Alexander M. Kytmanov

The Bochner-Martinelli Integral and Its Applications

Translated from the Russian
by Harold P. Boas

Birkhäuser Verlag
Basel · Boston · Berlin

Author:

Alexander M. Kytmanov
Krasnoyarsk State University
Institute of Physics
Akademgorodok
Krasnoyarsk 660036
Russia

Originally published in Russian under the title «Integral Bochnera-Martinelli i evo primeneniya» by Nauka, Novosibirsk branch, 1992.

A CIP catalogue record for this book is available from the Library of Congress, Washington D.C., USA

Deutsche Bibliothek Cataloging-in-Publication Data
Kytmanov, Aleksandr M.:
The Bochner-Martinelli integral and its applications / Alexander M. Kytmanov.
Transl. from the Russian by Harold P. Boas. - Basel ; Boston ; Berlin : Birkhäuser, 1995
Einheitssacht.: Integral Bochnera-Martinelli i ego primenija <engl.>
ISBN-13:978-3-0348-9904-8 e-ISBN-13:978-3-0348-9094-6
DOI: 10.1007/978-3-0348-9094-6

Softcover reprint of the hardcover 1st edition 1995

Printed on acid-free paper produced of chlorine-free pulp ∞

ISBN-13:978-3-0348-9904-8

9 8 7 6 5 4 3 2 1

Contents

Preface IX

Preface to the English Edition XI

1 The Bochner-Martinelli Integral

1 The Bochner-Martinelli integral representation 1
 1.1 Green's formula in complex form 1
 1.2 The Bochner-Martinelli formula for smooth functions . . . 4
 1.3 The Bochner-Martinelli representation for holomorphic functions 4
 1.4 Some integral representations 6
2 Boundary behavior 13
 2.1 The Sokhotskiĭ-Plemelj formula for functions satisfying a Hölder condition 13
 2.2 Analogue of Privalov's theorem for integrable functions . . 17
 2.3 Further results 20
3 Jump theorems 21
 3.1 Integrable and continuous functions 21
 3.2 Functions of class $\mathcal{L}^p$ 24
 3.3 Distributions 27
 3.4 Differential forms 30
4 Boundary behavior of derivatives 33
 4.1 Formulas for finding derivatives 33
 4.2 Jump theorem for derivatives 36
 4.3 Jump theorem for the "normal" derivative 40
5 The Bochner-Martinelli integral in the ball 44
 5.1 The spectrum of the Bochner-Martinelli operator 44
 5.2 Computation of the Bochner-Martinelli integral in the ball 49
 5.3 Some applications 51
 5.4 Characterization of the ball using the Bochner-Martinelli operator 53

2 CR-Functions Given on a Hypersurface

6 Analytic representation of CR-functions 55
6.1 Currents 55
6.2 The problem of analytic representation 61
6.3 The theorem on analytic representation 64
6.4 Some corollaries 69
6.5 Further results and generalizations 70
7 The Hartogs-Bochner extension theorem 70
7.1 The Hartogs-Bochner theorem 70
7.2 Weinstock's extension theorem 72
7.3 The theorem of Harvey and Lawson 73
8 Holomorphic extension from a part of the boundary 75
8.1 Statement of the problem 75
8.2 Lupacciolu's theorem 75
8.3 The $\overline{\partial}$-problem for the Bochner-Martinelli kernel 77
8.4 Proof of Lupacciolu's theorem 81
8.5 Extension of the class of compact sets 84
8.6 The case of a hypersurface 88
8.7 Further results and generalizations 90
9 Removable singularities of CR-functions 91
9.1 Bounded CR-functions 91
9.2 Integrable CR-functions 94
9.3 Further results 96
10 Analogue of Riemann's theorem for CR-functions 97
10.1 Statement of the problem and results 97
10.2 Auxiliary results 99
10.3 Analogue of Smirnov's theorem 101
10.4 Proof of the main result 102
10.5 Further results 103

3 Distributions Given on a Hypersurface

11 Harmonic representation of distributions 105
11.1 Statement of the problem 105
11.2 Boundary values of harmonic functions of finite order of growth 107
11.3 Corollaries 109
11.4 Theorems on harmonic extension 111
12 Multiplication of distributions 115
12.1 Different approaches to multiplication of distributions 115
12.2 Definition of the product of distributions using harmonic representations 116
12.3 Properties of the product of distributions given on a hypersurface 118
12.4 Properties of products of distributions in $\mathcal{D}'(\mathbf{R}^n)$ 119

12.5 Multiplication of hyperfunctions with compact support . . . 122
12.6 Multiplication in the sense of Mikusiński 124
12.7 Multipliable distributions 126
12.8 Boundary values of polyharmonic functions of finite order of growth 128
12.9 The class of homogeneous multipliable distributions 129
12.10 Further results 135
13 The generalized Fourier transform 137
13.1 Functions of slow growth 137
13.2 Distributions of slow growth 141
13.3 The inversion formula 142
13.4 Analogue of Vladimirov's theorem 144
13.5 Determination of the Fourier transform of some distributions 149

4 The $\overline{\partial}$-Neumann Problem
14 Statement of the $\overline{\partial}$-Neumann problem 155
14.1 The Hodge operator 155
14.2 Statement of the problem 157
14.3 The homogeneous $\overline{\partial}$-Neumann problem 159
15 Functions represented by Bochner-Martinelli 161
15.1 Smooth functions 161
15.2 Continuous functions 162
15.3 Functions with the one-dimensional holomorphic extension property 163
15.4 Generalizations for differential forms 165
16 Iterates of the Bochner-Martinelli integral 168
16.1 The theorem on iterates 168
16.2 Auxiliary results 170
16.3 Proof of the theorem on iterates and some corollaries 172
17 Uniqueness theorem for the $\overline{\partial}$-Neumann problem 174
17.1 Proof of the theorem 174
17.2 Corollaries of the uniqueness theorem 176
18 Solvability of the $\overline{\partial}$-Neumann problem 177
18.1 The tangential $\overline{\partial}_\tau$-equation 177
18.2 The $\overline{\partial}$-Neumann problem for smooth functions 179
18.3. The $\overline{\partial}$-Neumann problem for distributions 181
18.4 Generalization to differential forms 182
19 Integral representation in the ball 183
19.1 The $\overline{\partial}$-Neumann problem in the ball 183
19.2 Auxiliary results 185
19.3 Proof of the main theorem 187

5 Some Applications and Open Problems
20 Multidimensional logarithmic residues 189
20.1 The residue formula for smooth functions 189
20.2 The formula for logarithmic residues 191
20.3 The singular Bochner-Martinelli integral 192
20.4 The formula for logarithmic residues with singularities on the boundary 195
21 Multidimensional analogues of Carleman's formula 200
21.1 The classical Carleman-Goluzin-Krylov formula 200
21.2 Holomorphic extension from a part of the boundary 201
21.3 Yarmukhamedov's formula 203
21.4 Aĭzenberg's formula 205
22 The Poincaré-Bertrand formula 206
22.1 The singular Bochner-Martinelli integral depending on a parameter 206
22.2 Estimates of some integrals 208
22.3 Composition of the singular Bochner-Martinelli integral and an integral with a weak singularity 211
22.4 The Poincaré-Bertrand formula 214
23 Problems on holomorphic extension 220
23.1 Functions representable by the Cauchy-Fantappiè formula . 220
23.2 Differential criteria for holomorphicity of functions 222
23.3 The generalized $\overline{\partial}$-Neumann problem 229
23.4 The general form of integral representations in $\mathbf{C}^2$ 230

6 Holomorphic Extension of Functions
24 Holomorphic extension of hyperfunctions 233
24.1 Hyperfunctions as boundary values of harmonic functions . 233
24.2 Holomorphic extension of hyperfunctions into a domain . . 240
25 Holomorphic extension of functions 244
25.1 Holomorphic extension using the Bochner-Martinelli integral 244
25.2 Holomorphic extension using Cauchy-Fantappiè integrals . 253
26 The Cauchy problem for holomorphic functions 261
26.1 Statement of the problem 261
26.2 Some additional information on the Bochner-Martinelli integral 262
26.3 Weak boundary values of holomorphic functions of class $\mathcal{L}^q(D)$ 264
26.4 Doubly orthogonal bases in spaces of harmonic functions . . 267
26.5 Criteria for solvability of Problem 1 268

Bibliography 271

Index 289

Preface

The Bochner-Martinelli integral representation for holomorphic functions of several complex variables (which has already become classical) appeared in the works of Martinelli and Bochner at the beginning of the 1940's. It was the first essentially multidimensional representation in which the integration takes place over the whole boundary of the domain. This integral representation has a universal kernel (not depending on the form of the domain), like the Cauchy kernel in $\mathbf{C}^1$. However, in $\mathbf{C}^n$ when $n > 1$, the Bochner-Martinelli kernel is harmonic, but not holomorphic. For a long time, this circumstance prevented the wide application of the Bochner-Martinelli integral in multidimensional complex analysis. Martinelli and Bochner used their representation to prove the theorem of Hartogs (Osgood-Brown) on removability of compact singularities of holomorphic functions in $\mathbf{C}^n$ when $n > 1$.

In the 1950's and 1960's, only isolated works appeared that studied the boundary behavior of Bochner-Martinelli (type) integrals by analogy with Cauchy (type) integrals. This study was based on the Bochner-Martinelli integral being the sum of a double-layer potential and the tangential derivative of a single-layer potential. Therefore the Bochner-Martinelli integral has a jump that agrees with the integrand, but it behaves like the Cauchy integral under approach to the boundary, that is, somewhat worse than the double-layer potential. Thus, the Bochner-Martinelli integral combines properties of the Cauchy integral and the double-layer potential.

Interest in the Bochner-Martinelli representation grew in the 1970's in connection with the increased attention to integral methods in multidimensional complex analysis. Moreover, it turned out that the very general Cauchy-Fantappiè integral representation found by Leray is easily obtained from the Bochner-Martinelli representation. Koppelman's representation for exterior differential forms, which has the Bochner-Martinelli representation as a special case, appeared at the same time. The kernels in Koppelman's formula are constructed (like the Bochner-Martinelli kernel) by using derivatives of the fundamental solution of Laplace's equation.

The Cauchy-Fantappiè and Koppelman representations found significant applications in multidimensional complex analysis: constructing good integral representations for holomorphic functions, an explicit solution of the $\overline{\partial}$-equation and estimates of this solution, uniform approximation of holomorphic functions on compact sets, etc.

At the beginning of the 1970's, it was shown that, notwithstanding the non-holomorphicity of the kernel, the Bochner-Martinelli representation holds only for holomorphic functions. In 1975, Harvey and Lawson obtained a result for odd-dimensional manifolds on spanning by complex chains; the Bochner-Martinelli formula lies at its foundation. In the 1980's, the Bochner-Martinelli formula was successfully exploited in the theory of functions of several complex variables: in multidimensional residues, in complex (algebraic) geometry, in questions of rigidity of holomorphic mappings, in finding analogues of Carleman's formula, etc. Since these questions were not reflected in any monograph in the literature, it seemed necessary to write this book.

The exposition is relatively elementary and self-contained. For example, many properties of the singular Bochner-Martinelli integral can be deduced from the general theory of singular integral operators, but in view of the concrete form of the Bochner-Martinelli kernel, we obtain them by using ordinary properties of improper integrals (and Stokes's formula). In sum, one may say that the Bochner-Martinelli formula is the connection between complex and harmonic analysis in $\mathbf{C}^n$. This becomes especially apparent in the solution of the $\overline{\partial}$-Neumann problem: any function that is orthogonal to the holomorphic functions is the "normal" derivative of a harmonic function.

The proofs given in the book either are proofs of results that can be found only in journal articles, or else are different from earlier proofs. Therefore there is no proof in the book of the theorem of Harvey and Lawson, nor applications of the Bochner-Martinelli integral in algebraic geometry, which have been previously treated in monograph form.

The bibliography contains mainly works that are connected with the Bochner-Martinelli integral.

I thank L. A. Aĭzenberg, who provided the impetus for writing this book; the participants in the Krasnoyarsk seminar in the theory of functions; and especially N. N. Tarkhanov, who helped improve the proofs of a number of results.

Preface to the English Edition

Nearly four years have passed since the Russian version of the book was written. Therefore it is natural for the author to want to introduce many changes and additions in the text. This especially applies to Chapter 2. A great many new results have appeared on these problems, often not directly connected with the Bochner-Martinelli integral, but dealing with removable singularities of CR-functions given on a hypersurface. In fact, a new book would be required on this subject. Moreover, a large survey by Chirka and Stout [39] has appeared, in which these results find a thorough treatment.

In sum, the text has basically been modified only as necessary to correct misprints and the statements of some theorems. The changes are mainly connected with the question of iterates of the Bochner-Martinelli integral. As remarked by Professor Straube, Theorem 16.1 cannot hold for all the Sobolev spaces $\mathcal{W}_2^s(D)$, $s \geq 1$, and for all domains. Therefore, in the English version of the book, Theorem 16.1 has been left in the form in which it was given by Romanov [169], that is, for the space $\mathcal{W}_2^1(D)$. Corresponding changes were made in Theorems 17.1 and 17.2 as well.

In addition, a supplementary Chapter 6 was written for the book. It is devoted to the question of describing functions given on a hypersurface that can be extended holomorphically into a fixed domain. Here it is not assumed that the domain is the envelope of holomorphy of the hypersurface. It turns out that the Bochner-Martinelli integral is the ideal instrument for solving such problems. We also see here a connection between the theory of holomorphic functions and the theory of harmonic functions in $\mathbf{C}^n$.

December 1993

Preface to the English Edition

Chapter 1

The Bochner-Martinelli Integral

1 The Bochner-Martinelli integral representation

1.1 Green's formula in complex form

We consider n-dimensional complex space $\mathbf{C}^n$ with variables $z = (z_1, z_2, \dots, z_n)$. If z and w are points in $\mathbf{C}^n$, then we write $\langle z, w\rangle = z_1w_1 + \cdots + z_nw_n$, and $|z| = \sqrt{\langle z, \bar{z}\rangle}$, where $\bar{z} = (\bar{z}_1, \dots, \bar{z}_n)$. The topology in $\mathbf{C}^n$ is given by the metric $(z, w) \mapsto |z - w|$. If $z \in \mathbf{C}^n$, then $\operatorname{Re} z = (\operatorname{Re} z_1, \dots, \operatorname{Re} z_n) \in \mathbf{R}^n$, where we write $\operatorname{Re} z_j = x_j$, and $\operatorname{Im} z = (\operatorname{Im} z_1, \dots, \operatorname{Im} z_n)$ with $\operatorname{Im} z_j = y_j$; that is, $z_j = x_j + iy_j$ for $j = 1, \dots, n$. Thus $\mathbf{C}^n \cong \mathbf{R}^{2n}$. The orientation of $\mathbf{C}^n$ is determined by the coordinate order $(x_1, \dots, x_n, y_1, \dots, y_n)$. Accordingly, the volume form dv is given by $dv = dx_1 \wedge \cdots \wedge dx_n \wedge dy_1 \wedge \cdots \wedge dy_n = dx \wedge dy = (i/2)^n\, dz \wedge d\bar{z} = (-i/2)^n\, d\bar{z} \wedge dz$, where $dz = dz_1 \wedge \cdots \wedge dz_n$.

As usual, a function f on an open set $U \subset \mathbf{C}^n$ belongs to the space $\mathcal{C}^k(U)$ if f is k times continuously differentiable in U. (Here $0 \le k \le \infty$, and $\mathcal{C}^0(U) = \mathcal{C}(U)$.) If M is a closed set in $\mathbf{C}^n$, then f belongs to $\mathcal{C}^k(M)$ when f extends to some neighborhood U of M as a function of class $\mathcal{C}^k(U)$. We will also consider the space $\mathcal{C}^r(U)$ (or $\mathcal{C}^r(M)$) when $r \ge 0$ is not necessarily an integer. A function f belongs to $\mathcal{C}^r(U)$ if it lies in the class $\mathcal{C}^{[r]}(U)$ (where $[r]$ is the integral part of r), and all its derivatives of order $[r]$ satisfy a Hölder condition on U with exponent $r - [r]$.

The space $\mathcal{O}(U)$ consists of those functions f that are holomorphic on the open set U; when M is a closed set, $\mathcal{O}(M)$ consists of those functions f that are holomorphic in some neighborhood of M (a different neighborhood for each function). A function f belongs to $\mathcal{A}(U)$ if f is holomorphic in U and continuous on the closure $\overline{U}$ (that is, $f \in \mathcal{O}(U) \cap \mathcal{C}(\overline{U})$).

A domain D in $\mathbf{C}^n$ has boundary of class $\mathcal{C}^k$ (we write $\partial D \in \mathcal{C}^k$) if $D = \{\, z : \rho(z) < 0 \,\}$, where ρ is a real-valued function of class $\mathcal{C}^k$ on some neighborhood of the closure of D, and the differential $d\rho \ne 0$ on ∂D. If $k = 1$, then we say that D is a domain with smooth boundary. We will call the function ρ a defining function for the domain D. The orientation of the boundary ∂D is induced by the orientation of D.

By a domain with piecewise-smooth boundary ∂D we will understand a smooth polyhedron, that is, a domain of the form $D = \{ z : \rho_j(z) < 0, j = 1, \dots, m \}$, where the real-valued functions ρ_j are class $\mathcal{C}^1$ in some neighborhood of the closure $\overline{D}$, and for every set of distinct indices $j_1, \dots, j_s$ we have $d\rho_{j_1} \wedge \cdots \wedge d\rho_{j_s} \neq 0$ on the set $\{ z : \rho_{j_1}(z) = \cdots = \rho_{j_s}(z) = 0 \}$. It is well known that Stokes's formula holds for such domains D and surfaces ∂D.

We denote the ball of radius $\epsilon > 0$ with center at the point $z \in \mathbf{C}^n$ by $B(z, \epsilon)$, and we denote its boundary by $S(z, \epsilon)$ (that is, $S(z, \epsilon) = \partial B(z, \epsilon)$).

Consider the exterior differential form $U(\zeta, z)$ of type $(n, n-1)$ given by

$$U(\zeta, z) = \frac{(n-1)!}{(2\pi i)^n} \sum_{k=1}^{n} (-1)^{k-1} \frac{\bar{\zeta}_k - \bar{z}_k}{|\zeta - z|^{2n}} \, d\bar{\zeta}[k] \wedge d\zeta,$$

where $d\bar{\zeta}[k] = d\bar{\zeta}_1 \wedge \cdots \wedge d\bar{\zeta}_{k-1} \wedge d\bar{\zeta}_{k+1} \wedge \cdots \wedge d\bar{\zeta}_n$. When $n = 1$, the form $U(\zeta, z)$ reduces to the Cauchy kernel $(2\pi i)^{-1}(\zeta - z)^{-1} \, d\zeta$. The form $U(\zeta, z)$ clearly has coefficients that are harmonic in $\mathbf{C}^n \setminus \{z\}$, and it is closed with respect to ζ (that is, $d_\zeta U(\zeta, z) = 0$).

Let $g(\zeta, z)$ be the fundamental solution to the Laplace equation:

$$g(\zeta, z) = \begin{cases} -\dfrac{(n-2)!}{(2\pi i)^n} |\zeta - z|^{2-2n} & \text{for } n > 1, \\ (2\pi i)^{-1} \ln |\zeta - z|^2 & \text{for } n = 1. \end{cases}$$

Then

$$\begin{aligned} U(\zeta, z) &= \sum_{k=1}^{n} (-1)^{k-1} \frac{\partial g}{\partial \zeta_k} \, d\bar{\zeta}[k] \wedge d\zeta \\ &= (-1)^{n-1} \, \partial_\zeta g \wedge \sum_{k=1}^{n} d\bar{\zeta}[k] \wedge d\zeta[k], \end{aligned}$$

where the operator $\partial = \sum_{k=1}^{n} (d\zeta_k)(\partial / \partial \zeta_k)$. We will write the Laplace operator Δ in the following form:

$$\Delta = \sum_{k=1}^{n} \frac{\partial^2}{\partial \zeta_k \partial \bar{\zeta}_k} = \frac{1}{4} \sum_{k=1}^{n} \left(\frac{\partial^2}{\partial x_k^2} + \frac{\partial^2}{\partial y_k^2} \right)$$

(if $\zeta_k = x_k + i y_k$, then $\partial / \partial \zeta_k = (1/2)(\partial / \partial x_k - i \partial / \partial y_k)$, and $\partial / \partial \bar{\zeta}_k = \overline{\partial / \partial \zeta_k}$). When $f \in \mathcal{C}^1(U)$, we define the differential form μ_f via

$$\mu_f = \sum_{k=1}^{n} (-1)^{n+k-1} \frac{\partial f}{\partial \bar{\zeta}_k} \, d\zeta[k] \wedge d\bar{\zeta}.$$

Theorem 1.1. *Let D be a bounded domain in $\mathbf{C}^n$ with piecewise-smooth boundary, and let $f \in \mathcal{C}^2(\overline{D})$. Then*

$$\int_{\partial D} f(\zeta)\, U(\zeta, z) + \int_D g(\zeta, z)\Delta f(\zeta)\, d\bar{\zeta} \wedge d\zeta - \int_{\partial D} g(\zeta, z)\, \mu_f(\zeta) = \begin{cases} f(z), & \text{if } z \in D, \\ 0, & \text{if } z \notin \overline{D}. \end{cases} \tag{1.1}$$

(The integral in (1.1) converges absolutely.)

Formula (1.1) is Green's formula in complex form.

Proof. Since

$$d_\zeta(f(\zeta)\, U(\zeta, z) - g(\zeta, z)\, \mu_f(\zeta)) = -g(\zeta, z)\Delta f\, d\bar{\zeta} \wedge d\zeta, \tag{1.2}$$

Stokes's formula implies that (1.1) holds for $z \notin \overline{D}$. If $z \in D$, then for sufficiently small positive ϵ, we obtain from (1.2) and Stokes's formula that

$$\int_{\partial D} f\, U(\zeta, z) - \int_{\partial D} g(\zeta, z)\, \mu_f + \int_{D \setminus B(z,\epsilon)} g(\zeta, z)\Delta f\, d\bar{\zeta} \wedge d\zeta = \int_{S(z,\epsilon)} f(\zeta)\, U(\zeta, z) - \int_{S(z,\epsilon)} g(\zeta, z)\, \mu_f.$$

When $n > 1$,

$$\left| \int_{S(z,\epsilon)} g(\zeta, z)\, \mu_f \right| \leq \frac{(n-2)!}{(2\pi)^n \epsilon^{2n-2}} \int_{S(z,\epsilon)} |\mu_f| \leq C\epsilon,$$

that is,

$$\lim_{\epsilon \to 0+} \int_{S(z,\epsilon)} g(\zeta, z)\, \mu_f = 0.$$

(The argument for $n = 1$ is analogous.) However,

$$\begin{aligned} \int_{S(z,\epsilon)} f(\zeta)\, U(\zeta, z) &= \frac{(n-1)!}{(2\pi i)^n} \epsilon^{-2n} \int_{S(z,\epsilon)} f(\zeta) \sum_{k=1}^{n} (-1)^{k-1} (\bar{\zeta}_k - \bar{z}_k)\, d\bar{\zeta}[k] \wedge d\zeta \\ &= \frac{(n-1)!}{(2\pi i)^n} \epsilon^{-2n} \int_{B(z,\epsilon)} \left[n f(\zeta) + \sum_{k=1}^{n} \frac{\partial f}{\partial \bar{\zeta}_k} (\bar{\zeta}_k - \bar{z}_k) \right] d\bar{\zeta} \wedge d\zeta. \end{aligned}$$

Since $\lim_{\epsilon \to 0+} \epsilon^{-2n} \int_{B(z,\epsilon)} \sum_{k=1}^{n} (\partial f / \partial \bar{\zeta}_k)(\bar{\zeta}_k - \bar{z}_k)\, d\bar{\zeta} \wedge d\zeta = 0$, we have

$$\begin{aligned} \lim_{\epsilon \to 0+} \int_{S(z,\epsilon)} f(\zeta)\, U(\zeta, z) &= \lim_{\epsilon \to 0+} n!\, (2\pi i \epsilon^2)^{-n} \int_{B(z,\epsilon)} f(\zeta)\, d\bar{\zeta} \wedge d\zeta \\ &= \lim_{\epsilon \to 0+} n!\, (\pi \epsilon^2)^{-n} \int_{B(z,\epsilon)} f(\zeta)\, dv = f(z) \end{aligned}$$

(by the mean-value theorem). □

Corollary 1.2. *Let D be a bounded domain with piecewise-smooth boundary, and let f be a harmonic function in D of class $\mathcal{C}^1(\overline{D})$. Then*

$$\int_{\partial D} f(\zeta)\,U(\zeta,z) - \int_{\partial D} g(\zeta,z)\,\mu_f = \begin{cases} f(z), & z \in D, \\ 0, & z \notin \overline{D}. \end{cases} \tag{1.3}$$

Formula (1.3) was given byBochner [28] in deriving the Bochner-Martinelli integral representation.

1.2 The Bochner-Martinelli formula for smooth functions

Theorem 1.3. *Let D be a bounded domain with piecewise-smooth boundary, and let f be a function in $\mathcal{C}^1(\overline{D})$. Then*

$$f(z) = \int_{\partial D} f(\zeta)\,U(\zeta,z) - \int_D \overline{\partial} f(\zeta) \wedge U(\zeta,z), \qquad z \in D, \tag{1.4}$$

where $\overline{\partial} = \sum_{k=1}^{n}(d\bar{\zeta}_k)(\partial/\partial\bar{\zeta}_k)$, and the integral in (1.4) converges absolutely.

Formula (1.4) was obtained by Koppelman in [102]. When $n = 1$, it reduces to the Cauchy-Green formula.

Proof. Supposing at first that $f \in \mathcal{C}^2(\overline{D})$, we transform the integral:

$$\int_D \overline{\partial} f \wedge U(\zeta,z) = \int_D \sum_{k=1}^{n} \frac{\partial f}{\partial \bar{\zeta}_k}\frac{\partial g}{\partial \zeta_k}\, d\bar{\zeta} \wedge d\zeta = \int_D \partial_\zeta g \wedge \mu_f$$
$$= \int_D d_\zeta(g\mu_f) - \int_D g\,\Delta f\, d\bar{\zeta} \wedge d\zeta = \int_{\partial D} g\mu_f - \int_D g\,\Delta f\, d\bar{\zeta} \wedge d\zeta$$

(here we have applied Stokes's formula, since all the integrals converge absolutely). Then for $z \in D$, formula (1.1) implies that

$$\int_D \overline{\partial} f \wedge U(\zeta,z) = \int_{\partial D} f(\zeta)\,U(\zeta,z) - f(z).$$

Now if $f \in \mathcal{C}^1(\overline{D})$, we obtain (1.4) by approximating f (in the metric of $\mathcal{C}^1(\overline{D})$) by functions of class $\mathcal{C}^2(\overline{D})$. □

1.3 The Bochner-Martinelli representation for holomorphic functions

Theorem 1.4 (Martinelli, Bochner). *If D is a bounded domain in $\mathbf{C}^n$ with piecewise-smooth boundary, and f is a holomorphic function in D of class $\mathcal{C}(\overline{D})$, then*

$$f(z) = \int_{\partial D} f(\zeta)\,U(\zeta,z), \qquad z \in D. \tag{1.5}$$

Formula (1.5) was obtained in $\mathbf{C}^n$ by Martinelli [149], and then by Bochner [28] independently and by different methods. The history of the construction of this formula has been described in detail in [104]. It is the first integral representation for holomorphic functions in $\mathbf{C}^n$ in which the integration is carried out over the whole boundary of the domain. This formula is by now classical and has found a place in many textbooks on multidimensional complex analysis (see, for example, [51, 180, 209]).

Formula (1.5) reduces to Cauchy's formula when $n = 1$, but in contrast to Cauchy's formula, the kernel in (1.5) is not holomorphic (in z and ζ) when $n > 1$. By splitting the kernel $U(\zeta, z)$ into real and imaginary parts, it is easy to show that $U(\zeta, z)$ is the sum of a double-layer potential and a tangential derivative of a single-layer potential; consequently, the Bochner-Martinelli integral inherits some of the properties of the Cauchy integral and some of the properties of the double-layer potential. It differs from the Cauchy integral in not being a holomorphic function, and it differs from the double-layer potential in having somewhat worse boundary behavior. At the same time, it establishes a connection between harmonic and holomorphic functions in $\mathbf{C}^n$ when $n > 1$.

Later on we shall need formula (1.5) for the Hardy spaces $\mathcal{H}^p(D)$, so we now recall some definitions (see, for example, [75, 194]). Let D be a bounded domain, and suppose that ∂D is a (connected) Lyapunov surface, that is, $\partial D \in \mathcal{C}^1$ and the outer unit normal vector $\nu(\zeta)$ to the surface ∂D satisfies a Hölder condition with some exponent $\alpha > 0$ (that is, $|\nu(\zeta) - \nu(\xi)| \leq C|\zeta - \xi|^\alpha$ for ζ and ξ in ∂D). It is known that in such domains, the Green function $G(\zeta, z)$ has good boundary behavior: for fixed $z \in D$, the function $G(\zeta, z) \in \mathcal{C}^1(\overline{D})$, and its first derivatives satisfy a Hölder condition on $\overline{D}$.

We say that a holomorphic function f belongs to $\mathcal{H}^p(D)$ (where $p > 0$) if

$$\sup_{\epsilon>0} \int_{\partial D} |f(\zeta - \epsilon\nu(\zeta))|^p \, d\sigma < +\infty$$

(here $d\sigma$ is the surface area element on ∂D). A holomorphic function f belongs to $\mathcal{H}^\infty(D)$ if $\sup_D |f(z)| < \infty$.

The class $\mathcal{H}^p(D)$ may also be defined in the following way. Let $D = \{z : \rho(z) < 0\}$ with $d\rho \neq 0$ on ∂D, and let $D_\epsilon = \{z : \rho(z) < -\epsilon\}$ for $\epsilon > 0$. A holomorphic function $f \in \mathcal{H}^p(D)$ if

$$\sup_{\epsilon>0} \int_{\partial D_\epsilon} |f(\zeta)|^p \, d\sigma_\epsilon < +\infty.$$

As is shown in [194], this definition does not depend on the choice of the smooth function ρ defining the domain D.

Theorem 1.5. *If $p \geq 1$ and $f \in \mathcal{H}^p(D)$, then*

$$f(z) = \int_{\partial D} f(\zeta)\, U(\zeta, z), \qquad z \in D.$$

Proof. If $p \geq 1$ and $f \in \mathcal{H}^p(D)$, then f has normal boundary values almost everywhere on ∂D (see [75, 194]) that form a function of class $\mathcal{L}^p(\partial D)$ (we denote these boundary values again by f). Moreover, the function f can be reconstructed in D from its boundary values by Poisson's formula

$$f(z) = \int_{\partial D} f(\zeta) P(\zeta, z)\, d\sigma$$

(where $P(\zeta, z)$ is the Poisson kernel for D). Since the Green function $G(\zeta, z) = g(\zeta, z) + h(\zeta, z)$, where for fixed $z \in D$ the function $h(\zeta, z)$ is harmonic in D of class $\mathcal{C}^1(\overline{D})$, we have

$$P(\zeta, z)\, d\sigma = U(\zeta, z)\big|_{\partial D} + \sum_{k=1}^{n} \frac{\partial h}{\partial \zeta_k} (-1)^{k-1}\, d\bar{\zeta}[k] \wedge d\zeta\big|_{\partial D}.$$

Since the differential form $\sum_{k=1}^{n} (-1)^{k-1} (\partial h / \partial \zeta_k)\, d\bar{\zeta}[k] \wedge d\zeta$ is closed, we have

$$\int_{\partial D} f(\zeta) \sum_{k=1}^{n} (-1)^{k-1} \frac{\partial h}{\partial \zeta_k}\, d\bar{\zeta}[k] \wedge d\zeta = \int_D f\, d\left(\sum_{k=1}^{n} (-1)^{k-1} \frac{\partial h}{\partial \zeta_k}\, d\bar{\zeta}[k] \wedge d\zeta \right) = 0.$$

Consequently, formula (1.5) holds for $f \in \mathcal{H}^p(D)$. (This theorem can be obtained by applying Theorem 1.4 to the domains D_ϵ and letting ϵ tend to zero, but we have given a different proof to show the connection between the Poisson integral and the Bochner-Martinelli integral.) □

1.4 Some integral representations

We remark first of all that it is possible to derive from the Bochner-Martinelli formula (1.5) the Cauchy-Fantappiè formula that was obtained by Leray [135, 136].

Let D be a bounded domain with piecewise-smooth boundary, and suppose that for a point $z \in D$ there is defined on ∂D a continuously differentiable vector-valued function $\eta(\zeta) = (\eta_1(\zeta), \dots, \eta_n(\zeta))$ such that

$$\sum_{k=1}^{n} (\zeta_k - z_k) \eta_k(\zeta) = 1, \qquad \zeta \in \partial D.$$

Theorem 1.6 (Leray). *Every function $f \in \mathcal{C}(\overline{D})$ that is holomorphic in D satisfies the equation*

$$f(z) = \frac{(n-1)!}{(2\pi i)^n} \int_{\partial D} f(\zeta)\, \omega'(\eta) \wedge d\zeta, \tag{1.6}$$

where $\omega'(\eta) = \sum_{k=1}^{n} (-1)^{k-1} \eta_k\, d\eta[k]$.

Proof. Henkin's proof is as follows. Consider in the space $\mathbf{C}^{2n}$ of variables $(\eta,\zeta)=(\eta_1,\dots,\eta_n,\zeta_1,\dots,\zeta_n)$ the analytic hypersurface $M_z=\{(\eta,\zeta):\sum_{k=1}^n(\zeta_k-z_k)\eta_k=1\}$, on which the form $\omega'(\eta)\wedge d\zeta$ is closed. The two cycles $\Gamma_1=\{(\eta,\zeta):\zeta\in\partial D,\ \eta_j=(\bar\zeta_j-\bar z_j)|\zeta-z|^{-2}\}$ and $\Gamma_2=\{(\eta,\zeta):\zeta\in\partial D,\ \eta_j=\eta_j(\zeta),\ j=1,\dots,n\}$ in M_z are homotopic in M_z (the homotopy being given by the formula $\tilde\eta_j=t(\bar\zeta_j-\bar z_j)|\zeta-z|^{-2}+(1-t)\eta_j(\zeta)$), that is, they are homologous cycles. Consequently, $\int_{\Gamma_1}f(\zeta)\omega'(\eta)\wedge d\zeta=\int_{\Gamma_2}f(\zeta)\omega'(\eta)\wedge d\zeta$ when f is a holomorphic function. But $\omega'((\bar\zeta_j-\bar z_j)/|\zeta-z|^2)\wedge d\zeta=(2\pi i)^nU(\zeta,z)/(n-1)!$. Hence (1.6) follows. □

The Cauchy-Fantappiè representation has turned out to be very useful, and it has many applications in multidimensional complex analysis (see the books and surveys [3, 10, 74, 77, 104, 172, 175]). Formula (1.6) together with (1.4) makes it possible to obtain multidimensional analogues of the Cauchy-Green formula (see the surveys [74, 159], and also [18]).

Now we consider some generalizations of the Bochner-Martinelli formula (1.5). First we remark that Yarmukhamedov extended (1.5) to unbounded domains in [221]. Suppose given in a bounded domain D with piecewise-smooth boundary a function $h(\zeta,z)$ depending on the parameter $z\in D$ such that $h\in\mathcal{C}^1(\overline{D})$, and h is harmonic in D for each point $z\in D$. Then, as already remarked in the proof of Theorem 1.5, the differential form

$$\bar\mu_{\bar h}=\sum_{k=1}^n(-1)^{n+k-1}\frac{\partial h}{\partial\zeta_k}\,d\bar\zeta[k]\wedge d\zeta$$

is closed in $\overline{D}$, and so $\int_{\partial D}f(\zeta)\,\bar\mu_{\bar h}(\zeta)=0$ for each function $f\in\mathcal{A}(D)$. Consequently, the integral representation

$$f(z)=\int_{\partial D}f(\zeta)\,[U(\zeta,z)+\bar\mu_{\bar h}(\zeta)],\qquad z\in D,\tag{1.7}$$

holds for such functions f.

Yarmukhamedov [221] proposed the following construction for choosing the function h. Let $K(w)$ be an entire function of one complex variable $w=u+iv$ taking real values on the real axis (that is, for $v=0$), and such that $K(u)\neq0$ for all u. We define the real-valued function $\Phi(\zeta,z)$ for ζ and z in $\mathbf{C}^n$ by the formula

$$C_nK(\operatorname{Im}z_n)\Phi(\zeta,z)=\frac{\partial^{n-2}}{\partial s^{n-2}}\operatorname{Im}\frac{K(i\alpha+\operatorname{Im}z_n)}{\alpha(i\alpha+\operatorname{Im}(\zeta_n-z_n))},$$

where $s=\sum_{k=1}^{n-1}|\zeta_k-z_k|^2$, $\alpha^2=s+\operatorname{Re}^2(\zeta_n-z_n)$, and $C_n=(-1)^{n-1}(2\pi i)^n$.

Lemma 1.7. *When $n>1$, the function $\Phi(\zeta,z)$ can be represented in the form $\Phi(\zeta,z)=g(\zeta,z)+h(\zeta,z)$, where $g(\zeta,z)$ is the fundamental solution of the Laplace equation (see subsection 1), and $h(\zeta,z)$ is defined for all ζ and z in $\mathbf{C}^n$ and is harmonic in $\zeta\in\mathbf{C}^n$ for fixed z.*

Proof. The proof of this lemma in [221] consists in expanding $K(u+iv)$ in powers of v and directly verifying the harmonicity of Φ. □

We note that when $K \equiv 1$, we obtain $\Phi(\zeta, z) = g(\zeta, z)$.

Theorem 1.8 (Yarmukhamedov). *Suppose G is an unbounded domain in $\mathbf{C}^n$ (where $n > 1$) with piecewise-smooth boundary, and f is a function of class $\mathcal{A}(G)$. If $\lim_{R\to\infty} \int_{\partial(G\setminus B(0,R))} f(\zeta)\,\Omega(\zeta, z) = 0$ for each fixed $z \in G$, then*

$$f(z) = \int_{\partial G} f(\zeta)\,\Omega(\zeta, z), \qquad \text{where} \qquad \Omega(\zeta, z) = \sum_{k=1}^{n} (-1)^{k-1} \frac{\partial \Phi}{\partial \zeta_k}\, d\bar{\zeta}[k] \wedge d\zeta.$$

Proof. The proof of the theorem consists in applying formula (1.7) to the domain $G \cap B(0, R)$, and then letting R tend to infinity. □

Integral representations for holomorphic functions with various growth properties can be obtained by making appropriate choices of the domain G and the function K. Examples are given in [221, 222].

We have already remarked that the Bochner-Martinelli kernel is the sum of a double-layer potential and a tangential derivative of a single-layer potential. Martinelli [152] noted that if two continuous unit vector fields $\nu(\zeta)$ and $s(\zeta)$ are chosen on the boundary of a bounded domain with smooth boundary, and $\nu = is$ (where ν is the outer normal field to ∂D), then the restriction of the kernel $U(\zeta, z)$ to ∂D coincides with $((\partial/\partial\nu) + i(\partial/\partial s))g(\zeta, z)\, d\sigma$.

Henkin and Leiterer [76] (see also [77, Chapter 4]) extended formula (1.5) to domains D in Stein manifolds. Let X be a Stein manifold of dimension n, and let D be a relatively compact domain in X with smooth boundary. To obtain an analogue of the Bochner-Martinelli representation on X, we need to choose a mapping to replace $\zeta - z$. It would seem natural to choose a mapping $u(z, \zeta)$ from $X \times X$ into $\mathbf{C}^n$ with the following properties:

1. $u(z, \zeta) \neq 0$ if $\zeta \neq z$,
2. for each point $z \in X$, the mapping $u(z, \cdot)$ is biholomorphic in some neighborhood of z, where $u(z, z) = 0$.

However, since such a mapping $u(z, \zeta)$ does not exist in general, we need to take a mapping $s(z, \zeta)$ with values in the tangent bundle $T(X)$ such that $s(z, \zeta) \in T_z(X)$ for all $(z, \zeta) \in X \times X$ and such that conditions (1) and (2) hold. But in principle, even such a mapping may not exist (property (1) may fail). Therefore the following construction is proposed in [76]. First construct a mapping $s(z, \zeta)$ satisfying only condition (2). Then, using Cartan's Theorem B, find a holomorphic function $\varphi(z, \zeta)$ on $X \times X$ and an integer $\kappa \geq 0$ such that $\varphi(z, z) = 1$ for all $z \in X$ and $\varphi^\kappa(z, \zeta)/(\chi s(z, \zeta), s(z, \zeta))$ is a smooth function for all $z \neq \zeta$; here χ is a fiber-preserving mapping of class C^∞ from $T(X)$ into the cotangent bundle $T^*(X)$ such that the expression $(\chi a, a)^{1/2} = \|a\|$ defines the norm in the fibers of $T(X)$ (where (b, a) is the value of the covector $b \in T_z^*(X)$ on the vector $a \in T_z(X)$).

Theorem 1.9 (Dynin, Leiterer, Henkin). *The formula*

$$f(z) = \frac{(n-1)!}{(2\pi i)^n} \int_{\partial D} f(\zeta) \frac{\varphi^\nu(\zeta, z)\, \omega'(\chi(s(z,\zeta))) \wedge \omega(s(z,\zeta))}{(\chi s(z,\zeta), s(z,\zeta))^n},$$

holds for each integer $\nu \geq 2n\kappa$ and each function $f \in \mathcal{O}(\overline{D})$, where

$$\omega(s(z,\zeta)) = d_\zeta s_1 \wedge \cdots \wedge d_\zeta s_n.$$

Andreotti and Norguet obtained another generalization of (1.5) in [14] (it was proved by another method in [151]).

Suppose D is a bounded domain with piecewise-smooth boundary, $\alpha = (\alpha_1, \ldots, \alpha_n)$ is a multi-index, f is a function holomorphic in D and continuous on $\overline{D}$, and $\partial^\alpha f = \partial^{\|\alpha\|} f / \partial z_1^{\alpha_1} \ldots \partial z_n^{\alpha_n}$, where $\|\alpha\| = \alpha_1 + \cdots + \alpha_n$. Consider the following differential form:

$$\omega_\alpha(\zeta, z) = \frac{(n-1)!\, \alpha_1! \ldots \alpha_n!}{(2\pi i)^n} \sum_{k=1}^n \frac{(-1)^{k-1}(\bar\zeta_k - \bar z_k)\, d\bar\zeta^{\alpha+I}[k] \wedge d\zeta}{(|\zeta_1 - z_1|^{2(\alpha_1+1)} + \cdots + |\zeta_n - z_n|^{2(\alpha_n+1)})^n},$$

where

$$d\bar\zeta^{\alpha+I}[k] = d\bar\zeta_1^{\alpha_1+1} \wedge \cdots \wedge d\bar\zeta_{k-1}^{\alpha_{k-1}+1} \wedge d\bar\zeta_{k+1}^{\alpha_{k+1}+1} \wedge \cdots \wedge d\bar\zeta_n^{\alpha_n+1}.$$

Theorem 1.10 (Andreotti, Norguet). *The formula*

$$\partial^\alpha f(z) = \int_{\partial D} f(\zeta)\, \omega_\alpha(\zeta, z) \tag{1.8}$$

holds for every point $z \in D$ and every multi-index α.

Proof. The proof given in [10, page 60] goes as follows. First verify that ω_α is a closed form, so that integration over ∂D can be replaced by integration over the set $\{\zeta : |\zeta_1 - z_1|^{2\alpha_1+2} + \cdots + |\zeta_n - z_n|^{2\alpha_n+2} = \epsilon^2\}$. Expand the function f in powers of $\zeta - z$ in a neighborhood of z, and integrate the series termwise against the form $\omega_\alpha(\zeta, z)$. We obtain $\partial^\alpha f(z)$ as the result of a direct calculation. When $\alpha = (0, \ldots, 0)$, formula (1.8) reduces to (1.5). □

We note that (1.8) can be generalized in the spirit of the Cauchy-Fantappiè formula (see [10, page 61]). The Andreotti-Norguet formula was carried over to Stein manifolds in [233]. Analogues of the Bochner-Martinelli formula have also been considered in quaternionic analysis [206, 207] and in Clifford analysis [190]. Other generalizations of the Bochner-Martinelli formula may be found in [59, 101, 155, 191, 219, 220]. In conclusion, we give another generalization of the Bochner-Martinelli formula, to the case of differential forms, that was obtained by Koppelman [102].

Let $I = (i_1, \ldots, i_q)$ and $J = (j_1, \ldots, j_p)$ be increasing multi-indices, that is, $1 \le i_1 < \cdots < i_q \le n$ and $1 \le j_1 < \cdots < j_p \le n$, where $0 \le p \le n$ and $0 \le q \le n$. For $q \le n-1$, we consider the differential forms

$$U_{p,q}(\zeta, z) = (-1)^{p(n-q-1)} \frac{(n-1)!}{(2\pi i)^n} \times$$
$$\times \sum_{I,J}{}' \sum_{k \notin I} \sigma(I,k)\sigma(J) \frac{(\bar{\zeta}_k - \bar{z}_k)}{|\zeta - z|^{2n}}\, d\bar{\zeta}[I,k] \wedge d\zeta[J]\, d\bar{z}_I \wedge dz_J,$$

where $dz_J = dz_{j_1} \wedge \cdots \wedge dz_{j_p}$, the form $d\zeta[J]$ is obtained from $d\zeta$ by eliminating the differentials $d\zeta_{j_1}, \ldots, d\zeta_{j_p}$, and the prime on the summation sign indicates that the sum is taken over increasing multi-indices I and J. The symbols $\sigma(I,k)$ and $\sigma(J)$ are given by $\sigma(I,k)\, dz = dz_k \wedge dz_I \wedge dz[I,k]$ and $\sigma(J)\, dz = dz_J \wedge dz[J]$. Also, we set $U_{p,-1} \equiv U_{p,n} \equiv 0$. The forms $U_{p,q}$ are to be understood as double differential forms. If $\gamma = \sum'_{I,J} \gamma_{I,J}\, d\bar{z}_I \wedge dz_J$, then $\bar{\partial}\gamma = \sum_{k=1}^n \sum'_{I,J} (\partial\gamma_{I,J}/\partial\bar{z}_k)\, d\bar{z}_k \wedge d\bar{z}_I \wedge dz_J$.

Theorem 1.11 (Koppelman). *Suppose D is a bounded domain in $\mathbf{C}^n$ with piecewise-smooth boundary, and γ is a differential form of type (p,q) with coefficients of class $\mathcal{C}^1(\overline{D})$. Then*

$$\int_{\partial D} \gamma(\zeta) \wedge U_{p,q}(\zeta, z) - \int_D \bar{\partial}\gamma \wedge U_{p,q}(\zeta, z) - \bar{\partial} \int_D \gamma(\zeta) \wedge U_{p,q-1}(\zeta, z)$$
$$= \begin{cases} \gamma(z), & z \in D, \\ 0, & z \notin \overline{D}. \end{cases} \tag{1.9}$$

(The integrals over D converge absolutely.)

Lemma 1.12. *The following formulas hold:*

$$U_{p,q}(\zeta, z) = -U_{n-p,n-q-1}(z, \zeta),$$
$$\bar{\partial}_\zeta U_{p,q}(\zeta, z) = (-1)^{p+q}\, \bar{\partial}_z U_{p,q-1}(\zeta, z), \qquad p,q = 0,1,\ldots,n,$$

and in particular

$$\bar{\partial}_\zeta U_{p,0} = \bar{\partial}_z U_{p,n-1} = 0.$$

Proof. The lemma follows directly from the definition of the kernel $U_{p,q}(\zeta, z)$. □

Lemma 1.13. *If $f \in \mathcal{C}^1(B(z,r))$, then*

$$\lim_{\epsilon \to 0+} \int_{S(z,\epsilon)} f(\zeta)\, |\zeta - z|^{m-2n}\, d\bar{\zeta}[k] \wedge d\zeta = \begin{cases} (-1)^{k-1} \dfrac{(2\pi i)^n}{n!} \dfrac{\partial f}{\partial \bar{z}_k}, & m = 0, \\ 0, & m > 0, \end{cases}$$

$$\lim_{\epsilon \to 0+} \int_{S(z,\epsilon)} f(\zeta)\, |\zeta - z|^{m-2n}\, d\bar{\zeta} \wedge d\zeta[k] = \begin{cases} (-1)^{n+k-1} \dfrac{(2\pi i)^n}{n!} \dfrac{\partial f}{\partial z_k}, & m = 0, \\ 0, & m > 0. \end{cases}$$

Proof. The proof follows from Stokes's formula and the mean-value theorem. □

Lemma 1.14. *If the differential form γ of type (p,q) has coefficients of class C^1 in $B(z,r)$, then*

$$\lim_{\epsilon\to 0^+}\int_{S(z,\epsilon)}\gamma(\zeta)\wedge U_{p,q}(\zeta,z)=(1-q/n)\,\gamma(z).$$

Proof. It suffices to prove the lemma for γ of the form

$$\gamma(\zeta)=f(\zeta)\,d\bar\zeta_I\wedge d\zeta_J,\qquad I=(i_1,\dots,i_q),\quad J=(j_1,\dots,j_q). \tag{1.10}$$

For $K=(k_1,\dots,k_q)$, we have

$$\int_{S(z,\epsilon)}\gamma(\zeta)\wedge U_{p,q}(\zeta,z)=\epsilon^{-2n}(2\pi i)^{-n}(n-1)!\,\times$$
$$\times\int_{S(z,\epsilon)}f(\zeta)\,d\bar\zeta_I\wedge{\sum_K}'\sum_{k\notin K}\sigma(K,k)(\bar\zeta_k-\bar z_k)\,d\bar\zeta[K,k]\wedge d\zeta\,d\bar z_K\wedge dz_J.$$

By Lemma 1.13,

$$\lim_{\epsilon\to 0^+}\epsilon^{-2n}(2\pi i)^{-n}(n-1)!\int_{S(z,\epsilon)}f(\zeta)\sum_{k\notin I}(-1)^{k-1}(\bar\zeta_k-\bar z_k)\,d\bar\zeta[k]\wedge d\zeta$$
$$=((n-q)/n)f(z).$$

The integral

$$\epsilon^{-2n}\int_{S(z,\epsilon)}f(\zeta){\sum_{K\neq I}}'\sum_{k\notin K}\sigma(K,k)(\bar\zeta_k-\bar z_k)\,d\bar\zeta_I\wedge d\bar\zeta[K,k]\wedge d\zeta$$

consists of terms of the form $\epsilon^{-2n}\int_{S(z,\epsilon)}f(\zeta)(\bar\zeta_k-\bar z_k)\,d\bar\zeta[j]\wedge d\zeta$, where $j\neq k$, so (by Lemma 1.13) such integrals tend to zero when $\epsilon\to 0^+$. □

Lemma 1.15. *Suppose D and γ satisfy the hypotheses of Theorem 1.11. Then for $z\in D$, we have*

$$\bar\partial_z\int_D\gamma(\zeta)\,U_{p,q-1}(\zeta,z)=\mathrm{P.V.}\int_D\gamma(\zeta)\wedge\bar\partial_z U_{p,q-1}(\zeta,z)-(q/n)\,\gamma(z).$$

Proof. Suppose γ has the form (1.10). First we compute

$$\frac{\partial}{\partial\bar z_k}\int_D f(\zeta)\frac{\bar\zeta_j-\bar z_j}{|\zeta-z|^{2n}}\,d\bar\zeta\wedge d\zeta.$$

We have

$$\int_{B(z,\epsilon)}f(\zeta)\frac{\bar\zeta_j-\bar z_j}{|\zeta-z|^{2n}}\,d\bar\zeta\wedge d\zeta=\frac{1}{1-n}\int_{B(z,\epsilon)}f(\zeta)\frac{\partial}{\partial\zeta_j}\frac{1}{|\zeta-z|^{2n-2}}\,d\bar\zeta\wedge d\zeta$$
$$=\frac{(-1)^{n-j-1}}{1-n}\int_{S(z,\epsilon)}\frac{f(\zeta)}{|\zeta-z|^{2n-2}}\,d\bar\zeta\wedge d\zeta[j]-\frac{1}{1-n}\int_{B(z,\epsilon)}\frac{1}{|\zeta-z|^{2n-2}}\frac{\partial f}{\partial\zeta_j}\,d\bar\zeta\wedge d\zeta.$$

Then

$$\frac{\partial}{\partial \bar{z}_k} \int_{B(z,\epsilon)} f(\zeta) \frac{\bar{\zeta}_j - \bar{z}_j}{|\zeta - z|^{2n}} \, d\bar{\zeta} \wedge d\zeta$$
$$= (-1)^{n+j} \int_{S(z,\epsilon)} f(\zeta) \frac{\zeta_k - z_k}{|\zeta - z|^{2n}} \, d\bar{\zeta} \wedge d\zeta[j] + \int_{B(z,\epsilon)} \frac{\zeta_k - z_k}{|\zeta - z|^{2n}} \frac{\partial f}{\partial \zeta_j} \, d\bar{\zeta} \wedge d\zeta.$$

The second integral in this formula converges to zero as $\epsilon \to 0^+$ because of the absolute integrability of the integrand, and the first integral converges by Lemma 1.13 to $-((2\pi i)^n/n!)\delta_{kj} f(z)$, where δ_{kj} is the Kronecker symbol. Therefore

$$\frac{\partial}{\partial \bar{z}_k} \int_{D} f(\zeta) \frac{\bar{\zeta}_j - \bar{z}_j}{|\zeta - z|^{2n}} \, d\bar{\zeta} \wedge d\zeta$$
$$= \text{P.V.} \int_D f(\zeta) \frac{\partial}{\partial \bar{z}_k} \frac{\bar{\zeta}_j - \bar{z}_j}{|\zeta - z|^{2n}} \, d\bar{\zeta} \wedge d\zeta - \frac{(2\pi i)^n}{n!} \delta_{kj} f(z). \quad (1.11)$$

We obtain the assertion of the lemma by using (1.11) and the definition of $U_{p,q-1}$. □

The argument also shows that the integrals in Lemma 1.15 are functions of class $\mathcal{C}^1(\overline{D})$.

Remark. The proof of Lemma 1.15 shows that if $\partial D \in \mathcal{C}^1$ and $z \in \partial D$, then

$$\overline{\partial}_z \int_D \gamma(\zeta) \wedge U_{p,q-1}(\zeta, z) = \text{P.V.} \int_D \gamma(\zeta) \wedge \overline{\partial}_z U_{p,q-1}(\zeta, z) - (q/2n)\, \gamma(z).$$

Proof of Theorem 1.11. Let γ be a form of the type (1.10). If $z \notin \overline{D}$, then (1.9) is a consequence of Stokes's formula and Lemma 1.12. Suppose $z \in D$. The form $U_{p,q}(\zeta, z)$ has no singularities in the domain $D \setminus B(z, \epsilon)$, so (by Stokes's formula)

$$\int_{\partial(D \setminus B(z,\epsilon))} \gamma \wedge U_{p,q} = \int_{\partial D} \gamma \wedge U_{p,q} - \int_{S(z,\epsilon)} \gamma \wedge U_{p,q} = \int_{D \setminus B(z,\epsilon)} d(\gamma \wedge U_{p,q})$$
$$= \int_{D \setminus B(z,\epsilon)} \overline{\partial}\gamma \wedge U_{p,q} + (-1)^{p+q} \int_{D \setminus B(z,\epsilon)} \gamma \wedge \overline{\partial}_\zeta U_{p,q}$$
$$= \int_{D \setminus B(z,\epsilon)} \overline{\partial}\gamma \wedge U_{p,q} + \int_{D \setminus B(z,\epsilon)} \gamma \wedge \overline{\partial}_z U_{p,q-1}.$$

Now we pass to the limit as $\epsilon \to 0^+$ in these integrals. Using Lemmas 1.14 and 1.15, we have

$$\int_{\partial D} \gamma \wedge U_{p,q} - (1 - q/n)\, \gamma(z) = \int_D \overline{\partial}\gamma \wedge U_{p,q} + \overline{\partial}_z \int_D \gamma \wedge U_{p,q-1} + (q/n)\, \gamma(z).$$

This completes the proof. This proof is due to Tarkhanov. □

When $p = q = 0$, we obtain (1.4) from (1.9) by using that $U_{0,0} = U$ and $U_{0,-1} = 0$.

Formula (1.9) also has been generalized in the spirit of the Cauchy-Fantappiè formula and plays an important role in modern multidimensional complex analysis (see [3, 7, 10, 74, 77, 104, 130, 131, 172, 203]).

2 Boundary behavior of the Bochner-Martinelli integral

2.1 The Sokhotskiĭ-Plemelj formula for functions satisfying a Hölder condition

Let D be a bounded domain with piecewise-smooth boundary, and let f be an integrable function on ∂D (that is, $f \in \mathcal{L}^1(\partial D)$). We consider the Bochner-Martinelli (type) integral

$$F(z) = \int_{\partial D} f(\zeta)\, U(\zeta, z), \qquad z \notin \partial D. \tag{2.1}$$

It is a function that is harmonic both in D and in $\mathbf{C}^n \setminus \overline{D}$; moreover, $F(z) = O(|z|^{1-2n})$ as $|z| \to \infty$. Sometimes, when it is necessary to distinguish, we will write F^+ for the integral (2.1) when $z \in D$, and F^- when $z \notin \overline{D}$. When $z \in \partial D$, the integral (2.1) in general does not exist as an improper integral, since the integrand has the singularity $|\zeta - z|^{1-2n}$. Therefore, when $z \in \partial D$, we will consider the Cauchy principal value of the Bochner-Martinelli integral:

$$\text{P.V.} \int_{\partial D} f(\zeta)\, U(\zeta, z) = \lim_{\epsilon \to 0+} \int_{\partial D \setminus B(z,\epsilon)} f(\zeta)\, U(\zeta, z), \qquad z \in \partial D.$$

Subsequently we shall sometimes omit the principal value sign P.V., that is, we shall always assume that an integral of the form (2.1) is understood in the sense of a principal value when $z \in \partial D$.

In this section we are interested in analogues of the Sokhotskiĭ-Plemelj formula for the Bochner-Martinelli integral, that is, in the connection between the boundary values of the functions $F^{\pm}(z)$ and the singular integral. First we shall consider the simpler case when the density f satisfies a Hölder condition with exponent $\alpha > 0$, that is, $|f(\zeta) - f(\eta)| \leq C|\zeta - \eta|^{\alpha}$ for ζ and η in ∂D. Generally speaking, these formulas can be deduced from properties of potentials, but we will give a direct proof.

When $z \in \partial D$, we denote by $\tau(z)$ the expression

$$\tau(z) = \lim_{\epsilon \to 0+} \operatorname{vol}\{S(z, \epsilon) \cap D\} / \operatorname{vol} S(z, \epsilon).$$

In other words, $\tau(z)$ is the solid angle of the tangent cone to the surface ∂D at z. Since we are considering a domain D with piecewise-smooth boundary, the quantity $\tau(z)$ is defined and is not zero.

Lemma 2.1. P. V. $\int_{\partial D} U(\zeta, z) = \tau(z)$ *for* $z \in \partial D$.

Proof. By definition,

$$\text{P. V.} \int_{\partial D} U(\zeta, z) = \lim_{\epsilon \to 0+} \int_{\partial D \setminus B(z,\epsilon)} U(\zeta, z).$$

But

$$\int_{\partial D \setminus B(z,\epsilon)} U(\zeta, z) = \int_{\partial(D \setminus B(z,\epsilon))} U(\zeta, z) + \int_{S^+(z,\epsilon)} U(\zeta, z),$$

where $S^+(z,\epsilon)$ is the part of the sphere $S(z,\epsilon)$ lying in D, that is, $S^+(z,\epsilon) = D \cap S(z,\epsilon)$. The sign of the second term has been changed because the orientation of $S(z,\epsilon)$ (induced by the orientation of the ball $B(z,\epsilon)$) is opposite to the orientation of ∂D. For $z \notin D \setminus B(z,\epsilon)$, the integral $\int_{\partial(D\setminus B(z,\epsilon))} U(\zeta, z) = 0$, while the form $U(\zeta, z)$ is closed, so

$$\begin{aligned} \int_{\partial D \setminus B(z,\epsilon)} U(\zeta, z) &= \int_{S^+(z,\epsilon)} U(\zeta, z) \\ &= \frac{(n-1)!}{(2\pi i)^n \epsilon^{2n}} \int_{S^+(z,\epsilon)} \sum_{k=1}^{n} (-1)^{k-1} (\bar{\zeta}_k - \bar{z}_k)\, d\bar{\zeta}[k] \wedge d\zeta. \end{aligned}$$

It remains to note that the restriction

$$\sum_{k=1}^{n} (-1)^{k-1} (\bar{\zeta}_k - \bar{z}_k)\, d\bar{\zeta}[k] \wedge d\zeta \big|_{S(z,\epsilon)} = \epsilon\, 2^{n-1} i^n\, d\sigma, \tag{2.2}$$

where $d\sigma$ is the area element on the sphere. Indeed, the restriction of the form $dx[k]$ to a smooth surface S in $\mathbf{R}^n$ equals $(-1)^{k-1}\gamma_k\, d\sigma$, where γ_k is the kth direction cosine of the normal to S. Thus, for the unit sphere $S(0,1)$, the restriction of $dx[k]$ to $S(0,1)$ equals $(-1)^{k-1} x_k\, d\sigma$. Now passing to coordinates $(x_1, \ldots, x_n, y_1, \ldots, y_n)$ in $\mathbf{C}^n \cong \mathbf{R}^{2n}$, we express $d\bar{\zeta}[k] \wedge d\zeta$ in terms of the forms $dx[k] \wedge dy$ and $dx \wedge dy[k]$ and take their restrictions to $S(z,\epsilon)$ to obtain (2.2). (In §3, we will give expressions for the restrictions of the forms $d\bar{z}[k] \wedge dz$ and $dz[k] \wedge d\bar{z}$ to an arbitrary smooth surface.)

Thus

$$\int_{\partial D \setminus B(z,\epsilon)} U(\zeta, z) = \operatorname{vol} S^+(z,\epsilon) / \operatorname{vol} S(z,\epsilon) \to \tau(z)$$

as $\epsilon \to 0^+$. □

We extend $f(z)$ to a neighborhood $V(\partial D)$ as a function satisfying a Hölder condition on $V(\partial D)$ with the same exponent α, and we again denote it by $f(z)$. Consider the integral

$$\Phi(z) = \int_{\partial D} (f(\zeta) - f(z))\, U(\zeta, z). \tag{2.3}$$

If $z \notin \partial D$, then the integral (2.3) has no singularity, while if $z \in \partial D$, then

$$|f(\zeta) - f(z)| \cdot |U(\zeta, z)| \leq C|\zeta - z|^{\alpha+1-2n}\, d\sigma,$$

so the integral $\Phi(z)$ is absolutely convergent.

Lemma 2.2. *If f satisfies a Hölder condition in $V(\partial D)$ with exponent α, where $0 < \alpha < 1$, then $\Phi(z)$ satisfies a Hölder condition in $V(\partial D)$ with the same exponent α.*

Proof. Let z^1 and z^2 be points in $V(\partial D)$ with $|z^1 - z^2| = \delta$, where δ is sufficiently small. Consider the ball $B(z^1, 2\delta) \subset V(\partial D)$, and set $\sigma_\delta = \partial D \cap B(z^1, 2\delta)$. Then

$$\left| \int_{\sigma_\delta} (f(\zeta) - f(z^j))\, U(\zeta, z^j) \right| \leq C_1 \int_{\sigma_\delta} |\zeta - z^j|^{1+\alpha-2n}\, d\sigma \leq C_2 \delta^\alpha$$

for $j = 1, 2$. When σ_δ is a smooth surface, it is easy to obtain this inequality by replacing the z^j by their projections onto σ_δ and the integral over σ_δ by the integral over the $(2n-1)$-dimensional sphere of radius δ, and passing to polar coordinates in this sphere. If σ_δ is piecewise smooth, we estimate the integral over each smooth piece of σ_δ this way.

We consider the difference of the integrals (2.3) over $\partial D \setminus \sigma_\delta$ at the points z^1 and z^2, which equals

$$\int_{\partial D \setminus \sigma_\delta} (f(\zeta) - f(z^2))\, (U(\zeta, z^2) - U(\zeta, z^1)) + (f(z^1) - f(z^2)) \int_{\partial D \setminus \sigma_\delta} U(\zeta, z^1). \tag{2.4}$$

We have already dealt with the second integral in Lemma 2.1 (except that there $z^1 \in \partial D$), from which we obtain $\left|\int_{\partial D \setminus \sigma_\delta} U(\zeta, z^1)\right| \leq 1$. Consequently,

$$|f(z^1) - f(z^2)| \left| \int_{\partial D \setminus \sigma_\delta} U(\zeta, z^1) \right| \leq C_3 \delta^\alpha.$$

Now we estimate the first term in (2.4). If $\zeta \in \partial D \setminus \sigma_\delta$, then

$$\begin{aligned}
\left| \frac{(\bar{\zeta}_j - \bar{z}^1_j)}{|\zeta - z^1|^{2n}} - \frac{(\bar{\zeta}_j - \bar{z}^2_j)}{|\zeta - z^2|^{2n}} \right| &\leq \frac{|z^2_j - z^1_j|}{|\zeta - z^1|^{2n}} + |\zeta_j - z^2_j| \left| \frac{1}{|\zeta - z^1|^{2n}} - \frac{1}{|\zeta - z^2|^{2n}} \right| \\
&\leq \frac{|\zeta_j - z^2_j| \cdot |z^2 - z^1|}{|\zeta - z^1| \cdot |\zeta - z^2|} \sum_{s=0}^{2n-1} \frac{|\zeta - z^2|^{s+1-2n}}{|\zeta - z^1|^s} + \frac{|z^2 - z^1|}{|\zeta - z^1|^{2n}} \\
&\leq C_4 \delta |\zeta - z^1|^{-2n}
\end{aligned}$$

since $|\zeta - z^1| \leq 2|\zeta - z^2|$. Thus

$$|\Phi(z^1) - \Phi(z^2)| \leq C_5 \delta^\alpha + C_6 \delta \int_{\partial D \setminus \sigma_\delta} |\zeta - z^1|^{\alpha - 2n}\, d\sigma.$$

If σ_δ is a smooth surface, then by replacing the point z^1 by its projection onto σ_δ we obtain

$$\int_{\partial D\setminus\sigma_\delta} |\zeta - z^1|^{\alpha-2n}\, d\sigma \leq C_7\delta^{\alpha-1}.$$

□

Remark. As in the case of a Cauchy-type integral, when $\alpha = 1$ the function $\Phi(z)$ will satisfy the condition

$$|\Phi(z^1) - \Phi(z^2)| \leq C|z^1 - z^2| \cdot |\ln|z^1 - z^2||,$$

since $\int_{\partial D\setminus\sigma_\delta} |\zeta - z^1|^{1-2n}\, d\sigma \leq C_8|\ln\delta|$.

Theorem 2.3. *Let D be a bounded domain with piecewise-smooth boundary ∂D, and let $f \in \mathcal{C}^\alpha(\partial D)$, where $0 < \alpha < 1$. Then the Bochner-Martinelli integral F^+ extends continuously to $\overline{D}$ as a function of class $\mathcal{C}^\alpha(\overline{D})$, while F^- extends continuously to $\mathbf{C}^n\setminus D$ as a function of class $\mathcal{C}^\alpha(\mathbf{C}^n\setminus D)$. Moreover, the Sokhotskiĭ-Plemelj formulas are valid: for $z \in \partial D$,*

$$\begin{aligned} F^+(z) &= (1-\tau(z))f(z) + \text{P.V.}\int_{\partial D} f(\zeta)\, U(\zeta, z), \\ F^-(z) &= -\tau(z) + \text{P.V.}\int_{\partial D} f(\zeta)\, U(\zeta, z). \end{aligned} \tag{2.5}$$

Proof. The first part of the theorem follows from Lemma 2.2. We consider the integral

$$\text{P.V.}\int_{\partial D} f(\zeta)\, U(\zeta, z) = \int_{\partial D} (f(\zeta) - f(z))\, U(\zeta, z) + \tau(z)f(z)$$

(by Lemma 2.1). Since Φ is continuous in $V(\partial D)$ (by Lemma 2.2),

$$\int_{\partial D} (f(\zeta) - f(z))\, U(\zeta, z) = F^+(z) - f(z),$$

that is, $F^+(z) = (1 - \tau(z))f(z) + \text{P.V.}\int_{\partial D} f(\zeta)\, U(\zeta, z)$. On the other hand, $\int_{\partial D}(f(\zeta) - f(z))\, U(\zeta, z) = F^-(z)$. □

Remark. If we introduce the norm

$$\|f\|_{\mathcal{C}^\alpha} = \sup_{\partial D}|f| + \sup_{\zeta,\eta\in\partial D} |f(\zeta) - f(\eta)| \cdot |\zeta - \eta|^{-\alpha}$$

in the space $\mathcal{C}^\alpha(\partial D)$ of functions f satisfying a Hölder condition with exponent α, then Lemmas 2.1 and 2.2 show that the Bochner-Martinelli integral and the Bochner-Martinelli singular integral define bounded operators in this space for $0 < \alpha < 1$ (when $\partial D \in \mathcal{C}^1$).

Lemma 2.2 is contained in a paper of Chirka [36]. Many versions of Theorem 2.3 have been given. Lu Qi-Keng and Zhong Tongde [141] proved (2.5) for domains with boundary of class C^2. Kakichev [91] showed in this case that F^+ and F^- satisfy a Hölder condition on ∂D. Then these formulas were obtained by Harvey and Lawson [66] for domains with smooth boundary. The case of piecewise-smooth boundary was noted in [139, 161].

Corollary 2.4. *If $\partial D \in C^1$, then for $z \in \partial D$ formula (2.5) takes the form*

$$\begin{aligned} F^+(z) &= \frac{1}{2}f(z) + \mathrm{P.V.}\int_{\partial D} f(\zeta)\,U(\zeta, z), \\ F^-(z) &= -\frac{1}{2}f(z) + \mathrm{P.V.}\int_{\partial D} f(\zeta)\,U(\zeta, z), \end{aligned} \tag{2.6}$$

and therefore the Bochner-Martinelli singular integral also satisfies a Hölder condition with exponent α on ∂D.

Corollary 2.5. *If ∂D is piecewise smooth, then*

$$F^+(z) - F^-(z) = f(z), \qquad z \in \partial D.$$

2.2 Analogue of Privalov's theorem for integrable functions

In this subsection, we consider bounded domains D with boundary of class C^1 and functions f that are integrable on ∂D (that is, $f \in \mathcal{L}^1(\partial D)$). Let $z^0 \in \partial D$. Consider a right circular cone V_{z^0} with vertex at z^0 and axis that coincides with the normal to ∂D at z^0, the angle β between the axis and the generator of the cone being less than $\pi/2$. Let $z \in D \cap V_{z^0}$. Suppose that z^0 is a Lebesgue point for f, that is,

$$\lim_{\epsilon \to 0+} \epsilon^{1-2n} \int_{\partial D \cap B(z^0, \epsilon)} |f(\zeta) - f(z^0)|\, d\sigma = 0.$$

Theorem 2.6 (Kytmanov). *If $z \in D \cap V_{z^0}$, then*

$$\lim_{\substack{z \to z^0 \\ z \in V_{z^0}}} \left[\int_{\partial D} (f(\zeta) - f(z^0))\, U(\zeta, z) - \int_{\partial D \setminus B(z^0, |z - z_0|)} (f(\zeta) - f(z^0))\, U(\zeta, z^0) \right] = 0.$$

Proof. We make a unitary transformation of $\mathbf{C}^n$ and a translation so that z^0 goes to 0 and the tangent plane to ∂D at z^0 goes to the plane $T = \{ w \in \mathbf{C}^n : \operatorname{Im} w_n = 0 \}$. The surface ∂D will then be given in a neighborhood of the origin by equations $\zeta_1 = w_1, \ldots, \zeta_{n-1} = w_{n-1}, \zeta_n = u_n + i\varphi(w)$, where $w = (w_1, \ldots, w_{n-1}, u_n) \in T$; the function $\varphi \in C^1(W)$, where W is a neighborhood of the origin in the plane T; and $\varphi(w) = o(|w|)$ as $w \to 0$. We denote the projection of z onto the $\operatorname{Im} w_n$ axis by $\tilde{z}$. Then $|z - \tilde{z}| \le |\tilde{z}| \tan\beta$ and $|z| \le |\tilde{z}|/\cos\beta$.

Fix $\epsilon_0 > 0$, and choose a $(2n-1)$-dimensional ball B' in the plane T with center at 0 and radius ϵ_0 such that

1. $B' \subset W$;

2. $|w - \tilde{z}| \le C|\zeta(w) - z|$ for $w \in B'$, where C is a constant not depending on w and z.

Condition (2) is guaranteed by the relations

$$\begin{aligned}
|w - \zeta(w)| &= |\varphi(w)| = o(|w|), \qquad |w| \to 0;\\
|w| &\le |w - \tilde{z}|, \quad |\tilde{z}| \le |w - \tilde{z}|,\\
|w - \tilde{z}| &\le |w - \zeta(w)| + |\zeta(w) - z| + |z - \tilde{z}|\\
&\le |\varphi(w)| + |\zeta(w) - z| + |\tilde{z}| \tan\beta\\
&\le |\varphi(w)| + |\zeta(w) - z| + (\tan\beta)(|w - \zeta(w)| + |\zeta(w) - z|)\\
&= (1 + \tan\beta)(|\varphi(w)| + |\zeta(w) - z|)\\
&\le C|\zeta(w) - z|.
\end{aligned}$$

We note that the ball B' and the constant C may be taken to be independent of the point $z^0 = 0$. If 0 is a Lebesgue point for the function $f(\zeta)$, then 0 is also a Lebesgue point for the function $f(\zeta(w))$.

It is clear that the form of the kernel $U(\zeta, z)$ does not change under a translation.

Lemma 2.7. *The kernel $U(\zeta, z)$ is invariant with respect to unitary transformations.*

Proof. Suppose that the unitary transformation has the form $\zeta = A\zeta'$, where A is a unitary matrix $\|a_{jk}\|_{j,k=1}^n$. Then the distance $|\zeta - z|$ does not change, $d\zeta = (\det A)\, d\zeta' = e^{i\psi}\, d\zeta'$, and

$$\sum_{k=1}^n (-1)^{k-1}(\zeta_k - z_k)\, d\zeta[k] = \sum_{k=1}^n (-1)^{k-1} \sum_{j=1}^n a_{jk}(\zeta'_j - z'_j) \sum_{p=1}^n A_{pk}\, d\zeta'[p],$$

where A_{pk} is the minor of the matrix A corresponding to the element a_{pk}, so that

$$\sum_{k=1}^n (-1)^{k-1} a_{jk} A_{pk} = \begin{cases} 0, & j \ne p, \\ (-1)^{p-1} \det A, & j = p. \end{cases}$$

□

We now continue with the proof of Theorem 2.6. Let $|z| = \epsilon$. We transform the difference of the integrals

$$\begin{aligned}
&\int_{\partial D} (f(\zeta) - f(0))\, U(\zeta, z) - \int_{\partial D \setminus B(0,\epsilon)} (f(\zeta) - f(0))\, U(\zeta, 0)\\
&= \int_{\partial D \setminus B(0,\epsilon)} (f(\zeta) - f(0))\, (U(\zeta, z) - U(\zeta, 0)) + \int_{\partial D \cap B(0,\epsilon)} (f(\zeta) - f(0))\, U(\zeta, z).
\end{aligned}$$

Now $|\zeta - z|^{1-2n} \le C_1\epsilon^{1-2n}$, since $C|\zeta(w) - z| \ge |w - \tilde{z}| \ge |\tilde{z}| \ge \epsilon\cos\beta$, so

$$\left|\int_{\partial D\cap B(0,\epsilon)} (f(\zeta) - f(0))\, U(\zeta, z)\right| \le C_2\epsilon^{1-2n} \int_{\partial D\cap B(0,\epsilon)} |f(\zeta) - f(0)|\, d\sigma \to 0$$

as $\epsilon \to 0^+$. Now consider the difference

$$\frac{(\bar{\zeta}_j - \bar{z}_j)}{|\zeta - z|^{2n}} - \frac{\bar{\zeta}_j}{|\zeta|^{2n}} = \bar{\zeta}_j\left(\frac{1}{|\zeta - z|^{2n}} - \frac{1}{|\zeta|^{2n}}\right) - \frac{\bar{z}_j}{|\zeta - z|^{2n}}.$$

We have

$$\frac{|z_j|}{|\zeta - z|^{2n}} \le \frac{C_3|\tilde{z}|}{|w - \tilde{z}|^{2n}} = \frac{C_3|\tilde{z}|}{(|w|^2 + |\tilde{z}|^2)^n}.$$

However

$$|\zeta_j|\left|\frac{1}{|\zeta - z|^{2n}} - \frac{1}{|\zeta|^{2n}}\right| = \frac{|\zeta_j|\cdot\big|\,|\zeta| - |\zeta - z|\,\big|}{|\zeta|\cdot|\zeta - z|}\sum_{s=0}^{2n-1}\frac{|\zeta - z|^{1+s-2n}}{|\zeta|^s}$$
$$\le |z|\sum_{s=0}^{2n-1}\frac{|\zeta - z|^{s-2n}}{|\zeta|^s}.$$

We have $|\zeta(w)| \ge C_4|w| \ge C_5|w - \tilde{z}|$ since $\zeta \notin B(0,\epsilon)$, and the ratio $|w|/|w - \tilde{z}|$ is bounded below by a nonzero constant because it equals the cosine of the angle between the vectors w and $w - \tilde{z}$, and this angle cannot be greater than $\pi/4$. Thus

$$|U(\zeta, z) - U(\zeta, 0)| \le C_6|\tilde{z}|\,(|w|^2 + |\tilde{z}|^2)^{-n}\, d\sigma.$$

Since $d\sigma \le C_7\, dS$, where dS is the area element of the plane T, we obtain

$$\int_{B(0,\epsilon^0)\cap\partial D\setminus B(0,\epsilon)} |f(\zeta) - f(0)|\cdot|U(\zeta, z) - U(\zeta, 0)|$$
$$\le C_8\int_{B(0,\epsilon^0)\cap T\setminus B(0,\epsilon)} |f(\zeta(w)) - f(0)|\cdot|\tilde{z}|\,(|w|^2 + |\tilde{z}|^2)^{-n}\, dS$$
$$\le C_8\int_{T\cap B(0,\epsilon^0)} |f(\zeta(w)) - f(0)|\cdot|\tilde{z}|\,(|w|^2 + |\tilde{z}|^2)^{-n}\, dS.$$

If $\epsilon \to 0^+$, then $|\tilde{z}| \to 0$, while the expression $|\tilde{z}|(|w|^2 + |\tilde{z}|^2)^{-n}$ is the Poisson kernel for the half-space. Since 0 is a Lebesgue point of $f(\zeta(w))$, it is well known that this integral converges to zero as $\epsilon \to 0^+$ (see, for example, [195]). □

This theorem (see [122]) is an analogue of Privalov's theorem (see [163]) for an integral of Cauchy type.

Remark. It can be seen from the proof of the theorem that the hypothesis on ∂D can be weakened; it is enough that there exist a tangent plane T_{z^0} to ∂D at z^0.

Theorem 2.6 shows that the existence of the Bochner-Martinelli singular integral at z^0 is equivalent to the existence of the limit of $F^+(z)$ as $z \to z^0$ along nontangential paths. Therefore, if the singular integral exists, so does $\lim_{z \to z^0} F^+(z)$, and also the Sokhotskiĭ-Plemelj formula (2.6) holds.

We now give a sufficient condition for the existence of the singular integral. Let $\omega_f(\delta, z) = \sup_{\zeta \in \partial D \cap B(z,\delta)} |f(\zeta) - f(z)|$ denote the modulus of continuity of f at z.

Corollary 2.8. *If $f \in \mathcal{L}^1(\partial D)$, and if the Dini condition*

$$\int_0^1 \omega_f(\delta, z)\delta^{-1}\, d\delta < +\infty$$

holds at the point $z \in \partial D$, then the Bochner-Martinelli singular integral exists at z, and the Sokhotskiĭ-Plemelj formula (2.6) holds.

Here, as in Theorem 2.6, the point $\tilde{z}$ tends to z along nontangential paths.

Proof. First we note that if the Dini condition holds for f at z, then f is continuous at z (that is, $\omega_f(\delta, z) \to 0$ as $\delta \to 0^+$), so z is a Lebesgue point for f. Consequently, Theorem 2.6 holds for f and z. Consider the integral:

$$\begin{aligned}\int_{\partial D \cap B(z,\epsilon)} |f(\zeta) - f(z)| \cdot |U(\zeta, z)| &\leq C_1 \int_{\partial D \cap B(z,\epsilon)} |f(\zeta) - f(z)| \cdot |\zeta - z|^{1-2n}\, d\sigma \\ &\leq C_2 \int_0^\epsilon \omega_f(\delta, z)\delta^{1-2n}\, d\delta \int_{S(z,\delta)\cap\partial D} d\tau_\delta \\ &\leq C_3 \int_0^\epsilon \omega_f(\delta, z)\delta^{-1}\, d\delta < +\infty.\end{aligned}$$

□

Corollary 2.8 was given in [161]. A closely related question was considered by Gaziev [53, 54].

If $f \in \mathcal{L}^1(\partial D)$, the existence of the Bochner-Martinelli singular integral at almost all points of ∂D can be deduced from the general theory of singular integral operators (see, for example, [95, 153, 193]). The Bochner-Martinelli operator is a singular integral operator because it is defined on constant functions. Therefore the Sokhotskiĭ-Plemelj formula (2.6) holds at almost all points $z \in \partial D$. Moreover, this singular integral defines a bounded operator from $\mathcal{L}^p(\partial D)$ to $\mathcal{L}^p(\partial D)$ for $p > 1$.

2.3 Further results

First we mention that a Sokhotskiĭ-Plemelj formula was obtained in [231] for Yarmukhamedov's representation (see Theorem 1.8).

For the Cauchy-Fantappiè formula in strongly pseudoconvex domains with the Henkin-Ramirez kernels, the question of the boundary behavior of the Cauchy-Fantappiè integral with Hölder or integrable density has been studied in detail (see the surveys [3, 74, 75, 77, 95]). The Sokhotskiĭ-Plemelj formula for this case was obtained in [12, 96]. We should mention that the principal value of the Cauchy-Fantappiè singular integral is determined by deleting Hörmander "balls" rather than the usual balls $B(z, \epsilon)$. Generalizations of these results are given in [59]. The Sokhotskiĭ-Plemelj formula was considered for some other Cauchy-Fantappiè integrals in [35, 60, 138]. Formula (2.6) for integral representations on Stein manifolds was found in [35, 131]. For Koppelman integral representations for elliptic complexes, it was obtained in [203].

3 Jump theorems for the Bochner-Martinelli integral

3.1 Integrable and continuous functions

We saw in §2 that the Sokhotskiĭ-Plemelj formula implies a jump theorem (see Corollary 2.5). As a rule, a jump theorem is simpler to prove than a Sokhotskiĭ-Plemelj formula, and moreover the difference $F^+ - F^-$ may have a limit on ∂D even when the functions F^+ and F^- themselves do not. Therefore jump theorems hold for a wider class of functions than do Sokhotskiĭ-Plemelj formulas.

First we study the case when D is a bounded domain with boundary of class $\mathcal{C}^1$, and $f \in \mathcal{L}^1(\partial D)$. As in Theorem 2.6, we consider a right circular cone V_{z^0} with vertex at $z^0 \in \partial D$ whose axis coincides with the normal to ∂D at z^0, the angle β between the axis and the generator being less than $\pi/2$. We take two points $z^+ \in V_{z^0} \cap D$ and $z^- \in V_{z^0} \cap (\mathbf{C}^n \setminus \overline{D})$ such that $a|z^+ - z^0| \le |z^- - z^0| \le b|z^+ - z^0|$, where a and b are constants not depending on $z^\pm$, and $0 < a \le b < \infty$.

Theorem 3.1 (Dautov, Kytmanov). *If z^0 is a Lebesgue point of the function $f \in \mathcal{L}^1(\partial D)$, then*

$$\lim_{z^\pm \to z^0} (F(z^+) - F(z^-)) = f(z^0) \tag{3.1}$$

(where F is defined by (2.1)). If $f \in \mathcal{C}(\partial D)$, then the limit (3.1) exists for all points $z^0 \in \partial D$, and it is attained uniformly if the angle β and the constants a and b are fixed.

Proof. The proof is carried out in essentially the same way as the proof of Theorem 2.6. Using a unitary transformation and a translation, we take z^0 to 0 and the tangent plane to ∂D at z^0 to the plane $T = \{w \in \mathbf{C}^n : \operatorname{Im} w_n = 0\}$. The surface ∂D will then be given in a neighborhood of 0 by a system of equations $\zeta_1 = w_1, \ldots, \zeta_{n-1} = w_{n-1}, \zeta_n = u_n + i\varphi(w)$, where $w = (w_1, \ldots, w_{n-1}, u_n) \in T$, the function $\varphi(w)$ is of class $\mathcal{C}^1$ in a neighborhood W of 0 in the plane T, and $\varphi(w) = o(|w|)$ as $w \to 0$. We denote the projections of the points $z^\pm$ onto the

$\operatorname{Im} w_n$ axis by $\tilde{z}^{\pm}$. Then

$$|z^{\pm} - \tilde{z}^{\pm}| \le |\tilde{z}^{\pm}| \cdot \tan\beta, \qquad |z^{\pm}| \le |\tilde{z}^{\pm}|/\cos\beta, \qquad \text{and} \\ a|\tilde{z}^{+}| \cdot \cos\beta \le |\tilde{z}^{-}| \le b|\tilde{z}^{+}|/\cos\beta. \tag{3.2}$$

As in Theorem 2.6, we fix a ball B' in the plane T with center at 0 and radius ϵ such that

1. $B' \subset W$;
2. $|w - \tilde{z}^{\pm}| \le C|\zeta(w) - z^{\pm}|$ for $w \in B'$, where C is a constant independent of the point $z^0 = 0$. Here $B' = B(z^0, \varepsilon) \cap T$ and $\Gamma = B(z^0, \epsilon) \cap \partial D$.

Consider the difference

$$F(z^+) - F(z^-) = \int_{\partial D} (f(\zeta) - f(z^0))\, U(\zeta, z^+) - \int_{\partial D} (f(\zeta) - f(z^0))\, U(\zeta, z^-) \\ + f(z^0) \int_{\partial D} (U(\zeta, z^+) - U(\zeta, z^-)).$$

Since $\int_{\partial D}(U(\zeta, z^+) - U(\zeta, z^-)) = 1$, it is enough to show that

$$\lim_{z^{\pm} \to z^0} \int_{\partial D} (f(\zeta) - f(z^0))\, (U(\zeta, z^+) - U(\zeta, z^-)) = 0.$$

In the integral $\int_{\partial D \setminus \Gamma}(f(\zeta)-f(z^0))\,(U(\zeta, z^+)-U(\zeta, z^-))$, we can take the limit inside, since $z^0 \notin \partial D \setminus \Gamma$. It remains to consider this integral over the set Γ. From condition (2) on the choice of B' and the inequality $|\zeta(w)| \le C_1|w| \le C_1|w - \tilde{z}^{\pm}|$, we obtain

$$\left| \frac{\bar{\zeta}_k}{|\zeta - z^+|^{2n}} - \frac{\bar{\zeta}_k}{|\zeta - z^-|^{2n}} \right| = \left| \frac{1}{|\zeta - z^+|} - \frac{1}{|\zeta - z^-|} \right| \sum_{j=0}^{2n-1} \frac{|\zeta_k| \cdot |\zeta - z^-|^{j+1-2n}}{|\zeta - z^+|^j} \\ = \big||\zeta - z^+| - |\zeta - z^-|\big| \sum_{j=0}^{2n-1} \frac{|\zeta_k| \cdot |\zeta - z^-|^{j-2n}}{|\zeta - z^+|^{j+1}} \\ \le C_1 C^{2n} \sum_{j=0}^{2n-1} \frac{|w - \tilde{z}^-|^{j-2n}(|z^+| + |z^-|)}{|w - \tilde{z}^+|^j}. \tag{3.3}$$

We may assume that $a_1 = a\cos\beta < 1$; then $|w - \tilde{z}^{\pm}| \ge |w - a_1\tilde{z}^+|$ in view of (3.2). Therefore we have from (3.3) that

$$\left| \frac{\bar{\zeta}_k}{|\zeta - z^+|^{2n}} - \frac{\bar{\zeta}_k}{|\zeta - z^-|^{2n}} \right| \le \frac{d\,|\tilde{z}^+|}{|w - a_1\tilde{z}^+|^{2n}},$$

where d depends only on a, b, C, C_1, and β. In precisely the same way,

$$\left| \frac{\bar{z}_k^+}{|\zeta - z^+|^{2n}} - \frac{\bar{z}_k^-}{|\zeta - z^-|^{2n}} \right| \le \frac{|z_k^+|}{|\zeta - z^+|^{2n}} + \frac{|z_k^-|}{|\zeta - z^-|^{2n}} \le \frac{d_1\,|\tilde{z}|}{|w - a_1\tilde{z}^+|^{2n}}.$$

Finally, $d\sigma \le d_2\, dS$, where dS is the surface area element on the surface T, and d_2 is independent of z^0. Therefore

$$\left|\int_\Gamma (f(\zeta) - f(0))\,(U(\zeta, z^+) - U(\zeta, z^-))\right| \le d_3 \int_{B'} \frac{|f(\zeta(w)) - f(0)| \cdot |\tilde{z}^+|}{(|w|^2 + a_1^2|\tilde{z}^+|^2)^n}\, dS. \tag{3.4}$$

Since $|\tilde{z}|/(|w|^2 + a_1^2|\tilde{z}^+|^2)^n$ is the Poisson kernel for the half-space, and 0 is a Lebesgue point for $f(\zeta(w))$, the last expression tends to zero as $|\tilde{z}^+| \to 0$ (see [195, Theorem 1.25]).

If f is continuous on ∂D, then for each $\delta > 0$, we choose a ball B' of radius ϵ such that $|f(\zeta(w)) - f(0)| < \delta$ for $w \in B'$ (where ϵ may be taken independent of the point $z^0 = 0$). Then we obtain from (3.4) that

$$\begin{aligned}\left|\int_\Gamma (f(\zeta) - f(0))\,(U(\zeta, z^+) - U(\zeta, z^-))\right| &\le d_4\delta \int_{B'} \frac{a_1|\tilde{z}^+|\, dS}{(|w|^2 + a_1^2|\tilde{z}^+|^2)^n} \\ &\le d_4\delta \int_T \frac{a_1|\tilde{z}^+|\, dS}{(|w|^2 + a_1^2|\tilde{z}^+|^2)^n},\end{aligned}$$

and the last integral equals a constant not depending on $\tilde{z}^+$. □

For continuous functions, Theorem 3.1 was given in [41]; it is an analogue of the corresponding result for an integral of Cauchy type (see, for example, [156, pp. 55–56]).

Corollary 3.2. *Under the hypotheses of Theorem 3.1, if F^+ extends continuously to $\overline{D}$, then F^- extends continuously to $\mathbf{C}^n \setminus D$, and conversely (where $f \in C(\partial D)$).*

This corollary, given in [41], was also remarked by Harvey and Lawson in [66].

We now give an example showing that when f is continuous, the function F can fail to extend to certain points of the boundary ∂D. This example is contained in [41] and is based on a corresponding example from [64, Chap. II, §7].

Let D be a domain such that $\overline{D}$ is contained in the unit ball $B(0,1)$, and ∂D contains a $(2n-1)$-dimensional ball B' of radius $R < 1$ with center at the point 0 in the plane $T = \{\, z \in \mathbf{C}^n : \operatorname{Im} z_n = 0 \,\}$. We set $f(\zeta) = \zeta_n/(|\zeta| \ln |\zeta|)$ on ∂D, so that $f \in C(\partial D)$. We will show that $F(z)$ is unbounded in every neighborhood of the origin. Set $z = (0, \ldots, 0, iy_n)$, with $y_n > 0$. We need to show that the integral $I(z) = \int_{B'} f(\zeta)\, U(\zeta, z)$ is unbounded in every neighborhood of the origin. Now $d\bar{\zeta}[k] \wedge d\zeta = 0$ on the set B' for $k \ne n$, so

$$I(z) = \frac{(n-1)!\,(-1)^{n-1}}{(2\pi i)^n} \int_{B'} \frac{\eta_n(\eta_n - iy_n)\, d\bar{\zeta}[n] \wedge d\zeta}{|\zeta| \ln|\zeta|\, (|\zeta|^2 + y_n^2)^n},$$

where $\eta_n = \operatorname{Re} \zeta_n$. As in Theorem 3.1,

$$\left|\int_{B'} f(\zeta) \frac{\eta_n y_n\, d\bar{\zeta}[n] \wedge d\zeta}{(|\zeta|^2 + y_n^2)^n}\right| \le C \int_T \frac{y_n\, dS}{(|\zeta|^2 + y_n^2)^n} \le C_1.$$

If we introduce polar coordinates in B', then $dS = |\zeta|^{2n-2}\, d|\zeta| \wedge d\omega$, where $d\omega$ is the surface area element on the unit sphere in $\mathbf{R}^{2n-1}$. Integrating in ω, we obtain

$$I_1 = \int_{B'} \frac{\eta_n^2\, dS}{|\zeta| \cdot |\ln|\zeta|| \cdot (|\zeta|^2 + y_n^2)^n} = C_2 \int_0^R \frac{|\zeta|^{2n}\, d|\zeta|}{|\zeta| \cdot |\ln|\zeta|| \cdot (|\zeta|^2 + y_n^2)^n}$$
$$\geq C_2 \int_\epsilon^R \frac{|\zeta|^{2n-1}\, d|\zeta|}{|\ln|\zeta|| \cdot (|\zeta|^2 + y_n^2)^n}.$$

However

$$\lim_{y_n \to 0} \int_\epsilon^R \frac{|\zeta|^{2n-1}\, d|\zeta|}{|\ln|\zeta|| \cdot (|\zeta|^2 + y_n^2)^n} = \int_\epsilon^R \frac{d|\zeta|}{|\zeta| \cdot |\ln|\zeta||} = -\ln|\ln R| + \ln|\ln \epsilon|.$$

Fix $M > 0$. If we take ϵ sufficiently small, then $\ln|\ln\epsilon| - \ln|\ln R| > 2M$, and so $I_1 > C_2 M$ for y_n sufficiently small.

If $f \in \mathcal{C}(\partial D)$, then the limit in Theorem 2.6 is attained uniformly, so we obtain the following result from Theorems 2.6 and 3.1.

Corollary 3.3. *Suppose $\partial D \in \mathcal{C}^1$ and $f \in \mathcal{C}(\partial D)$. For the integral F^+ to extend continuously to $\overline{D}$, it is necessary and sufficient that the singular integral*

$$\text{P.V.} \int_{\partial D} f(\zeta)\, U(\zeta, z), \qquad z \in \partial D,$$

converge uniformly with respect to $z \in \partial D$.

This was noted by Gaziev in [55] (also see [56]) for domains with boundary of class $\mathcal{C}^2$. We remark that the Sokhotskiĭ-Plemelj formula for domains with piecewise-smooth boundary is given inaccurately in [53, 55] (the coefficient 1/2 is given everywhere in place of the coefficient in formula (2.5)).

3.2 Functions of class $\mathcal{L}^p$

Suppose to start with that D is a bounded domain in $\mathbf{C}^n$ with smooth boundary ∂D, and $f \in \mathcal{L}^p(\partial D)$ with $p \geq 1$. We denote the unit outer normal to ∂D at ζ by $\nu(\zeta)$.

Theorem 3.4 (Kytmanov). *If $F(z)$ is an integral of the form (2.1), then*

$$\lim_{\epsilon \to 0^+} \int_{\partial D} |F(z - \epsilon\nu(z)) - F(z + \epsilon\nu(z)) - f(z)|^p\, d\sigma = 0,$$

and in addition,

$$\int_{\partial D} |F(z - \epsilon\nu(z)) - F(z + \epsilon\nu(z))|^p\, d\sigma \leq C \int_{\partial D} |f|^p\, d\sigma, \tag{3.5}$$

where the constant C is independent of f and ϵ (for sufficiently small ϵ, the point $z - \epsilon\nu(z) \in D$, and $z + \epsilon\nu(z) \in \mathbf{C}^n \setminus \overline{D}$). If $f \in \mathcal{L}^\infty(\partial D)$, then

$$\sup_{\partial D} |F(z - \epsilon\nu(z)) - F(z + \epsilon\nu(z))| \leq C \operatorname*{ess\,sup}_{\partial D} |f|.$$

Proof. We write $z^+ = z - \epsilon\nu(z)$ and $z^- = z + \epsilon\nu(z)$. For each point $\zeta \in \partial D$, we take a ball $B(\zeta, r)$ of radius r not depending on ζ such that, for $z \in \partial D \cap B(\zeta, r)$, we have $|\zeta - z^\pm|^2 \geq k(|w - \zeta|^2 + \epsilon^2)$ for $\epsilon < r/2$ (here k is independent of ζ and ϵ), where w is the projection of z onto the tangent plane T_ζ to ∂D at ζ. This can always be done because $\big|\, |\zeta - w| - |\zeta - z| \,\big| \leq |w - z| = o(|\zeta - w|)$ as $w \to \zeta$ (see the proof of Theorems 2.6 and 3.1). We have

$$\begin{aligned}
&\int_{\partial D} |F(z^+) - F(z^-) - f(z)|^p \, d\sigma \\
&\qquad = \int_{\partial D} d\sigma(z) \left| \int_{\partial D} (f(\zeta) - f(z))\,(U(\zeta, z^+) - U(\zeta, z^-)) \right|^p \\
&\qquad \leq \int_{\partial D} d\sigma(z) \left(\int_{\partial D} |U(\zeta, z^+) - U(\zeta, z^-)| \right)^{p-1} \times \\
&\qquad\qquad \times \int_{\partial D} |f(\zeta) - f(z)|^p \, |U(\zeta, z^+) - U(\zeta, z^-)|
\end{aligned}$$

by Jensen's inequality (see, for example, [71, §2.2]) applied to the integral

$$\left(\int_{\partial D} |f(\zeta) - f(z)| \cdot |U(\zeta, z^+) - U(\zeta, z^-)| \, d\sigma \right)^p.$$

We estimated the integral $\int_{\partial D} |U(\zeta, z^+) - U(\zeta, z^-)|$ in Theorem 3.1 and showed that it is bounded by a constant not depending on ϵ, while the integral

$$\begin{aligned}
&\int_{\partial D} d\sigma(z) \int_{\partial D} |f(\zeta) - f(z)|^p \, |U(\zeta, z^+) - U(\zeta, z^-)| \\
&\quad \leq C_1 \sum_{m=1}^n \int_{\partial D} d\sigma(\zeta) \int_{\partial D} |f(\zeta) - f(z)|^p \left| \frac{\bar\zeta_m - \bar z_m^+}{|\zeta - z^+|^{2n}} - \frac{\bar\zeta_m - \bar z_m^-}{|\zeta - z^-|^{2n}} \right| d\sigma(z).
\end{aligned}$$

If $z \in B(\zeta, r) \cap \partial D$, then

$$\begin{aligned}
\left| \frac{\bar\zeta_m - \bar z_m}{|\zeta - z^+|^{2n}} - \frac{\bar\zeta_m - \bar z_m}{|\zeta - z^-|^{2n}} \right| &= |\bar\zeta_m - \bar z_m| \cdot \big| |\zeta - z^+| - |\zeta - z^-| \big| \sum_{j=0}^{2n-1} \frac{|\zeta - z^-|^{j-2n}}{|\zeta - z^+|^{j+1}} \\
&\leq \frac{6\epsilon n}{k^n (|w - \zeta|^2 + \epsilon^2)^n},
\end{aligned}$$

while $|\epsilon\nu_m|\zeta - z^+|^{-2n} + \epsilon\nu_m|\zeta - z^-|^{-2n}| \leq 2\epsilon k^{-n}(|w-\zeta|^2+\epsilon^2)^{-n}$. Then

$$\int_{\partial D\cap B(\zeta,r)} |f(\zeta)-f(z)|^p \left|\frac{\bar\zeta_m - \bar z_m^+}{|\zeta - z^+|^{2n}} - \frac{\bar\zeta_m - \bar z_m^-}{|\zeta - z^-|^{2n}}\right| d\sigma(z)$$
$$\leq d\int_{T_\zeta\cap B(\zeta,r)} \frac{|f(\zeta)-f(z(w))|^p\epsilon}{(|w-\zeta|^2+\epsilon^2)^n}\, dS(w) = d\cdot I_1.$$

Introducing the variable $t = \epsilon^{-1}(w-\zeta)$ in $\mathbf{R}^{2n-1}$, we obtain that

$$I_1 = \int_{\{\epsilon|t|<r\}} \frac{|f(\zeta)-f(z(\zeta+\epsilon t))|^p}{(|t|^2+1)^n}\, dS(t),$$

and the integral

$$I_\epsilon(t) = \int_{\partial D} |f(\zeta)-f(z(\zeta+\epsilon t))|^p\, d\sigma(\zeta)$$

converges to zero as $\epsilon \to 0^+$ for fixed t. Also $I_\epsilon(t) \leq A\|f\|^p_{\mathcal{L}^p}$. Therefore

$$\int_{\partial D} d\sigma(\zeta) \int_{\{\epsilon|t|<r\}} \frac{|f(\zeta)-f(z(\zeta+\epsilon t))|^p}{(|t|^2+1)^n}\, dS(t) = \int_{\{\epsilon|t|<r\}} \frac{I_\epsilon(t)}{(|t|^2+1)^n}\, dS(t)$$
$$\leq \int_{\mathbf{R}^{2n-1}} \frac{I^*_\epsilon(t)}{(|t|^2+1)^n}\, dS(t),$$

where $I^*_\epsilon(t) = I_\epsilon(t)$ in the ball $\{t : \epsilon|t| < r\}$, and $I^*_\epsilon(t) = 0$ outside this ball. In the last integral, we may take the limit as $\epsilon \to 0^+$ under the integral sign by Lebesgue's dominated convergence theorem.

It remains to consider the integral

$$\int_{\partial D} d\sigma(\zeta) \int_{\partial D\setminus B(\zeta,r)} |f(\zeta)-f(z)|^p \left|\frac{\bar\zeta_m - \bar z_m^+}{|\zeta - z^+|^{2n}} - \frac{\bar\zeta_m - \bar z_m^-}{|\zeta - z^-|^{2n}}\right| d\sigma(z).$$

Since $|\zeta - z| \geq r$, we have $|\zeta - z^\pm| \geq \big||\zeta - z| - |z - z^\pm|\big| \geq r - \epsilon > r/2$. Then

$$\left|\frac{\bar\zeta_m - \bar z_m}{|\zeta - z^+|^{2n}} - \frac{\bar\zeta_m - \bar z_m}{|\zeta - z^-|^{2n}}\right| \leq |\bar\zeta_m - \bar z_m|\cdot|z^+ - z^-| \sum_{j=0}^{2n-1} \frac{|\zeta - z^-|^{j-2n}}{|\zeta - z^+|^{j+1}} \leq d_1\epsilon,$$

$$\text{while} \quad \left|\frac{\epsilon\nu_m}{|\zeta - z^+|^{2n}} + \frac{\epsilon\nu_m}{|\zeta - z^-|^{2n}}\right| \leq d_2\epsilon, \quad \text{that is,}$$

$$\int_{\partial D} d\sigma(\zeta) \int_{\partial D\setminus B(\zeta,r)} |f(\zeta)-f(z)|^p \left|\frac{\bar\zeta_m - \bar z_m^+}{|\zeta - z^+|^{2n}} - \frac{\bar\zeta_m - \bar z_m^-}{|\zeta - z^-|^{2n}}\right| d\sigma(z)$$
$$\leq d_3\epsilon\left(\int_{\partial D} |f|^p\, d\sigma\right)^2.$$

Inequality (3.5) is proved analogously. Theorem 3.4 was given in [119] for $f \in \mathcal{L}^1(\partial D)$ and in [121] for functions of class $\mathcal{L}^p(\partial D)$. □

3.3 Distributions

Now we consider a bounded domain D with boundary of class C^∞. As usual, $\mathcal{E}(\partial D) = C^\infty(\partial D)$ is the space of infinitely differentiable functions on ∂D with the topology of uniform convergence of the functions and all their derivatives, while $\mathcal{E}'(\partial D)$ is the space of distributions (with compact support), that is, the continuous linear functionals on $\mathcal{E}(\partial D)$.

Now we want to define the action of these functionals on the Bochner-Martinelli kernel $U(\zeta, z)$. To do this, we need to compute the restriction to ∂D of the differential forms $d\bar{\zeta}[k] \wedge d\zeta$ and $d\zeta[k] \wedge d\bar{\zeta}$ in terms of the Lebesgue surface measure $d\sigma$. Suppose $D = \{ z : \rho(z) < 0 \}$, where $\rho \in C^\infty(\mathbf{C}^n)$ and $d\rho \neq 0$ on ∂D.

Lemma 3.5. *The restriction of the form $d\bar{\zeta}[k] \wedge d\zeta$ to the boundary ∂D equals $2^{n-1} i^n (-1)^{k-1} (\partial\rho/\partial\bar{\zeta}_k)\, d\sigma / |\operatorname{grad} \rho|$, and the restriction of $d\zeta[k] \wedge d\bar{\zeta}$ to ∂D equals $2^{n-1} i^n (-1)^{n+k-1} (\partial\rho/\partial\zeta_k)\, d\sigma/|\operatorname{grad}\rho|$, where $\operatorname{grad}\rho = (\partial\rho/\partial\zeta_1, \ldots, \partial\rho/\partial\zeta_n)$.*

Proof. It is well known that

$$\begin{aligned} dx[k] \wedge dy\big|_{\partial D} &= (-1)^k \gamma_k \, d\sigma, \\ dx \wedge dy[k]\big|_{\partial D} &= (-1)^{n+k-1} \gamma_{k+n}\, d\sigma, \end{aligned} \tag{3.6}$$

where $x = \operatorname{Re}\zeta$ and $y = \operatorname{Im}\zeta$. Here the γ_k are the direction cosines of the normal vector to ∂D: namely,

$$\gamma_k = \frac{\partial\rho}{\partial x_k} \Bigg/ \sqrt{\sum_{j=1}^n \left[\left(\frac{\partial\rho}{\partial x_j}\right)^2 + \left(\frac{\partial\rho}{\partial y_j}\right)^2 \right]}$$

$$\gamma_{n+k} = \frac{\partial\rho}{\partial y_k} \Bigg/ \sqrt{\sum_{j=1}^n \left[\left(\frac{\partial\rho}{\partial x_j}\right)^2 + \left(\frac{\partial\rho}{\partial y_j}\right)^2 \right]}.$$

We obtain the assertion of the lemma by using (3.6) and the formulas $(\partial\rho/\partial z_k) = ((\partial\rho/\partial x_k) - i(\partial\rho/\partial y_k))/2$, $(\partial\rho/\partial\bar{z}_k) = \overline{(\partial\rho/\partial z_k)}$,

$$|\operatorname{grad}\rho| = \frac{1}{2}\sqrt{\sum_{j=1}^n \left[\left(\frac{\partial\rho}{\partial x_j}\right)^2 + \left(\frac{\partial\rho}{\partial y_k}\right)^2 \right]},$$

and $dz_k \wedge d\bar{z}_k = -2i\, dx_k \wedge dy_k$. □

From Lemma 3.5, we have

$$U(\zeta, z)\big|_{\partial D} = \frac{(n-1)!}{2\pi^n} \sum_{k=1}^n \frac{(\bar{\zeta}_k - \bar{z}_k)}{|\zeta - z|^{2n}} \frac{\partial\rho}{\partial\bar{z}_k} \frac{d\sigma}{|\operatorname{grad}\rho|} = M(\zeta, z)\, d\sigma.$$

When $S \in \mathcal{E}'(\partial D)$, we define $\widehat{S}(z) = S_\zeta(M(\zeta, z))$ for $z \notin \partial D$. If S is given by an integrable function f, then $\widehat{f}$ is the Bochner-Martinelli integral (2.1). Obviously $\widehat{S}(z)$ is harmonic off ∂D.

Theorem 3.6 (Chirka). *For every function* $\varphi \in \mathcal{E}(\partial D)$,

$$\lim_{\epsilon\to 0+}\int_{\partial D}|\widehat{S}(z-\epsilon\nu(z))-\widehat{S}(z+\epsilon\nu(z))|\varphi(z)\,d\sigma = S(\varphi)=\langle S,\varphi\rangle, \tag{3.7}$$

where, as usual, $\nu(z)$ *is the unit outer normal vector to* ∂D, *and* $S(\varphi)=\langle S,\varphi\rangle$ *is the value of the functional* S *on the the element* φ.

Proof. Each distribution S in $\mathcal{E}'(\partial D)$ has a singularity of finite order, so $S = X_1\dots X_N f$, where $f\in C^1(\partial D)$, and the X_j are vector fields tangent to ∂D with coefficients of class $C^\infty(\partial D)$, that is, $X_j=\sum_{k=1}^n (a_k(z)(\partial/\partial x_k)+b_k(z)(\partial/\partial y_k))$ with $\sum_{k=1}^n (a_k(\partial\rho/\partial x_k)+b_k(\partial\rho/\partial y_k))=0$ on ∂D.

We can extend distributions S of order N as currents: continuous linear functionals on the space of differential forms β of degree $(2n-1)$ with coefficients of class C^∞ with compact support in a neighborhood $V(\partial D)$. This extension can be realized in the following way: if β is a form of the indicated class, then there is a function $\widetilde{\beta}(z)$, infinitely differentiable with compact support in $V(\partial D)$, such that the form $(\beta-\widetilde{\beta}\sigma)\wedge d\rho$ has a zero on ∂D of order at least N (here σ is a form of type $(2n-1)$ representing Lebesgue measure on the level surfaces of ρ). We set $\widetilde{S}(\beta)=S(\widetilde{\beta})$. It is clear that this definition does not depend on the choice of the function $\widetilde{\beta}$.

Lemma 3.7. *Suppose* $S\in\mathcal{E}'(\partial D)$ *has the property that for every form* β *of type* $(2n-1)$ *with compact support in* $V(\partial D)$ *and coefficients of class* C^∞,

$$\lim_{\epsilon\to 0+}\left(\int_{\partial D_{-\epsilon}}\widehat{S}\beta-\int_{\partial D_\epsilon}\widehat{S}\beta\right)=\widetilde{S}(\beta) \tag{3.8}$$

(here $D_\epsilon=\{\,z:\rho(z)<\epsilon\,\}$*). Then for every tangential vector field* X *on* ∂D, *the distribution* XS *has the same property:*

$$\lim_{\epsilon\to 0+}\left(\int_{\partial D_{-\epsilon}}\widehat{XS}\beta-\int_{\partial D_\epsilon}\widehat{XS}\beta\right)=\widetilde{XS}(\beta).$$

Proof. Let $\widetilde{X}$ be an arbitrary vector field in $V(\partial D)$ that is tangent to every ∂D_ϵ and that coincides with X on ∂D (the coefficients of $\widetilde{X}$ are class C^∞). We denote the chain $\partial D_{-\epsilon}-\partial D_\epsilon$ by Γ_ϵ. By hypothesis,

$$\widetilde{XS}(\beta)=\widetilde{X}\widetilde{S}(\beta)=-\widetilde{S}(\widetilde{X}\beta)=-\lim_{\epsilon\to 0+}\int_{\Gamma_\epsilon}\widehat{S}\widetilde{X}\beta=\lim_{\epsilon\to 0+}\int_{\Gamma_\epsilon}(\widetilde{X}\widehat{S})\beta.$$

We need to prove this equation not for $\widetilde{X}\widehat{S}$, but for $\widehat{XS}$. To do this we choose $\widetilde{X}$ so that $\int_{\Gamma_\epsilon}(\widehat{XS}-\widetilde{X}\widehat{S})\beta=0$ for all ϵ that are sufficiently small (in modulus). Let $\widetilde{X} = \sum_{j=1}^{n}(a_j(\partial/\partial\xi_j)+b_j(\partial/\partial\eta_j))$, where $\xi_j = \operatorname{Re}\zeta_j$ and $\eta_j = \operatorname{Im}\zeta_j$. Then

$$\widehat{XS}(z) = -\widetilde{S}_\zeta(\widetilde{X}_\zeta U(\zeta,z)) = -\widetilde{S}_\zeta\left(\sum_{j=1}^{n}\left(a_j(\zeta)\frac{\partial}{\partial\xi_j}+b_j(\zeta)\frac{\partial}{\partial\eta_j}\right)U(\zeta,z)\right).$$

Since $(\partial/\partial\xi_j)U(\zeta,z) = -(\partial/\partial x_j)U(\zeta,z)$, we have

$$X_\zeta U(\zeta,z) = \\ -\widetilde{X}_z U(\zeta,z) + \sum_{j=1}^{n}\left[(a_j(z)-a_j(\zeta))\frac{\partial}{\partial x_j} + (b_j(z)-b_j(\zeta))\frac{\partial}{\partial y_j}\right]U(\zeta,z).$$

We need to arrange that the a_j and b_j satisfy the equations

$$\int_{\Gamma_\epsilon}(\widehat{XS} - \widetilde{X}\widehat{S})\beta \\ = \int_{\Gamma_\epsilon}\widetilde{S}\Big(\sum_{j=1}^{n}\left[(a_j(z)-a_j(\zeta))\frac{\partial}{\partial x_j} + (b_j(z)-b_j(\zeta))\frac{\partial}{\partial y_j}\right]\Big)U(\zeta,z)\beta(z) = 0$$

for all ϵ, that is,

$$\sum_{j=1}^{n}\int_{\Gamma_\epsilon}\left[a_j(z)\widetilde{S}\left(\frac{\partial}{\partial x_j}U(\zeta,z)\right) + b_j(z)\widetilde{S}\left(\frac{\partial}{\partial y_j}U(\zeta,z)\right)\right]\beta(z) = \psi(\epsilon),$$

where $\psi(\epsilon)$ is obtained by integrating the terms with $a_j(\zeta)$ and $b_j(\zeta)$. Since

$$\widetilde{S}_\zeta\left(\frac{\partial}{\partial x_j}U(\zeta,z)\right) = \frac{\partial}{\partial x_j}\widetilde{S}_\zeta(U(\zeta,z)) \quad\text{and}\quad \widetilde{S}_\zeta\left(\frac{\partial}{\partial y_j}U(\zeta,z)\right) = \frac{\partial}{\partial y_j}\widetilde{S}_\zeta(U(\zeta,z)),$$

and these functions cannot be simultaneously zero on Γ_ϵ (otherwise $\widehat{S} \equiv C_\epsilon$ on Γ_ϵ and the jump of $\widehat{S}$ on ∂D equals a constant, but not the function S, which may be assumed to be different from a constant), so the required extension of X to $\widetilde{X}$ exists. □

We now complete the proof of the theorem. Since S has a singularity of finite order, we write $S = X_1 \dots X_N f$, where $f \in \mathcal{C}^1(\partial D)$, and the X_j are tangential vector fields. Formula (3.7) holds for the function f (see Theorem 2.3), and this means that (3.8) holds, so by Lemma 3.7, formula (3.8) also holds for the distribution S. Hence we obtain (3.7). □

Remark. Theorem 3.6 holds for surfaces with a finite degree N of smoothness and for distributions S on ∂D with singularity no greater than N.

Theorem 3.6 and Lemma 3.7 were given in [36]. These assertions were proved for the case of hypersurfaces in [7, §15].

3.4 Differential forms

Let D be a bounded domain in $\mathbf{C}^n$ with boundary of class $\mathcal{C}^2$, so $D = \{ z : \rho(z) < 0 \}$, with $\rho \in \mathcal{C}^2(\overline{D})$ and $d\rho \neq 0$ on ∂D. If γ is a differential form of type (p,q) with coefficients of class $\mathcal{C}(\overline{D})$, we say that the tangential part γ_τ of γ equals zero on ∂D if

$$\int_{\partial D} \gamma \wedge \varphi = 0 \tag{3.9}$$

for all forms φ of type $(n-p, n-q-1)$ with coefficients of class $\mathcal{C}^\infty(\overline{D})$. This means that the form $\gamma \wedge \overline{\partial}\rho$ equals zero on ∂D. Indeed, let

$$\gamma = \sum_I{}' \sum_J{}' a_{I,J}(z)\, d\bar{z}_I \wedge dz_J,$$

where $I = (i_1, \ldots, i_q)$ and $J = (j_1, \ldots, j_p)$. Multiplying γ by φ of the form $\varphi = \psi\, dz[J] \wedge d\bar{z}[K]$, where $K = (k_1, \ldots, k_{q+1})$, we obtain

$$\begin{aligned}\gamma \wedge \varphi &= \sigma(J)(-1)^{n(n-q-1)}\psi \sum_{I\cup k=K}{}' a_{I,J}(z)\, d\bar{z}_I \wedge d\bar{z}[K] \wedge dz \\ &= \sigma(J)(-1)^{n(n-q-1)}\psi \sum_{I\cup k=K}{}' (-1)^{k-1}\sigma(I,k) a_{I,J}\, d\bar{z}[k] \wedge dz.\end{aligned}$$

Applying Lemma 3.5, we have

$$\gamma \wedge \varphi\big|_{\partial D} = 2^{n-1} i^n \sigma(J)(-1)^{n(n-q-1)} |\operatorname{grad}\rho|^{-1} \psi \sum_{I\cup k}{}' \sigma(I,k) a_{I,J} \frac{\partial \rho}{\partial \bar{z}_k}\, d\sigma.$$

Since ψ is arbitrary, it follows from (3.9) that $\sum'_{I\cup k=K} \sigma(I,k) a_{I,J}(\partial\rho/\partial\bar{z}_k) = 0$ on ∂D. But this means precisely that $\gamma \wedge dz[J] \wedge d\bar{z}[K] \wedge \overline{\partial}\rho$ equals zero on ∂D, that is, $\gamma \wedge \overline{\partial}\rho = 0$ on ∂D.

If γ is of type (n,q), then the condition $\gamma \wedge \overline{\partial}\rho = 0$ on ∂D means that the restriction of γ to ∂D equals zero, since in this case $\gamma \wedge d\rho = \gamma \wedge \overline{\partial}\rho = 0$.

Suppose γ is a form of type (p,q), where $0 \le p, q \le n$, with coefficients of class $\mathcal{C}^1(\overline{D})$. We write

$$\gamma^{\pm}(z) = \int_{\partial D} \gamma(\zeta) \wedge U_{p,q}(\zeta, z), \qquad z \notin \partial D,$$

where $U_{p,q}$ is the kernel of the Koppelman integral representation (1.9).

Theorem 3.8 (Aĭzenberg, Dautov). *The forms $\gamma^{\pm}$ extend continuously to the boundary ∂D, and $\gamma_\tau^+ - \gamma_\tau^- = \gamma_\tau$ on ∂D.*

Proof. Since the coefficients of the kernels $U_{p,q}(\zeta, z)$ are derivatives of the fundamental solution $g(\zeta, z)$, the form γ^+ (respectively γ^-) has coefficients of class $\mathcal{C}^\alpha(\overline{D})$ (respectively $\mathcal{C}^\alpha(\mathbf{C}^n \setminus D)$) for every α with $0 < \alpha < 1$ (see the proof of Theorem 2.3).

We will show that

$$\int_{\partial D} \gamma \wedge \varphi = \int_{\partial D} (\gamma^+ - \gamma^-) \wedge \varphi$$

for a form φ of type $(n-p, n-q-1)$ with coefficients of class $\mathcal{C}^\infty(\mathbf{C}^n)$.

The boundary ∂D_ϵ of the domain $D_\epsilon = \{z : \rho(z) < -\epsilon\}$ is smooth for sufficiently small $|\epsilon|$, while $D_{-\epsilon} \supset \overline{D}$ and $D \supset \overline{D}_\epsilon$ for $\epsilon > 0$. We have

$$\begin{aligned}
\int_{\partial D} &(\gamma^+ - \gamma^-) \wedge \varphi \\
&= \lim_{\epsilon \to 0+} \left(\int_{\partial D_\epsilon} \gamma^+ \wedge \varphi - \int_{\partial D_{-\epsilon}} \gamma^- \wedge \varphi \right) \\
&= \lim_{\epsilon \to 0+} \left[\int_{\partial D_\epsilon} \varphi(z) \wedge \int_{\partial D} \gamma(\zeta) \wedge U_{p,q}(\zeta, z) \right. \\
&\qquad \left. - \int_{\partial D_{-\epsilon}} \varphi(z) \wedge \int_{\partial D} \gamma(\zeta) \wedge U_{p,q}(\zeta, z) \right] \\
&= \lim_{\epsilon \to 0+} \int_{\partial D} \gamma(\zeta) \wedge \left[\int_{\partial D_\epsilon} \varphi(z) \wedge U_{p,q}(\zeta, z) - \int_{\partial D_{-\epsilon}} \varphi(z) \wedge U_{p,q}(\zeta, z) \right].
\end{aligned}$$

But $U_{p,q}(\zeta, z) = -U_{n-p,n-q-1}(z, \zeta)$ (this is easy to deduce from the definition of $U_{p,q}(\zeta, z)$), so by applying (1.9) to the domain $D_{-\epsilon} \setminus D_\epsilon$ we obtain

$$\begin{aligned}
\int_{\partial D} &(\gamma^+ - \gamma^-) \wedge \varphi \\
&= \lim_{\epsilon \to 0+} \int_{\partial D} \gamma(\zeta) \wedge \int_{\partial(D_{-\epsilon} \setminus D_\epsilon)} \varphi(z) \wedge U_{n-p,n-q-1}(z, \zeta) \\
&= \lim_{\epsilon \to 0+} \int_{\partial D} \gamma(\zeta) \wedge \left[\varphi(\zeta) + \int_{D_{-\epsilon} \setminus D_\epsilon} \bar\partial \varphi(z) \wedge U_{n-p,n-q-1}(z, \zeta) \right. \\
&\qquad \left. + \bar\partial \int_{D_{-\epsilon} \setminus D_\epsilon} \varphi(z) \wedge U_{n-p,n-q-2}(z, \zeta) \right] \\
&= \int_{\partial D} \gamma \wedge \varphi + \lim_{\epsilon \to 0+} \left[\int_{\partial D} \gamma(\zeta) \wedge \int_{D_{-\epsilon} \setminus D_\epsilon} \bar\partial \varphi \wedge U_{n-p,n-q-1}(z, \zeta) \right. \\
&\qquad \left. + (-1)^{p+q} \int_{\partial D} \bar\partial \gamma(\zeta) \wedge \int_{D_{-\epsilon} \setminus D_\epsilon} \varphi(z) \wedge U_{n-p,n-q-2}(z, \zeta) \right].
\end{aligned}$$

We have used Stokes's formula in the second term and the equality $\bar\partial(\gamma \wedge \varphi) = d(\gamma \wedge \varphi)$ (it is here that we use the condition that $\gamma \wedge \varphi$ has type $(n, n-1)$). It remains to show that

$$\lim_{\epsilon\to 0+}\int_{\partial D} d\bar{\zeta}[j]\wedge d\zeta\int_{D_{-\epsilon}\setminus D_\epsilon} f(\zeta,z)\frac{\bar{\zeta}_k-\bar{z}_k}{|\zeta-z|^{2n}}\,d\bar{z}\wedge dz=0 \tag{3.10}$$

for $f\in\mathcal{C}(\partial D\times\overline{D_{-\epsilon}\setminus D_\epsilon})$ (since the coefficients of the forms under the integral sign in the last formula are of this type). We will show that

$$\int_{\partial D\times(D_{-\epsilon}\setminus D_\epsilon)}|\zeta-z|^{1-2n}\,d\sigma(\zeta)\,dv(z)$$

exists as an improper integral. Consider the limit

$$\lim_{\epsilon\to 0+}\int_{[\partial D\times(D_{-\epsilon}\setminus D_\epsilon)]\setminus\{(\zeta,z):|\zeta-z|<\epsilon\}}|\zeta-z|^{1-2n}\,d\sigma(\zeta)\,dv(z). \tag{3.11}$$

Since the integral in (3.11) is proper, we may treat it as an iterated integral. Consequently, the limit in (3.11) exists if and only if the limit

$$\lim_{\epsilon\to 0+}\int_{\partial D} h_\epsilon(\zeta)\,d\sigma(\zeta) \tag{3.12}$$

exists, where $h_\epsilon(\zeta)=\int_{(D_{-\epsilon}\setminus D_\epsilon)\setminus B(\zeta,\epsilon)}|\zeta-z|^{1-2n}\,dv(z)$.

Since $0\le h_\epsilon(\zeta)\le\int_{D_{-\epsilon}\setminus D_\epsilon}|\zeta-z|^{1-2n}\,dv(z)$, and the latter integral converges, it follows by Lebesgue's dominated convergence theorem that the limit in (3.12) exists. Hence the function

$$\int_{D_{-\epsilon}\setminus D_\epsilon} f(\zeta,z)\frac{\bar{\zeta}_k-\bar{z}_k}{|\zeta-z|^{2n}}\,d\bar{z}\wedge dz$$

is integrable in ζ, and since the volume of $(D_{-\epsilon}\setminus D_\epsilon)$ converges to zero as $\epsilon\to 0^+$, equation (3.10) holds. □

Theorem 3.8 was obtained, in essence, in [7, §2].

Corollary 3.9. *If γ is a form of type (n,q) with $0\le q\le n$, or of type $(p,0)$ with $0\le p\le n$, then $(\gamma^+-\gamma^-)|_{\partial D}=\gamma|_{\partial D}$.*

In this case, as we have remarked, $\gamma|_{\partial D}=\gamma_\tau$.

In general, Corollary 3.9 does not hold for forms of other bidegrees. As an example, consider a domain D such that ∂D contains a $(2n-1)$-dimensional ball B' of radius $R<1$ with center at the origin, lying in the plane $\{\operatorname{Im} z_n=0\}$. Let $\alpha=d\bar{z}_{n-q+1}\wedge\cdots\wedge d\bar{z}_{n-1}\wedge dz_1\wedge\cdots\wedge dz_p$, and $\gamma=d\bar{z}_n\wedge\alpha$. Then γ has type (p,q). Put $\gamma_1^\pm=\int_{B'}\gamma(\zeta)\wedge U_{p,q}(\zeta,z)$. Since the difference $\gamma^\pm-\gamma_1^\pm$ has no jump on B', we have

$$(\gamma^+-\gamma^-)|_{\partial D}=(\gamma_1^+-\gamma_1^-)|_{\partial D}$$

for $z\in B'$.

The form $\alpha(\zeta)\wedge U_{p,q}(\zeta,z)$ has type $(n,n-2)$ in ζ, so the restriction of $\gamma\wedge U_{p,q}$ to B' is zero, that is, $\gamma_1^{\pm}\equiv 0$. Consequently, $(\gamma^+-\gamma^-)|_{\partial D}=0$, but $\gamma|_{\partial D}\neq 0$ for $z\in B'$.

Generalizations of the jump theorem for other integral representations (such as Cauchy-Fantappiè, Yarmukhamedov, etc.) may be found in [18, 33, 35, 128, 132, 203, 231].

4 Boundary behavior of derivatives of the Bochner-Martinelli integral

4.1 Formulas for finding derivatives

Suppose D is a bounded domain with piecewise-smooth boundary, $f\in C^1(\partial D)$, and F is the Bochner-Martinelli integral (2.1).

Lemma 4.1. *Derivatives of F may be found by the formulas*

$$\frac{\partial F}{\partial z_m}=\int_{\partial D}\frac{\partial f}{\partial \zeta_m}U(\zeta,z)+(-1)^{n+m}\int_{\partial D}\sum_{s=1}^{n}\frac{\partial f}{\partial\bar\zeta_s}\frac{\partial g}{\partial\zeta_s}\,d\bar\zeta\wedge d\zeta[m], \tag{4.1}$$

$$\frac{\partial F}{\partial \bar z_m}=\int_{\partial D}\frac{\partial f}{\partial \bar\zeta_m}U(\zeta,z)+(-1)^{m}\int_{\partial D}\sum_{s=1}^{n}\frac{\partial f}{\partial\bar\zeta_s}\frac{\partial g}{\partial\zeta_s}\,d\bar\zeta[m]\wedge d\zeta, \tag{4.2}$$

where $g=g(\zeta,z)$ is the fundamental solution to Laplace's equation (see §1).

Proof. We prove, for example, formula (4.1); formula (4.2) is proved analogously. Recall that

$$U(\zeta,z)=\sum_{s=1}^{n}(-1)^{s-1}\frac{\partial g}{\partial\zeta_s}(\zeta,z)\,d\bar\zeta[s]\wedge d\zeta.$$

Now

$$\frac{\partial F}{\partial z_m}=-\int_{\partial D}f(\zeta)\frac{\partial}{\partial\zeta_m}U(\zeta,z)=-\int_{\partial D}\frac{\partial}{\partial\zeta_m}(fU)+\int_{\partial D}\frac{\partial f}{\partial\zeta_m}U(\zeta,z),$$

but

$$(-1)^s\int_{\partial D}\frac{\partial}{\partial\zeta_m}\left(f\frac{\partial g}{\partial\zeta_s}\right)d\bar\zeta[s]\wedge d\zeta=(-1)^{n+m}\int_{\partial D}\frac{\partial}{\partial\bar\zeta_s}\left(f\frac{\partial g}{\partial\zeta_s}\right)d\bar\zeta\wedge d\zeta[m],$$

since

$$d\left(f\frac{\partial g}{\partial\zeta_s}\right)d\bar\zeta[s]\wedge d\zeta[m]=(-1)^{s-1}\frac{\partial}{\partial\bar\zeta_s}\left(f\frac{\partial g}{\partial\zeta_s}\right)d\bar\zeta\wedge d\zeta[m]$$
$$+(-1)^{m+n}\frac{\partial}{\partial\zeta_m}\left(f\frac{\partial g}{\partial\zeta_s}\right)d\bar\zeta[s]\wedge d\zeta.$$

Consequently

$$\begin{aligned}\frac{\partial F}{\partial z_m} &= \int_{\partial D} \frac{\partial f}{\partial \zeta_m} U(\zeta, z) + (-1)^{m+n} \sum_{s=1}^{n} \int_{\partial D} \frac{\partial}{\partial \bar{\zeta}_s} \left(f \frac{\partial g}{\partial \zeta_s} \right) d\bar{\zeta} \wedge d\zeta[m] \\ &= \int_{\partial D} \frac{\partial f}{\partial \zeta_m} U(\zeta, z) + (-1)^{m+n} \sum_{s=1}^{n} \int_{\partial D} \frac{\partial f}{\partial \bar{\zeta}_s} \frac{\partial g}{\partial \zeta_s} d\bar{\zeta} \wedge d\zeta[m],\end{aligned}$$

since g is a harmonic function. □

Now consider a domain D with boundary of class $\mathcal{C}^2$, and suppose that $f \in \mathcal{C}^1(\partial D)$. If $D = \{ z : \rho(z) < 0 \}$ and $\rho \in \mathcal{C}^2(\overline{D})$ with $d\rho \neq 0$ on ∂D, we denote $(\partial \rho / \partial z_k)/|\operatorname{grad} \rho|$ by ρ_k and $\overline{\rho_k}$ by $\rho_{\bar{k}}$. The surface area element is then

$$\begin{aligned}d\sigma &= i^{-n} 2^{1-n} \sum_{k=1}^{n} (-1)^{n+k-1} \rho_{\bar{k}} \, d\zeta[k] \wedge d\bar{\zeta}\big|_{\partial D} \\ &= i^{-n} 2^{1-n} \sum_{k=1}^{n} (-1)^{k-1} \rho_k \, d\bar{\zeta}[k] \wedge d\zeta\big|_{\partial D}\end{aligned}$$

(see Lemma 3.5).

Lemma 4.2. *For $z \notin \partial D$, let $\Phi(z) = i^n 2^{n-1} \int_{\partial D} f(\zeta) g(\zeta, z) \, d\sigma(\zeta)$ be a single-layer potential. Then*

$$\begin{aligned}\frac{\partial \Phi}{\partial z_m} &= - \int_{\partial D} f \rho_m U(\zeta, z) \\ &\quad + i^n 2^{n-1} \sum_{k=1}^{n} \int_{\partial D} \left[\rho_k \frac{\partial}{\partial \zeta_m} (f \rho_{\bar{k}}) - \rho_m \frac{\partial}{\partial \zeta_k} (f \rho_{\bar{k}}) \right] g(\zeta, z) \, d\sigma(\zeta),\end{aligned} \tag{4.3}$$

$$\begin{aligned}\frac{\partial \Phi}{\partial \bar{z}_m} &= - \int_{\partial D} f \rho_{\overline{m}} U(\zeta, z) \\ &\quad + i^n 2^{n-1} \sum_{k=1}^{n} \int_{\partial D} \left[\rho_k \frac{\partial}{\partial \bar{\zeta}_m} (f \rho_{\bar{k}}) - \rho_{\overline{m}} \frac{\partial}{\partial \zeta_k} (f \rho_{\bar{k}}) \right] g(\zeta, z) \, d\sigma(\zeta).\end{aligned} \tag{4.4}$$

Proof. We have

$$\begin{aligned}\frac{\partial \Phi}{\partial z_m} &= - \int_{\partial D} f(\zeta) \frac{\partial g}{\partial \zeta_m} \sum_{k=1}^{n} \rho_{\bar{k}} (-1)^{n+k-1} \, d\zeta[k] \wedge d\bar{\zeta} \\ &= \sum_{k=1}^{n} (-1)^{n+k-1} \int_{\partial D} \frac{\partial}{\partial \zeta_m} (f \rho_{\bar{k}}) g(\zeta, z) \, d\zeta[k] \wedge d\bar{\zeta} \\ &\quad - \sum_{k=1}^{n} (-1)^{n+k-1} \int_{\partial D} \frac{\partial}{\partial \zeta_m} (f \rho_{\bar{k}} g) \, d\zeta[k] \wedge d\bar{\zeta}.\end{aligned}$$

Just as in Lemma 4.1, we obtain

$$(-1)^{n+k}\int_{\partial D}\frac{\partial}{\partial\zeta_m}(f\rho_{\bar k}g)\,d\zeta[k]\wedge d\bar\zeta=(-1)^{n+m}\int_{\partial D}\frac{\partial}{\partial\zeta_k}(f\rho_{\bar k}g)\,d\zeta[m]\wedge d\bar\zeta.$$

Therefore

$$\begin{aligned}\frac{\partial\Phi}{\partial z_m}&=\sum_{k=1}^{n}(-1)^{n+k-1}\int_{\partial D}\frac{\partial}{\partial\zeta_m}(f\rho_{\bar k})g(\zeta,z)\,d\zeta[k]\wedge d\bar\zeta\\&\quad+(-1)^{n+m}\sum_{k=1}^{n}\int_{\partial D}\frac{\partial}{\partial\zeta_k}(f\rho_{\bar k})g(\zeta,z)\,d\zeta[m]\wedge d\bar\zeta\\&\quad-i^n2^{n-1}\sum_{k=1}^{n}\int_{\partial D}f\rho_{\bar k}\frac{\partial g}{\partial\zeta_k}\rho_m\,d\sigma\\&=\sum_{k=1}^{n}\int_{\partial D}\Big[(-1)^{n+k-1}\frac{\partial}{\partial\zeta_m}(f\rho_{\bar k})\,d\zeta[k]\wedge d\bar\zeta\\&\quad+(-1)^{n+m}\frac{\partial}{\partial\zeta_k}(f\rho_{\bar k})\,d\zeta[m]\wedge d\bar\zeta\Big]\,g(\zeta,z)\\&\quad-\int_{\partial D}f\rho_m\,U(\zeta,z).\end{aligned}$$

Formula (4.4) is proved analogously. □

Theorem 4.3. *If $\partial D\in\mathcal{C}^2$ and $f\in\mathcal{C}^2(\partial D)$, then the integral F extends to $\overline{D}$ and to $\mathbf{C}^n\setminus D$ as a function of class $\mathcal{C}^{1+\alpha}$ for $0<\alpha<1$. Moreover*

$$\frac{\partial F}{\partial z_m}=\int_{\partial D}\left(\frac{\partial f}{\partial\zeta_m}-\rho_m\sum_{k=1}^{n}\rho_k\frac{\partial f}{\partial\bar\zeta_k}\right)U(\zeta,z)+i^n2^{n-1}\int_{\partial D}\psi_1(\zeta)g(\zeta,z)\,d\sigma(\zeta),\tag{4.5}$$

where

$$\psi_1=\sum_{s,k=1}^{n}\left[\rho_k\frac{\partial}{\partial\zeta_s}\left(\rho_m\rho_{\bar k}\frac{\partial f}{\partial\bar\zeta_s}\right)-\rho_m\frac{\partial}{\partial\zeta_k}\left(\rho_m\rho_{\bar k}\frac{\partial f}{\partial\bar\zeta_s}\right)\right],$$

and

$$\frac{\partial F}{\partial\bar z_m}=\int_{\partial D}\left(\frac{\partial f}{\partial\bar\zeta_m}-\rho_{\overline{m}}\sum_{k=1}^{n}\rho_k\frac{\partial f}{\partial\bar\zeta_k}\right)U(\zeta,z)+i^n2^{n-1}\int_{\partial D}\psi_2(\zeta)g(\zeta,z)\,d\sigma(\zeta),\tag{4.6}$$

where

$$\psi_2=\sum_{s,k=1}^{n}\left[\rho_k\frac{\partial}{\partial\bar\zeta_s}\left(\rho_{\overline{m}}\rho_{\bar k}\frac{\partial f}{\partial\bar\zeta_s}\right)-\rho_{\overline{m}}\frac{\partial}{\partial\zeta_k}\left(\rho_{\overline{m}}\rho_{\bar k}\frac{\partial f}{\partial\bar\zeta_s}\right)\right].$$

Proof. Formulas (4.5) and (4.6) follow from Lemmas 4.1 and 4.2, while the boundary behavior of the integral F follows from Theorem 2.3 and properties of the single-layer potential. □

Formulas (4.1)–(4.6) essentially are classical formulas of potential theory (see [64]).

Corollary 4.4. *If $\partial D \in \mathcal{C}^k$ and $f \in \mathcal{C}^m(\partial D)$, where $k \geq m$, then the Bochner-Martinelli integrals extend to $\overline{D}$ and to $\mathbf{C}^n \setminus D$ as functions of class $\mathcal{C}^{m-\epsilon}$, where ϵ is any number such that $0 < \epsilon < 1$.*

Proof. The proof follows from Theorem 4.3 by induction on m. This assertion was noted in [36]. □

Corollary 4.5. *If $\partial D \in \mathcal{C}^2$ and $f \in \mathcal{C}^2(\partial D)$, then the jumps of the derivatives of F are given by*

$$\begin{aligned}
\frac{\partial F^+}{\partial z_m} - \frac{\partial F^-}{\partial z_m} &= \frac{\partial f}{\partial z_m} - \rho_m \sum_{k=1}^{n} \frac{\partial f}{\partial \bar{z}_k}\rho_k, \qquad z \in \partial D,\\
\frac{\partial F^+}{\partial \bar{z}_m} - \frac{\partial F^-}{\partial \bar{z}_m} &= \frac{\partial f}{\partial \bar{z}_m} - \rho_{\overline{m}} \sum_{k=1}^{n} \frac{\partial f}{\partial \bar{z}_k}\rho_k, \qquad z \in \partial D.
\end{aligned} \tag{4.7}$$

4.2 Jump theorem for derivatives

If we are concerned only with the jump of derivatives (that is, with formula (4.7)), then we can weaken the conditions on ∂D and on f.

Let D be a bounded domain with boundary of class $\mathcal{C}^1$: $D = \{ z : \rho(z) < 0 \}$, where $\rho \in \mathcal{C}^1(\mathbf{C}^n)$, and $d\rho \neq 0$ on ∂D. If $z \in \partial D$, then we denote by $z^+ \in D$ and $z^- \notin \overline{D}$ points on the normal to ∂D at z such that $|z^+ - z| = |z^- - z|$.

Lemma 4.6. *Let*

$$\Phi_{m,\bar{k}}(z) = \int_{\partial D} \frac{\partial g(\zeta, z)}{\partial \zeta_m}\, d\bar{\zeta}[k] \wedge d\zeta, \qquad z \notin \partial D, \qquad \textit{and}$$

$$\Phi_{\overline{m},k}(z) = \int_{\partial D} \frac{\partial g(\zeta, z)}{\partial \bar{\zeta}_m}\, d\zeta[k] \wedge d\bar{\zeta}, \qquad z \notin \partial D.$$

Then

$$\lim_{z^{\pm} \to z} (\Phi_{m,\bar{k}}(z^+) - \Phi_{m,\bar{k}}(z^-)) = (-1)^{k-1}\rho_{\bar{k}}\rho_m, \tag{4.8}$$

$$\lim_{z^{\pm} \to z} (\Phi_{\overline{m},k}(z^+) - \Phi_{\overline{m},k}(z^-)) = (-1)^{k-1}\rho_k\rho_{\overline{m}}, \tag{4.9}$$

and these limits are attained uniformly in z.

Proof. Since $\Phi_{m,\bar{k}} = i^n 2^{n-1}(-1)^{k-1}\int_{\partial D}(\partial g/\partial\zeta_m)\rho_{\bar{k}}\,d\sigma$, and the jump of the integral $\int_{\partial D}\rho_{\bar{k}}\rho_m U(\zeta,z)$ equals $\rho_{\bar{k}}\rho_m$, to prove (4.8) it is enough to show that the jump of the integral $\int_{\partial D}\rho_{\bar{k}} d_m g(\zeta,z)\,d\sigma$ is zero, where $d_m = (\partial/\partial\zeta_m) - \rho_m(\sum_{l=1}^n \rho_{\bar{l}}\partial/\partial\zeta_l))$ (recall that $U(\zeta,z)\big|_{\partial D} = i^n 2^{n-1}\sum_{l=1}^n(\partial g/\partial\zeta_l)\rho_{\bar{l}}\,d\sigma)$. Notice that d_m is a tangential vector field. Since $|\rho_{\bar{k}}| \le C$ on ∂D, it is enough to show that

$$\lim_{z^\pm\to z}\int_{\partial D}\left|d_m g(\zeta,z^+) - d_m g(\zeta,z^-)\right|\,d\sigma = 0, \tag{4.10}$$

the limit being attained uniformly in z. Since the points $z^\pm$ lie on the normal to ∂D at z, we can write $z^\pm - z = \pm\overline{\operatorname{grad}\rho}\cdot t/|\operatorname{grad}\rho|$ with $t\in\mathbf{R}$. Writing $\alpha(z) = \overline{\operatorname{grad}\rho}$ and $w = \zeta - z$, we obtain

$$\begin{aligned}&d_m(g(\zeta,z^+) - g(\zeta,z^-))\\ &\quad = a(\zeta,z)\left(\frac{1}{|w-\alpha t|^{2n}} - \frac{1}{|w+\alpha t|^{2n}}\right) - b(\zeta,z)t\left(\frac{1}{|w-\alpha t|^{2n}} + \frac{1}{|w+\alpha t|^{2n}}\right),\end{aligned}$$

where

$$a(\zeta,z) = \bar{w}_m - \rho_m(\zeta)\sum_{k=1}^n \bar{w}_k\rho_{\bar{k}}(\zeta), \qquad \text{and}$$

$$b(\zeta,z) = \rho_m(z) - \rho_m(\zeta)\sum_{k=1}^n \rho_{\bar{k}}(\zeta)\rho_k(z).$$

Since

$$\begin{aligned}b(\zeta,z) &= b(\zeta,z) - b(\zeta,\zeta)\\ &= (\rho_m(z) - \rho_m(\zeta)) - \rho_m(\zeta)\sum_{k=1}^n \rho_{\bar{k}}(\zeta)(\rho_k(\zeta) - \rho_k(z)),\end{aligned}$$

we have $|b(\zeta,z)| \le \omega(|w|) \to 0$ as $|w| \to 0$ (where $\omega(|w|)$ is the modulus of continuity of $b(\zeta,z)$). Moreover, it is clear that $|a(\zeta,z)| \le 2|w|$ (by the Schwarz inequality).

Now let T_z be the hyperplane tangent to ∂D at z. We choose a coordinate system $u = (u_1,\dots,u_{2n-1})$ in T_z with center at z. We fix $\epsilon > 0$ and choose a $(2n-1)$ dimensional ball B' in T_z with center at z and such that

1. each real line parallel to the normal to T_z and passing through a point $u\in B'$ intersects ∂D at one point, which we denote by $\zeta(u)$;
2. $|\sin\beta| \ge 1/2$ when $u\in B'$, where β is the angle between the normal to ∂D at z and the line passing through z and $\zeta(u)$;
3. $|\zeta(u) - z| < \epsilon$ for $u\in B'$;

4. $\omega(|\zeta(u) - z|) < \epsilon$ for $u \in B'$;

5. $|\zeta(u) - u| < \epsilon|u|$ for $u \in B'$;

6. $|u - z^{\pm}| \leq 2|\zeta(u) - z^{\pm}|$ for $u \in B'$.

These conditions are essentially the same as in Theorem 3.1, and the radius d of the ball B' can be chosen to be independent of z.

Splitting the integral in (4.10) into the two parts over the set $\partial D \setminus B(z, r)$ and over the set $\partial D \cap B(z, r)$, we see that we can take the limit under the integral sign in the integral over $\partial D \setminus B(z, r)$. It remains to estimate the integral over $\partial D \cap B(z, r)$. Since $|u - z^+| = |u - z^-|$ for $u \in B'$, we have $|w \pm \alpha t| \geq |u - z^+|/2 = \sqrt{|u|^2 + t^2}/2$ by condition (6). Using condition (4), we obtain

$$|b(\zeta, z)| \cdot |t| \left(\frac{1}{|w - \alpha t|^{2n}} + \frac{1}{|w + \alpha t|^{2n}} \right) \leq \frac{\epsilon |t| 2^{2n+1}}{(|u|^2 + t^2)^n}.$$

Furthermore, since

$$\left| \frac{1}{|w - \alpha t|^{2n}} - \frac{1}{|w + \alpha t|^{2n}} \right| = \left| \frac{1}{|w - \alpha t|} - \frac{1}{|w + \alpha t|} \right| \sum_{j=0}^{2n-1} \frac{|w + \alpha t|^{j+1-2n}}{|w - \alpha t|^j},$$

we have

$$\begin{aligned} |a(\zeta, z)| \left| \frac{1}{|w - \alpha t|^{2n}} - \frac{1}{|w + \alpha t|^{2n}} \right| &\leq \frac{2^{2n+1} |w| (2n-1) \big| |w - \alpha t| - |w + \alpha t| \big|}{|w - \alpha t| (|u|^2 + t^2)^n} \\ &\leq \frac{n 2^{2n+3} \big| |w - \alpha t| - |w + \alpha t| \big|}{(|u|^2 + t^2)^n}. \end{aligned}$$

Consider the function $\lambda(w, t) = (|w - \alpha t| - |w + \alpha t|)/t$. Since $|w + \alpha t|^2 + |w - \alpha t|^2 = 4t \operatorname{Re}\langle \bar{\alpha}, w \rangle$, and $|w + \alpha t| - |w - \alpha t| \geq 2|w|$, we have $|\lambda(w, t)| \leq 2|\operatorname{Re}\langle \bar{\alpha}, w \rangle|/|w|$. By condition (5), we have

$$|\operatorname{Re}\langle \bar{\alpha}, w \rangle| = |\operatorname{Re}\langle \operatorname{grad} \rho, \zeta(u) - u \rangle|/|\operatorname{grad} \rho| \leq |\zeta(u) - u| \leq \epsilon |w|.$$

Hence

$$|a(\zeta, z)| \left| \frac{1}{|w - \alpha t|^{2n}} - \frac{1}{|w + \alpha t|^{2n}} \right| \leq \frac{C_1 \epsilon |t|}{(|u|^2 + t^2)^n}.$$

Since $d\sigma \leq C_2 dS$, where dS is the area element of the plane T_z, we have finally

$$\begin{aligned} &\int_{\partial D \cap B(z,r)} |d_m g(\zeta, z^+) - d_m g(\zeta, z^-)| \, d\sigma \\ &\qquad \leq C_3 \epsilon \int_{B'} \frac{|t|}{(|u|^2 + t^2)^n} \, dS \leq C_3 \epsilon \int_{T_z} \frac{|t|}{(|u|^2 + t^2)^n} \, dS = C_4 \epsilon. \end{aligned}$$

□

The proof of this lemma is essentially the same as the proof of Theorem 3.1.

Theorem 4.7 (Aronov). *Suppose $f \in \mathcal{C}^1(\partial D)$, and F is the Bochner-Martinelli integral (2.1). Then*

$$\lim_{z^\pm \to z} \left(\frac{\partial F(z^+)}{\partial \bar{z}_k} - \frac{\partial F(z^-)}{\partial \bar{z}_k} \right) = \frac{\partial f}{\partial \bar{z}_k} - \rho_{\bar{k}} \sum_{m=1}^n \rho_m \frac{\partial f}{\partial \bar{z}_m}, \tag{4.11}$$

$$\lim_{z^\pm \to z} \left(\frac{\partial F(z^+)}{\partial z_k} - \frac{\partial F(z^-)}{\partial z_k} \right) = \frac{\partial f}{\partial z_k} - \rho_k \sum_{m=1}^n \rho_m \frac{\partial f}{\partial \bar{z}_m}, \tag{4.12}$$

and these limits are attained uniformly in $z \in \partial D$.

Proof. By (4.2), we have

$$\frac{\partial F}{\partial \bar{z}_k} = \int_{\partial D} \frac{\partial f}{\partial \bar{\zeta}_k} U(\zeta, z) + (-1)^k \int_{\partial D} \sum_{m=1}^n \frac{\partial f}{\partial \bar{\zeta}_m} \frac{\partial g}{\partial \zeta_m} \, d\bar{\zeta}[k] \wedge d\zeta.$$

By Theorem 3.1, the jump of the first integral equals $\partial f / \partial \bar{z}_k$, and we represent the second integral in the form

$$\begin{aligned} &\int_{\partial D} \frac{\partial f}{\partial \bar{\zeta}_m} \frac{\partial g}{\partial \zeta_m} \, d\bar{\zeta}[k] \wedge d\zeta \\ &\qquad = \frac{\partial f}{\partial \bar{z}_m} \int_{\partial D} \frac{\partial g}{\partial \zeta_m} \, d\bar{\zeta}[k] \wedge d\zeta + \int_{\partial D} \left(\frac{\partial f}{\partial \bar{\zeta}_m} - \frac{\partial f}{\partial \bar{z}_m} \right) \frac{\partial g}{\partial \zeta_m} \, d\bar{\zeta}[k] \wedge d\zeta. \end{aligned}$$

By Lemma 4.6, the jump of the first integral equals $(-1)^{k-1} (\partial f / \partial \bar{z}_m) \rho_{\bar{k}}(z) \rho_m(z)$, and the jump of the second integral is zero (this is proved the same way as in Theorem 3.1). □

Theorem 4.7 was given in [17].

We obtain the following assertion from Theorem 4.7 and Corollary 4.4 by induction.

Corollary 4.8. *If $\partial D \in \mathcal{C}^k$ and $f \in \mathcal{C}^m(\partial D)$, where $m \leq k$, and $F^+ \in \mathcal{C}^m(\overline{D})$, then $F^- \in \mathcal{C}^m(\mathbf{C}^n \setminus D)$. Conversely, if $F^- \in \mathcal{C}^m(\mathbf{C}^n \setminus D)$, then $F^+ \in \mathcal{C}^m(\overline{D})$.*

Remark. Just as for Theorem 3.1, Theorem 4.7 can be obtained when f is differentiable on ∂D and all its derivatives are integrable on ∂D. The jump formulas (4.11) and (4.12) for derivatives will then hold at Lebesgue points of the derivatives of f.

Corollary 4.9. *If $\partial D \in \mathcal{C}^1$ and $f \in \mathcal{C}^1$, then the jump of the derivative $\bar{\partial}_n F = \sum_{k=1}^n \rho_k (\partial F / \partial \bar{z}_k)$ is zero.*

4.3 Jump theorem for the "normal" derivative

Corollary 4.9 shows that the jump of the "normal" derivative $\overline{\partial}_n F$ of the Bochner-Martinelli integral is zero. It turns out that this assertion is valid even for continuous functions f if the boundary of the domain is assumed to be class $\mathcal{C}^2$ smooth.

In this case, we may take as defining function

$$\rho(z) = \begin{cases} -\inf_{\zeta\in\partial D} |\zeta - z|, & z \in \overline{D}; \\ \inf_{\zeta\in\partial D} |\zeta - z|, & z \in \mathbf{C}^n \setminus \overline{D}. \end{cases}$$

Then $D = \{ z : \rho(z) < 0 \}$. Moreover, when $\partial D \in \mathcal{C}^2$ we have the following (see §2 of [213], and also [61]):

(a) there is a neighborhood V of ∂D such that $\rho \in \mathcal{C}^2(V)$;

(b) $|\operatorname{grad} \rho| = 1/2$ in V;

(c) if $z^\pm \in V$ are points on the normal to ∂D at z such that $|z^+ - z| = |z^- - z|$, then $(\partial\rho/\partial z_k)(z^\pm) = (\partial\rho/\partial z_k)(z)$ and $(\partial\rho/\partial \bar{z}_k)(z^\pm) = (\partial\rho/\partial \bar{z}_k)(z)$ for $k = 1, 2, \dots, n$.

In this case $\rho_k = 2(\partial\rho/\partial z_k)$ and $\rho_{\bar{k}} = 2(\partial\rho/\partial \bar{z}_k)$. Consequently $\overline{\partial}_n F = \sum_{k=1}^n \rho_k(\partial F/\partial \bar{z}_k) = 2\sum_{k=1}^n (\partial F/\partial \bar{z}_k)(\partial\rho/\partial z_k)$.

Theorem 4.10 (Kytmanov, Aĭzenberg). *If $f \in C(\partial D)$, then the integral F of the form (2.1) satisfies*

$$\lim_{z^\pm \to z} (\overline{\partial}_n F(z^+) - \overline{\partial}_n F(z^-)) = 0.$$

This limit is attained uniformly with respect to $z \in \partial D$. If $\overline{\partial}_n F(z^+)$ extends continuously to $\overline{D}$, then $\overline{\partial}_n F(z^-)$ extends continuously to $\mathbf{C}^n \setminus D$, and conversely.

Proof. If f is constant, then $\overline{\partial}_n F \equiv 0$. Thus, we may assume that $f(z) = 0$ at the point $z \in \partial D$. The restriction of the kernel $U(\zeta, z)$ to ∂D has the form

$$\frac{(n-1)!}{\pi^n} \sum_{k=1}^n \frac{\partial\rho}{\partial\zeta_k} \frac{(\bar{\zeta}_k - \bar{z}_k)}{|\zeta - z|^{2n}}\, d\sigma.$$

Consequently,

$$\begin{aligned} &\overline{\partial}_n F(z^+) - \overline{\partial}_n F(z^-) \\ &= \frac{-(n-1)!}{\pi^n} \int_{\partial D} f(\zeta) \sum_{k=1}^n \frac{\partial\rho(z)}{\partial z_k} \frac{\partial\rho}{\partial\bar{\zeta}_k} \left(\frac{1}{|\zeta - z^+|^{2n}} - \frac{1}{|\zeta - z^-|^{2n}} \right) d\sigma \\ &\quad + \frac{n!}{\pi^n} \int_{\partial D} f(\zeta) \left(\left[\sum_{k=1}^n \frac{\partial\rho}{\partial z_k}(\zeta_k - z_k^+) \sum_{m=1}^n \frac{\partial\rho}{\partial\bar{\zeta}_m}(\bar{\zeta}_m - \bar{z}_m^+) \right] \frac{1}{|\zeta - z^+|^{2+2n}} \right. \\ &\quad \left. - \left[\sum_{k=1}^n \frac{\partial\rho}{\partial z_k}(\zeta_k - z_k^-) \sum_{m=1}^n \frac{\partial\rho}{\partial\bar{\zeta}_m}(\bar{\zeta}_m - \bar{z}_m^-) \right] \frac{1}{|\zeta - z^-|^{2+2n}} \right) d\sigma. \end{aligned}$$

We denote the first integral by I_1 and the second by I_2.

Just as in Theorem 3.1, we make a unitary transformation and a translation so that z is taken to 0 and the tangent plane to ∂D at z is taken to the plane $T = \{ w \in \mathbf{C}^n : \operatorname{Im} w_n = 0 \}$. In a neighborhood of the origin, the boundary ∂D will be given by a system of equations $\zeta_1 = w_1, \ldots, \zeta_{n-1} = w_{n-1}, \zeta_n = u_n + i\varphi(w)$, where $w = (w_1, \ldots, w_{n-1}, u_n) \in T$. The function $\varphi(w)$ is class C^2 in a neighborhood W of the origin, and $z^{\pm} = (0, \ldots, 0, \pm i y_n)$. The surface ∂D is a Lyapunov surface with Hölder exponent equal to 1, so the following estimates hold (see [211, §22] and [192, §7]):

$$\begin{aligned} &|\varphi(w)| \le C|w|^2, && w \in W, \\ &\left|\frac{\partial \varphi}{\partial u_j}\right| \le C_1|w|, && j = 1, \ldots, n, \\ &\left|\frac{\partial \varphi}{\partial v_j}\right| \le C_1|w|, && j = 1, \ldots, n-1, \end{aligned} \tag{4.13}$$

where $u_j = \operatorname{Re} w_j$ and $v_j = \operatorname{Im} w_j$. Since $(\partial\varphi/\partial w_j) = -(\partial\rho/\partial w_j)/(\partial\rho/\partial y_n)$, and $|\partial\rho/\partial y_n| \ge C_2 > 0$ for $w \in W$, it follows that

$$\left|\frac{\partial \rho}{\partial \zeta_k}(\zeta(w))\right| \le C_3|w|, \qquad \left|\frac{\partial \rho}{\partial \bar{\zeta}_k}(\zeta(w))\right| \le C_3|w|, \tag{4.14}$$

for $w \in W$ and $k = 1, \ldots, n-1$.

We note that the constants do not depend on the point z under consideration. Finally,

$$|\zeta(w)| \le C_4|w|. \tag{4.15}$$

We fix $\epsilon > 0$, take a ball B' in the plane T with center at the origin, and choose $a > 0$ such that

1. $B' \subset W$;
2. $|f(\zeta(w))| < \epsilon$ for $w \in B'$;
3. $\{ z \in \mathbf{C}^n : (z_1, \ldots, z_{n-1}, \operatorname{Re} z_n) \in B', |\operatorname{Im} z_n| < a \} \subset W$;
4. $C(2|y_n| + C|w|^2) \le d < 1$ for $|y_n| < a$ and $w \in B'$ (the constant C being the one from (4.13)).

Since $z^{\pm} = (0, \ldots, 0, \pm i y_n)$, the identity $|\zeta(w) - z^{\pm}|^2 = |w|^2 + (\pm y_n - \varphi(w))^2$ holds. Hence

$$|\zeta - z^{\pm}|^2 = |w - z^{\pm}|^2 \left(1 - (\pm 2\varphi y_n - \varphi^2)/|w - z^{\pm}|^2\right).$$

But

$$\frac{|\pm 2\varphi y_n - \varphi^2|}{|w - z^{\pm}|^2} \le \frac{C|w|^2(2|y_n| - C|w|^2)}{|w|^2 + y_n^2} \le C(2|y_n| + C|w|^2) \le d < 1$$

for $|y_n| \le a$ and $w \in B'$. Consequently

$$\frac{1}{1-(\pm 2\varphi y_n - \varphi^2)/|w-z^\pm|^2} = \sum_{k=0}^{\infty} \frac{(\pm 2\varphi y_n - \varphi^2)^k}{|w-z^\pm|^{2k}} = 1 + \frac{(\pm 2\varphi y_n - \varphi^2)h(w,z)}{|w-z^\pm|^2},$$

and the function $h(w,z)$ is uniformly bounded for $w \in B'$ and $|y_n| \le a$. Therefore

$$\frac{1}{|\zeta - z^\pm|^{2n}} = \frac{1+(\pm 2\varphi y_n - \varphi^2)h_1(z,w)/|w-z^\pm|^2}{|w-z^\pm|^{2n}}, \tag{4.16}$$

$$\frac{1}{|\zeta - z^\pm|^{2+2n}} = \frac{1+(\pm 2\varphi y_n - \varphi^2)h_2(z,w)/|w-z^\pm|^2}{|w-z^\pm|^{2+2n}}, \tag{4.17}$$

and the functions h_1 and h_2 are uniformly bounded for $w \in B'$ and $|y_n| \le a$.

We set $\Gamma = \{\, \zeta \in \partial D : \zeta = \zeta(w),\ w \in B' \,\}$ and estimate the integral I_1 over the surface Γ. Using (4.16) and (4.17), we obtain

$$\left| \frac{1}{|\zeta - z^+|^{2n}} - \frac{1}{|\zeta - z^-|^{2n}} \right| \le \frac{2(|2\varphi y_n| + \varphi^2)|h_1|}{|w-z^+|^{2+2n}} \le C_5 \frac{2|y_n| + C|w|^2}{(|w|^2 + y_n^2)^n}.$$

Supposing that $d\sigma \le C_6\, dS$, where dS is the surface area element of the plane T, we have

$$|I_{1,\Gamma}| = \frac{(n-1)!}{\pi^n} \left| \int_\Gamma f(\zeta) \sum_{k=1}^{n} \frac{\partial \rho}{\partial z_k} \frac{\partial \rho}{\partial \bar{\zeta}_k} \left(\frac{1}{|\zeta - z^+|^{2n}} - \frac{1}{|\zeta - z^-|^{2n}} \right) d\sigma \right|$$
$$\le \epsilon\, C_7 \int_{B'} \frac{2|y_n| + C|w|^2}{(|w|^2 + y_n^2)^n}\, dS.$$

Now

$$\int_{B'} \frac{|y_n|}{(|w|^2 + y_n^2)^n}\, dS \le \int_T \frac{|y_n|}{(|w|^2 + y_n^2)^n}\, dS = \text{const},$$

while

$$\int_{B'} \frac{|w|^2}{(|w|^2 + y_n^2)^n}\, dS \le \int_{B'} \frac{1}{(|w|^2 + y_n^2)^{n-1}}\, dS.$$

Introducing polar coordinates in the ball B', we have $dS = |w|^{2n-2}\, d|w| \wedge d\omega$, where $d\omega$ is the surface area element in the unit sphere in $\mathbf{R}^{2n-1}$, so

$$\int_{B'} \frac{1}{(|w|^2 + y_n^2)^{n-1}}\, dS = \sigma_{2n-1} \int_0^R \frac{|w|^{2n-2}}{(|w|^2 + y_n^2)^{n-1}}\, d|w| \le R\, \sigma_{2n-1}.$$

Here R is the radius of the ball B', and σ_{2n-1} is the area of the unit sphere in $\mathbf{R}^{2n-1}$. Therefore $|I_{1,\Gamma}| \le C_8\epsilon$, where the constant C_8 is independent of z and y_n.

Obviously the integral I_1 over the surface $\partial D \setminus \Gamma$ can be made as small as desired as $z^{\pm} \to 0$.

We now show that the form of the integral I_2 does not change under a unitary transformation. Indeed, distance does not change, so the functions ρ, $d\sigma$, and $|\zeta - z|$ do not change. Consider the expression $\sum_{k=1}^n (\partial\rho/\partial z_k)(\zeta_k - z_k)$. Suppose the unitary transformation is given by the matrix $A = \|a_{jk}\|_{j,k=1}^n$, that is, by $z'_k = \sum_{j=1}^n a_{jk}z_j$ for $k = 1, \ldots, n$, and the inverse transformation is given by the matrix $B = \|b_{jk}\|_{j,k=1}^n$. Then $\sum_{k=1}^n a_{kj}b_{sk} = \delta_{js}$, where δ_{js} is the Kronecker symbol. Therefore

$$\sum_{k=1}^n \frac{\partial\rho}{\partial z_k}(\zeta_k - z_k) = \sum_{k,j,s=1}^n \frac{\partial\rho}{\partial z'_j} a_{kj}b_{sk}(\zeta'_s - z'_s) = \sum_{j,s=1}^n \frac{\partial\rho}{\partial z'_j}\delta_{js}(\zeta'_s - z'_s)$$
$$= \sum_{j=1}^n \frac{\partial\rho}{\partial z'_j}(\zeta'_j - z'_j).$$

It can be shown in the same way that the sum $\sum_{k=1}^n (\partial\rho/\partial\bar\zeta_k)(\bar\zeta_k - \bar z_k)$ does not change. Thus, the form of the integral I_2 is invariant under unitary transformations. Then

$$\sum_{k=1}^n \frac{\partial\rho}{\partial z_k}(0)(\zeta_k - z_k^{\pm}) \sum_{m=1}^n \frac{\partial\rho}{\partial\bar\zeta_m}(\bar\zeta_m - \bar z_m^{\pm})$$
$$= -\frac{i}{2}(\zeta_n - z_n^{\pm}) \sum_{m=1}^n \frac{\partial\rho}{\partial\bar\zeta_m}(\bar\zeta_m - \bar z_m^{\pm})$$
$$= -\frac{i}{2}\sum_{m=1}^{n-1} \frac{\partial\rho}{\partial\bar\zeta_m}\bar\zeta_m(\zeta_n - z_n^{\pm}) - \frac{i}{2}\frac{\partial\rho}{\partial\bar\zeta_n}(u_n^2 + \varphi^2 + y_n^2 \mp 2\varphi y_n).$$

We split the integral I_2 over the surface Γ into three integrals:

$$I'_2 = \frac{in!}{\pi^n}\int_{B'} \frac{f(\zeta(w))}{|w - z^+|^{2+2n}}\left(2\varphi y_n \frac{\partial\rho}{\partial\bar\zeta_n} + \sum_{m=1}^{n-1}\frac{\partial\rho}{\partial\bar\zeta_m}\bar\zeta_m y_n\right) d\sigma'.$$
$$I_2^{\pm} = \frac{\pm in!}{2\pi^n}\int_{B'} \frac{f(\zeta(w))}{|w - z^+|^{4+2n}}\left[\frac{\partial\rho}{\partial\bar\zeta_n}(u_n^2 + \varphi^2 + y_n^2 \pm 2\varphi y_n) + \right.$$
$$\left. + \sum_{m=1}^{n-1}\frac{\partial\rho}{\partial\bar\zeta_m}\bar\zeta_m(\zeta_n - z_n^{\pm})\right](\pm 2\varphi y_n - \varphi^2)h_2\, d\sigma'$$

where $d\sigma'$ is the image of $d\sigma$ under the mapping $w \to \zeta(w)$, and h_2 is defined in (4.17). Using (4.13)–(4.17), we find that

$$\begin{aligned}|I_2'| &\leq M_1\epsilon \int_{B'} \frac{M_2|w|^2|y_n| + M_3|w|^2|y_n|}{|w-z^+|^{2+2n}}\, dS \\ &\leq M_4\epsilon \int_T \frac{|y_n|}{(|w|^2+y_n^2)^n}\, dS = M_5\epsilon.\end{aligned}$$

Now

$$\begin{aligned}|I_2^{\pm}| &\leq M_6\epsilon \int_{B'} \frac{|w|^2}{(|w|^2+y_n^2)^{2+n}}(2|y_n| + C|w|^2)(M_7|w|^2 + M_8|w|^2|y_n| + M_9 y_n^2)\, dS \\ &\leq M_{10}\epsilon \int_{B'} \frac{|y_n|}{(|w|^2+y_n^2)^n}\, dS + M_{11}\epsilon \int_{B'} \frac{1}{(|w|^2+y_n^2)^{n-1}}\, dS \leq M_{12}\epsilon.\end{aligned}$$

The integral I_2 over $\partial D \setminus \Gamma$ also tends to zero. □

Theorem 4.10, given in [123], is an analogue of Lyapunov's theorem on the jump of the normal derivative of a double-layer potential (see [64, 192]). Just as in Theorem 3.1, it can be shown that for $f \in \mathcal{L}^1(\partial D)$, the difference $\overline{\partial}_n F(z^+) - \overline{\partial}_n F(z^-) \to 0$ as $z^{\pm} \to z$ at Lebesgue points of f. We also remark that Theorem 4.10 does not hold for the derivative $\sum_{k=1}^n (\partial F/\partial z_k)\rho_{\bar{k}}$.

5 The Bochner-Martinelli integral in the ball

5.1 The spectrum of the Bochner-Martinelli operator

In the previous sections we saw that the behavior of the Bochner-Martinelli integral is analogous to the behavior of a Cauchy-type integral or of a double-layer potential. Here we will find the spectral decomposition of the Bochner-Martinelli operator in the ball. It will differ significantly from the corresponding decomposition of the Cauchy operator in the disc or of the double-layer potential in the ball.

Let $B = B(0,1)$ be the unit ball in $\mathbf{C}^n$ with center at the origin, and let $S = S(0,1)$ be its boundary. We recall that the scalar product (f,g) of two functions f and g in $\mathcal{L}^2(S)$ is given by the integral $(f,g) = \int_S f\bar{g}\, d\sigma$.

We will identify the space $\mathcal{L}^2(S)$ with the space of harmonic extensions of such functions from S into B, that is, with the space of harmonic functions f in B for which

$$\sup_{0\leq r<1} \int_S |f(rz)|^2\, d\sigma(z) < \infty.$$

Recall that the Poisson kernel $P(\zeta, z)$for the ball B has the form

$$P(\zeta, z) = \frac{(n-1)!}{2\pi^n} \frac{(1-|z|^2)}{|\zeta - z|^{2n}}, \qquad z \in B, \quad \zeta \in S.$$

Lemma 5.1. *The restriction of the kernel* $U(\zeta, z)$ *to* S *equals*

$$\frac{1 - \langle \zeta, \bar{z} \rangle}{1 - |z|^2} P(\zeta, z)\, d\sigma, \qquad z \in B, \quad \zeta \in S.$$

Proof. Since the defining function ρ for B may be chosen as $\rho = |z|^2 - 1$, we obtain from Lemma 3.5 that

$$U(\zeta, z)\big|_{\partial B} = \frac{(n-1)!}{2\pi^n} \sum_{k=1}^{n} \frac{\zeta_k(\bar{\zeta}_k - \bar{z}_k)\, d\sigma(\zeta)}{|\zeta - z|^{2n}}.$$

Hence Lemma 5.1 follows. □

We remark that from Lemma 5.1 we can obtain another proof of the Bochner-Martinelli formula for holomorphic functions in the ball B. If f is holomorphic in B and $f \in \mathcal{C}(\overline{B})$, then $f(\zeta)(1 - \langle \zeta, \bar{z} \rangle)(1 - |z|^2)^{-1}$ is also holomorphic in B, and consequently is harmonic. Therefore it can be represented by Poisson's formula, that is (for $z \in B$),

$$\int_S f(\zeta)\, U(\zeta, z) = \int_S f(\zeta)(1 - \langle \zeta, \bar{z} \rangle)(1 - |z|^2)^{-1} P(\zeta, z)\, d\sigma = f(z).$$

Let $P_k(z)$ be a homogeneous harmonic polynomial of degree k of the form

$$P_k(z) = \sum_{\|\alpha + \beta\| = k} a_{\alpha,\beta} z^\alpha \bar{z}^\beta,$$

where $\alpha = (\alpha_1, \dots, \alpha_n)$ and $\beta = (\beta_1, \dots, \beta_n)$ are multi-indices, $z^\alpha = z_1^{\alpha_1} \dots z_n^{\alpha_n}$ and $\bar{z}^\beta$ are monomials, and $\|\alpha\| = \alpha_1 + \dots + \alpha_n$. Then

$$P_k(z) = \sum_{s+t=k} P_{s,t}(z),$$

where $P_{s,t} = \sum_{\|\alpha\|=s} \sum_{\|\beta\|=t} a_{\alpha,\beta} z^\alpha \bar{z}^\beta$.

It is clear that the $P_{s,t}$ are harmonic polynomials, homogeneous of degree s in z and degree t in $\bar{z}$. We denote the set of homogeneous harmonic polynomials $P_{s,t}$ by $\mathcal{P}_{s,t}$. Since the set of harmonic polynomials is dense in $\mathcal{L}^2(S)$, so is $\bigcup_{s,t} \mathcal{P}_{s,t}$. Moreover, if $P_{s,t} \in \mathcal{P}_{s,t}$ and $P_{l,m} \in \mathcal{P}_{l,m}$, then the scalar product $(P_{s,t}, P_{l,m}) = 0$ when $s \neq l$ or $t \neq m$. Indeed, since $d\bar{\zeta}[k] \wedge d\zeta\big|_S = (-1)^{k-1} 2^{n-1} i^n \zeta_k\, d\sigma$, and $d\zeta[k] \wedge d\bar{\zeta}\big|_S = (-1)^{n+k-1} 2^{n-1} i^n \bar{\zeta}_k\, d\sigma$, we have

$$\sum_{k=1}^{n} \frac{\partial P_{s,t}}{\partial \bar{\zeta}_k} (-1)^{n+k-1}\, d\zeta[k] \wedge d\bar{\zeta}\big|_S = i^n 2^{n-1} \sum_{k=1}^{n} \frac{\partial P_{s,t}}{\partial \bar{\zeta}_k} \bar{\zeta}_k\, d\sigma = i^n 2^{n-1} t P_{s,t}\, d\sigma.$$

Moreover, the form $\sum_{k=1}^{n}(\partial P_{s,t}/\partial\bar\zeta_k)(-1)^{n+k-1}\,d\zeta[k]\wedge d\bar\zeta$ is closed by the harmonicity of $P_{s,t}$, so by Stokes's formula

$$\begin{aligned} t2^{n-1}i^n(P_{s,t},P_{l,m}) &= \sum_{k=1}^{n}\int_S \overline{P}_{l,m}\frac{\partial P_{s,t}}{\partial\bar\zeta_k}(-1)^{n+k-1}\,d\zeta[k]\wedge d\bar\zeta \\ &= \int_S P_{s,t}\sum_{k=1}^{n}(-1)^{k-1}\frac{\partial\overline{P}_{l,m}}{\partial\zeta_k}\,d\bar\zeta[k]\wedge d\zeta \\ &= m2^{n-1}i^n(P_{s,t},P_{l,m}). \end{aligned}$$

Thus we have shown that we can always choose an orthonormal basis in $\mathcal{L}^2(S)$ consisting of polynomials $P_{s,t}$.

Since we are interested in the operator giving the Bochner-Martinelli integral, we write

$$(Mf)(z)=\int_S f(\zeta)U(\zeta,z),\qquad z\in B,$$

for $f\in\mathcal{L}^2(S)$.

Lemma 5.2. *If $P_{s,t}\in\mathcal{P}_{s,t}$, then*

$$MP_{s,t}=\frac{(n+s-1)}{(n+s+t-1)}P_{s,t}. \tag{5.1}$$

Proof. First we show that the harmonic extension of the polynomial $\zeta_k P_{s,t}$ from S to B is given by the formula

$$f(\zeta)=\zeta_k P_{s,t}(\zeta)+\frac{(1-|\zeta|^2)}{(n+s+t-1)}\frac{\partial P_{s,t}}{\partial\bar\zeta_k},\qquad \zeta\in\overline{B}.$$

In fact,

$$\begin{aligned} \Delta f &= \sum_{m=1}^{n}\frac{\partial^2 f}{\partial\zeta_m\partial\bar\zeta_m} \\ &= \frac{\partial}{\partial\bar\zeta_k}P_{s,t}-\frac{n}{(n+s+t-1)}\frac{\partial P_{s,t}}{\partial\bar\zeta_k}-\frac{1}{(n+s+t-1)}\sum_{m=1}^{n}\zeta_m\frac{\partial}{\partial\zeta_m}\left(\frac{\partial P_{s,t}}{\partial\bar\zeta_k}\right) \\ &\quad -\frac{1}{(n+s+t-1)}\sum_{m=1}^{n}\bar\zeta_m\frac{\partial}{\partial\bar\zeta_m}\left(\frac{\partial P_{s,t}}{\partial\bar\zeta_k}\right) \\ &= \frac{\partial P_{s,t}}{\partial\bar\zeta_k}-\frac{n}{(n+s+t-1)}\frac{\partial P_{s,t}}{\partial\bar\zeta_k}-\frac{s}{(n+s+t-1)}\frac{\partial P_{s,t}}{\partial\bar\zeta_k} \\ &\quad -\frac{(t-1)}{(n+s+t-1)}\frac{\partial P_{s,t}}{\partial\bar\zeta_k} \\ &= 0. \end{aligned}$$

Here we have used the formula

$$\Delta(fg) = g\Delta f + f\Delta g + \sum_{m=1}^{n} \left(\frac{\partial f}{\partial \zeta_m} \frac{\partial g}{\partial \bar{\zeta}_m} + \frac{\partial f}{\partial \bar{\zeta}_m} \frac{\partial g}{\partial \zeta_m} \right).$$

By Lemma 5.1,

$$\begin{aligned} MP_{s,t}(z) &= \int_S P_{s,t}(\zeta) \frac{(1 - \langle \zeta, \bar{z} \rangle)}{(1 - |z|^2)} P(\zeta, z)\, d\sigma \\ &= \frac{P_{s,t}}{(1 - |z|^2)} - \sum_{k=1}^{n} \frac{\bar{z}_k}{(1 - |z|^2)} \left(z_k P_{s,t} + \frac{(1 - |z|^2)}{(n + s + t - 1)} \frac{\partial P_{s,t}}{\partial \bar{z}_k} \right) \\ &= P_{s,t}(z) - \frac{t}{(n + s + t - 1)} P_{s,t}(z) = \frac{(n + s - 1)}{(n + s + t - 1)} P_{s,t}(z). \end{aligned}$$

□

We remark that when $n = 1$, the polynomial $P_{s,t}(z)$ is either $\bar{z}^t$ or z^s. Therefore we obtain that $Mz^s = z^s$, and $M\bar{z}^t = 0$ (in this case Mf is a Cauchy-type integral).

For $z \in S$, let $(M_\sigma f)(z) = 2 \int_S f(\zeta)\, U(\zeta, z)$ be twice the Bochner-Martinelli singular integral.

Corollary 5.3. *If $P_{s,t} \in \mathcal{P}_{s,t}$, then*

$$M_\sigma P_{s,t} = \frac{(n + s - t - 1)}{(n + s + t - 1)} P_{s,t}.$$

Proof. It follows from Lemma 5.2 and the Sokhotskiĭ-Plemelj formula (2.6) that $Mf = f/2 + (M_\sigma f)/2$ on S. □

When $n = 1$, we obtain the well-known equality $M_\sigma z^s = z^s$, while $M_\sigma \bar{z}^t = -\bar{z}^t$.

Theorem 5.4 (Romanov). *When $n > 1$, the operator M is a bounded self-adjoint operator $\mathcal{L}^2(S) \to \mathcal{L}^2(S)$ with $\|M\| = 1$. Every rational number in the interval $(0, 1]$ is an eigenvalue of M of infinite multiplicity. The spectrum of M coincides with the interval $[0, 1]$.*

Proof. The proof follows from Lemma 5.2, since $(n + s - 1)(n + s + t - 1)^{-1} \leq 1$, and consequently $\|Mf\|_{\mathcal{L}^2} \leq \|f\|_{\mathcal{L}^2}$ for $f \in \mathcal{L}^2(S)$, while $\|Mf\|_{\mathcal{L}^2} = \|f\|_{\mathcal{L}^2}$ for holomorphic f. □

Theorem 5.4 and Lemma 5.2 were obtained by Romanov in [168]. Our proof is taken from [114, 115].

Corollary 5.5. *When $n > 1$, the operator M_σ is bounded in $\mathcal{L}^2(S)$ with $\|M_\sigma\| = 1$. Every rational number in the interval $(-1, 1]$ is an eigenvalue of M_σ of infinite multiplicity.*

Thus, the operators M and M_σ are essentially different when $n = 1$ and when $n > 1$. For example, M_σ^2 is the identity operator when $n = 1$ (sometimes this equality is called the Poincaré-Bertrand formula (see [154, p. 125])), while $M_\sigma^2 f$ does not equal f in general when $n > 1$. In fact, $M_\sigma^2 f = f$ only for holomorphic functions; for example, $M_\sigma \bar{z}_k = (n-2)\bar{z}_k/n$. Moreover, $M_\sigma^l P_{s,t} \to 0$ when $l \to \infty$ if $t > 0$. In this sense, the Poincaré-Bertrand formula for M_σ fails when $n > 1$ (we will consider the question of the Poincaré-Bertrand formula later on, in chapter 5).

Theorem 5.6 (Romanov). *Suppose $n > 1$, and let P_H be the operator of projection from $\mathcal{L}^2(S)$ onto the subspace of holomorphic functions in $\mathcal{L}^2(S)$. Then $M^k \to P_H$ as $k \to \infty$ in the strong operator topology of $\mathcal{L}^2(S)$.*

Theorem 5.6 was given in [168]. Thus, although $M \neq P_H$ when $n > 1$ (it is true that $M = P_H$ when $n = 1$), its iterates converge to P_H. In chapter 4, we will consider generalizations of this theorem to other domains.

We can also use Lemma 5.2 to compute the Bochner-Martinelli integral of every polynomial.

Corollary 5.7. *If $Q_{s,t}$ is an arbitrary polynomial of the form*

$$Q_{s,t} = \sum_{\|\alpha\|=s} \sum_{\|\beta\|=t} a_{\alpha,\beta} z^\alpha \bar{z}^\beta,$$

then $MQ_{s,t} = \sum_{p\geq 0} Z_{s-p,t-p}$, where the polynomials $Z_{s-p,t-p}$ are given by the formula

$$Z_{s-p,t-p} = \frac{s-p+n-1}{p!\,(s+t+n-p-1)!} \sum_{j\geq 0} (-1)^j \frac{(s+t-j-2p+n-2)!}{j!} |z|^{2j} \Delta^{j+p} Q_{s,t}(z).$$

Proof. For the proof, we extend $Q_{s,t}$ into B as a harmonic function by Poisson's formula (see, for example, [189, chap. 11]) and then apply Lemma 5.2. □

If we consider the Bochner-Martinelli integral on the exterior of the ball, we can easily obtain the spectral resolution of this operator from the jump theorem and Lemma 5.2.

We now compare the the spectral decompositions of the single- and double-layer operators with the spectral decomposition of the Bochner-Martinelli operator.

Let $B \subset \mathbf{R}^n$, $x = (x_1, \ldots, x_n)$, and $y \in \mathbf{R}^n$, while σ_n is the surface area of the unit sphere in $\mathbf{R}^n$ (where $n \geq 3$). We define the single-layer operator T and the double-layer operator W via

$$(Tf)(x) = \frac{1}{\sigma_n} \int_S \frac{f(y)}{|x-y|^{n-2}}\, d\sigma(y),$$
$$(Wf)(x) = \frac{1}{\sigma_n} \int_S \frac{f(y)(1-\langle x,y\rangle)}{|x-y|^n}\, d\sigma(y),$$
$$\langle x,y\rangle = x_1y_1 + \cdots + x_ny_n, \qquad x \in B.$$

By expressing the kernels of these operators in terms of the Poisson kernel $P(x,y) = \sigma_n^{-1}(1-|x|^2)|x-y|^{-n}$, and writing the harmonic extension of $x_kP_m(x)$ from S to B for the homogeneous harmonic polynomial P_m of degree m in the form $x_kP_m + (1-|x|^2)(n+2m-2)^{-1}(\partial P_m/\partial x_k)$, we obtain as in Lemma 5.2 that $TP_m = (n-2)(n+2m-2)^{-1}P_m$ and $WP_m = (n+m-2)(n+2m-2)^{-1}P_m$; that is, we find the eigenvalues and eigenfunctions of T and W. Therefore T and W are bounded in $\mathcal{L}^2(S)$ with $\|T\| = \|W\| = 1$. The spectra of T and W are discrete (in contrast to the spectrum of M), with the single limit point 0 for T and 1/2 for W.

We remark that if we denote by W_σ twice the singular double-layer potential:

$$W_\sigma f = \text{P.V.}\, \frac{2}{\sigma_n} \int_S \frac{f(y)(1-\langle x,y\rangle)}{|x-y|^n}\, d\sigma(y), \qquad x \in S,$$

then, by the jump formula for this potential, $W_\sigma f = 2Wf - f$ on S, that is,

$$W_\sigma f = \frac{1}{\sigma_n} \int_S \frac{f(y)(1-2\langle x,y\rangle + |y|^2)}{|x-y|^n}\, d\sigma = Tf.$$

These assertions are classical (for $n=3$ they are given in [192, chap. 5, §1]).

5.2 Computation of the Bochner-Martinelli integral in the ball

We have seen that the Bochner-Martinelli integral of a polynomial can be computed in closed form (see Corollary 5.7). The computation of Mf for an arbitrary function f in $\mathcal{L}^2(S)$ can be reduced to the calculation of a one-dimensional integral.

Theorem 5.8 (Kytmanov). *Suppose $n > 1$ and $f \in \mathcal{L}^2(S)$. Then*

$$Mf = f - \sum_{k=1}^n \bar{z}_k \frac{\partial \psi}{\partial \bar{z}_k}, \qquad \text{or} \tag{5.2}$$

$$Mf = f - \sum_{k=1}^n \bar{z}_k \psi_k, \tag{5.3}$$

where $\psi = |z|^{1-n}\int_0^{|z|} |\zeta|^{n-2} f(\zeta)\, d|\zeta|$, and $\psi_k = |z|^{-n}\int_0^{|z|} |\zeta|^{n-1}(\partial f/\partial\bar{\zeta}_k)\, d|\zeta|$. The functions ψ and ψ_k are harmonic in B. To compute them, it is convenient to pass to polar coordinates. Formula (5.3) also holds for $n=1$ (we recall that $\mathcal{L}^2(S)$ is identified with the space of its harmonic extensions into B).

Proof. We prove, for example, (5.2). We decompose f into a series in harmonic polynomials $P_{s,t} \in \mathcal{P}_{s,t}$:

$$f = \sum_{s,t\geq 0} P_{s,t}. \tag{5.4}$$

This series converges uniformly and absolutely inside B, and it converges on S in the metric of $\mathcal{L}^2(S)$. Applying the operator M to (5.4), we obtain

$$Mf = \sum_{s,t\geq 0} \frac{(n+s-1)}{(n+s+t-1)} P_{s,t}.$$

On the other hand,

$$\begin{aligned}
\psi &= |z|^{1-n} \int_0^{|z|} |\zeta|^{n-2} \sum_{s,t\geq 0} P_{s,t}(\zeta)\, d|\zeta| \\
&= |z|^{1-n} \sum_{s,t\geq 0} \int_0^{|z|} |\zeta|^{n+s+t-2} P_{s,t}\left(\frac{\zeta}{|\zeta|}\right) d|\zeta| \\
&= \sum_{s,t\geq 0} \frac{1}{(n+s+t-1)} P_{s,t}(z).
\end{aligned}$$

Hence

$$f - \sum_{k=1}^{n} \bar{z}_k \frac{\partial \psi}{\partial \bar{z}_k} = \sum_{s,t\geq 0} \left(P_{s,t} - \frac{t}{(n+s+t-1)} P_{s,t} \right) = Mf.$$

Formula (5.3) is proved analogously. □

Theorem 5.8 was obtained in [114, 115].

If f is polyharmonic or real-analytic in $\overline{B}$, then to compute Mf we can use Almansi's representation (see [189, chap. II, §5]):

$$f(z) = \sum_{k\geq 0} f_k(z)|z|^{2k},$$

where the f_k are harmonic functions in B.

There are also formulas analogous to (5.2) and (5.3) for the potentials T and W:

$$Tf = \frac{(n-2)}{2} |x|^{1-n/2} \int_0^{|x|} |y|^{-2+n/2} f(y)\, d|y|, \qquad \text{and}$$

$$Wf = \frac{f}{2} + \frac{(n-2)}{4} |x|^{1-n/2} \int_0^{|x|} |y|^{-2+n/2} f(y)\, d|y|, \qquad f \in \mathcal{L}^2(S).$$

Consequently, Tf and Wf satisfy the equations

$$Tf + \frac{2|x|}{(n-2)} \frac{\partial Tf}{\partial |x|} = f,$$
$$\frac{2|x|}{(n-2)} \frac{\partial Wf}{\partial |x|} + Wf = f + \frac{|x|}{(n-2)} \frac{\partial f}{\partial |x|}.$$

We compute, for example, the Bochner-Martinelli integral for the function $f(\zeta) = |\zeta - w|^{-2}$, where $w \in B$ and $n = 2$. If we make the Kelvin transformation (in w), then $f(\zeta) = |w|^{-2}\left|\zeta - w|w|^{-2}\right|^{-2}$ for $\zeta \in S$, and this function has no singularities in B. Then $f(\zeta) = (|\zeta|^2|w|^2 + 1 - \langle \zeta, \bar{w} \rangle - \langle \bar{\zeta}, w \rangle)^{-1} = (|\zeta|^2|w|^2 + 1 - 2|\zeta||w|\cos\varphi)^{-1}$, where φ is the angle between the vectors ζ and w. By (5.2),

$$\begin{aligned}
\psi(z) &= \frac{1}{|z|} \int_0^{|z|} \frac{d|\zeta|}{(|\zeta||w| - \cos\varphi)^2 + \sin^2\varphi} \\
&= \frac{1}{|z||w||\sin\varphi|} \arctan\left(\frac{|\zeta||w| - \cos\varphi}{|\sin\varphi|}\right)\Bigg|_0^{|z|} \\
&= \frac{1}{|z||w||\sin\varphi|}\left[\arctan\left(\frac{|z||w| - \cos\varphi}{|\sin\varphi|}\right) + \arctan\left(\frac{\cos\varphi}{|\sin\varphi|}\right)\right] \\
&= \frac{1}{|z||w||\sin\varphi|} \arctan\left(\frac{|z||w||\sin\varphi|}{1 - |z||w|\cos\varphi}\right).
\end{aligned}$$

Since $\cos\varphi = (\langle z, \bar{w} \rangle + \langle \bar{z}, w \rangle)/(2|z||w|)$, we have

$$|\sin\varphi| = \sqrt{4|z|^2|w|^2 - (\langle z, \bar{w} \rangle + \langle \bar{z}, w \rangle)^2}/(2|z||w|),$$

that is,

$$\psi(z) = \frac{2\arctan\sqrt{4|z|^2|w|^2 - (\langle z, \bar{w} \rangle + \langle \bar{z}, w \rangle)^2}/(2 - \langle z, \bar{w} \rangle - \langle \bar{z}, w \rangle)}{\sqrt{4|z|^2|w|^2 - (\langle z, \bar{w} \rangle + \langle \bar{z}, w \rangle)^2}}.$$

Therefore

$$\begin{aligned}
(Mf)(z) &= \frac{(|z|^2|w|^2 - \langle z, \bar{w} \rangle)(\langle z, \bar{w} \rangle + \langle \bar{z}, w \rangle)}{(|z|^2|w|^2 + 1 - \langle z, \bar{w} \rangle - \langle \bar{z}, w \rangle)(4|z|^2|w|^2 - (\langle z, \bar{w} \rangle + \langle \bar{z}, w \rangle)^2)} + \\
&+ 2\frac{2|z|^2|w|^2 - (\langle z, \bar{w} \rangle + \langle \bar{z}, w \rangle)\langle \bar{z}, w \rangle}{[4|z|^2|w|^2 - (\langle z, \bar{w} \rangle + \langle \bar{z}, w \rangle)^2]^{3/2}} \arctan\frac{\sqrt{4|z|^2|w|^2 - (\langle z, \bar{w} \rangle + \langle \bar{z}, w \rangle)^2}}{2 - \langle z, \bar{w} \rangle - \langle \bar{z}, w \rangle}.
\end{aligned}$$

5.3 Some applications

Suppose $B \subset \mathbf{R}^n$, where $n \geq 3$, and $h \in \mathcal{L}^2(S)$. Consider the integral equation $f + \lambda Tf = h$. From the form of Tf, we find that in B,

$$(n-2)(1+\lambda)f + 2|x|\frac{\partial f}{\partial |x|} = (n-2)h + 2|x|\frac{\partial h}{\partial |x|}, \qquad |\lambda| < 1.$$

Hence

$$f = h - \frac{\lambda}{2}(n-2)|x|^{-(n-2)(1+\lambda)/2} \int_0^{|x|} |y|^{-1+(n-2)(\lambda+1)/2} h(y)\, d|y|.$$

If the values of h are known only on S, then we must replace $h(y)$ in these formulas with the Poisson integral of h. We then obtain an integral representation for the solution f. We also note that the homogeneous equation $(n-2)(1+\lambda)f + 2|x|(\partial f/\partial|x|) = 0$ has (for $|\lambda| < 1$) the nontrivial solution $f = c|x|^{-(n-2)(1+\lambda)/2}$, but this solution is singular at $x = 0$, so we may exclude it from consideration.

Consider the equation $f - \lambda Wf = h$ for $h \in \mathcal{L}^2(S)$. Then in B we have

$$(n-2)(1+\lambda)f + (2+\lambda)|x|\frac{\partial f}{\partial|x|} = (n-2)h + 2|x|\frac{\partial h}{\partial|x|}, \qquad |\lambda| < 1.$$

Therefore

$$f = \frac{2}{2+\lambda}h - \frac{(n-2)\lambda}{(2+\lambda)^2}|x|^{-(n-2)(1+\lambda)/(2+\lambda)} \int_0^{|x|} |y|^{-1+(n-2)(1+\lambda)/(2+\lambda)} h(y)\, d|y|.$$

Now suppose $B \subset \mathbf{C}^n$. From the Sokhotskiĭ-Plemelj formula and Theorem 5.8, we have

$$M_\sigma f = 2Mf - f = f - 2\sum_{k=1}^n \bar{z}_k \frac{\partial \psi}{\partial \bar{z}_k}, \qquad f \in \mathcal{L}^2(S).$$

Consider the singular integral equation $f + \lambda M_\sigma f = h$, where $h \in \mathcal{L}^2(S)$. From (5.2) we obtain that

$$(1-\lambda)\sum_{k=1}^n \bar{z}_k \frac{\partial f}{\partial \bar{z}_k} + (1+\lambda)\sum_{k=1}^n z_k \frac{\partial f}{\partial z_k} + (n-1)(1+\lambda)f$$
$$= (n-1)h + |z|\frac{\partial h}{\partial|z|}, \qquad |\lambda| < 1.$$

If h is decomposed into a series of the form $h = \sum_{s\geq 0} P_{ts,qs}$, where $t, q \geq 0$, then

$$f = \frac{(t+q)h}{(1-\lambda)q + (1+\lambda)t} - \frac{2\lambda q(t+q)(n-1)}{[(1-\lambda)q + (1+\lambda)t]^2}|z|^{-\beta} \int_0^{|z|} |\zeta|^{\beta-1} h(\zeta)\, d|\zeta|,$$

where $\beta = (n-1)(1+\lambda)(t+q)/[(1-\lambda)q + (1+\lambda)t]$.

We may also consider more complicated integral equations, for example $f + \lambda x_k Tf = h$. To do this we would apply Theorem 5.8, extending the function $x_k Tf$ to be harmonic in B. This may be done by Almansi's formula, since $\Delta^2(x_k Tf) = 0$. We obtain in B an ordinary second-order Euler equation.

5.4 Characterization of the ball using the Bochner-Martinelli operator

We have seen that the operator M_σ giving the Bochner-Martinelli singular integral is self-adjoint in $\mathcal{L}^2(S(0,1))$. It turns out that this property completely characterizes the ball.

Theorem 5.9 (Boas). *Let D be a bounded domain in $\mathbf{C}^n$ with boundary of class C^1. Suppose that the operator M_σ giving the Bochner-Martinelli singular integral is self-adjoint in $\mathcal{L}^2(\partial D)$. Then D is a ball in $\mathbf{C}^n$.*

Proof. We have already noted (see §2) that the Bochner-Martinelli integral is singular. By the general theory of singular integrals (see, for example, [95, 153, 193]), it gives a bounded operator in $\mathcal{L}^p(\partial D)$ when $p > 1$. It follows from Lemma 3.5 that $U(\zeta, z)\big|_{\partial D} = M(\zeta, z)\, d\sigma$ on ∂D, where

$$M(\zeta, z) = \frac{(n-1)!}{2\pi^n} \sum_{k=1}^{n} (\bar{\zeta}_k - \bar{z}_k)\rho_{\bar{k}}(\zeta).$$

Now the operator adjoint to the Bochner-Martinelli operator has the kernel $\overline{M(z,\zeta)}$. The equality of these kernels gives

$$\sum_{k=1}^{n} (\bar{\zeta}_k - \bar{z}_k)\rho_{\bar{k}}(\zeta) = \sum_{k=1}^{n} (z_k - \zeta_k)\rho_k(z) \tag{5.5}$$

for all ζ and z in ∂D.

Let $\operatorname{Re} z_j = x_j$, $\operatorname{Im} z_j = y_j$, $\operatorname{Re} \zeta_j = \xi_j$, and $\operatorname{Im} \zeta_j = \eta_j$ for $j = 1, \dots, n$. Taking the real part in (5.5), we obtain

$$\begin{aligned}\sum_{k=1}^{n} (\xi_k - x_k)\frac{\partial \rho}{\partial \xi_k}|\operatorname{grad} \rho(\zeta)|^{-1} + \sum_{k=1}^{n} (\eta_k - y_k)\frac{\partial \rho}{\partial \eta_k}|\operatorname{grad} \rho(\zeta)|^{-1} \\ = \sum_{k=1}^{n} (x_k - \xi_k)\frac{\partial \rho}{\partial x_k}|\operatorname{grad} \rho(z)|^{-1} + \sum_{k=1}^{n} (y_k - \eta_k)\frac{\partial \rho}{\partial y_k}|\operatorname{grad} \rho(z)|^{-1}.\end{aligned}$$

Hence the angles between the chord joining z and ζ and the outer normals to ∂D at z and ζ are equal.

The proof of Theorem 5.9 is now a consequence of the following assertion. □

Lemma 5.10. *A ball is the only bounded domain D in $\mathbf{R}^n$ with smooth boundary such that the chord joining any two boundary points makes equal angles with the outer normals to the boundary at the two points.*

Proof. Since D is a bounded domain, its chords have a maximal length. Moreover, a maximal chord is normal at both endpoints. Making a dilation and a rotation, we may assume that a maximal chord has length 2 and intersects the boundary at the points $\pm e^1$ (where $e^j = (0, \dots, 1, \dots, 0)$ with a 1 in the jth place).

The proof proceeds by induction on k. Suppose that the coordinates are chosen such that the points $\pm e^j$, $j = 1, \dots, k$, lie on the boundary ∂D, and the unit normal at $\pm e^j$ is $\pm e^j$, $j = 1, \dots, k$.

Suppose b lies on the boundary ∂D and has maximal distance from the k-dimensional plane generated by the vectors $e^1, \dots, e^k$. Then the normal at b is necessarily orthogonal to this plane. We may assume (after a rotation leaving the first k coordinates unchanged) that $b = (b_1, \dots, b_{k+1}, 0, \dots, 0)$, and the normal at b is e^{k+1}. Applying the hypothesis of the lemma at the points b and $\pm e^j$, we obtain $\langle e^j - b, e^j\rangle = \langle b - e^j, e^{k+1}\rangle$ and $\langle -e^j - b, -e^j\rangle = \langle b + e^j, e^{k+1}\rangle$, $j = 1, \dots, k$ (here $\langle x, y\rangle = x_1y_1 + \dots + x_ny_n$). Consequently, $1 - b_j = b_{k+1}$ and $1 + b_j = b_{k+1}$, that is, $b_j = 0$, $j = 1, \dots, k$, and $b_{k+1} = 1$. The axis Ox_{k+1} intersects the boundary ∂D at a second point b'. The hypothesis of the lemma applied to b and b' implies that the normal at b' is $-e^{k+1}$. In view of the preceding considerations, $b' = -e^{k+1}$, that is, the induction step is completed.

Consequently, the coordinates may be chosen so that the points $\pm e^j$, $j = 1, \dots, n$, lie on the boundary ∂D, and the normals (to ∂D) at the points $\pm e^j$ are the vectors $\pm e^j$. Let $(x_1, \dots, x_n)$ be an arbitrary point of ∂D, and let $\nu = (\nu_1, \dots, \nu_n)$ be the unit normal to ∂D at x. From the hypothesis of the lemma, applied to the points x and $\pm e^j$, we have $\langle e^j - x, e^j\rangle = \langle x - e^j, \nu\rangle$ and $\langle -e^j - x, -e^j\rangle = \langle x + e^j, \nu\rangle$, so $\langle x, e^j\rangle = \langle e^j, \nu\rangle$, $j = 1, \dots, n$. Then $x_j = \nu_j$, so that $|x| = |\nu| = 1$. □

Theorem 5.9 and Lemma 5.10 were proved in [26].

Chapter 2

CR-Functions Given on a Hypersurface

6 Analytic representation of CR-functions

6.1 Currents

Subsequently we shall need the language of De Rham currents. We now give the necessary definitions, statements of theorems, and examples. All the information in this subsection may be found in [37, 65].

Let Ω be an open set in $\mathbf{R}^n$, and let $\mathcal{E}^p(\Omega) = \mathcal{E}^p$ be the space of exterior differential forms of degree p with coefficients of class $\mathcal{C}^\infty(\Omega)$. The topology in $\mathcal{E}^p$ is the usual one: the topology of uniform convergence of all coefficients together with all their derivatives on every compact set $K \subset \Omega$. The space $\mathcal{D}^p(\Omega) = \mathcal{D}^p$ is the space of p-forms in $\mathcal{E}^p$ with compact support. A sequence of forms in $\mathcal{D}^p$ converges to zero if the supports of all the forms lie in some fixed compact subset of Ω and the coefficients of these forms converge uniformly with all derivatives to zero on this compact set. Every differential form in $\mathcal{E}^p$ or $\mathcal{D}^p$ can be written as

$$\varphi = \sum_{|I|=p}' \varphi_I(x)\, dx_I, \tag{6.1}$$

where $I = (i_1, \ldots, i_p)$, $|I|$ is the number of components of the increasing multi-index I, $dx_I = dx_{i_1} \wedge \cdots \wedge dx_{i_p}$, and the prime on the summation sign means that the sum is taken over increasing multi-indices $(1 \leq i_1 < i_2 < \cdots < i_p \leq n)$.

A current T in Ω of dimension p means a continuous linear functional on $\mathcal{D}^p(\Omega)$. The space of currents of dimension p is denoted by $\mathcal{D}'_p(\Omega) = \mathcal{D}'_p$. If $T \in \mathcal{D}'_p$, then the value of T on the form $\varphi \in \mathcal{D}^p$ is denoted by $T(\varphi)$ or by $\langle T, \varphi \rangle$. We usually consider the weak topology in $\mathcal{D}'_p$: a sequence T_k converges (weakly) to zero as $k \to \infty$ if $T_k(\varphi) \to 0$ as $k \to \infty$ for every $\varphi \in \mathcal{D}^p$.

A current T has singularity m if it extends to the space of forms (with compact support) with coefficients of class $\mathcal{C}^m(\Omega)$. The currents of singularity 0 are called currents of measure type.

By definition, $\mathcal{D}'_p = \{0\}$ if $p > n$ or $p < 0$.

The space of currents $\mathcal{D}'_p$ may be identified with the space $\mathcal{D}'^{n-p}(\Omega) = \mathcal{D}'^{n-p}$, that is, with the space of exterior differential forms of degree $(n-p)$ having distribution coefficients. Thus, each current $T \in \mathcal{D}'_p$ can be written as

$$T = \sum_{|J|=n-p}{}' T_J \, dx_J, \tag{6.2}$$

where $J = (j_1, \dots, j_{n-p})$, and $T_J \in \mathcal{D}'$. If $\varphi \in \mathcal{D}^p$ has the form (6.1), then to find $T(\varphi)$ we need to multiply the current (6.2) by φ, take the coefficient of dx, and compute the value of the resulting distribution on the resulting function from $\mathcal{D}(\Omega)$. For example, if $T = T_J \, dx_J$, where $J = (j_1, \dots, j_{n-p})$, and $\varphi = \varphi_J(x) \, dx[J]$ (where $dx[J] = dx_{i_1} \wedge \dots \wedge dx_{i_p}$ with $i_1, \dots, i_p \notin J$ and $i_1 < \dots < i_p$), then $T(\varphi) = \sigma(J) T_J(\varphi_J)$, where the symbol $\sigma(J)$ is defined via $dx_J \wedge dx[J] = \sigma(J) \, dx$. Consequently, the currents of dimension 0 or n can be identified with the space of distributions $\mathcal{D}'(\Omega)$.

The support $\operatorname{supp} T$ of a current $T \in \mathcal{D}'_p$ of the form (6.2) is the union of the supports of the distributions T_J. The currents with compact support are denoted by $\mathcal{E}'_p = \mathcal{E}'^{n-p}$. Each current in $\mathcal{E}'_p$ has finite singularity.

The operator d of exterior differentiation is defined as usual: if T is the current (6.2), then

$$dT = \sum_{k=1}^{n} \sum_{|J|=n-p}{}' \frac{\partial T_J}{\partial x_k} \, dx_k \wedge dx_J.$$

For $\varphi \in \mathcal{D}^{p-1}$ of the form (6.1), we obtain

$$\begin{aligned}
(dT)(\varphi) &= \sum_{k=1}^{n} \sum_{|J|=n-p}{}' \sum_{|I|=p-1}{}' \frac{\partial T_J}{\partial x_k}(\varphi_I) \, dx_k \wedge dx_J \wedge dx_I \\
&= -\sum_{k=1}^{n} \sum_{|J|=n-p}{}' \sum_{|I|=p-1}{}' T_J \left(\frac{\partial \varphi_I}{\partial x_k} \right) dx_k \wedge dx_J \wedge dx_I \\
&= (-1)^{n-p+1} \sum_{|J|=n-p}{}' \sum_{|I|=p-1}{}' \sum_{k=1}^{n} T_J \left(\frac{\partial \varphi_I}{\partial x_k} \right) dx_J \wedge dx_k \wedge dx_I \\
&= (-1)^{n-p+1} T(d\varphi).
\end{aligned}$$

Thus $(dT)(\varphi) = (-1)^{n-p+1} T(d\varphi)$; in particular, $dT \in \mathcal{D}'_{p-1} = \mathcal{D}'^{n-p+1}$. It is clear that $d^2 T = 0$.

A current T is called closed if $dT = 0$. A current T is called exact if $T = dS$.

We now give some basic examples of currents.

(1) Let M be a smooth, oriented, closed submanifold of Ω of dimension p. We define a current $[M]$ by setting

$$[M](\varphi) = \int_M \varphi, \qquad \varphi \in \mathcal{D}^p.$$

Clearly $[M]$ is a current of measure type, and the support of $[M]$ coincides with the set M. If M is an oriented manifold with smooth boundary N, then Stokes's formula shows that

$$\begin{aligned} d[M](\varphi) &= (-1)^{n-p+1}[M](d\varphi) = (-1)^{n-p+1}\int_M d\varphi \\ &= (-1)^{n-p+1}\int_N \varphi = (-1)^{n-p+1}[N](\varphi), \qquad \varphi \in \mathcal{D}^{p-1}, \end{aligned}$$

that is,

$$d[M] = (-1)^{n-p+1}[dM] = (-1)^{n-p+1}[N]. \tag{6.3}$$

It is not necessary that M be closed to define the current $[M]$. It suffices if the intersection of M with each compact subset K of Ω has finite Hausdorff p-measure (that is, $\Lambda_p(M \cap K) < \infty$). This is the case, for example, if $M = G$ is an open subset of Ω. We can find $d[G]$ if G has smooth (or piecewise-smooth) boundary Γ (the orientation of Γ is assumed to be compatible with the orientation of G). By formula (6.3),

$$d[G] = -[\Gamma].$$

We can consider, instead of submanifolds in Ω, locally finite singular chains Σ of dimension p, and define the current $[\Sigma]$.

Let Γ be a smooth, closed, oriented hypersurface in Ω defined by a function ρ, that is, $\Gamma = \{ x \in \Omega : \rho(x) = 0 \}$ with $\rho \in \mathcal{C}^1(\Omega)$ and $d\rho \neq 0$ on Γ. Then

$$[\Gamma] = \sum_{i=1}^n m_i \, dx_i, \tag{6.4}$$

where the measures m_i have the form $dm_i = (\partial\rho/\partial x_i)|d\rho|^{-1}\, d\sigma$, the surface area element on Γ being $d\sigma = d\Lambda_{n-1}$, and $|d\rho| = (\sum_{i=1}^n (\partial\rho/\partial x_i)^2)^{1/2}$. Indeed, suppose $\varphi \in \mathcal{D}^{n-1}$ is given by $\varphi = \sum_{i=1}^n \varphi_i \, dx[i]$. Then by definition,

$$\begin{aligned} \left(\sum_{i=1}^n m_i \, dx_i\right)(\varphi) &= \sum_{i=1}^n m_i(\varphi_i) \, dx_i \wedge dx[i] = \sum_{i=1}^n (-1)^{i-1} \int \varphi_i \, dm_i \\ &= \sum_{i=1}^n (-1)^{i-1} \int_\Gamma \varphi \frac{\partial \rho}{\partial x_i} |d\rho|^{-1} \, d\sigma = \sum_{i=1}^n \int_\Gamma \varphi_i \, dx[i] = \int_\Gamma \varphi \\ &= [\Gamma](\varphi). \end{aligned}$$

(2) Let ψ be an arbitrary p-form with locally integrable coefficients. We define a current $[\psi]$ of dimension $(n-p)$ via

$$[\psi](\varphi) = \int_\Omega \psi \wedge \varphi, \qquad \varphi \in \mathcal{D}^{n-p}(\Omega).$$

Obviously $[\psi]$ is a current of measure type, and the support of $[\psi]$ coincides with the essential support of the form ψ.

If ψ has smooth coefficients in Ω, then

$$\begin{aligned} d[\psi](\varphi) &= (-1)^{p+1}[\psi](d\varphi) = (-1)^{p+1}\int_\Omega \psi \wedge d\varphi \\ &= \int_\Omega d\psi \wedge \varphi - \int_\Omega d(\psi \wedge \varphi) = [d\psi](\varphi). \end{aligned}$$

Thus $d[\psi] = [d\psi]$.

We can introduce, besides the operation of exterior differentiation of currents, the operation of exterior product of a current and a form ψ, as follows. If $T \in \mathcal{D}'_p$ has singularity m, and ψ is a k-form with coefficients of class $C^m(\Omega)$, then the current $T \wedge \psi$ acts by the rule

$$(T \wedge \psi)(\varphi) = T(\psi \wedge \varphi), \qquad \varphi \in \mathcal{D}^{p-k}.$$

The conditions on ψ can be weakened in many cases. For example, if $T = [M]$ is a smooth oriented manifold, then the form ψ can be taken to have coefficients that are locally integrable on M.

Now suppose Ω is an open set in $\mathbf{C}^n$. The complex structure of $\mathbf{C}^n$ induces a decomposition of $\mathcal{D}^p$ into a direct sum of spaces $\mathcal{D}^{r,s}(\Omega) = \mathcal{D}^{r,s}$ (where $r+s=p$) of forms of (bi)degree (r,s). A form $\varphi \in \mathcal{D}^{r,s}$ can be written as

$$\varphi = {\sum_{|I|=r}}' {\sum_{|J|=s}}' \varphi_{I,J}(z)\, dz_I \wedge d\bar{z}_J, \tag{6.5}$$

where $I = (i_1, \dots, i_r)$ and $J = (j_1, \dots, j_s)$ are increasing multi-indices, and $dz_I = dz_{i_1} \wedge \cdots \wedge dz_{i_r}$.

The space of currents $\mathcal{D}'_p$ accordingly decomposes into a direct sum of subspaces $\mathcal{D}'_{r,s} = \mathcal{D}'^{n-r,n-s}$, where $r+s=p$. If $T \in \mathcal{D}'_p$, then

$$T = \sum_{r+s=p} T_{r,s}, \tag{6.6}$$

where $T_{r,s} \in \mathcal{D}'_{r,s}$ satisfies

$$T_{r,s}(\varphi) = \begin{cases} T(\varphi), & \varphi \in \mathcal{D}^{r,s}; \\ 0, & \varphi \notin \mathcal{D}^{r,s}. \end{cases}$$

The decomposition (6.6) is the Dolbeault decomposition. Elements of $\mathcal{D}'_{r,s}$ are called currents of (bi)dimension (r,s) or (bi)degree $(n-r,n-s)$. As usual, we will indicate the dimension of a current by a lower index, and the degree by an upper index: $T_{r,s}=T^{n-r,n-s}$.

If a current $T\in\mathcal{D}'_p$ is written as a differential form, then by expressing dx_j and dy_j in terms of dz_j and $d\bar z_j$, we obtain the decomposition of T as the direct sum of currents $T_{r,s}$. We recall that we introduced a volume form dv in $\mathbf{C}^n$ as

$$dv=dx\wedge dy=(i/2)^n\,dz\wedge d\bar z=(-i/2)^n d\bar z\wedge dz.$$

Therefore, if a current T of dimension (r,s) is given by a differential form, and we want to find its value on φ of the form (6.5), after formal multiplication $T\wedge\varphi$ we still need to pass from the form $dz\wedge d\bar z$ to the volume form, that is, we need to divide by $(i/2)^n$.

The Dolbeault decomposition is obviously invariant under holomorphic mappings $f:\Omega\to\Omega'$.

The operator d also decomposes into a sum of two operators $d=\partial+\bar\partial$, where $\partial:\mathcal{D}'_{r,s}\to\mathcal{D}'_{r-1,s}$ and $\bar\partial:\mathcal{D}'_{r,s}\to\mathcal{D}'_{r,s-1}$. If a current T has the form

$$T=\sum_{|I|=n-r}{}'\sum_{|J|=n-s}{}' T_{I,J}\,dz_I\wedge d\bar z_J,$$

then

$$\bar\partial T=\sum_{k=1}^{n}\sum_{|I|=n-r}{}'\sum_{|J|=n-s}{}'\frac{\partial T_{I,J}}{\partial\bar z_k}\,d\bar z_k\wedge dz_I\wedge d\bar z_J.$$

In particular, $\bar\partial\bar\partial=0$, $\partial\partial=0$, $\partial\bar\partial+\bar\partial\partial=0$, and $(\bar\partial T)(\varphi)=(-1)^{r+s+1}T(\bar\partial\varphi)$ for $T\in\mathcal{D}'_{r,s}$.

For example, let us find the Dolbeault decomposition for a current $[\Gamma]$, where Γ is a smooth, closed, oriented hypersurface in Ω defined by a function $\rho\in\mathcal{C}^1(\Omega)$. We have $[\Gamma]=[\Gamma]^{1,0}+[\Gamma]^{0,1}$, where $[\Gamma]^{0,1}=\sum_{j=1}^n\bar\mu_j\,d\bar z_j$ and $[\Gamma]^{1,0}=\sum_{j=1}^n\mu_j\,dz_j$, the measure $d\mu_j$ being $(1/2)\rho_j\,d\sigma$. We recall (see §3) that

$$\rho_j=\frac{\partial\rho}{\partial z_j}|\operatorname{grad}\rho|^{-1}=\frac{\partial\rho}{\partial z_j}\left(\sum_{k=1}^{n}\left|\frac{\partial\rho}{\partial z_k}\right|^2\right)^{-1/2}.$$

Indeed, let $\varphi=\sum_{j=1}^n\varphi_j(z)\,dz\wedge d\bar z[j]$. Then by definition,

$$\begin{aligned}\left(\sum_{j=1}^{n}\bar\mu_j\,d\bar z_j\right)(\varphi)&=\sum_{j=1}^{n}\bar\mu_j(\varphi_j)\,d\bar z_j\wedge dz\wedge d\bar z[j]=2^n i^{-n}\sum_{j=1}^{n}(-1)^{n+j-1}\int\varphi_j\,d\bar\mu_j\\&=2^{n-1}i^{-n}\sum_{j=1}^{n}(-1)^{n+j-1}\int_\Gamma\varphi_j\rho_{\bar j}\,d\sigma=\sum_{j=1}^{n}\int_\Gamma\varphi_j\,dz\wedge d\bar z[j]\\&=\int_\Gamma\varphi=[\Gamma](\varphi)\end{aligned}$$

(by Lemma 3.5). Thus

$$[\Gamma]^{0,1} = \lambda\, d\sigma\, \overline{\partial}\rho, \tag{6.7}$$

where $\lambda = (1/2)|\operatorname{grad}\rho|^{-1} = |d\rho|^{-1}$ (we recall that $|d\rho|^2 = \sum_{j=1}^n [(\partial\rho/\partial x_j)^2 + (\partial\rho/\partial y_j)^2]$). In particular, if $|d\rho| = 1$, then $[\Gamma]^{0,1} = d\sigma\,\overline{\partial}\rho$. In exactly the same way, we see that $[\Gamma]^{1,0} = |d\rho|^{-1}\, d\sigma\, \partial\rho$.

Now consider the Bochner-Martinelli kernel $U(\zeta, z)$. This form has locally integrable coefficients, so it determines a current $[U(\zeta, z)]$ in $\mathcal{D}'_{0,1}(\mathbf{C}^n)$ (where z is a fixed point in $\mathbf{C}^n$). Let us find $\overline{\partial}\,[U(\zeta, z)]$.

If $\varphi \in \mathcal{D}(\mathbf{C}^n)$, then

$$\begin{aligned}
\overline{\partial}\,[U(\zeta,z)](\varphi) &= [U(\zeta,z)](\overline{\partial}\varphi) = \int U(\zeta,z)\wedge\overline{\partial}\varphi(\zeta)\\
&= \lim_{\epsilon\to 0+}\int_{\mathbf{C}^n\setminus B(z,\epsilon)} U(\zeta,z)\wedge\overline{\partial}\varphi(\zeta) = -\lim_{\epsilon\to 0+}\int_{\mathbf{C}^n\setminus B(z,\epsilon)} d(\varphi U(\zeta,z))\\
&= \lim_{\epsilon\to 0+}\int_{S(\zeta,\epsilon)} \varphi(\zeta)\, U(\zeta,z)\\
&= \frac{(n-1)!}{(2\pi i)^n}\lim_{\epsilon\to 0+}\epsilon^{-2n}\int_{S(z,\epsilon)}\varphi(\zeta)\sum_{k=1}^n(-1)^{k-1}(\bar\zeta_k-\bar z_k)\,d\bar\zeta[k]\wedge d\zeta\\
&= \frac{(n-1)!}{(2\pi i)^n}\lim_{\epsilon\to 0+}\epsilon^{-2n}\int_{B(z,\epsilon)}\left(n\varphi(\zeta)+\sum_{k=1}^n\frac{\partial\varphi}{\partial\bar\zeta_k}(\bar\zeta_k-\bar z_k)\right)d\bar\zeta\wedge d\zeta\\
&= \frac{n!}{\pi^n}\lim_{\epsilon\to 0+}\epsilon^{-2n}\int_{B(z,\epsilon)}\varphi(\zeta)\,dv = \varphi(z)
\end{aligned}$$

by the mean-value theorem. We have obtained that

$$\overline{\partial}\,[U(\zeta, z)] = \delta_z\, dv, \tag{6.8}$$

where $\delta_z \in \mathcal{D}'(\mathbf{C}^n)$ is the delta function at z.

From this, it is easy to deduce the Bochner-Martinelli formula (1.4) for smooth functions. Let G be a bounded domain in $\mathbf{C}^n$ with piecewise-smooth boundary Γ, and let χ_G be the characteristic function of G. Then the current $\chi_G[U] \in \mathcal{D}'_{0,1}$. Suppose $\varphi \in \mathcal{C}^1(\overline{G})$. Then (using (6.8))

$$\begin{aligned}
\overline{\partial}(\chi_G[U])(\varphi) &= \chi_G[U](\overline{\partial}\varphi) = [U](\overline{\partial}\varphi) - (1-\chi_G)[U](\overline{\partial}\varphi)\\
&= \varphi(z) - \int_{\mathbf{C}^n\setminus G} U(\zeta,z)\wedge\overline{\partial}\varphi(\zeta) = \varphi(z) - \int_\Gamma \varphi(\zeta)\,U(\zeta,z),
\end{aligned}$$

$$\chi_G[U](\overline{\partial}\varphi) = -\int_G \overline{\partial}\varphi(\zeta)\wedge U(\zeta,z), \qquad z\in G.$$

(Here it is assumed that the function $\varphi \in \mathcal{C}^1(\overline{G})$ has been extended to a function of class $\mathcal{C}^1(\mathbf{C}^n)$ with compact support.)

6.2 The problem of analytic representation

In the theory of functions of one complex variable, every function given on an arc is the difference of boundary values of a Cauchy-type integral of the function (Sokhotskiĭ-Plemelj formula). The smoothness properties of arcs and functions complement each other: the worse the arc, the better we need the function to be, and conversely; on good arcs we may even take distributions. The analytic representation of a distribution on a line segment as the jump of the boundary values of holomorphic functions has numerous applications (see, for example, [30]). In Chapter 1, we saw that we can solve the jump problem in several complex variables by using the Bochner-Martinelli integral, and the smoothness properties of functions and boundaries of domains are the same as for the Cauchy integral. Generally speaking, however, this representation is not holomorphic. Our aim in this section is to construct an analytic representation of functions given on a hypersurface in $\mathbf{C}^n$ as the difference of boundary values of holomorphic functions. Our exposition basically follows [36].

Let Ω be an open set (or domain) in $\mathbf{C}^n$, and let Γ be a relatively closed, piecewise-smooth, oriented hypersurface in Ω. The surface Γ is defined by a function $\rho \in \mathcal{C}^1(\Omega)$ such that $\Gamma = \{ z \in \Omega : \rho(z) = 0 \}$, and $d\rho \neq 0$ at the smooth points of Γ. Then Γ divides Ω into two open sets: $\Omega^+ = \{ z : \rho(z) > 0 \}$, and $\Omega^- = \{ z : \rho(z) < 0 \}$. The orientation of Γ is supposed to be compatible with the orientation of Ω^+, and $\nu(z)$ is the unit normal vector to Γ directed to the side of increasing ρ. If Γ is a smooth surface, then for every compact set $K \subset \Gamma$ there is a suitably small positive ϵ such that each point of the form $\zeta + t\nu(\zeta)$ is in Ω^+, and $\zeta - t\nu(\zeta)$ is in Ω^-, when $\zeta \in K$ and $0 < t \leq \epsilon$.

If h is a function defined in $\Omega \setminus \Gamma$, we set $h^{\pm}(\zeta) = \lim_{t\to 0^+} h(\zeta \pm t\nu(\zeta))$ (if these limits exist).

Our problem is to find, given a function f on Γ, a holomorphic function in $\Omega \setminus \Gamma$ such that $h^+(\zeta) - h^-(\zeta) = f(\zeta)$ for $\zeta \in \Gamma$. When $n = 1$, no conditions on f and Ω are needed for such a representation to exist (other than compatible smoothness of f and Γ). When $n > 1$, there are such conditions.

Let h be a holomorphic function in Ω^+, and suppose that h is continuous up to the surface Γ (where $\Gamma \in \mathcal{C}^1$). Let us consider the current $h[\Omega^+] \in \mathcal{D}'_{n,n}(\Omega)$ and find $\overline{\partial}(h[\Omega^+])$. If $\varphi \in \mathcal{D}^{n,n-1}$, then

$$\overline{\partial}(h[\Omega^+])(\varphi) = -h[\Omega^+](\overline{\partial}\varphi) = -\int_{\Omega^+} h\,\overline{\partial}\varphi = -\int_{\Omega^+} d(h\varphi) = -\int_{\Gamma} h^+\varphi.$$

Thus, $\overline{\partial}(h[\Omega^+]) = -h^+[\Gamma]^{0,1}$ (we assume that h has been extended into Ω^- by zero).

Since $\overline{\partial}\,\overline{\partial} = 0$, we obtain that the current $h^+[\Gamma]^{0,1}$ is $\overline{\partial}$-closed: $\overline{\partial}h^+[\Gamma]^{0,1} = 0$. In the same way, we see that if h is holomorphic in Ω^-, then $h^-[\Gamma]^{0,1}$ is $\overline{\partial}$-closed. Hence, if we can solve the jump problem for a function f, then $\overline{\partial}f[\Gamma]^{0,1} = 0$.

We make the following definition: *if f is locally integrable on Γ (that is, $f \in \mathcal{L}^1_{\text{loc}}(\Gamma)$) and the current $f[\Gamma]^{0,1}$ is $\overline{\partial}$-closed, then f is called a CR-function* *on* Γ. When $n = 1$, the condition that $f[\Gamma]^{0,1}$ be closed is always satisfied, that is, every function f on Γ is a CR-function. When $n > 1$, the condition may be written as follows: if $\varphi \in \mathcal{D}^{n,n-2}(\Omega)$, then

$$\overline{\partial}(f[\Gamma]^{0,1})(\varphi) = f[\Gamma]^{0,1}(\overline{\partial}\varphi) = \int_\Gamma f\,\overline{\partial}\varphi = 0.$$

We say that a function $f \in \mathcal{L}^1_{\mathrm{loc}}(\Gamma)$ *satisfies the (weak) tangential Cauchy-Riemann equations on* Γ *if*

$$\int_\Gamma f\,\overline{\partial}\varphi = 0 \tag{6.9}$$

for all $\varphi \in \mathcal{D}^{n,n-2}(\Omega)$. Thus, f is a CR-function on Γ if and only if f satisfies the tangential Cauchy-Riemann equations (6.9) (when $n > 1$).

If $\Gamma \in \mathcal{C}^1$ and $f \in \mathcal{C}^1(\Gamma)$, it is easy to derive differential equations for f from (6.9). Indeed,

$$\int_\Gamma f\,\overline{\partial}\varphi = \int_\Gamma d(f\varphi) - \int_\Gamma \overline{\partial}f \wedge \varphi = -\int_\Gamma \overline{\partial}f \wedge \varphi$$

since φ has compact support in Ω. If $\int_\Gamma \overline{\partial}f \wedge \varphi = 0$ for every form $\varphi \in \mathcal{D}^{n,n-2}(\Omega)$, then the tangential part of the form $\overline{\partial}f$ (that is, $(\overline{\partial}f)_\tau$) is zero on Γ (see (3.9)). Since the normal part of $\overline{\partial}f$ is proportional to $\overline{\partial}\rho$ (since $\overline{\partial}\rho \wedge \varphi = d\rho \wedge \varphi$ and the restriction of $d\rho$ to Γ is zero),

$$\overline{\partial}f = (\overline{\partial}f)_\tau + \lambda\,\frac{\overline{\partial}\rho}{|\overline{\partial}\rho|}$$

(we have normalized $\overline{\partial}\rho$, where $|\overline{\partial}\rho| = \left(\sum_{k=1}^n |\partial\rho/\partial\bar{z}_k|^2\right)^{1/2} = |\operatorname{grad}\rho|$).

It follows that the form $\overline{\partial}f - \lambda\,\overline{\partial}\rho|\overline{\partial}\rho|^{-1}$ must be orthogonal to the form $\overline{\partial}\rho|\overline{\partial}\rho|^{-1}$, that is,

$$\sum_{k=1}^n \left(\frac{\partial f}{\partial \bar{z}_k} - \lambda\rho_{\bar{k}}\right)\frac{\partial\rho}{\partial z_k} = 0.$$

Therefore $\lambda = \sum_{k=1}^n (\partial f/\partial\bar{z}_k)(\partial\rho/\partial z_k)|\operatorname{grad}\rho|^{-1} = \sum_{k=1}^n (\partial f/\partial\bar{z}_k)\rho_k$. In §4, we denoted this expression by $\overline{\partial}_n f$ (see Corollary 4.9). Thus,

$$\overline{\partial}f = (\overline{\partial}f)_\tau + \overline{\partial}_n f\,\frac{\overline{\partial}\rho}{|\overline{\partial}\rho|}. \tag{6.10}$$

The condition $(\overline{\partial}f)_\tau = 0$ may be written alternatively as

$$\overline{\partial}f \wedge \overline{\partial}\rho = 0 \qquad \text{on } \Gamma. \tag{6.11}$$

Indeed, from (6.10) we have $\overline{\partial} f \wedge \overline{\partial}\rho = (\overline{\partial} f)_\tau \wedge \overline{\partial}\rho$. Since the forms $(\overline{\partial} f)_\tau$ and $\overline{\partial}\rho$ are orthogonal, $(\overline{\partial} f)_\tau \wedge \overline{\partial}\rho = 0$ if and only if $(\overline{\partial} f)_\tau = 0$ or $\overline{\partial}\rho = 0$, but $\overline{\partial}\rho \neq 0$ (see also §3). Condition (6.11) may be rewritten as a system of differential equations

$$\frac{\partial f}{\partial \bar{z}_j}\frac{\partial \rho}{\partial \bar{z}_k} - \frac{\partial f}{\partial \bar{z}_k}\frac{\partial \rho}{\partial \bar{z}_j} = 0 \qquad \text{on } \Gamma \tag{6.12}$$

for all $j, k = 1, \ldots, n$. The equations (6.12) are not independent. If, for example, $\partial\rho/\partial z_n \neq 0$, then it is enough to write these equations for $j = n$ and $k = 1, \ldots, n-1$.

We recall that the complex tangent space $T_\zeta^c(\Gamma)$ at the point $\zeta \in \Gamma$ consists of the vectors $w = \sum_{k=1}^n a_k(\partial/\partial z_k)$ such that $w(\rho) = \sum_{k=1}^n a_k(\partial\rho/\partial z_k) = 0$. If $(\partial\rho/\partial z_n) \neq 0$, then the vectors $w_k = (\partial\rho/\partial z_k)(\partial/\partial z_n) - (\partial\rho/\partial z_n)(\partial/\partial z_k)$, for $k = 1, \ldots, n-1$, are a basis for T_ζ^c. Therefore the $\bar{w}$ form a basis for $\bar{T}_\zeta^c$. Consequently, condition (6.12) is equivalent to

$$\bar{w}(f) = 0 \tag{6.13}$$

for each point $\zeta \in \Gamma$ and each vector $w \in T_\zeta^c$.

Thus, if the function $f \in \mathcal{C}^1(\Gamma)$ satisfies the tangential Cauchy-Riemann equations on Γ, then it satisfies the equivalent conditions (6.11)–(6.13) (see, for example, [180, §20]).

We also define CR-distributions on Γ. If Γ is a hypersurface of class $\mathcal{C}^m$ in Ω with defining function ρ, then we may consider distributions f on Γ with singularity no greater than m. For every form $\varphi \in \mathcal{D}^{n,n-1}(\Omega)$, there is a compactly supported function β of class $\mathcal{C}^{m-1}(\Omega)$ such that the form $(\varphi - \beta\omega) \wedge \overline{\partial}\rho$ has a zero on Γ of order at least $m-1$ (here ω is an $(n, n-1)$-form of class $\mathcal{C}^{m-1}$ representing the Euclidean volume on the level sets of ρ; for example, we may take ω to be $c\sum_{k=1}^n(-1)^{k-1}\rho_{\bar{k}}\, d\zeta \wedge d\bar{\zeta}[k]$). The function β may always be taken as the coefficient of $d\sigma$ in the restriction of φ to a level surface of ρ.

We set $f[\Gamma]^{0,1}(\varphi) = f(\beta)$. It is clear that this definition does not depend on the choice of ρ. *A distribution f of singularity not greater than $m-1$ on Γ is called a CR-distribution if the current $f[\Gamma]^{0,1}$ is $\overline{\partial}$-closed in Ω.*

Locally, condition (6.9) is also sufficient for f to be represented as the jump of holomorphic functions. However, an additional condition on Ω is needed for a global representation. We consider an example (see [36]).

Let $\Omega = \{(z_1, z_2) : |z_1| < 1, |z_2| < 1\} \setminus \{y_2 = z_1 = 0\}$ be the bidisc in $\mathbf{C}^2$ with a segment deleted. For Γ we take the intersection of Ω with the hypersurface $\{y_2 = 0\}$, and we consider on Γ the function $f = 1/z_1$. It is holomorphic in a neighborhood of Γ, and so it satisfies (6.9). Suppose that f were represented in the form $h^+ - h^-$, where $h^\pm$ are the boundary values of some function h that is holomorphic in $\Omega \setminus \Gamma$. Since

$$\lim_{t\to 0^+}\int_{|z_1|=1/2} h(z_1, \pm it)\, dz_1 = 0$$

by Cauchy's theorem (the functions $h(z_1, \pm it)$ are holomorphic in z_1 in the unit disc when $t \neq 0$), then $\int_{|z_1|=1/2} f(z_1, 0)\, dz_1$ also must be zero; but it equals $\int_{|z_1|=1/2} z_1^{-1}\, dz_1 = 2\pi i \neq 0$, and we have obtained a contradiction.

This example suggests viewing the problem of representing a CR-function as the difference of boundary values of holomorphic functions as a limiting case of the first Cousin problem. Therefore, we will subsequently require that the first cohomology group

$$H^1(\Omega, \mathcal{O}) = 0. \tag{6.14}$$

The result is that conditions (6.9) and (6.14) are sufficient for solvability of the problem.

Here is the idea of the proof: since $\overline{\partial} f[\Gamma]^{0,1} = 0$ and $H^1(\Omega, \mathcal{O}) = 0$, the current $f[\Gamma]^{0,1}$ is $\overline{\partial}$-exact, that is, $f[\Gamma]^{0,1} = \overline{\partial} h$, where h is a distribution in Ω. Then $\overline{\partial} h = 0$ outside Γ, since $\operatorname{supp} f[\Gamma]^{0,1} \subset \Gamma$, so h is a holomorphic function outside Γ. We need to show that the jump of h on Γ equals f.

We consider the one-dimensional case as an illustration. As a solution of the $\overline{\partial}$-equation, we consider the distribution h acting by the formula

$$h(\varphi) = -\frac{1}{2\pi i} \int_\Gamma f(\zeta) \int_\Omega \frac{\varphi(z)}{\zeta - z}\, d\zeta,$$

where $\varphi \in \mathcal{D}^{1,1}(\Omega)$. Indeed, if $f \in \mathcal{L}^1(\Gamma)$, and $\varphi = \overline{\partial}\alpha$, $\alpha = \psi\, dz$, then

$$\frac{1}{2\pi i} \int_\Omega \frac{\varphi(\zeta)}{\zeta - z} = \frac{1}{2\pi i} \int_\Omega \frac{\overline{\partial}\alpha(\zeta)}{\zeta - z} = -\psi(z)$$

by the Cauchy-Green formula, that is,

$$\overline{\partial} h(\alpha) = -h(\overline{\partial}\alpha) = \int_\Gamma f(\zeta)\psi(\zeta)\, d\zeta = f[\Gamma]^{0,1}(\alpha).$$

It follows from Fubini's theorem that the function $(2\pi i)^{-1} \int_\Gamma f(\zeta)(\zeta - z)^{-1}\, d\zeta$, locally integrable on the whole plane, coincides with h. Therefore the jump of h on Γ equals f. Any other solution h' of the $\overline{\partial}$-equation differs from h by a holomorphic function in Ω, and so has the same jump.

6.3 The theorem on analytic representation

Let Ω be a domain in $\mathbf{C}^n$ satisfying the condition $H^1(\Omega, \mathcal{O}) = 0$, and let Γ be a relatively closed, piecewise-smooth, orientable hypersurface in Ω.

Theorem 6.1 (Andreotti, Hill, Chirka). *Let f be a CR-function on Γ. There exists a (generalized) function $h \in \mathcal{D}'_{n,n}(\Omega)$ such that $\overline{\partial} h = f[\Gamma]^{0,1}$. It is holomorphic in $\Omega \setminus \Gamma$ and has the following boundary properties:*

1. *if Γ is piecewise smooth, and f satisfies a Hölder condition on Γ with exponent $\alpha > 0$, then h extends continuously to $\overline{\Omega}^+$ and to $\overline{\Omega}^-$;*
2. *if Γ is class $\mathcal{C}^k$, with $1 \leq k \leq \infty$, and $f \in \mathcal{C}^k(\Gamma)$, then h extends to a function of class $\mathcal{C}^{k-\epsilon}(\overline{\Omega}^+)$ and $\mathcal{C}^{k-\epsilon}(\overline{\Omega}^-)$ for every positive ϵ, and in these cases*
$$h^+(z) - h^-(z) = f(z), \qquad z \in \Gamma; \tag{6.15}$$
3. *if $f \in \mathcal{L}^1_{\text{loc}}(\Gamma)$ and $\Gamma \in \mathcal{C}^1$, then (6.15) holds almost everywhere for the nontangential limits of h;*
4. *if $\Gamma \in \mathcal{C}^1$, and $f \in \mathcal{L}^p_{\text{loc}}(\Gamma)$, where $p \geq 1$, then*
$$\lim_{\epsilon \to 0+} \int_K |h(\zeta + \epsilon\nu(\zeta)) - h(\zeta - \epsilon\nu(\zeta)) - f(\zeta)|^p \, d\sigma = 0$$
for every compact set $K \subset \Gamma$;
5. *if $\Gamma \in \mathcal{C}^\infty$ and f is an arbitrary CR-distribution, then*
$$f[\Gamma]^{0,1}(\varphi) = \lim_{\epsilon \to 0+} \int_\Gamma |h(\zeta + \epsilon\nu(\zeta)) - h(\zeta - \epsilon\nu(\zeta))| \varphi(\zeta) \, d\sigma$$
for every function $\varphi \in \mathcal{D}(\Gamma)$.

For the case when Ω is a complex manifold, this theorem was proved in [13, 36] (see also [37]). To prove the theorem, we first construct a suitable integral representation for solving the $\overline{\partial}$-problem in $\mathbf{C}^n$.

Let $\Omega = \mathbf{C}^n$, and let $M = \sum_{j=1}^n M_j(\partial/\partial\bar{z}_j)$ be a vector field, where $M_j \in \mathcal{D}'(\mathbf{C}^n)$. Let $T \in \mathcal{E}'^{p,q}(\mathbf{C}^n)$ be a current of the form
$$T = \sum_{|I|=p}{}' \sum_{|J|=q}{}' T_{I,J} \, dz_I \wedge d\bar{z}_J.$$
We introduce the operation of convolution contraction:
$$M \sharp T = \sum_{|I|=p}{}' \sum_{|J|=q}{}' \sum_{j \in J} (M_j * T_{I,J}) \sigma_{I,J}(j) \, dz_I \wedge d\bar{z}_{J \setminus j},$$
where $J \setminus j$ is the multi-index obtained from J by removing the index j, and the symbol $\sigma_{I,J}(j)$ is defined by
$$d\bar{z}_j \wedge dz_I \wedge d\bar{z}_{J\setminus j} = \sigma_{I,J}(j) \, dz_I \wedge d\bar{z}_J.$$
Since $\operatorname{supp} T_{I,J}$ is compact, the convolution of the distributions M_j and $T_{I,J}$ is well defined, and moreover $M_j * T_{I,J} \in \mathcal{D}'(\mathbf{C}^n)$ (see, for example, [211, §7]). Therefore $M \sharp T \in \mathcal{D}'^{p,q-1}(\mathbf{C}^n)$. When $q = 0$, we set $M \sharp T = 0$. We denote $\overline{\partial}\text{iv}\, M = \sum_{j=1}^n (\partial M_j/\partial\bar{z}_j)$.

Theorem 6.2 (Harvey, Lawson). *The following $\bar\partial$-homotopy formula holds:*

$$\bar\partial(M\sharp T)+M\sharp\bar\partial T=(\bar\partial\mathrm{iv}\,M)*T. \tag{6.16}$$

Proof. Since (6.16) is linear in M and T, it suffices to prove it for the case $M=M_j(\partial/\partial\bar z_j)$ and $T=T_{I,J}\,dz_I\wedge d\bar z_J$. Since the degree in z does not vary in (6.16), we may assume that $p=0$, that is, $T=T_J\,d\bar z_J$. Then

$$M\sharp T=M_j*T_J\sigma_J(j)\,d\bar z_{J\setminus j},$$
$$\bar\partial T=\sum_{k\notin J}\frac{\partial T_J}{\partial\bar z_k}\sigma_{J\cup k}(k)\,d\bar z_{J\cup k}.$$

Here $J\cup k$ is the increasing multi-index obtained by augmenting the multi-index J by k. Furthermore,

$$M\sharp\bar\partial T=\sum_{k\notin J}\left(M_J*\frac{\partial T_J}{\partial\bar z_k}\right)\sigma_{J\cup k}(k)\sigma_{J\cup k}(j)\,d\bar z_{J\cup k\setminus j}, \tag{6.17}$$
$$\bar\partial(M\sharp T)=\sum_{k\notin J\setminus j}\frac{\partial}{\partial\bar z_k}(M_j*T_J)\,\sigma_J(j)\sigma_{J\cup k\setminus j}(k)\,d\bar z_{J\cup k\setminus j}, \tag{6.18}$$
$$(\bar\partial\mathrm{iv}\,M)*T=\left(\frac{\partial M_j}{\partial\bar z_j}*T_J\right)\,d\bar z_J. \tag{6.19}$$

If $j\notin J$, then $M\sharp T=0$, so $\bar\partial(M\sharp T)=0$ in (6.18), and if $k\neq j$, then $\sigma_{J\cup k}(j)=0$, so only one term $M_j*(\partial T_J/\partial\bar z_j)\,d\bar z_J$ remains on the right-hand side of (6.17). For every derivative $\partial/\partial\bar z_j$, we have

$$\frac{\partial}{\partial\bar z_j}(u*v)=\left(\frac{\partial}{\partial\bar z_j}u\right)*v=u*\left(\frac{\partial v}{\partial\bar z_j}\right),$$

so when $j\notin J$, the right-hand sides of (6.17) and (6.19) are equal, and the expression in (6.18) is zero.

Suppose $j\in J$. The terms in (6.17) and (6.18) corresponding to $k\notin J$ differ only in sign. This sign is $\sigma_{J\cup k}(k)\sigma_{J\cup k}(j)$ in (6.17) and $\sigma_J(j)\sigma_{J\cup k\setminus j}(k)$ in (6.18). If $k<j$, then $\sigma_{J\cup k}(k)=\sigma_{J\cup k\setminus j}(k)$, and $\sigma_{J\cup k}(j)=-\sigma_J(j)$. If $k>j$, then $\sigma_{J\cup k}(j)=\sigma_J(j)$, and $\sigma_{J\cup k}(k)=-\sigma_{J\cup k\setminus j}(k)$. Thus, in each case the terms in (6.17) and (6.18) corresponding to $k\notin J$ have opposite signs.

If $k\in J$, then necessarily $k=j$. There are no such terms in (6.17), and in (6.18) this term is

$$\frac{\partial}{\partial\bar z_j}(M_j*T_J)\,d\bar z_J=(\bar\partial\mathrm{iv}\,M)*T.$$

□

Theorem 6.2 was obtained in [66] (see also [65]).

Corollary 6.3. *If* $\overline{\partial}\mathrm{iv}\, M = \delta = \delta_0$*, then for* $T \in \mathcal{E}'^{p,q}$ *we have*

$$T = \overline{\partial}(M \sharp T) + M \sharp \overline{\partial} T. \tag{6.20}$$

If $\overline{\partial} T = 0$ *and* $q \geq 1$*, then the current* $S = M \sharp T$ *is a solution to the* $\overline{\partial}$*-equation* $\overline{\partial} S = T$.

We can take $M = (n-1)!\pi^{-n} \sum_{j=1}^{n} \bar{\zeta}_j |\zeta|^{-2n} (\partial / \partial \bar{\zeta}_j)$, the Bochner-Martinelli field. The coefficients $\bar{\zeta}_j |\zeta|^{-2n}$ are locally integrable functions in $\mathbf{C}^n$, so that $\bar{\zeta}_j |\zeta|^{-2n} \in \mathcal{D}'(\mathbf{C}^n)$, and the equation $\overline{\partial}\mathrm{iv}\, M = \delta$ is verified in the same way as (6.8). Moreover, $M \sharp \delta_z\, dv = U(\zeta, z)$. The coefficients of the current $M \sharp T$ are harmonic functions outside the support of T.

By using the Bochner-Martinelli field, we can obtain from Theorem 6.2 Grothendieck's lemma for currents (see [37]).

Corollary 6.4. *Let* T *be a* $\overline{\partial}$*-closed current defined in the polydisc (that is,* $T \in \mathcal{D}'^{p,q}(U)$ *and* $\overline{\partial} T = 0$*). When* $q \geq 1$*, for every polydisc* $U' \Subset U$ *there is a current* $S \in \mathcal{D}'^{p,q-1}(U')$ *such that* $\overline{\partial} S = T$ *in* U'*, and when* $q = 0$*, the coefficients of* T *are holomorphic functions in* U.

Proof. If $q \geq 1$ and $U' \Subset U$, we consider the current $T' = \psi T$, where $\psi \in \mathcal{D}(U)$ and $\psi = 1$ in a neighborhood of $\overline{U'}$. Evidently $T' \in \mathcal{E}'^{p,q}(\mathbf{C}^n)$, and we can apply formula (6.20), where M is the Bochner-Martinelli field.

Since $\overline{\partial} T' = 0$ in U', the coefficients of the current $S' = M \sharp \overline{\partial} T'$ are harmonic in U'. Since $\overline{\partial} S' = \overline{\partial} T' = 0$ in U', the usual Grothendieck lemma implies $S' = \overline{\partial} \varphi$ in U', where φ is a form of class $\mathcal{C}^\infty(U')$. We obtain from (6.20) that

$$T = \overline{\partial}(M \sharp T' + \varphi) \qquad \text{in } U'.$$

If $q = 0$ and $p = 0$, then as above the distribution $S' = M \sharp \overline{\partial} T'$ is a harmonic function in U', and $\overline{\partial} S' = \overline{\partial} T' = 0$ in U', so S is a holomorphic function in U'. Since $M \sharp T' = 0$, we have $T' = S'$ in U', that is, T is a holomorphic function in U (since U' is an arbitrary polydisc compactly contained in U).

Using these assertions, we can obtain Dolbeault's theorem that when $p \geq 1$, the cohomology group $H^p(\Omega, \mathcal{O})$ coincides with the quotient group of the $\overline{\partial}$-closed currents $T \in \mathcal{D}'^{0,p}(\Omega)$ by the $\overline{\partial}$-exact currents (see, for example, [37]). In particular, if $H^p(\Omega, \mathcal{O}) = 0$, then the equation $\overline{\partial} S = T$ (where $\overline{\partial} T = 0$) is solvable for currents, with the singularity of S no worse than that of T.

Finally, if formula (6.20) is applied to a current of the form $\gamma[G]$, where G is a bounded domain in $\mathbf{C}^n$ with piecewise-smooth boundary, γ is a form of type (p,q) with coefficients of class $\mathcal{C}^1(\overline{G})$, and we take the Bochner-Martinelli field as M, then we obtain none other than the Bochner-Martinelli-Koppelman integral representation (1.9) (see [37]). □

Proof of Theorem 6.1. Suppose $H^1(\Omega, \mathcal{O}) = 0$, and let f be a CR-function on Γ, that is, $\overline{\partial} f[\Gamma]^{0,1} = 0$. By Dolbeault's theorem, $f[\Gamma]^{0,1} = -\overline{\partial} h$, where $h \in \mathcal{D}'(\Omega)$.

Since $\operatorname{supp} f[\Gamma]^{0,1} \subset \Gamma$, we have $\overline{\partial} h = 0$ in $\Omega \setminus \Gamma$. By Grothendieck's lemma (see Corollary 6.4), h is a holomorphic function in $\Omega \setminus \Gamma$. If h extends to a continuous function $h^{\pm}$ in $\overline{\Omega}^{\pm}$, then we have seen (see subsection 2) that

$$\overline{\partial} h^{+}[\Omega^{+}] = -h^{+}[\Gamma]^{0,1} \qquad \text{and} \qquad \overline{\partial} h^{-}[\Omega^{-}] = h^{-}[\Gamma]^{0,1}$$

(taking account of the orientation of Γ), that is, $-f[\Gamma]^{0,1} = \overline{\partial} h = \overline{\partial}(h^{+}[\Omega^{+}] + h^{-}[\Omega^{-}]) = -h^{+}[\Gamma]^{0,1} + h^{-}[\Gamma]^{0,1}$, so that $f = h^{+} - h^{-}$ on Γ.

Now let a be an arbitrary point of Γ, and let U be a polydisc of sufficiently small radius centered at a and contained in Ω, while U' is any polydisc centered at a and compactly contained in U. Let ψ be a function in $\mathcal{D}(U)$ that is equal to 1 on a neighborhood of the closure of U'. The current $T = \psi f[\Gamma]^{0,1}$ then belongs to $\mathcal{E}'^{0,1}(\mathbf{C}^n)$, and $\overline{\partial} T = 0$ in U'. We rewrite T by formula (6.20), using the Bochner-Martinelli field M:

$$T = \overline{\partial}(M \sharp T) + M \sharp \overline{\partial} T.$$

Now $\overline{\partial} T = \overline{\partial}(M \sharp \overline{\partial} T) = 0$ in U', and in addition $M \sharp \overline{\partial} T$ has harmonic coefficients in U' (since $\operatorname{supp} \overline{\partial} T \cap U' = \varnothing$), so by the usual Grothendieck lemma, $M \sharp \overline{\partial} T = \overline{\partial} \varphi$, where φ is a function of class $\mathcal{C}^{\infty}(U')$. Thus

$$T = \overline{\partial}(M \sharp T + \varphi).$$

Consequently, φ has no jump on Γ. Moreover, $\overline{\partial} h = f[\Gamma]^{0,1} = T$ in U', so by the Grothendieck lemma (see Corollary 6.4), the function $h - M \sharp T - \varphi$ is holomorphic in U'. Thus, the jump of h on $\Gamma \cap U'$ is completely determined by $M \sharp T$. Now $[\Gamma]^{0,1} = d\sigma\, \overline{\partial}\rho |\overline{\partial}\rho|^{-1}/2$ by (6.7), and the definition of the convolution contraction implies that

$$M \sharp T = \int_{\Gamma \cap U} \psi(\zeta) f(\zeta)\, U(\zeta, z), \qquad z \notin \Gamma. \tag{6.21}$$

Indeed, if $T = \sum_{k=1}^{n} T_k\, d\bar{z}_k$, then $M \sharp T = \sum_{j=1}^{n} (M_j * T_j)$, while

$$T_j = \psi(\zeta) f(\zeta) \frac{\partial \rho}{\partial \bar{\zeta}_j} \frac{d\sigma}{2|\overline{\partial}\rho|}.$$

Then

$$\begin{aligned} M \sharp T &= \sum_{j=1}^{n} \frac{(n-1)!}{2\pi^n} \int_{\Gamma} \frac{\bar{\zeta}_j - \bar{z}_j}{|\zeta - z|^{2n}} \psi(\zeta) f(\zeta) \frac{\partial \rho}{\partial \bar{\zeta}_j} \frac{d\sigma}{|\overline{\partial}\rho|} \\ &= \frac{(n-1)!}{(2\pi i)^n} \int_{\Gamma} \psi(\zeta) f(\zeta) \sum_{j=1}^{n} \frac{\bar{\zeta}_j - \bar{z}_j}{|\zeta - z|^{2n}} (-1)^{j-1}\, d\bar{\zeta}[j] \wedge d\zeta. \end{aligned}$$

(Here we have used Lemma 3.5 again).

Therefore the boundary behavior of h is the same as that of the Bochner-Martinelli integral (6.21). To complete the proof of Theorem 6.1, it suffices to refer to Theorems 2.3, 2.5, 3.1, 3.4, 3.6, and 4.4. These assertions were proved for the case when $\Gamma \cap U$ is the boundary of a domain, but we can always extend $\Gamma \cap U$ to the boundary ∂G of some domain G, and the function ψf extends to this part of the boundary with the same smoothness. Then the Bochner-Martinelli integral over the whole boundary has the required properties, and this integral over $\partial G \setminus (\Gamma \cap U)$ has no jump on $\Gamma \cap U$ (it is a function that is harmonic in $\mathbf{C}^n \setminus (\partial G \setminus (\Gamma \cap U))$). □

6.4 Some corollaries

Suppose the hypotheses of Theorem 6.1 hold for a domain Ω, a surface Γ, and a function f. We take a smooth point $\zeta^0 \in \Gamma$ and a neighborhood U of ζ^0 such that $U \subset \Omega$, the surface $U \cap \Gamma$ is smooth, and the domain $U_\epsilon = \{\zeta + t\nu(\zeta^0) : \zeta \in U^-, t \geq \epsilon\} \cap U$ contains U^- when $\epsilon > 0$, while it is contained in U^- when $\epsilon < 0$. We recall that $\nu(\zeta^0)$ is the unit normal directed to the side on which the defining function ρ is increasing, and $U^- = \{\zeta \in U : \rho(\zeta) < 0\}$. Then the domains U^- and U^+ expand semi-continuously into U, so (see [52, §3]) the class of functions $\mathcal{O}(U)$ is dense in $\mathcal{O}(U^\pm)$ (and also in $\mathcal{O}(U_\epsilon)$).

Theorem 6.5 (Chirka). *The following assertions hold:*

1. *if $p \geq 1$, and if the CR-function $f \in \mathcal{L}^p(U \cap \Gamma)$, then f can be approximated in the metric of $\mathcal{L}^p(U \cap \Gamma)$ by functions in $\mathcal{O}(U)$;*

2. *if $k \geq 1$, $U \cap \Gamma \in C^k$, and the CR-function $f \in C^k(U \cap \Gamma)$, then f can be approximated in the metric of $C^k(U \cap \Gamma)$ by functions in $\mathcal{O}(U)$;*

3. *if $U \cap \Gamma \in C^\infty$, and f is a CR-distribution, then f can be approximated in the weak topology on $U \cap \Gamma$ by functions in $\mathcal{O}(U)$, that is, there is a sequence of functions f_j in $\mathcal{O}(U)$ such that*

$$f[\Gamma]^{0,1}(\varphi) = \lim_{j \to \infty} \int_\Gamma f_j \varphi$$

for every $\varphi \in \mathcal{D}^{n,n-1}(U)$.

Proof. Suppose (by Theorem 6.1) that $h \in \mathcal{O}(\Omega \setminus \Gamma)$ has jump on Γ equal to f. Then

$$\lim_{\epsilon \to 0+} \int_{\Gamma \cap \overline{U}} |h(\zeta + \epsilon\nu(\zeta)) - h(\zeta - \epsilon\nu(\zeta)) - f(\zeta)|^p \, d\sigma = 0.$$

Then (1) follows from the density of $\mathcal{O}(U)$ in $\mathcal{O}(U_\epsilon)$. Parts (2) and (3) are proved in the same way. □

Cases (2) and (3) were given in [36].

Corollary 6.6. *Each CR-function f on Γ can be locally approximated by polynomials, in the metric of $\mathcal{L}^p$ if $f \in \mathcal{L}^p_{\text{loc}}(\Gamma)$ with $p \geq 1$, in the metric of $\mathcal{C}^k$ if $\Gamma \in \mathcal{C}^k$ and $f \in \mathcal{C}^k(\Gamma)$ with $k \geq 1$, and in the weak topology if f is a CR-distribution and $\Gamma \in \mathcal{C}^\infty$.*

Theorem 6.1 on the analytic representation of CR-functions can also be applied to the problem of local extension. Suppose, for example, Ω and Γ are such that every holomorphic function in Ω^+ extends holomorphically to some fixed neighborhood U of the point $\zeta \in \Gamma$. By applying Theorem 6.1, we obtain that every CR-function f on Γ extends into this neighborhood as a holomorphic function. More precise results are discussed in [36].

6.5 Further results and generalizations

In place of a hypersurface, we may consider a CR-manifold Γ (that is, a smooth manifold for which the dimension of the complex tangent space $T^c_\zeta(\Gamma)$ is constant on Γ). The problem would be to show that a CR-function on Γ (or a CR-form) can be represented as a sum of values of a finite number of holomorphic functions (or $\overline{\partial}$-closed forms) given in wedge domains abutting at Γ. Such a representation has been obtained only in special cases: for totally real manifolds by Henkin (see [72]), and for manifolds with some restriction on the Levi form (see [3, 74]).

An example of Trépreau given in a paper of Baouendi and Rothschild [24] shows that in general this problem is not solvable. The CR-manifold Γ in this example has the form $\Gamma = \{ (z, w_1, w_2) \in \mathbf{C}^3 : \operatorname{Im} w_1 - (\operatorname{Re} z)^2 \operatorname{Re} w_2 = 0, \operatorname{Im} w_2 + (\operatorname{Re} z)^2 \operatorname{Re} w_1 = 0 \}$. It turns out that any CR-function f given on $\Gamma \cap B(0,R)$ has a wave front set $WF_0 f$ that is either empty or equal to $\mathbf{R}^2 \setminus \{0\}$. On the other hand, there exist smooth CR-functions f which do not extend to be holomorphic on any wedge adjoining Γ. Hence, using results of [24], it can be shown that there is a CR-function f on $\Gamma \cap B(0,R)$ which cannot be decomposed as a sum of values of a finite number of holomorphic functions given on wedge domains with edge $\Gamma \cap B(0,R)$. This example is discussed in more detail in [24].

7 The Hartogs-Bochner extension theorem

7.1 The Hartogs-Bochner theorem

Let G be a bounded domain in $\mathbf{C}^n$ (where $n > 1$) with connected smooth boundary Γ.

Theorem 7.1. *If f is a CR-function on Γ, then f extends holomorphically into G as a function F. This extension is given by the Bochner-Martinelli integral (2.1), and it has the following boundary properties:*

1. *if $\Gamma \in \mathcal{C}^k$, where $k \geq 1$, and $f \in \mathcal{C}^r(\Gamma)$, where $r \in \mathbf{R}$ and $r \leq k$, then $F \in \mathcal{C}^r(\overline{G})$, and $F(z) = f(z)$ for $z \in \Gamma$;*

2. *if* $\Gamma \in \mathcal{C}^1$, *and* $f \in \mathcal{L}^p(\Gamma)$, *where* $p \geq 1$, *then* $F \in \mathcal{H}^p(G)$, *the nontangential limit of* F *agrees with* f, *and*

$$\lim_{\epsilon \to 0^+} \int_\Gamma |F(\zeta - \epsilon\nu(\zeta)) - f(\zeta)|^p \, d\sigma = 0;$$

3. *if* f *is a CR-distribution, then* $F(\zeta - \epsilon\nu(\zeta))$ *converges weakly to* $f(\zeta)$ *on* Γ.

Proof. The proof can be obtained from Theorem 6.1, or we can again use (6.20). Indeed, since $\overline{\partial} f[\Gamma]^{0,1} = 0$, and $\operatorname{supp} f[\Gamma]^{0,1}$ is compact, formula (6.20) implies

$$f[\Gamma]^{0,1} = \overline{\partial}(M \natural f[\Gamma]^{0,1}) = \overline{\partial} F,$$

while we have seen that

$$F(z) = \int_\Gamma f(\zeta)\, U(\zeta, z), \qquad z \notin \Gamma.$$

Moreover, $M \natural f[\Gamma]^{0,1}$ is holomorphic in $\mathbf{C}^n \setminus \Gamma$, and this function tends to zero when $|z| \to \infty$. By Hartogs's theorem, the function $F^-(z)$ for $z \notin \overline{G}$ continues holomorphically into G, and so by Liouville's theorem $F^- \equiv 0$. Since the boundary Γ is connected, the boundary value of F^+ agrees with f. The boundary properties of F^+ follow from the properties of the Bochner-Martinelli integral (see Theorems 2.3, 3.1, 3.4, 3.6, and 4.8). □

Theorem 7.1 (in contrast to Theorem 6.1) is not true when $n = 1$.

Bochner [28] and Martinelli [150] independently in 1943 found a rigorous proof of Hartogs's theorem on removability of compact singularities of holomorphic functions. Moreover, Bochner essentially proved Theorem 7.1 for the case $F \in \mathcal{C}^1$ and $f \in \mathcal{C}^1(\Gamma)$ (although of course he did not have the concepts of CR-functions, the operator $\overline{\partial}_\tau$, and so on). In 1970, Weinstock [214, 215] proved it for continuous functions f (when $\Gamma \in \mathcal{C}^\infty$).

Fichera [47] was able to replace the differential conditions on f by integral conditions (see also the survey [48]).

At the beginning of the 1970's, papers were published on this subject by Hörmander [79],[1] Vinogradov [208], Baĭkov [23], Aronov and Dautov [19]. In the survey [75], Chirka gave a proof for integrable functions f and Lyapunov surfaces. Finally, the theorem was carried over to noncompact complex manifolds Ω with the condition $H_c^1(\Omega, \mathcal{O}) = 0$ (where $H_c^1(\Omega, \mathcal{O})$ is the group of compact cohomologies) in the work of Harvey and Lawson [66] and Chirka [36] (see also [37, 65]). It was considered on Stein manifolds in [128, 129, 131].

We give a simple generalization of Theorem 7.1 to unbounded domains in $\mathbf{C}^n$ from [158].

[1] The order of citation is connected with the chronology.

Theorem 7.2 (Naser Shafii). *Let D be an unbounded domain in $\mathbf{C}^n$, where $n \geq 2$, with connected smooth boundary Γ, and let f be a CR-function on Γ. If the envelope of holomorphy of $\mathbf{C}^n \setminus \overline{D}$ coincides with $\mathbf{C}^n$, then f extends holomorphically into D. (The boundary properties of the extension are the same as in Theorem 7.1, except that convergence in the norm of $\mathcal{L}^p(\Gamma)$ must be replaced with convergence in $\mathcal{L}^p(\Gamma \cap K)$, where K is a compact set in $\mathbf{C}^n$.)*

Proof. The proof follows at once from Theorem 6.1. Since $\overline{\partial} f[\Gamma]^{0,1} = 0$, there is a function $h \in \mathcal{O}(\mathbf{C}^n \setminus \Gamma)$ with jump on Γ equal to f. Since h^- is holomorphic in $\mathbf{C}^n \setminus \overline{D}$, it extends to a holomorphic function $\tilde{h}$ in $\mathbf{C}^n$. Then $h^+ - \tilde{h}$ is holomorphic in D, and its boundary value agrees with f. □

We will give another generalization of Theorem 7.1 to unbounded domains in §8.

7.2 Weinstock's extension theorem

If the boundary ∂G of the domain is not connected, then the Hartogs-Bochner theorem becomes false. For example, if $G = B(0,2) \setminus \overline{B(0,1)}$, then by taking $f = 1$ on $S(0,2)$ and $f = 0$ on $S(0,1)$, we obtain a CR-function f on ∂G that does not extend holomorphically into G (for this is forbidden by the uniqueness theorem).

For domains with disconnected boundary, the condition of orthogonality to exact forms can be replaced by the condition of orthogonality to $\overline{\partial}$-closed forms, and the Hartogs-Bochner theorem becomes true.

Theorem 7.3 (Weinstock). *Suppose D is a bounded domain in $\mathbf{C}^n$, and f is a function on ∂D satisfying the condition*

$$\int_{\partial D} f\, \alpha = 0$$

for every form α of type $(n, n-1)$ with coefficients of class $\mathcal{C}^\infty(\overline{D})$ and such that $\overline{\partial}\alpha = 0$ in $\overline{D}$. Then f extends holomorphically into D. The smoothness of f and ∂D are the same as in Theorem 7.1. The holomorphic extension of f is given by the Bochner-Martinelli integral.

Proof. Let

$$F(z) = F^{\pm}(z) = \int_{\partial D} f(\zeta)\, U(\zeta, z), \qquad z \notin \partial D.$$

Since $\overline{\partial}_\zeta U(\zeta, z) = 0$, and the form $U(\zeta, z)$ has coefficients of class $\mathcal{C}^\infty(\overline{D})$ if $z \notin \overline{D}$, we have $F^- = 0$ outside $\overline{D}$. Then by the jump theorem for the Bochner-Martinelli integral given in §3, the boundary values of $F^+(z)$ on ∂D coincide with f. We will show that F^+ is a holomorphic function. It is clear that

$$\frac{\partial U}{\partial \bar{z}_k} = \overline{\partial}_\zeta \omega_k,$$

where

$$\omega_k = (-1)^k \left[\sum_{j=1}^{k-1} (-1)^j \frac{\partial g}{\partial \zeta_j}\, d\bar{\zeta}[j,k] + \sum_{j=k+1}^{n} (-1)^{j-1} \frac{\partial g}{\partial \zeta_j}\, d\bar{\zeta}[j,k] \right] \wedge d\zeta.$$

The form ω_k has a singularity only at $\zeta = z$. Therefore

$$\frac{\partial F^+}{\partial \bar{z}_k} = \int_{\partial D} f(\zeta)\, \overline{\partial}_\zeta \omega_k(\zeta, z).$$

Replacing the form ω_k by the form $\psi\omega_k$, where $\psi \in \mathcal{D}(\mathbf{C}^n)$ and $\psi = 1$ in a neighborhood of ∂D, while $\psi = 0$ in a neighborhood of z, we have

$$\frac{\partial F^+}{\partial \bar{z}_k} = \int_{\partial D} f(\zeta)\, \overline{\partial}_\zeta (\psi\omega_k) = 0$$

by the hypothesis of the theorem. Thus, F^+ is holomorphic in D. □

We will show in Chapter 4 that to extend f holomorphically into D, it suffices to know that f is orthogonal to the forms $U(\zeta, z)$ for $z \notin \overline{D}$, and not to all $\overline{\partial}$-closed forms.

Theorem 7.3 was proved by Weinstock [215] for continuous functions f and $\partial D \in C^\infty$. The proof we have given, valid for all functions f and $\partial D \in C^1$, is due to Aronov and Dautov (see [7, §3], and [19]). Theorem 7.3 was proved for functions f of class $\mathcal{L}^1(\partial D)$ in [75, §1].

7.3 The theorem of Harvey and Lawson

We now give a very remarkable application of the Hartogs-Bochner theorem: a result of Harvey and Lawson [65, 66] characterizing boundaries of complex manifolds.

Let M be a smooth manifold in $\mathbf{C}^n$. It is called maximally complex if the dimension of the complex tangent space $T^c_\zeta(M)$ is as large as possible. For even-dimensional manifolds M ($\dim_\mathbf{R} M = 2p$), this means that $T^c_\zeta(M) = T_\zeta(M)$ for every point ζ. A theorem of Levi-Civita (see, for example, [180]) shows that M is a complex manifold. If M is odd-dimensional ($\dim_\mathbf{R} M = 2p+1$), then the complex dimension of T^c_ζ equals p.

Examples of such manifolds are boundaries of complex manifolds: if X is a complex manifold of dimension $(p+1)$ with smooth boundary M, then in a neighborhood of a point $\zeta \in M$ the manifold X is biholomorphic to an open set in $\mathbf{C}^{p+1}$ in which the image of M is a hypersurface; so the complex dimension of $T^c_\zeta(M)$ equals p. Moreover, $d[M] = 0$ by Stokes's formula.

If M is a smooth curve in $\mathbf{C}^n$, then of course M is maximally complex, but boundaries of complex curves satisfy an additional moment condition. Let

X be a complex one-dimensional manifold with compact boundary curve M. If $\omega = \sum_{j=1}^{n} \omega_j(z)\, dz_j$ is a holomorphic form in $\mathbf{C}^n$, then

$$\int_M \omega = \int_X d\omega = \int_X \bar\partial\omega = 0,$$

since there are no nontrivial forms on X of degree $(0,2)$.

If M is a smooth compact curve in $\mathbf{C}^n$ satisfying the condition $\int_M \omega = 0$ for every holomorphic 1-form in $\mathbf{C}^n$, then we say that M satisfies the moment condition. Of course, the moment condition also arises in the case of manifolds M of arbitrary dimension. However, when $p \geq 1$, it follows from the maximal complexity of M (see [66]).

Using the idea of maximal complexity, we can give another definition of the tangential Cauchy-Riemann equations: *suppose $f \in \mathcal{C}^1(\Gamma)$ and Γ is a smooth hypersurface in $\mathbf{C}^n$. For f to be a CR-function on Γ, it is necessary and sufficient that the graph of f, that is, the manifold $M = \{ (z_1, \dots, z_n, w) : z \in \Gamma,\ w = f(z) \}$, be maximally complex.*

Indeed, if $\Gamma = \{z : \rho(z) = 0\}$, $\rho \in \mathcal{C}^1(\mathbf{C}^n)$, and $d\rho \neq 0$ on Γ, then $M = \{ (z, w) : \rho(z) = 0,\ w - f(z) = 0 \}$. The vector $u = \sum_{k=1}^{n} a_k(\partial/\partial z_k) + b(\partial/\partial w)$ is a complex tangent vector ($u \in T^c_{(\zeta,\eta)}(M)$) if $u(\rho) = 0$, $u(\mathrm{Re}(w - f(z))) = 0$, and $u(\mathrm{Im}(w - f(z))) = 0$. But $u(\rho) = \sum_{k=1}^{n} a_k(\partial\rho/\partial z_k)$, that is, the vector $u' = \sum_{k=1}^{n} a_k(\partial/\partial z_k) \in T^c_\zeta(\Gamma)$. If f is a CR-function, then $u'(\bar f) = 0$ (see (6.13)), and we obtain that

$$b = \sum_{k=1}^{n} a_k \frac{\partial f}{\partial z_k},$$

so the complex dimension of $T^c_{(\zeta,\eta)}(M)$ equals $n-1$. If $u'(\bar f) \neq 0$ for some $u' \in T^c_\zeta(\Gamma)$, then obviously the dimension of $T^c_{(\zeta,\eta)}(M)$ is less than $n-1$.

Thus, Theorem 7.1 may be interpreted as a theorem on the spanning of a complex manifold by the graphs of functions.

Theorem 7.4 (Harvey, Lawson). *Suppose M is a compact maximally complex manifold of dimension $2p-1$ in $\mathbf{C}^n$ (or in a Stein manifold), and $d[M] = 0$ (when $p = 1$, we assume that M satisfies the moment condition). Then there exists a unique holomorphic p-chain T in $\mathbf{C}^n \setminus M$ with compact support such that $dT = [M]$. Moreover, there is a compact subset $A \subset M$ of Hausdorff $(2p-1)$-dimensional measure zero such that for each point of $M \setminus A$ in a neighborhood of which M is of class $\mathcal{C}^k$, where $1 \leq k \leq \infty$, there is a neighborhood in which* $(\operatorname{supp} T) \cup M$ *is a regular $\mathcal{C}^k$ submanifold with boundary.*

In particular, if M is connected, then there exists a unique, irreducible, complex p-dimensional analytic subset $V \subset \mathbf{C}^n \setminus M$ with compact closure in $\mathbf{C}^n$ such that $d[V] = \pm[M]$, with the boundary regularity indicated above.

Proof. The proof may be found in [65, 66]. The idea is to represent M (locally) as the graph of a CR-function and to apply the Hartogs-Bochner theorem. □

8 Holomorphic extension from a part of the boundary of a domain

8.1 Statement of the problem

In this section, we assume that the dimension $n > 1$. There is an extensive literature on removable singularities of holomorphic functions of one and several variables. Recently, there has been interest in the problem of removable singularities of CR-functions. We mention first of all the work of Harvey and Polking [67], in which this problem was studied for solutions of arbitrary linear differential equations. They obtained results in terms of the smoothness of solutions and the metric dimension of compact singularities. For example, in our case, if $f \in \mathcal{L}^\infty(\Gamma)$ and f is a CR-function on $\Gamma \setminus K$, where $\Lambda_{2n-2}(K) = 0$, then f is a CR-function on Γ.

As Henkin showed [73], there is a phenomenon of removability of compact singularities for CR-functions, but the CR-manifold must be 1-concave (for the case of a hypersurface Γ, this means that the Levi form of Γ must have at each point $\zeta \in \Gamma$ both a negative and a positive eigenvalue; it cannot, for example, be the boundary of a domain).

We will be interested in two questions:

1. If Γ is the boundary of a bounded domain D, and f is a CR-function on $\Gamma \setminus K$, where K is a compact subset of Γ, what condition must be imposed on K so that f extends into D as a holomorphic function?

2. Under what conditions on K and f does the extension F have "good" boundary behavior near K? In other words, when does F determine on Γ a CR-function $\tilde{f}$? This function also will serve as an extension of f from $\Gamma \setminus K$ to Γ.

In this section we will consider the first question. We first give a number of results of Lupacciolu [143, 144] related to this problem. Most of them were obtained for continuous CR-functions, and we will generalize them somewhat to the case of functions of class $\mathcal{L}^p$ (using properties of the Bochner-Martinelli integral from Chapter 1).

8.2 Lupacciolu's theorem

Suppose $n > 1$ and D is a bounded domain $\mathbf{C}^n$ such that $\overline{D}$ has a schlicht envelope of holomorphy. If $\mathcal{O}(\overline{D})$ is the space of functions holomorphic on $\overline{D}$, then for a compact set $K \subset \overline{D}$ we set

$$\widehat{K}_{\overline{D}} = \left\{ z \in \overline{D} : |h(z)| \le \max_K |h|, \quad h \in \mathcal{O}(\overline{D}) \right\}.$$

We assume that the hypersurface $\Gamma = \partial D \setminus K$ is a smooth (class C^1), connected manifold in $\mathbf{C}^n \setminus K$.

As was noted in [198], the hypothesis of schlichtness of the envelope of $\overline{D}$ was left out in [143, 144].

Suppose $K = \widehat{K}_{\overline{D}}$; for example, K could be polynomially convex, or $K = \widehat{K}_\Omega$, where $\widehat{K}_\Omega$ is the envelope of K with respect to $\mathcal{O}(\Omega)$, with Ω an open set containing $\overline{D}$.

Theorem 8.1 (Lupacciolu). *If $K = \widehat{K}_{\overline{D}}$, $\Gamma = \partial D \setminus K$, and f is a CR-function of class $\mathcal{L}^p_{\mathrm{loc}}(\Gamma)$, where $p \geq 1$, then there exists a holomorphic function F in $D \setminus K$ whose boundary value agrees with f, that is, for every point $z \in \Gamma$ there is a ball $B(z,r)$ such that*

$$\lim_{\epsilon \to 0+} \int_{B(z,r)\cap\Gamma} |F(\zeta - \epsilon\nu(\zeta)) - f(\zeta)|^p \, d\sigma = 0.$$

Here, as usual, $\nu(\zeta)$ is the outer unit normal vector to Γ at $\zeta \in \Gamma$, and $d\sigma$ is the surface area element of Γ. Moreover, if $\zeta \in \Gamma$ is a Lebesgue point of f, then the nontangential limit of F at ζ coincides with $f(\zeta)$.

The hypothesis that f be a CR-function on Γ means that f is a CR-function on Γ with respect to the open set $G = \mathbf{C}^n \setminus K$.

This theorem was proved by Lupacciolu in [143] for the case $f \in \mathcal{C}(\Gamma)$. It was carried over to the class $\mathcal{L}^1$ in [119].

Before proving the theorem, we discuss some preliminaries. Let V be an open neighborhood of K in $\mathbf{C}^n$, and let β be a function of class $\mathcal{C}^\infty$ in $\mathbf{C}^n$ that is equal to 1 on K, satisfies $0 \leq \beta < 1$ off K, and has compact support in V. For each positive ϵ, we write $D_\epsilon = D \cap \{z : \beta(z) < 1-\epsilon\}$, and $\Gamma_\epsilon = \partial D \cap \{z : \beta(z) < 1-\epsilon\}$. By Sard's theorem, the set Γ_ϵ is a smooth manifold with smooth boundary $\partial\Gamma_\epsilon$ for almost all positive ϵ (we may apply Sard's theorem to the restriction of β to $\Gamma = \partial D \setminus K$). Since ∂D is compact, the image of the set of critical points of β is compact, that is, the set of positive ϵ for which $\partial\Gamma_\epsilon$ is a smooth manifold is open. Now $D \setminus K = \bigcup_{\epsilon>0} D_\epsilon$, and $\Gamma = \partial D \setminus K = \bigcup_{\epsilon>0} \Gamma_\epsilon$.

If $p \geq 1$ and $f \in \mathcal{L}^p_{\mathrm{loc}}(\Gamma)$, then by Fubini's theorem, $f \in \mathcal{L}^p(\partial\Gamma_\epsilon)$ for almost all positive ϵ. Indeed, if $\epsilon \in [a,b]$, where $0 < a \leq b$, the surface Γ_ϵ is smooth, and $\Gamma_{a,b} = \{z \in \partial D : 1-b \leq \beta(z) \leq 1-a\}$, then

$$\int_{\Gamma_{a,b}} |f|^p \, d\sigma = \int_a^b d\epsilon \int_{\partial\Gamma_\epsilon} |f|^p \, \psi \, d\alpha,$$

where $d\alpha$ is the surface area element of $\partial\Gamma_\epsilon$ (generated by Hausdorff measure Λ_{2n-2}), and $\psi = |d\sigma/d\epsilon|$; since $\psi \neq 0$ on $\Gamma_{a,b}$, we obtain that $\int_{\partial\Gamma_\epsilon} |f|^p \, d\alpha$ is finite for almost all $\epsilon \in [a,b]$.

Lemma 8.2. *For every form $\varphi \in \mathcal{D}^{n,n-2}(\mathbf{C}^n \setminus K)$, and for almost all positive ϵ, we have*

$$\int_{\Gamma_\epsilon} f \, \overline{\partial}\varphi = \int_{\partial\Gamma_\epsilon} f \, \varphi.$$

Proof. Applying Corollary 6.6 to f, we obtain that for every point $z \in \Gamma$, there is a ball $B(z,r)$ disjoint from K and a sequence of polynomials P_k such that

$$\int_{\Gamma \cap B(z,r)} |f - P_k|^p \, d\sigma \to 0 \qquad \text{as } k \to \infty.$$

Suppose $z \in \Gamma_{a,b}$, $\operatorname{supp} \varphi \subset B(z,r)$, and $B(z,r) \cap \Gamma \subset \Gamma_{a,b}$. By Fubini's theorem,

$$\int_{\Gamma_{a,b} \cap B(z,r)} |f - P_k|^p \, d\sigma = \int_a^b d\epsilon \int_{\partial\Gamma_\epsilon \cap B(z,r)} |f - P_k|^p \, \psi \, d\alpha \to 0$$

as $k \to \infty$, so there exists a subsequence k_s such that

$$\int_{\partial\Gamma_\epsilon \cap B(z,r)} |f - P_{k_s}|^p \, \psi \, d\alpha \to 0$$

as $s \to \infty$ for almost all $\epsilon \in [a,b]$. Therefore

$$\lim_{s \to \infty} \int_{\partial\Gamma_\epsilon \cap B(z,r)} |f - P_{k_s}|^p \, d\alpha = 0.$$

But in view of Stokes's formula,

$$\int_{\Gamma_\epsilon} P_{k_s} \overline{\partial} \varphi = \int_{\partial \Gamma_\epsilon} P_{k_s} \varphi.$$

We obtain the required result by passing to the limit as $s \to \infty$. □

Lemma 8.2 was given in [119].

Subsequently we will assume that the sequence $\epsilon_s \to 0$ is decreasing as $s \to \infty$ and is chosen so that $\partial\Gamma_{\epsilon_s}$ is a smooth manifold, $f \in \mathcal{L}^p(\partial\Gamma_{\epsilon_s})$, and Lemma 8.2 holds for Γ_{ϵ_s}. We write $D_{\epsilon_s} = D_s$ and $\Gamma_{\epsilon_s} = \Gamma_s$.

8.3 The $\overline{\partial}$-problem for the Bochner-Martinelli kernel

Suppose G is an open neighborhood of $\overline{D}$, and $h \in \mathcal{O}(G)$ (we may assume that G is a domain of holomorphy). For each positive ϵ, we consider the set

$$G_\epsilon(h) = \left\{ z \in G : |h(z)| > \max_{\overline{D \setminus D_\epsilon}} |h| \right\}.$$

Then $G_\epsilon(h) \subset G \setminus (\overline{D \setminus D_\epsilon})$, and for every $z \in G_\epsilon(h)$ the level set $L_z(h) = \{\zeta \in G : h(\zeta) = h(z)\} \subset G_\epsilon(h)$. We write $G(h) = \{z \in G : |h(z)| > \max_K |h|\}$. Since $K = \bigcap_{\epsilon>0} \overline{D \setminus D_\epsilon}$, we have $G(h) = \bigcup_{\epsilon>0} G_\epsilon(h)$. Since $K = \widehat{K}_{\overline{D}}$, we have $\overline{D} \setminus K \subset \bigcup_{G \supset \overline{D}} \bigcup_{h \in \mathcal{O}(G)} G(h)$.

By Hefer's theorem, for each $h \in \mathcal{O}(G)$ there are holomorphic functions $h_1(\zeta, z), \ldots, h_n(\zeta, z)$ on $G \times G$ such that

$$h(z) - h(\zeta) = \sum_{k=1}^{n} h_k(\zeta, z)(z_k - \zeta_k). \tag{8.1}$$

We can explicitly compute a solution of the $\overline{\partial}$-problem $\overline{\partial}\Phi_k(\zeta) = U(\zeta, z)$ on the set $G \setminus L_z(h)$. Let

$$U_k(\zeta, z) = \frac{(-1)^k (n-2)!}{(\zeta_k - z_k)(2\pi i)^n} \left[\sum_{j=1}^{k-1} (-1)^j \frac{\bar{\zeta}_j - \bar{z}_j}{|\zeta - z|^{2n-2}} \, d\bar{\zeta}[j, k] \right.$$
$$\left. + \sum_{j=k+1}^{n} (-1)^{j-1} \frac{\bar{\zeta}_j - \bar{z}_j}{|\zeta - z|^{2n-2}} \, d\bar{\zeta}[k, j] \right] \wedge d\zeta.$$

(The forms $U_k(\zeta, z)$ were first considered by Martinelli [150] in the proof of the Hartogs extension theorem.)

Lemma 8.3. *Set*

$$\Phi_h(\zeta) = (h(\zeta) - h(z))^{-1} \sum_{k=1}^{n} h_k(\zeta, z)(\zeta_k - z_k) U_k(\zeta, z).$$

Then $\Phi_h(\zeta)$ is defined in $G \setminus L_z(h)$, and $\overline{\partial}_\zeta \Phi_h(\zeta) = U(\zeta, z)$.

Proof. It is easy to check that $\overline{\partial}_\zeta U_k(\zeta, z) = U(\zeta, z)$ outside $L_z(\zeta_k)$. Therefore

$$\begin{aligned} \overline{\partial}\Phi_h(\zeta) &= \sum_{k=1}^{n} h_k(\zeta, z)(h(\zeta) - h(z))^{-1} \overline{\partial} U_k(\zeta, z)(\zeta_k - z_k) \\ &= \sum_{k=1}^{n} h_k(\zeta, z)(\zeta_k - z_k)(h(\zeta) - h(z))^{-1} U(\zeta, z) = U(\zeta, z). \end{aligned}$$

□

Now suppose that G and G' are open sets in $\mathbf{C}^n$ with nontrivial intersection, $h \in \mathcal{O}(G)$, and $h' \in \mathcal{O}(G')$. We consider the Hefer decomposition (8.1) for h and h' and a point $z \in G \cap G'$.

If $n \geq 3$, we consider the forms $U_{k,j}(\zeta, z)$ given for $1 \leq k < j \leq n$ by

$$U_{k,j}(\zeta, z) = \frac{(-1)^{n+j+k}(n-3)! \, |\zeta - z|^{4-2n}}{(2\pi i)^n (\zeta_k - z_k)(\zeta_j - z_j)} \left\{ \sum_{m=1}^{k-1} (-1)^m (\bar{\zeta}_m - \bar{z}_m) \, d\bar{\zeta}[m, k, j] \right.$$
$$\left. + \sum_{m=k+1}^{j-1} (-1)^{m-1} (\bar{\zeta}_m - \bar{z}_m) \, d\bar{\zeta}[k, m, j] + \sum_{m=j+1}^{n} (-1)^m (\bar{\zeta}_m - \bar{z}_m) \, d\bar{\zeta}[k, j, m] \right\} \wedge d\zeta.$$

For $k > j$, we set $U_{k,j} = -U_{j,k}$. We further denote

$$\chi_{h,h'}(\zeta) = \sum_{1\le k<j\le n} \frac{(h_k h'_j - h'_k h_j)(\zeta_k - z_k)(\zeta_j - z_j)U_{k,j}(\zeta, z)}{(h(\zeta) - h(z))(h'(\zeta) - h'(z))}.$$

Lemma 8.4. *The following equalities hold on the set* $(G \setminus L_z(h)) \cap (G' \setminus L_z(h'))$:

$$\Phi_h(\zeta) - \Phi_{h'}(\zeta) = \bar\partial \chi_{h,h'}(\zeta), \qquad \text{if } n \ge 3, \tag{8.2}$$

$$\Phi_h(\zeta) - \Phi_{h'}(\zeta) = \frac{(h_1 h'_2 - h_2 h'_1)\, d\zeta_1 \wedge d\zeta_2}{(2\pi i)^2 (h(\zeta) - h(z))(h'(\zeta) - h'(z))}, \qquad \text{if } n = 2. \tag{8.3}$$

Proof. When $n = 2$, we have

$$\begin{aligned}
&\Phi_h(\zeta) - \Phi_{h'}(\zeta)\\
&= \frac{d\zeta}{(2\pi i)^2 |\zeta - z|^2} \left[\frac{h_1(\bar\zeta_2 - \bar z_2) - h_2(\bar\zeta_1 - \bar z_1)}{h(\zeta) - h(z)} - \frac{h'_1(\bar\zeta_2 - \bar z_2) - h'_2(\bar\zeta_1 - \bar z_1)}{h'(\zeta) - h'(z)} \right]\\
&= \frac{(h_1 h'_2 - h_2 h'_1)\, d\zeta_1 \wedge d\zeta_2}{(2\pi i)^2 (h(\zeta) - h(z))(h'(\zeta) - h'(z))}.
\end{aligned}$$

When $n \ge 3$, it is easy to check that $U_k(\zeta, z) - U_j(\zeta, z) = \bar\partial U_{k,j}(\zeta, z)$, so

$$\begin{aligned}
\bar\partial \chi_{h,h'}(\zeta) &= \sum_{1\le k<j\le n} \frac{(h_k h'_j - h'_k h_j)(\zeta_k - z_k)(\zeta_j - z_j)(U_k(\zeta, z) - U_j(\zeta, z))}{(h(\zeta) - h(z))(h'(\zeta) - h'(z))}\\
&= \frac{1}{(h(\zeta) - h(z))(h'(\zeta) - h'(z))} \Big[\sum_{k<j} (h_k h'_j - h'_k h_j)(\zeta_k - z_k)(\zeta_j - z_j) U_k\\
&\qquad - \sum_{k<j} (h_k h'_j - h'_k h_j)(\zeta_k - z_k)(\zeta_j - z_j) U_j \Big]\\
&= \frac{1}{(h(\zeta) - h(z))(h'(\zeta) - h'(z))} \Big[\sum_{k<j} (h_k h'_j - h'_k h_j)(\zeta_k - z_k)(\zeta_j - z_j) U_k\\
&\qquad + \sum_{k>j} (h_k h'_j - h'_k h_j)(\zeta_k - z_k)(\zeta_j - z_j) U_k \Big]\\
&= \sum_{k=1}^{n} \frac{(\zeta_k - z_k) h_k U_k}{(h(\zeta) - h(z))} - \sum_{k=1}^{n} \frac{(\zeta_k - z_k) h'_k U_k}{(h'(\zeta) - h'(z))}\\
&= \Phi_h(\zeta) - \Phi_{h'}(\zeta).
\end{aligned}$$

□

It is clear that

$$\frac{\partial U_k}{\partial \bar{z}_k} = \frac{(-1)^k (n-1)!}{(2\pi i)^n} \left[\sum_{j=1}^{k-1} \frac{(-1)^j (\bar{\zeta}_j - \bar{z}_j)\, d\bar{\zeta}[j,k]}{|\zeta - z|^{2n}} + \sum_{j=k+1}^{n} \frac{(-1)^{j-1} (\bar{\zeta}_j - \bar{z}_j)\, d\bar{\zeta}[k,j]}{|\zeta - z|^{2n}} \right] \wedge d\zeta$$

$$= \frac{(n-1)(\zeta_k - z_k)\, U_k(\zeta, z)}{|\zeta - z|^2},$$

that is, $\partial U_k / \partial \bar{z}_k$ has a singularity only at $\zeta = z$, but not on the plane $L_z(\zeta_k)$. Therefore we obtain

$$\frac{\partial U}{\partial \bar{z}_k} = \bar{\partial}_\zeta \left(\frac{\partial U_k}{\partial \bar{z}_k} \right)$$

in $\mathbf{C}^n \setminus \{z\}$.

When $n \geq 3$,

$$\frac{\partial U_{k,j}}{\partial \bar{z}_k} = \frac{(n-2)(\zeta_k - z_k)\, U_{k,j}}{|\zeta - z|^2},$$

that is, this form has singularities on the plane $L_z(\zeta_j)$, but not on $L_z(\zeta_k)$. Consequently,

$$\frac{\partial U_k}{\partial \bar{z}_k} - \frac{\partial U_j}{\partial \bar{z}_k} = \bar{\partial}_\zeta \left(\frac{\partial U_{k,j}}{\partial \bar{z}_k} \right)$$

on the set $\mathbf{C}^n \setminus L_z(\zeta_j)$.

When $n = 2$, it is obvious that

$$\frac{\partial U_1}{\partial \bar{z}_j} - \frac{\partial U_2}{\partial \bar{z}_j} = 0, \qquad j = 1, 2.$$

If for $n \geq 3$ we consider the form

$$\Psi_h^k(\zeta) = (h(\zeta) - h(z))^{-1} \sum_{j \neq k} h_j(\zeta_j - z_j) \frac{\partial U_{k,j}}{\partial \bar{z}_k}$$

on the set $G \setminus L_z(h)$, then we have the following.

Lemma 8.5. *On the set $G \setminus L_z(h)$, we have*

$$\frac{\partial \Phi_h}{\partial \bar{z}_k} = \frac{\partial U_k}{\partial \bar{z}_k} - \bar{\partial}_\zeta \Psi_h^k, \qquad k = 1, \ldots, n, \qquad \textit{for } n \geq 3, \textit{ and}$$

$$\frac{\partial \Phi_h}{\partial \bar{z}_k} = \frac{\partial U_k}{\partial \bar{z}_k}, \qquad k = 1, 2, \qquad \textit{for } n = 2.$$

Lemmas 8.3–8.5 were given in [143].

8.4 Proof of Lupacciolu's theorem

Suppose $G \supset \overline{D}$, $h \in \mathcal{O}(G)$, and $G(h) = \bigcup_{s=1}^{\infty} G_s(h)$, where $G_s(h) = \{ z \in G : |h(z)| > \max_{\overline{D \setminus D_s}} |h| \}$. The sequences Γ_s and D_s are chosen the same as in subsection 2. Consider the function

$$F_h^s(z) = \int_{\Gamma_s} f(\zeta)\, U(\zeta, z) - \int_{\partial\Gamma_s} f(\zeta)\, \Phi_h(\zeta). \tag{8.4}$$

The idea of the proof is to show that F_h^s is holomorphic in $G_s(h) \setminus \Gamma$ and independent of s and h, while outside $\overline{D}$ it equals zero, and so (8.4) gives a holomorphic extension of f into $D \setminus K$ (by the theorem on the jump of the Bochner-Martinelli integral).

The function (8.4) is defined for $z \notin \Gamma$ and $z \in G_s(h)$, since $\partial\Gamma_s \subset \overline{D \setminus D_s}$. We first show that it does not depend on s. Indeed, if $s' \geq s$, then the function $F_h^{s'}$ also is defined in $G_s(h) \setminus \Gamma$, so

$$F_h^{s'}(z) - F_h^s(z) = \int_{\Gamma_{s'} \setminus \Gamma_s} f\, U(\zeta, z) - \int_{\partial\Gamma_{s'}} f\, \Phi_h + \int_{\partial\Gamma_s} f\, \Phi_h.$$

Since $|h(z)| > |h(\zeta)|$ if $z \in G_s(h)$, and $\zeta \in \overline{\Gamma_{s'} \setminus \Gamma_s}$ (because $\Gamma_{s'} \setminus \Gamma_s \subset \overline{D \setminus D_s}$), the form Φ_h has no singularities on $\overline{\Gamma_{s'} \setminus \Gamma_s}$, and $\bar\partial \Phi_h = U(\zeta, z)$ by Lemma 8.3. The coefficients of Φ_h are class C^∞, and we can extend them outside $\overline{\Gamma_{s'} \setminus \Gamma_s}$ as functions of class C^∞ with compact support, so that Lemma 8.2 is applicable. Consequently, $F_h^{s'}(z) = F_h^s(z)$. From now on we shall omit the symbol s on the function F_h^s.

We now show the uniqueness of the extension.

Lemma 8.6. *If F satisfies the hypotheses of Theorem 8.1, then for $z \in G_s(h)$ we have*

$$F(z) = F_h(z), \qquad z \in D \setminus K.$$

Proof. The Bochner-Martinelli formula (1.5) can be applied to the function F in the domain D_s. The boundary $\partial D_s = \Gamma_s \cup K_s$, where $K_s = D \cap \{ z : 1 - \beta(z) = \epsilon_s \}$, and we may assume that K_s is also a smooth manifold. For $z \in G_s(h) \cap D$, we then have

$$F(z) = \int_{\partial D_s} F(\zeta)\, U(\zeta, z) = \int_{\Gamma_s} f\, U(\zeta, z) + \int_{K_s} F\, U(\zeta, z).$$

But $|h(z)| > |h(\zeta)|$ for $\zeta \in K_s \subset \overline{D \setminus D_s}$, so $U(\zeta, z) = \bar\partial \Phi_h(\zeta)$, and by Stokes's formula

$$\int_{K_s} F\, U(\zeta, z) = \int_{\partial K_s} f\, \Phi_h = - \int_{\partial\Gamma_s} f\, \Phi_h.$$

□

This lemma is essentially unnecessary, for the uniqueness of the extension of f follows from the usual uniqueness theorem for functions of class $\mathcal{H}^p(D)$.

Lemma 8.7. *The function $F_h(z)$ is holomorphic in $G(h) \setminus \Gamma$.*

Proof. If $z \in G_s(h) \setminus \Gamma$, then by Lemmas 8.5 and 8.2 we have

$$\frac{\partial F_h}{\partial \bar{z}_k} = \int_{\Gamma_s} f \frac{\partial U}{\partial \bar{z}_k} - \int_{\partial\Gamma_s} f \frac{\partial \Phi_h}{\partial \bar{z}_k} = \int_{\Gamma_s} f \frac{\partial U}{\partial \bar{z}_k} - \int_{\partial\Gamma_s} f \left(\frac{\partial U_k}{\partial \bar{z}_k} - \overline{\partial}\Psi_h^k\right)$$
$$= \int_{\Gamma_s} f \,\overline{\partial}_\zeta \left(\frac{\partial U_k}{\partial \bar{z}_k}\right) - \int_{\partial\Gamma_s} f \frac{\partial U_k}{\partial \bar{z}_k}.$$

But since the form $\partial U_k/\partial \bar{z}_k$ has no singularities on Γ_s (if $z \notin \Gamma$), we obtain that $\partial F_h/\partial \bar{z}_k = 0$ for $k = 1, \ldots, n$ by again applying Lemma 8.2. □

Lemma 8.8. *If $h' \in \mathcal{O}(G')$, where $G' \supset \overline{D}$, then $F_h(z) = F_{h'}(z)$ for $z \in G'_s(h') \cap G_s(h) \setminus \Gamma$.*

Proof. First consider the case $n \geq 3$. Using (8.2), we have

$$F_h(z) - F_{h'}(z) = -\int_{\partial\Gamma_s} f(\zeta)\,\overline{\partial}_\zeta \chi_{h,h'}(\zeta).$$

The form $\chi_{h,h'}$ has no singularities on $\partial\Gamma_s$, and outside $\partial\Gamma_s$ we can extend the coefficients of this form to functions of class $\mathcal{C}^\infty$ with compact support in $\mathbf{C}^n \setminus K$, so $F_h(z) = F_{h'}(z)$ by Lemma 8.2.

If $n = 2$, then by using (8.3) we obtain

$$F_h(z) - F_{h'}(z) = \frac{1}{(2\pi i)^2} \int_{\partial\Gamma_s} \frac{f(\zeta)(h_2 h_1' - h_1 h_2')\,d\zeta}{(h(\zeta) - h(z))(h'(\zeta) - h'(z))}. \tag{8.5}$$

Since $z \in G'_s(h') \cap G_s(h) \setminus \Gamma$, we have $|h(\zeta)| < |h(z)|$ and $|h'(\zeta)| < |h'(z)|$ for $\zeta \in \partial\Gamma_s$, so

$$\frac{1}{(h(\zeta) - h(z))(h'(\zeta) - h'(z))} = \sum_{k,j\geq 0} \frac{(h(\zeta))^k (h'(\zeta))^j}{(h(z))^{1+k}(h'(z))^{1+j}},$$

and this series converges absolutely and uniformly on $\partial\Gamma_s$. Substituting it into (8.5) and integrating term by term, we find

$$F_h(z) - F_{h'}(z) = \frac{1}{(2\pi i)^2} \sum_{k,j\geq 0} \frac{1}{(h(z))^{1+k}(h'(z))^{1+j}} \int_{\partial\Gamma_s} f\,\mu_{k,j},$$

where $\mu_{k,j} = (h_2 h_1' - h_1 h_2')(h(\zeta))^k (h'(\zeta))^j\,d\zeta$. The form $\mu_{k,j}$ is $\overline{\partial}$-closed on Γ_s, so by Lemma 8.2,

$$\int_{\partial\Gamma_s} f\,\mu_{k,j} = \int_{\Gamma_s} f\,\overline{\partial}\mu_{k,j} = 0.$$

□

Thus, the integral (8.4) defines a function F that is holomorphic in $D \setminus K$ and holomorphic outside $\overline{D}$. Let W be a neighborhood of $\partial D \setminus K = \Gamma$ contained in $\bigcup_{G \supset \overline{D}} \bigcup_{h \in \mathcal{O}(G)} G(h)$ such that $W \setminus \Gamma = W_+ \cup W_-$, where $W_+ \subset D$, while $W_- \subset \mathbf{C}^n \setminus \overline{D}$ and $W_+ \cup W_-$ are connected sets. We will show that $F = 0$ in W_-.

Consider a neighborhood $G \supset \overline{D}$ and $h \in \mathcal{O}(G)$, and suppose $z \in G$ is a point such that $|h(z)| > \max_{\overline{D}} |h|$. Such a point exists since $\overline{D} \subset G$ and $|h|$ does not attain a maximum inside G. This point $z \notin \overline{D}$. Let $G_1(h) = \{ z : |h(z)| > \max_{\overline{D}} |h| \}$. Then

$$F(z) = F_h(z) = \int_{\Gamma_s} f\, U(\zeta, z) - \int_{\partial \Gamma_s} f\, \Phi_h(\zeta).$$

But Φ_h does not have singularities on Γ_s because $L_z(h)$ does not intersect ∂D, while $\bar{\partial} \Phi_h = U(\zeta, z)$. By Lemma 8.2, $F(z) = 0$.

Since $F = 0$ in $G_1(h) \subset G(h)$, while $G_1(h) \cap W_- \neq \varnothing$ (so G_1 will abut $\partial D \setminus K$) and W_- is a connected set, $F \equiv 0$ in W_-.

Now we consider the boundary behavior of $F(z)$. Suppose $z^0 \in \Gamma$ and $B(z^0, r)$ is a ball with center at z^0, $B(z^0, r) \cap \overline{D \setminus D_s} = \varnothing$, and $\nu(z)$ is the outer unit normal to Γ. Then

$$\begin{aligned} F(z - \zeta\nu(z)) &= F(z - \zeta\nu(z)) - F(z + \zeta\nu(z)) \\ &= \int_{\Gamma_s} f(\zeta) \big(U(\zeta, z - \epsilon\nu(z)) - U(\zeta, z + \epsilon\nu(z)) \big) \\ &\qquad + \int_{\partial \Gamma_s} f(\zeta) \big(\Phi_h(\zeta, z + \epsilon\nu(z)) - \Phi_h(\zeta, z - \epsilon\nu(z)) \big). \end{aligned}$$

The second integral in this formula has no jump, since we can choose the ball $B(z^0, r)$ so that $|h(z)| > |h(\zeta)|$ for $z \in \Gamma \cap B(z^0, r)$ and $\zeta \in \partial \Gamma_s$, which means that $|h(z \pm \epsilon\nu(z))| > |h(\zeta)|$ for sufficiently small positive ϵ. Moreover, this integral is a function of class C^∞ in $B(z^0, r)$. We split the integral over Γ_s into the two pieces over $\Gamma_s \setminus B(z, r)$ and over $\Gamma \cap B(z, r)$. The integral over $\Gamma_s \setminus B(z, r)$ also has no jump. Now we obtain Theorem 8.1 by applying Theorem 3.4 to the integral over $\Gamma \cap B(z, r)$.

Theorem 8.1 generalizes a number of results from [147, 197].

Remark. As can be seen from the proof, we do not need to require connectedness of Γ in Theorem 8.1, but only, for example, the following: the compact set $K \subset \overline{D}$, and K is convex with respect to the class $\mathcal{O}(G)$, where G is a domain of holomorphy containing $\overline{D}$, while the complement $\mathbf{C}^n \setminus \overline{D}$ is connected. Then every CR-function $f \in \mathcal{L}^p_{\text{loc}}(\Gamma)$ extends holomorphically into $D \setminus K$. Indeed, in this case F will be holomorphic in $G \setminus \overline{D}$, and consequently will extend holomorphically into D by Hartogs's theorem (see Theorem 8.14 below).

The result was obtained for connected Γ in Stein manifolds by Laurent-Thiébaut [133].

Corollary 8.9. *If $K \subset \partial D$, and $\partial D \setminus \widehat{K}_{\overline{D}}$ is a smooth connected hypersurface, then every CR-function $f \in \mathcal{L}^p_{\text{loc}}(\partial D \setminus \widehat{K}_{\overline{D}})$ extends holomorphically into $D \setminus \widehat{K}_{\overline{D}}$ (in the sense of Theorem 8.1). If $\widehat{K}_{\overline{D}} \subset \partial D$, then f extends holomorphically into D.*

For domains in $\mathbf{C}^2$, the $\mathcal{O}(\overline{D})$-convexity of K is also a necessary condition for the existence of a holomorphic extension.

Theorem 8.10 (Stout). *Suppose D is a bounded strongly pseudoconvex domain in $\mathbf{C}^2$ with $\partial D \in \mathcal{C}^2$. If K is a compact subset of ∂D, then every CR-function in $\mathcal{C}(\partial D \setminus K)$ extends holomorphically into D if and only if $K = \widehat{K}_{\overline{D}}$.*

For the proof, see [198].

8.5 Extension of the class of compact sets

If N is an arbitrary set in $\mathbf{C}^n$, and $K \subset N$, we set

$$\widetilde{K}_N = \bigcap_{h \in \mathcal{O}(N)} h^{-1}(h(K)) = \bigcap_{G \supset N} \bigcap_{h \in \mathcal{O}(G)} h^{-1}(h(K)),$$

where G is an open set. In other words, $\widetilde{K}_N$ is the complement in N of the set of points z for which there exists a function $h \in \mathcal{O}(N)$ with $h(z) = 0$ but $h \neq 0$ in K. In particular, if $N = \mathbf{C}^n$, then $\widehat{K}_{\mathbf{C}^n}$ is the rational envelope of K.

Theorem 8.11 (Lupacciolu). *Suppose $n \geq 3$. Let D be a bounded domain such that $\overline{D}$ has a schlicht envelope of holomorphy. Suppose $K \subset \overline{D}$ is a compact set, and $\Gamma \subset \partial D \setminus K$ is a smooth connected hypersurface in $\mathbf{C}^n \setminus K$. Every CR-function $f \in \mathcal{L}^p_{\text{loc}}(\Gamma)$, where $p \geq 1$, extends holomorphically into $D \setminus K$ if one of the following two conditions holds:*

1. *there is a neighborhood N of K in $\overline{D}$ such that $\widetilde{K}_N = K$, and each component of the hypersurface $\Gamma \cap N$ contains peak points for $\mathcal{O}(\overline{D})$;*
2. *there is a neighborhood N of K in $\overline{D}$ such that $\widetilde{K}_N = K$, and $\widehat{K}_{\overline{D}} \cap \partial D = K \cap \partial D$.*

A point $z^0 \in \partial D$ is a peak point for $\mathcal{O}(\overline{D})$ if there exists a function $h \in \mathcal{O}(\overline{D})$ for which $|h(z_0)| > |h(z)|$, $z \in \overline{D} \setminus \{z^0\}$.

Theorem 8.11 was proved by Lupacciolu in [144] for continuous CR-functions.

An example in [144] shows that this theorem is false when $n = 2$. Indeed, consider the domain $D = \{ (z_1, z_2) \in \mathbf{C}^2 : |z_1|^2 + |z_2|^2 < 2,\ |z_2| < 1 \}$ and the set $K = \{ (z_1, z_2) : |z_1|^2 + |z_2|^2 \leq 2,\ |z_2| = 1 \}$. Then $\Gamma = \{ (z_1, z_2) : |z_1|^2 + |z_2|^2 = 2,\ |z_2| < 1 \}$ is a smooth connected hypersurface. Let N be an arbitrary neighborhood of K in $\overline{D}$, and $h(z_1, z_2) = z_2$. Then $h^{-1}(h(K)) = K$ in N, that is, $\widetilde{K}_N = K$. Every point of Γ is a peak point for functions holomorphic in $\overline{B(0, \sqrt{2})}$, and moreover $\widehat{K}_{\overline{D}} \cap \partial D = K$. Consequently, both hypotheses of Theorem 8.11

hold. The function $f = 1/z_1$ is holomorphic in a neighborhood of Γ, but it does not extend holomorphically into D.

When $n \geq 3$, Theorem 8.1 follows from Theorem 8.11, for if $K = \widehat{K}_{\overline{D}}$, then $\widetilde{K}_{\overline{D}} = \widehat{K}_N = K$.

Proof. The proof of Theorem 8.11 is essentially a repetition of the proof of Theorem 8.1. It has been shown that the function

$$F(z) = \int_{\Gamma_s} f\,U(\zeta, z) - \int_{\partial\Gamma_s} f\,\Phi_h(\zeta), \qquad z \notin \Gamma, \tag{8.6}$$

does not depend on s and h. We only need to prove that F gives a holomorphic extension of f into $D \setminus K$. In case (2), this follows because F gives a holomorphic extension of f into $D \setminus \widetilde{K}_{\overline{D}}$. In case (1), it can be shown in the same way as in Theorem 8.1, by using a peak function h for the point $z^0 \in N \cap \Gamma$, that $F = 0$ for points in $\mathbf{C}^n \setminus \overline{D}$ near z^0. □

We will now give a result which follows from Theorem 6.1 on the analytic representation of CR-functions and which generalizes the preceding theorem. Suppose G is a domain in $\mathbf{C}^n$ and K is a compact set in G. We denote by $H^1_\Phi(G \setminus K, \mathcal{O})$ the quotient group of the group of $\overline{\partial}$-closed forms α in $G \setminus K$ of type $(0,1)$ such that $\operatorname{supp}\alpha \Subset G$ by the group of $\overline{\partial}$-exact forms $\overline{\partial}\gamma$, where γ is a function on $G \setminus K$ and $\operatorname{supp}\gamma \Subset G$. Thus, as the family of supports Φ we consider sets in $G \setminus K$ that are relatively compact in G. The regularization theorem (see, for example, [37]) shows that in this definition, we may take forms with coefficients of class $C^\infty(G \setminus K)$, or currents in $G \setminus K$.

Lemma 8.12. *If G is a domain of holomorphy in $\mathbf{C}^n$, where $n \geq 2$, and K is a holomorphically convex compact set in G (that is, $K = \widehat{K}_G$), then $H^1_\Phi(G \setminus K, \mathcal{O}) = 0$.*

Proof. Let α be a $\overline{\partial}$-closed form of type $(0,1)$ with coefficients of class $C^\infty(G \setminus K)$ and $\operatorname{supp}\alpha \Subset G$. By Theorem 5.1.6 from [79], there is a strongly plurisubharmonic function $\varphi \in C^\infty(G)$ such that $\varphi < 0$ on K, but $\varphi > 0$ outside any neighborhood ω of the set $\widehat{K}_G = K$, and moreover the sets $K_c = \{z \in G : \varphi(z) < c\}$ are relatively compact in G and $(\widehat{\overline{K_c}})_G = \overline{K_c}$. By choosing a sequence of neighborhoods ω_s approximating K, we obtain a sequence of strongly pseudoconvex domains K_s with the following properties: $\overline{K}_{s+1} \subset K_s$, $\bigcap_s K_s = K$, $\partial K_s \in C^\infty$, $K_s \Subset G$, and $(\widehat{\overline{K_s}})_G = \overline{K_s}$. The last equality allows us to assert by the Oka-Weil theorem (Corollary 5.2.9 in [79]) that every function in $\mathcal{O}(\overline{K_s})$ can be uniformly approximated on $\overline{K_s}$ by functions in $\mathcal{O}(G)$.

We consider on the boundary ∂K_s the tangential Cauchy-Riemann equation

$$\overline{\partial}_\tau v = \alpha. \tag{8.7}$$

(We say that $\overline{\partial}_\tau v = \alpha$ on ∂K_s if $\overline{\partial} v \wedge \overline{\partial}\rho = \alpha \wedge \overline{\partial}\rho$ on ∂K_s, where ρ is a defining function for ∂K_s; see also equation (3.9).) Then $\overline{\partial}_\tau \alpha = 0$ on ∂K_s, since $\overline{\partial}\alpha = 0$

in $G \setminus K \supset \partial K_s$. When $n \geq 3$, this condition is sufficient for solvability of (8.7) (see, for example, the survey [74, §8.3]). For solvability of (8.7) when $n = 2$, it is necessary that α be orthogonal to all forms γ of type $(2,0)$ with coefficients holomorphic in $\overline{K_s}$ (see [74] or [10]), that is,

$$\int_{\partial K_s} \alpha \wedge \gamma = 0. \tag{8.8}$$

Since any function $h \in \mathcal{O}(\overline{K_s})$ can be uniformly approximated on $\overline{K_s}$ by functions in $\mathcal{O}(G)$, it is enough to verify (8.8) on forms γ with coefficients holomorphic in G.

Consider in G a domain $G_1 \supset \operatorname{supp} \alpha$ with $\partial G_1 \in \mathcal{C}^\infty$. Then by Stokes's formula,

$$\int_{\partial K_s} \alpha \wedge \gamma = \int_{\partial G_1} \alpha \wedge \gamma = 0.$$

Thus, condition (8.8) holds, so equation (8.7) is indeed solvable when $n = 2$.

We denote by $v_s \in \mathcal{C}^\infty(G)$ a solution of (8.7) in K_s. Equation (8.7) means that $\overline{\partial} v_s - \alpha = \beta\, \overline{\partial}\rho$ on ∂K_s, where ρ is a defining function for K_s. Consequently, the function $u_s = v_s - \rho\beta$ satisfies the condition $\overline{\partial} u_s = \overline{\partial} v_s - \rho\, \overline{\partial}\beta - \beta\, \overline{\partial}\rho = \overline{\partial} v_s - \beta\overline{\partial}\rho = \alpha$ on ∂K_s.

We now define a current α_s on G via $\alpha_s = \alpha$ in $G \setminus K_s$, and $\alpha_s = \overline{\partial} u_s$ on $\overline{K_s}$. Let us find $\overline{\partial}\alpha_s$. If ψ is a form of type $(n, n-2)$ with coefficients in $\mathcal{D}(G)$, then

$$\begin{aligned}
\overline{\partial}\alpha_s(\psi) = -\alpha_s(\overline{\partial}\psi) &= -\int_{K_s} \overline{\partial} u_s \wedge \overline{\partial}\psi - \int_{G\setminus K_s} \alpha \wedge \overline{\partial}\psi \\
&= -\int_{\partial K_s} \overline{\partial} u_s \wedge \psi + \int_{\partial K_s} \alpha \wedge \psi = 0.
\end{aligned}$$

We note that $\operatorname{supp} \alpha_s \Subset G$. Since the compact cohomology group $H^1_c(G, \mathcal{O}) = 0$, there is a distribution $\varphi_s \in \mathcal{E}'(G)$ such that $\overline{\partial}\varphi_s = \alpha_s$.

Consider the difference $\varphi_{s+1} - \varphi_s$. Outside K_s we have $\overline{\partial}(\varphi_{s+1} - \varphi_s) = 0$, so $\varphi_{s+1} - \varphi_s$ is holomorphic outside K_s and $\operatorname{supp}(\varphi_{s+1} - \varphi_s) \Subset G$, whence $\varphi_{s+1} = \varphi_s$ outside K_s. Thus, by setting $\varphi = \varphi_s$, we obtain that $\overline{\partial}\varphi = \alpha$ in $G \setminus K$, and $\operatorname{supp} \varphi \Subset G$. Again applying the regularization theorem from [37], we may assume that φ is class $\mathcal{C}^\infty(G \setminus K)$. □

The proof of Lemma 8.12 follows the scheme of [174], which goes back to Hörmander (see Theorem 2.3.2 of [79]). For a Stein manifold G, it was carried out in [121].

Remark. Lemma 8.12 holds for a wider class of domains G and compact sets K. For example, it is enough to require that $H^1_c(G, \mathcal{O}) = 0$ and K is approximated by a sequence of domains K_s in which problem (8.7) is solvable for $\overline{\partial}$-closed forms α of type $(0, 1)$.

Lemma 8.13. *Suppose D is a bounded domain in $\mathbf{C}^n$, and $\mathbf{C}^n \setminus \overline{D}$ is connected. Let $K \subset \overline{D}$ be a compact set with $H^1_\Phi(G \setminus K, \mathcal{O}) = 0$ for some domain of holomorphy $G \supset \overline{D}$, while $\Gamma = \partial D \setminus K$ is a smooth manifold in $\mathbf{C}^n \setminus K$. Then for every CR-function f on Γ there is a holomorphic function F in $D \setminus K$ with the following boundary properties:*

1. *if $\Gamma \in \mathcal{C}^m$ and $f \in \mathcal{C}^r(\Gamma)$, where $r \leq m$, then $F \in \mathcal{C}^r((D \setminus K) \cup \Gamma)$, and $F = f$ on Γ;*

2. *if $f \in \mathcal{L}^p_{\mathrm{loc}}(\Gamma)$, where $p \geq 1$, then for every point $z \in \Gamma$ there is a ball $B(z,r)$ for which*

$$\lim_{\epsilon \to 0+} \int_{\Gamma \cap B(z,r)} |F(\zeta - \epsilon\nu(\zeta)) - f(\zeta)|^p \, d\sigma = 0,$$

   and the nontangential limit of F on Γ agrees with f almost everywhere;

3. *if $\Gamma \in \mathcal{C}^\infty$ and f is a CR-distribution on Γ, then $F(z - \epsilon\nu(z)) \to f(z)$ on Γ in the weak topology as $\epsilon \to 0^+$.*

In cases (1) and (2), the holomorphic extension of f is given by (8.6).

Proof. Consider the current $f[\Gamma]^{0,1}$, which is $\overline{\partial}$-closed in $\Omega = G \setminus K$ and has $\operatorname{supp} f[\Gamma]^{0,1} \Subset G$. By the hypothesis of the lemma, there is a distribution h in Ω such that $\operatorname{supp} h \Subset G$ and $\overline{\partial} h = f[\Gamma]^{0,1}$, that is, $h \in \mathcal{O}(\Omega \setminus \Gamma)$. One component of $\Omega \setminus \Gamma$ coincides with $G \setminus \overline{D}$. We denote the function h in $G \setminus \overline{D}$ by h^+. Then $h^+ = 0$ since $\operatorname{supp} h^+ \Subset G$. Now in each component of $D \setminus K$, the function h^- has the boundary values indicated in the lemma in view of Theorems 6.1 and 7.1. □

Theorem 8.14 (Kytmanov). *Let D be a bounded domain in $\mathbf{C}^n$, $n \geq 2$, with $\mathbf{C}^n \setminus \overline{D}$ connected. Suppose $K \subset \overline{D}$ is a compact set that is holomorphically convex with respect to some domain of holomorphy $G \supset \overline{D}$ (that is, $K = \widehat{K}_G$), and $\partial D \setminus K = \Gamma \in \mathcal{C}^1$ in $\mathbf{C}^n \setminus K$. If f is a CR-function on Γ, then there exists a holomorphic extension F of f into $D \setminus K$ with the boundary properties listed in Lemma 8.13.*

Proof. The proof is immediate from Lemmas 8.12 and 8.13. Theorem 8.14 was given in [121] for domains D in Stein manifolds. □

If K is not contained in $\overline{D}$, then we must require $\mathbf{C}^n \setminus (\overline{D} \cup K)$ to be a connected set.

Theorem 8.14 gives an answer to problem 18 in [38].

We remark that if $K = \widehat{K}_G$, then $K = \widehat{K}_{\overline{D}}$. Therefore Theorem 8.14 sharpens Theorem 8.1 for the given class of compact sets, since we do not require connectedness of Γ and schlichtness of the envelope of holomorphy of $\overline{D}$, while the CR-function on Γ may be arbitrary (it is clear that if Γ is connected and $K = \widehat{K}_G$, then $\mathbf{C}^n \setminus (K \cup \overline{D})$ is connected).

Theorem 8.15 (Kytmanov). *Suppose G is a domain of holomorphy in $\mathbf{C}^n$, where $n > 2$, and D is a domain such that $D \Subset G$ and $\mathbf{C}^n \setminus \overline{D}$ is connected. Let $K \subset \overline{D}$ be a compact set that is approximated from outside by a sequence of domains $K_s \Subset G$. Suppose $\partial K_s \in \mathcal{C}^\infty$ is connected, and equation (8.7) is solvable for ∂K_s for $\bar\partial$-closed forms of type $(0,1)$. Then every CR-function f given on $\Gamma = \partial D \setminus K$ extends holomorphically into $D \setminus K$ in the sense of Theorem 8.14.*

Proof. The proof follows from Lemma 8.13 and the remark preceding it. Theorem 8.15 was given in [121]; it sharpens a result of Lupacciolu from [142, Theorem 2.3]. □

Corollary 8.16. *Let G, D, and K satisfy the hypotheses of Theorem 8.14. Then the number of connected components of $\partial D \setminus K$ coincides with the number of connected components of $D \setminus K$, that is, each connected component W of $D \setminus K$ corresponds to a unique connected component W' of $\partial D \setminus K$. Moreover, $W' \subset \overline{W}$, and there are no other connected components in $D \setminus K$.*

For the proof, we set $f = C_j$ in $W_j' \subset \partial D \setminus K$, and $C_j \neq C_l$ if $j \neq l$. By Theorem 8.14, this CR-function f extends holomorphically into $D \setminus K$, and in view of the uniqueness theorem, the components $W_j \subset D \setminus K$ corresponding to the W_j' must not intersect. There are no other components of $D \setminus K$ since $G \setminus K$ is connected.

Corollary 8.16 was given in [121]. It sharpens a theorem of Alexander and Stout (see, for example, [198]), since in [198] the case of strongly pseudoconvex domains D is considered.

8.6 The case of a hypersurface

We now give a local version of Theorem 8.14. Let Ω be a domain of holomorphy in $\mathbf{C}^n$. Let $K \subset \Omega$ be a compact set, and suppose the hypersurface Γ is a smooth, oriented, relatively closed manifold in $\Omega \setminus K$. We will assume that $\Gamma = \{ z \in \Omega \setminus K : \rho(z) = 0 \}$, $\rho \in \mathcal{C}^1(\Omega \setminus K)$, and $d\rho \neq 0$ on Γ. Then $\Omega \setminus (\Gamma \cup K) = \Omega^+ \cup \Omega^-$, where $\Omega^+ = \{ z \in \Omega \setminus K : \rho(z) > 0 \}$ and $\Omega^- = \{ z \in \Omega \setminus K : \rho(z) < 0 \}$.

We are interested in the question of whether every CR-function f on Γ can be represented as the difference of boundary values of functions $h^\pm$ that are holomorphic in $\Omega^\pm$. If $n = 2$, then even in the simplest situations the answer is negative. Consider the following example.

Suppose Ω is the unit bidisc, that is, $\Omega = \{ z \in \mathbf{C}^2 : |z_1| < 1, |z_2| < 1 \}$, let $K = \{(0,0)\}$, and let $\Gamma = \{ z \in \Omega \setminus K : |z_1| = |z_2| \}$. Then Γ is a smooth hypersurface in $\Omega \setminus K$. The set $\Omega \setminus (\Gamma \cup K) = \Omega^+ \cup \Omega^-$, where $\Omega^+ = \{ z \in \Omega : |z_1| < |z_2| \}$, and $\Omega^- = \{ z \in \Omega : |z_1| > |z_2| \}$. We take the function $1/(z_1 z_2)$ as the CR-function f on Γ. If $f = h^+ - h^-$ on Γ, where $h^\pm \in \mathcal{O}(\Omega^\pm)$, then

$$\int_{\substack{|z_1|=a-\varepsilon \\ |z_2|=a}} h^+ \, dz_1 \wedge dz_2 = 0,$$

since for fixed z_2 the function h^+ is holomorphic in the disc $\{|z_1| < |z_2|\}$. For precisely the same reason,

$$\int_{\substack{|z_1|=a+\epsilon \\ |z_2|=a}} h^- \, dz_1 \wedge dz_2 = 0,$$

and then $\int_{\substack{|z_1|=a \\ |z_2|=a}} f \, dz_1 \wedge dz_2$ must be zero; but it equals $(2\pi i)^2$. This example shows that in $\mathbf{C}^2$, we cannot remove even a point from Ω.

It turns out that it is more natural to consider our problem when K is a closed set in Ω (and not compact). But the example from subsection 2 in §6 shows that this problem too cannot be solved in general.

If we recognize that $H^1(\Omega \setminus K, \mathcal{O}) \neq 0$ in these examples, then they become understandable.

Theorem 8.17 (Kytmanov). *Let Ω be a domain of holomorphy in $\mathbf{C}^n$, where $n \geq 3$. Consider a compact set $K = \widehat{K}_\Omega \subset \Omega$ and a smooth, oriented, relatively closed hypersurface Γ in $\Omega \setminus K$. If f is a CR-function on Γ, then f can be represented on Γ as the difference of boundary values of functions $h^\pm \in \mathcal{O}(\Omega^\pm)$. This means that*

1. *if $\Gamma \in C^k$, and $f \in C^r(\Gamma)$, where $0 < r \leq k$, then $h^\pm \in C^{r-\epsilon}(\Gamma \cup \Omega^\pm)$ for $\epsilon > 0$, and $f = h^+ - h^-$ on Γ;*

2. *if $\Gamma \in C^1$, and $f \in \mathcal{L}^p_{\text{loc}}(\Gamma)$, where $p \geq 1$, then for every point $z \in \Gamma$ there is a neighborhood W_z in which*

$$\lim_{\epsilon \to 0+} \int_{\Gamma \cap W_z} |f(\zeta) - h^+(\zeta + \epsilon\nu(\zeta)) + h^-(\zeta - \epsilon\nu(\zeta))|^p \, d\sigma = 0,$$

and moreover the difference of the nontangential limits of the functions $h^\pm$ agrees with f;

3. *if $\Gamma \in C^\infty$, and $f \in \mathcal{D}'(\Gamma)$, then for every function $\alpha \in \mathcal{D}(\Gamma)$,*

$$\lim_{\epsilon \to 0+} \int_\Gamma \big(h^+(\zeta + \epsilon\nu(\zeta)) - h^-(\zeta - \epsilon\nu(\zeta))\big)\alpha(\zeta) \, d\sigma = f(\alpha).$$

Proof. The proof consists in applying Theorem 6.1 concerning analytic representation to the domain $\Omega \setminus K$. We only need to show that $H^1(\Omega \setminus K, \mathcal{O}) = 0$. □

Lemma 8.18. $H^1(\Omega \setminus K, \mathcal{O}) = 0$ *under the hypotheses of Theorem 8.17.*

Proof. The proof is carried out exactly the same as in Lemma 8.12: we choose a sequence of strongly pseudoconvex domains K_s approximating K, and in $\Omega \setminus K$ we consider a $\bar\partial$-closed form α of type $(0,1)$. Since $n \geq 3$, we can solve (8.7), the tangential Cauchy-Riemann equation $\bar\partial_\tau v_s = \alpha$, on ∂K_s (when $n = 2$, this is not so, since the form α may not be orthogonal to holomorphic forms of type $(2,0)$). We further define the current α_s, which is $\overline{\partial}$-closed in Ω, and therefore there is a distribution φ_s in $\mathcal{D}'(\Omega)$ such that $\overline{\partial}\varphi_s = \alpha_s$.

Consider the difference $\varphi_{s+1} - \varphi_s$. In $\Omega \setminus \overline{K}_s$, we have $\overline{\partial}(\varphi_{s+1} - \varphi_s) = 0$, that is, $\varphi_{s+1} - \varphi_s = h_s$, where h_s is holomorphic in $\Omega \setminus \overline{K}_s$. By Hartogs's theorem, h_s extends holomorphically into Ω. If we consider the current $\psi_s = \varphi_s - \sum_{j=1}^{s-1} h_j$, then $\overline{\partial}\psi_s = \overline{\partial}\varphi_s = \alpha$ in $\Omega \setminus \overline{K}_s$, and $\psi_{s+1} - \psi_s = \varphi_{s+1} - \varphi_s - h_s = 0$ in $\Omega \setminus \overline{K}_s$. Thus, the sequence of currents ψ_s converges in $\Omega \setminus K$ to some current ψ. It is evident that $\overline{\partial}\psi = \alpha$ in $\Omega \setminus K$. □

Remark. The triviality of the group $H^1(\Omega \setminus K, \mathcal{O})$ may also be deduced from the general Alexander-Pontryagin duality theorem for cohomology groups (see [58]), but the proof we have given makes it possible to extend the class of compact sets for which $H^1(\Omega \setminus K, \mathcal{O}) = 0$. For example, it is enough to require that $H^1(\Omega, \mathcal{O}) = 0$ and that the compact set K can be approximated by a sequence of domains K_s in which (8.7) is solvable for $\overline{\partial}$-closed forms of type $(0,1)$. Therefore we have the following.

Theorem 8.19 (Kytmanov). *Let Ω be a domain in $\mathbf{C}^n$, where $n \geq 3$, such that $H^1(\Omega, \mathcal{O}) = 0$, and suppose that the compact set $K \subset \Omega$ can be approximated by the same sequence of domains as in Theorem 8.15. If Γ is a smooth, oriented, relatively closed hypersurface in $\Omega \setminus K$, then every CR-function on Γ is the difference of boundary values of holomorphic functions $h^\pm$ in $\Omega^\pm$ in the sense of Theorem 8.17.*

Theorems 8.17 and 8.18 were given in [121].

Corollary 8.20. *Suppose Γ is a smooth, oriented hypersurface in $\mathbf{C}^n$, where $n \geq 3$, and Ω is a domain of holomorphy in $\mathbf{C}^n$ such that $\Omega \setminus \Gamma$ consists of two connected components (Ω^+ and Ω^-). Suppose $z^0 \in \Gamma$, and the envelope of holomorphy of Ω^+ contains some neighborhood W of z^0. If K is a compact subset of $\Gamma \cap W$ containing z^0, and $\widehat{K}_\Omega = K$, then every CR-function f on $\Gamma \setminus K$ extends holomorphically into $W \cap \Omega^-$.*

Proof. The proof follows immediately from Theorem 8.17, since $f = h^+ - h^-$ on $\Gamma \setminus K$ with $h^\pm \in \mathcal{O}(\Omega^\pm)$, and h^+ extends holomorphically into W. □

In particular, Corollary 8.20 holds for hypersurfaces with nondegenerate Levi form.

8.7 Further results and generalizations

First of all, we give an application of Theorem 8.1. Consider complex projective space $\mathbf{CP}^n$ represented in the form of the union of $\mathbf{C}^n$ and the hyperplane Π at infinity. If $\Omega \subset \mathbf{C}^n$, we write $\widetilde{\Omega}$ for the closure of Ω in $\mathbf{CP}^n$.

Theorem 8.21 (Lupacciolu). *Suppose D is an unbounded domain in $\mathbf{C}^n$, where $n \geq 2$, with connected boundary $\partial D \in C^1$. If there exists an algebraic hypersurface T such that $\widetilde{T} \cap \widetilde{D} = \varnothing$, then every CR-function $f \in \mathcal{C}(\partial D)$ extends holomorphically into D as a function $F \in \mathcal{C}(\overline{D})$.*

We illustrate the proof for the example $T = \{ z_n = 0 \}$. Let $(\zeta_0, \zeta_1, \dots, \zeta_n)$ be homogeneous coordinates in $\mathbf{CP}^n$. Then $\Pi = \{ \zeta : \zeta_0 = 0 \}$, and $T = \{ \zeta : \zeta_n = 0 \}$. Consider the biholomorphic mapping $\tau : \mathbf{CP}^n \setminus T \to \mathbf{C}^n$ defined by

$$\tau(\zeta_0, \dots, \zeta_n) = (\zeta_0/\zeta_n, \dots, \zeta_{n-1}/\zeta_n).$$

Then $G = \tau(D)$ is a bounded domain in $\mathbf{C}^n$, $K = \tau(\widetilde{D} \cap \Pi) \subset \partial G$, $\partial G \setminus K$ is a smooth surface, $f \circ \tau^{-1}$ is a CR-function on $\partial G \setminus K$, and $\widehat{K}_{\overline{G}} = K$ since $\widetilde{D} \cap \Pi$ lies in the zero set of the holomorphic function ζ_0. We obtain the result by applying Theorem 8.1.

The proof goes analogously in the general case (see [145]). The hypothesis of schlichtness of $\overline{G}$ is superfluous here because $\widehat{K}_{\mathbf{C}^n} \subset \tau(\Pi)$, since $K \subset \tau(\Pi)$.

Recently a very large number of works have been devoted to the question of holomorphic extension of CR-functions from a part of the boundary of a domain. A survey of these results may be found in the articles of Stout [198] and Chirka and Stout [39]. Here we have mentioned only those results in which the Bochner-Martinelli integral is used explicitly. We mention one more fact of this kind from [146].

Theorem 8.22 (Lupacciolu, Stout). *Let D be a bounded domain in $\mathbf{C}^n$ with boundary of class C^2. Let K be a compact subset of ∂D with $\Lambda_{2n-2}(K) = 0$. If $f \in \mathcal{C}(\partial D \setminus K)$, and $\int_{\partial D} f \varphi = 0$ for every $\bar\partial$-closed differential form $\varphi \in \mathcal{E}^{n,n-1}(\overline{D})$ with* $\operatorname{supp} \varphi \cap K = \varnothing$, *then f extends holomorphically into D.*

9 Removable singularities of CR-functions

9.1 Bounded CR-functions

In this section, $n > 1$.

Theorem 9.1 (Kytmanov). *Under the hypotheses of Theorems 8.1, 8.11, 8.14, and 8.15, if the CR-function $f \in \mathcal{L}^\infty(\Gamma)$, then its holomorphic extension F also is bounded, and*

$$\|f\|_{\mathcal{L}^\infty} = \sup_{D \setminus K} |F|.$$

Proof. Suppose $F(z^0) = 1$ for some point $z^0 \in D \setminus K$, yet $\|f\|_{\mathcal{L}^\infty} < 1$. Since the f^k are also CR-functions, and their holomorphic extensions are given by the functions F^k, we have by (8.6) that

$$1 = F^k(z^0) = \int_{\Gamma_s} f^k(\zeta)\, U(\zeta, z) - \int_{\partial \Gamma_s} f^k(\zeta)\, \Phi(\zeta).$$

The integral on the right-hand side tends to zero as $k \to \infty$ since $f^k(\zeta) \to 0$ as $k \to \infty$, and we can apply the Lebesgue dominated convergence theorem. □

Corollary 9.2. *Let D be a bounded domain in $\mathbf{C}^n$ with connected complement and piecewise-smooth boundary ∂D. Let $K \subset \partial D$ and $\partial D \setminus K = \Gamma \in \mathcal{C}^1$. Then every CR-function $f \in \mathcal{L}^\infty(\Gamma)$ extends to a CR-function $\tilde{f}$ on the whole boundary ∂D in each of the following cases: (a) the surface Γ is connected and $K = \widehat{K}_{\overline{D}}$, (b) $K = \widehat{K}_G$, where G is a domain of holomorphy containing $\overline{D}$.*

Proof. The proof follows from Theorem 9.1, since f extends holomorphically into D as a bounded function F. □

Corollary 9.3 (Analogue of Riemann's theorem). *Suppose $\overline{D}$ is a holomorphically convex compact set, $\partial D \in \mathcal{C}^1$, and $\mathcal{O}(\overline{D})$ is dense in the class of functions $\mathcal{A}(D)$. If $h \in \mathcal{A}(D)$, f is a CR-function on $\partial D \setminus K$, where $K = \{\, z \in \partial D : h(z) = 0 \,\}$, and $f \in \mathcal{L}^\infty(\partial D)$, then f is a CR-function on ∂D.*

Since $\Lambda_{2n-1}(K) = 0$, we may assume here that f is defined on the whole boundary ∂D.

Proof. Let $S = \{\, z \in \overline{D} : h(z) = 0 \,\}$. Since $\mathcal{O}(\overline{D})$ is dense in $\mathcal{A}(D)$, we have $\widehat{K}_{\overline{D}} = \widehat{K}_{\mathcal{A}(D)}$, so $\widehat{K}_{\overline{D}} = S$. In view of Theorem 9.1, the function f extends into $D \setminus S$ as a holomorphic function F that is bounded in $D \setminus S$. By Riemann's theorem, F extends holomorphically into D. Consequently, f is a CR-function on ∂D. □

This result is contained in [198] for the case when D is a strongly pseudoconvex domain; our generalization is given in [119, 121].

Corollary 9.4. *Under the hypotheses of Corollary 9.3, the zero set K of $h \in \mathcal{A}(D)$ does not separate ∂D.*

Proof. If $\partial D \setminus K$ is disconnected, we choose a function f that is equal to 1 on one component of $\partial D \setminus K$ and equal to 0 on the remaining components of $\partial D \setminus K$. Then f is a CR-function on $\partial D \setminus K$. By the previous corollary, it extends holomorphically into D as a function F, which is impossible. □

This result is contained in [198] for strongly pseudoconvex domains.

We now consider the case of arbitrary hypersurfaces Γ.

Corollary 9.5. *Suppose Γ is a smooth oriented hypersurface in $\mathbf{C}^n$, where $n \geq 3$, and Ω is a domain of holomorphy such that Γ divides Ω into two connected components Ω^+ and Ω^-. Suppose $z \in \Gamma$ and the envelopes of holomorphy of $\Omega^\pm$ contain a neighborhood W of z. Let K be a compact subset of $W \cap \Gamma$ such that either $K = \widehat{K}_\Omega$ or the hypotheses of Theorem 8.15 are satisfied. Then every CR-function f on $\Gamma \setminus K$ extends holomorphically into W.*

Corollary 9.5 was obtained in [121]. The hypotheses on Γ hold, for example, if the restriction of the Levi form of Γ to $T_z^c(\Gamma)$ has at least one pair of eigenvalues of opposite signs (that is, Γ is a 1-concave manifold). In this case, the compact set K can be arbitrary (we only need to require that $\Gamma \cap W \setminus K$ be connected). If we denote by K' the intersection of all domains containing K in which the tangential $\overline{\partial}$-problem (8.7) is solvable, then $K' \subset \Gamma$. Thus, Corollary 9.5 reduces in this case to a theorem of Henkin [73] for hypersurfaces.

If we impose some additional requirement on f, boundedness near K for example, then no hypothesis on Γ is required.

Theorem 9.6 (Kytmanov). *Let Ω be a domain of holomorphy in $\mathbf{C}^n$, where $n \geq 3$, and let Γ be a smooth oriented hypersurface in Ω with $K = \widehat{K}_\Omega \subset \Gamma$. If the CR-function $f \in \mathcal{L}^\infty(\Gamma \setminus K)$, then f extends to Γ as a CR-function $\tilde{f}$.*

Proof. We modify the construction given in Theorem 8.1. We suppose first that $f \in \mathcal{L}^p_{\text{loc}}(\Gamma)$, where $p \geq 1$. We take a sequence of smooth hypersurfaces Γ_s with smooth compact boundaries $\partial\Gamma_s$ such that $\Gamma_s \subset \Gamma$, $\bigcup_s \Gamma_s = \Gamma \setminus K$, and $f \in \mathcal{L}^p(\partial\Gamma_s)$. Since we are interested in the behavior of f in a neighborhood of K, we may assume, by shrinking Ω, that $f \in \mathcal{L}^p(\Gamma_s)$. Moreover, we have

$$\int_{\Gamma_s} f\,\overline{\partial}\alpha = \int_{\partial\Gamma_s} f\,\alpha \tag{9.1}$$

for every form α of type $(n, n-2)$ with coefficients in $\mathcal{D}(\Omega \setminus K)$ (see Lemma 8.2).

Consider the function

$$F_s^h(z) = \int_{\Gamma_s} f\,U(\zeta, z) - \int_{\partial\Gamma_s} f\,\Phi_h(\zeta), \qquad z \notin \Gamma, \tag{9.2}$$

where $h \in \mathcal{O}(\Omega)$, and the form Φ_h is defined in Lemma 8.3. This function is defined on the set $\Omega_s(h) = \{\, z : |h(z)| > \max_{\Gamma\setminus\Gamma_s} |h| \,\}$. The function F_s^h does not depend on s because

$$F_{s+1}^h - F_s^h = \int_{\Gamma_{s+1}\setminus\Gamma_s} f\,U(\zeta, z) - \int_{\partial\Gamma_{s+1}} f\,\Phi_h(\zeta) + \int_{\partial\Gamma_s} f\,\Phi_h(\zeta).$$

Since $z \in \Omega_s(h) \setminus \Gamma$, the singularities of Φ_h do not intersect $\Gamma_{s+1} \setminus \Gamma_s$. Using (9.1) and the equality $\overline{\partial}_\zeta \Phi_h = U(\zeta, z)$, we obtain that $F_{s+1}^h = F_s^h$. We denote F_s^h by F^h.

The function F^h does not depend on h. This means that for $z \in \Omega_s(h) \cap \Omega_s(h')$, the functions F^h and $F^{h'}$ are equal. Indeed, Lemma 8.4 shows that $\Phi_h - \Phi_{h'} = \overline{\partial}_\zeta \chi_{h,h'}$ for $n \geq 3$, and $\chi_{h,h'}$ is defined in $\Omega_s(h) \cap \Omega_s(h')$. Therefore it follows from (9.1) that

$$F^h(z) - F^{h'}(z) = -\int_{\partial\Gamma_s} f\,\overline{\partial}_\zeta \chi_{h,h'} = 0.$$

Consequently, (9.2) determines a function $F = F^h$ in $\Omega \setminus \Gamma$, since $K = \widehat{K}_\Omega$.

Now consider $\partial F/\partial \bar{z}_k$. The derivative $\partial U(\zeta,z)/\partial \bar{z}_k = \bar{\partial}_\zeta(\partial U_k/\partial \bar{z}_k)$, and the forms $\partial U_k/\partial \bar{z}_k$ have singularities only at $\zeta = z$, while $(\partial \Phi_h/\partial \bar{z}_k) = (\partial U_k/\partial \bar{z}_k) - \bar{\partial}_\zeta \Psi_h^k$ when $n \geq 3$, where Ψ_h^k has singularities on $L_z(h)$ (see Lemma 8.5). Then

$$\frac{\partial F}{\partial \bar{z}_k} = \int_{\Gamma_s} f\,\bar{\partial}_\zeta\left(\frac{\partial U_k}{\partial \bar{z}_k}\right) - \int_{\partial\Gamma_s} f\,\frac{\partial U_k}{\partial \bar{z}_k}, \qquad z \notin \Gamma.$$

We will show that $\partial F/\partial \bar{z}_k \in \mathcal{C}^\infty(\Omega)$. Indeed, suppose $z^0 \in \Gamma$, and take a ball $B = B(z^0, r)$. Then

$$\begin{aligned}\frac{\partial F}{\partial \bar{z}_k} &= \int_{\Gamma_s\setminus B} f\,\bar{\partial}_\zeta\left(\frac{\partial U_k}{\partial \bar{z}_k}\right) - \int_{\partial\Gamma_s} f\,\frac{\partial U_k}{\partial \bar{z}_k} + \int_{B\cap\Gamma_s} f\,\bar{\partial}_\zeta\left(\frac{\partial U_k}{\partial \bar{z}_k}\right)\\ &= \int_{\Gamma_s\setminus B} f\,\bar{\partial}_\zeta\left(\frac{\partial U_k}{\partial \bar{z}_k}\right) - \int_{\partial\Gamma_s} f\,\frac{\partial U_k}{\partial \bar{z}_k} + \int_{\partial B\cap\Gamma_s} f\,\frac{\partial U_k}{\partial \bar{z}_k}.\end{aligned}$$

Thus, $\partial F/\partial \bar{z}_k$ extends into B as a function of class $\mathcal{C}^\infty$. We denote $F_k = \partial F/\partial \bar{z}_k$. Then the form $\omega = \sum_{k=1}^n F_k\, d\bar{z}_k$ is $\bar{\partial}$-closed in Ω (in $\Omega\setminus\Gamma$ this holds by hypothesis, and in Ω by continuity). Since Ω is a domain of holomorphy, the form ω is $\bar{\partial}$-exact, that is, $\omega = \bar{\partial}\eta$, where $\eta \in \mathcal{C}^\infty(\Omega)$. Then the functions $h^\pm = F - \eta$ are solutions to our problem, since the $h^\pm$ are holomorphic in $\Omega^\pm$, and $h^+ - h^- = f$ on $\Gamma \setminus K$ by (9.2). Consequently, the boundary behavior of the functions $h^\pm$ is determined by the behavior of $F(z)$. Now if $f \in \mathcal{L}^\infty(\Gamma \setminus K)$, then Lemma 9.7 and Theorem 3.4 show that the difference $h^+ - h^-$ defines a CR-function on Γ. □

9.2 Integrable CR-functions

Suppose D is a bounded domain in $\mathbf{C}^n$ with class $\mathcal{C}^1$ boundary, $K \subset \partial D$, and the CR-function $f \in \mathcal{L}^1(\Gamma \setminus K)$. The difference of the two integrals in (8.6) gives a holomorphic extension of f into $D \setminus \widehat{K}_{\overline{D}}$. One of these integrals is the Bochner-Martinelli integral, and the other is

$$I_2(z) = \int_{\partial\Gamma_s} f(\zeta)\,\Phi_h(\zeta),$$

the form Φ_h being defined on the set $G(h) = \{\, z \in G : |h(z)| > \max_K |h| \,\}$, where $G \supset \overline{D}$ and $h \in \mathcal{O}(G)$. The integral I_2 has singularities on the set $\bigcup_{z\in\partial\Gamma_s} L_z(h)$, where $L_z(h)$ is a level set of h, that is, $L_z(h) = \{\, \zeta \in G : h(\zeta) = h(z) \,\}$. The dimension of this set may be rather large.

We now show that the integral $I_2(z)$ converges for almost all points of some neighborhood of ∂D. Let V_z be a right circular cone with center at $z \in \partial D$ and axis along the normal to ∂D at z, with the angle at the vertex being $\beta < \pi/2$. Take points $z^\pm \in V_z$ with $z^+ \in D \cap V_z$ and $z^- \in (\mathbf{C}^n \setminus \overline{D}) \cap V_z$.

Lemma 9.7. *Suppose that for every $z \in \partial\Gamma_s$, the level set $L_z(h)$ intersects ∂D transversely. Then $I_2(z)$ converges for almost all points in a neighborhood of ∂D,* *and*

$$\int_{\partial D} |I_2(z^{\pm})|\, d\sigma(z) \le C \int_{\partial\Gamma_s} |f|\, d\alpha, \tag{9.3}$$

where $z^{\pm} \in V_z$ are such that, for z in a neighborhood W of the set $\bigcup_{z\in\partial\Gamma_s} L_z(h) \cap \partial D$, the values $h(z^{\pm})$ and $h(z)$ are equal, and $|z^{\pm} - z| < \epsilon$.

Proof. Since I_2 is defined outside $W \cap \partial D$, we may replace the integration over ∂D by integration over $W \cap \partial D$; changing the order of integration, we obtain estimate (9.3) when ∂D is replaced by $\partial D \setminus W$.

We consider the integral

$$\int_{\partial\Gamma_s}\int_{W\cap\partial D} |f|\, d\sigma(z)\, |\Phi_h(\zeta)| \le C_1 \int_{\partial\Gamma_s} |f|\, d\alpha \int_{W\cap\partial D} \frac{|\zeta - z^+|^{3-2n}}{|h(\zeta) - h(z^+)|}\, d\sigma(z).$$

We introduce a new coordinate system $(u_1, \dots, u_n)$ in a neighborhood of $\zeta \in \partial\Gamma_s$ such that $u_1(z) = h(z)$. In this neighborhood, $C_2|u(\zeta) - u(z)| \le |\zeta - z| \le C_3|u(\zeta) - u(z)|$. Similar inequalities will hold for the measures $d\alpha(\zeta)$ and $d\alpha(u)$, $d\sigma(z)$ and $d\sigma(u)$.

We will show that the integral

$$\int_{B(\zeta,r)\cap\partial D} \frac{|\zeta - u^+|^{3-2n}}{|\zeta_1 - u_1^+|}\, d\sigma(u)$$

is bounded. We denote the point $u(\zeta)$ again by ζ, and the image of W by W. Furthermore $|\zeta_1 - u_1^+| = |\zeta_1 - u_1| \le C_4\epsilon$. We choose the ball $B(\zeta, r)$ so that $|\zeta - u^+|^2 \ge K(|w - \zeta|^2 + \epsilon^2)$, where w is the projection of u onto the tangent plane T_ζ (see the proof of Theorem 3.1), and $|\zeta_1 - u_1^+| = |\zeta_1 - u_1| \ge k|\zeta_1 - w_1|$ (in view of the smoothness of ∂D), where the constant k depends only on r. Then

$$\int_{B(\zeta,r)\cap\partial D} \frac{|\zeta - u^+|^{3-2n}}{|\zeta_1 - u_1^+|}\, d\sigma(u) \le C_5 \int_{T_\zeta\cap B(\zeta,r)} \frac{|\zeta - w|^{3-2n}}{|\zeta_1 - w_1|}\, d\sigma(w).$$

Let $|w - \zeta| = |x|$, and $\zeta_1 - w_1 = x_{2n-2} + ix_{2n-1}$, $x = (x_1, \dots, x_{2n-1}) \in \mathbf{R}^{2n-1}$ (here we also use the transversality of the plane $\{\zeta_1 = w_1\}$ and ∂D). Then

$$\int_{T_\zeta\cap B(\zeta,r)} \frac{|\zeta - w|^{3-2n}}{|\zeta_1 - w_1|}\, d\sigma(w) = \int_{\{|x|<r\}} \frac{|x|^{3-2n}}{(x_{2n-1}^2 + x_{2n-2}^2)^{1/2}}\, dx.$$

We pass to polar coordinates

$$x_1 = R\cos\varphi_1,$$

$$\vdots$$

$$x_{2n-2} = R\sin\varphi_1 \dots \sin\varphi_{2n-3}\cos\varphi_{2n-2},$$
$$x_{2n-1} = R\sin\varphi_1 \dots \sin\varphi_{2n-3}\sin\varphi_{2n-2},$$

and taking into account that the Jacobian of the transformation is

$$J = R^{2n-2} \sin^{2n-3} \varphi_1 \ldots \sin \varphi_{2n-3},$$

we have

$$\begin{aligned}\int_{\{|x|<r\}} \frac{|x|^{3-2n}}{(x_{2n-1}^2 + x_{2n-2}^2)^{1/2}}\, dx &= \int_0^r R^{2-2n}\, dR \int_{\{|x|=R\}} \frac{|J|}{|\sin \varphi_1 \ldots \sin \varphi_{2n-3}|}\, d\omega \\ &= r \int_{\{|x|=R\}} |\sin^{2n-4} \varphi_1 \ldots \sin \varphi_{2n-4}|\, d\omega \\ &= rC_6.\end{aligned}$$

The integral over the set $\partial D \setminus B(\zeta, r)$ is also easy to estimate, since $|\zeta - u^+|$ is bounded away from zero. Thus

$$\int_{\partial\Gamma_s} |f|\, d\alpha(\zeta) \int_{\partial D} |\Phi_h / d\alpha|\, d\sigma(z) \leq C_7 \int_{\partial\Gamma_s} |f|\, d\alpha.$$

By Fubini's theorem, $I_2(z)$ converges for almost all z in a neighborhood of ∂D, which proves the lemma. □

Theorem 9.8 (Kytmanov). *Let D be a bounded domain in $\mathbf{C}^n$, where $n \geq 2$, and suppose that ∂D is a Lyapunov surface. Suppose K is a compact subset of ∂D, and $f \in \mathcal{L}^1(\partial D \setminus K)$ is a CR-function such that*

$$\int_{\partial\Gamma_s} |f|\, d\alpha = O(1)$$

as $s \to \infty$. Then f extends to ∂D as a CR-function $\tilde{f} \in \mathcal{L}^1(\partial D)$ in each of the following cases: (a) $\partial D \setminus K$ is connected and $K = \widehat{K}_{\overline{D}}$; (b) $K = \widehat{K}_G$, where G is a domain of holomorphy containing $\overline{D}$.

Proof. It is not hard to obtain from Theorems 8.1, 8.14, and 3.4 and Lemma 9.7 that the holomorphic extension of f into D is a function of class $\mathcal{H}^1(D)$. The conclusion of the theorem follows. □

9.3 Further results

First of all, we mention an analogue of Radó's theorem for CR-functions which was obtained by Rosay and Stout [174].

Theorem 9.9 (Rosay, Stout). *Let Γ be a hypersurface of class C^2 in an open set $\Omega \subset \mathbf{C}^n$, and suppose that the Levi form of Γ is nondegenerate on Γ. Then every function $f \in C(\Gamma)$ that is a CR-function outside its zero set K is a CR-function on Γ.*

Proof. The scheme of the proof is the following. First consider the case $n = 2$. Since the statement is local, we may assume that Γ bounds a strongly pseudoconvex domain D in $\mathbf{C}^2$. By Theorem 8.14, the function f extends to a holomorphic function F in $D \setminus \widehat{K}$. Using a result of Słodkowski [188], the authors show that F can be extended to a continuous function on $\overline{D}$ that equals zero on $\widehat{K}$. Applying Radó's theorem for holomorphic functions gives the result. The case $n > 2$ is reduced to the above case by using sections of Ω by suitable two-dimensional complex planes. □

Jöricke [89, 90] has considered a generalization of Hartogs's theorem in the following sense: let M be a closed set in the domain D; what conditions should be imposed on $M \cap \partial D$ so that every function holomorphic in $D \setminus M$ extends holomorphically to D?

10 Analogue of Riemann's theorem for CR-functions

10.1 Statement of the problem and results

As in the previous section, we assume that the dimension of $\mathbf{C}^n$ is strictly greater than 1.

We have already seen that an analogue of Riemann's theorem holds for bounded CR-functions. Here we will show that it holds in some cases for integrable functions. We start with the following.

Lemma 10.1. *Let Γ be a smooth oriented hypersurface in an open set $\Omega \subset \mathbf{C}^n$. Suppose $K = \{ z \in \Gamma : h(z) = 0 \}$, where h is a CR-function on Γ that satisfies a Hölder condition with exponent $\alpha > 0$ in a neighborhood of K, and suppose $f \in \mathcal{L}^1_{\mathrm{loc}}(\Gamma)$ is a CR-function on $\Gamma \setminus K$. Then there is a positive integer n_0 such that fh^{n_0} is a CR-function on Γ.*

Proof. If $K_d = \{ z \in \Omega : \inf_{\zeta \in K} |\zeta - z| \le d \}$, then

$$\sup_{K_d \cap \Gamma} |h| = \sup_{\substack{z \in K_d \cap \Gamma \\ \zeta \in K}} |h(z) - h(\zeta)| \le C d^{\alpha}.$$

Therefore $\sup_{K_d \cap \Gamma} |h^{n_0}| \le C_1 d^{1+\epsilon}$ for some n_0 and for $\epsilon > 0$. If φ is a differential form in $\mathcal{D}^{n,n-2}(\Omega)$, then

$$\int_\Gamma f h^{n_0}\, \overline{\partial}\varphi = \int_\Gamma f h^{n_0}\, \overline{\partial}\, [(1 - \psi_d)\varphi] + \int_\Gamma f h^{n_0}\, \overline{\partial}(\psi_d \varphi),$$

where $\psi_d \in \mathcal{D}(\Omega)$ equals 1 on $K_{d/3} \cap \operatorname{supp}\varphi$ and equals zero outside $K_{2d/3}$ (we may take as ψ_d the convolution of the characteristic function of a neighborhood of $K \cap \operatorname{supp}\varphi$ with the standard kernel). Then

$$\left| \frac{\partial}{\partial z_j} \psi_d \right| \le \frac{C_2}{d}, \qquad j = 1, \dots, n,$$

so

$$\left|\int_\Gamma f h^{n_0}\,\overline{\partial}(\psi_d\varphi)\right| \le C_3 d^\epsilon \int_{\Gamma\cap\operatorname{supp}\varphi} |f|\,d\sigma \to 0$$

as $d \to 0$. The support of $(1-\psi_d)\varphi$ does not intersect K, and h can be uniformly approximated by polynomials P_k in a neighborhood of each point of Γ (see Corollary 6.6). Therefore

$$\begin{aligned}\int_\Gamma f h^{n_0}\,\overline{\partial}\,[(1-\psi_d)\varphi] &= \lim_{k\to\infty}\int_\Gamma f P_k^{n_0}\,\overline{\partial}\,[(1-\psi_d)\varphi]\\ &= \lim_{k\to\infty}\int_\Gamma f\,\overline{\partial}[P_k^{n_0}(1-\psi_d)\varphi] = 0.\end{aligned}$$

□

This lemma was given in [116, 118].

Remark. If $d(z)$ denotes the distance from the point z to the compact set K, then Lemma 10.1 will hold for those CR-functions $f \in \mathcal{L}^1_{\text{loc}}(\Gamma\setminus K)$ for which

$$\int_{W(K)\cap\Gamma} |f(z)|d^{k_0}(z)\,d\sigma < +\infty$$

for a fixed k_0, where $W(K)$ is a neighborhood of K.

Theorem 10.2 (Kytmanov). *If D is a bounded domain in $\mathbf{C}^n$ with class C^1 boundary, $f \in \mathcal{L}^\infty(\partial D\setminus K)$ is a CR-function, where $K = \{\,z\in\partial D : h(z)=0\,\}$, and $h \in C^\alpha(\overline{D})$ is holomorphic in D, where $0<\alpha\le 1$, then f is a CR-function on ∂D.*

Proof. We may assume that $f\in\mathcal{L}^\infty(\partial D)$, since K has zero $(2n-1)$-dimensional Hausdorff measure. Now fh^{n_0} is a bounded CR-function on ∂D for some n_0 by Lemma 10.1. Moreover, fh^{n_0} is the boundary value of a function $F\in\mathcal{O}(D)$, by Theorem 7.1, where F is given by the Bochner-Martinelli integral. Therefore $\sup_D|F|\le\|f\|_{\mathcal{L}^\infty}$, that is, $F\in\mathcal{H}^\infty(D)$. The inequality $|fh^{n_0}|\le C|h^{n_0}|$ holds on ∂D, so the same inequality holds in D (see [175, Theorem 14.3.3]; although this assertion is given there only for $F\in\mathcal{A}(D)$, it is easy to carry over to the case $F\in\mathcal{H}^\infty(D)$). Hence $|F|\le C|h^{n_0}|$ in D. Therefore $Fh^{-n_0}\in\mathcal{H}^\infty(D)$. □

Theorem 10.2 was proved in [116, 118]. No additional hypothesis on D is required, as in Corollary 9.3, but the function h must satisfy a Hölder condition.

If, in the situation of Theorem 10.2, the CR-function $f\in\mathcal{L}^1(\partial D\setminus K)$, then it might not be a CR-function on ∂D. For example, suppose $f=1/z_1$, $D=B(0,1)$, and $K=\{\,z\in S(0,1) : z_1=0\,\}$. Then f is a CR-function on $S(0,1)\setminus K$, and we can show that $f \in \mathcal{L}^1(S(0,1))$. Indeed,

$$\begin{aligned}\int_{S(0,1)} |z_1|^{-1}\, d\sigma &\le C \int_{|z_1|\le 1} |z_1|^{-1}\, dx_1 \wedge dy_1 \int_{|z_2|^2+\cdots+|z_n|^2=1-|z_1|^2} d\sigma' \\ &\le C_1 \int_{|z_1|\le 1} |z_1|^{-1}(1-|z_1|^2)^{(2n-3)/2}\, dx_1 \wedge dy_1 \\ &= C_1 2\pi \int_0^1 (1-r^2)^{(2n-3)/2}\, dr < \infty.\end{aligned}$$

But f is not a CR-function on $S(0,1)$, for otherwise it would extend by the Hartogs-Bochner theorem to $B(0,1)$ as a function $F \in \mathcal{H}^1(D)$, and so by the uniqueness theorem would coincide with $1/z_1$ in $B(0,1)$.

However, if we replace the zero set K in Theorem 10.2 by a peak set, then the theorem remains true for functions of class $\mathcal{L}^1(\partial D)$.

Theorem 10.3 (Kytmanov). *Let D be a bounded domain in $\mathbf{C}^n$ such that ∂D is a Lyapunov surface. Suppose that $K \subset \partial D$ is a peak set for the class of holomorphic functions in D of class $\mathcal{C}^\alpha(\overline{D})$ (that is, $K = \{z \in \partial D : \psi(z) = 1\}$, where ψ is a holomorphic function in D of class $\mathcal{C}^\alpha(\overline{D})$, and $|\psi| < 1$ on $\partial D \setminus K$). If $f \in \mathcal{L}^1(\partial D)$, and f is a CR-function on $\partial D \setminus K$, then f is a CR-function on ∂D.*

Theorem 10.3 was proved in [118].

10.2 Auxiliary results

We recall some results from [75, 194]. The Green function $G(\zeta, z)$ for a domain D has the form $G(\zeta,z) = g(\zeta,z) + h(\zeta,z)$, where $g(\zeta,z)$ is the fundamental solution of Laplace's equation (see §1), and $h(\zeta,z)$ is a harmonic function in z for fixed $\zeta \in \overline{D}$. Moreover, for each $z \in D$ the function $h(\zeta,z)$ is of class $\mathcal{C}^{1+\beta}(\overline{D})$, where $0 < \beta \le 1$. Also, $h(\zeta,z) = h(z,\zeta)$, and $G(\zeta,z) = 0$ if $\zeta \in \partial D$ and $z \in D$. The function G differs in sign from the classical Green function.

If, as usual, $\nu(\zeta)$ is the outer unit normal to ∂D at ζ, then the Poisson kernel is

$$P(\zeta,z) = \frac{\partial G(\zeta,z)}{\partial \nu(\zeta)}, \qquad z \in D,$$

and $P(\zeta,z) \ge c > 0$ for fixed $z \in D$. For every function $u \in \mathcal{C}(\overline{D})$ that is harmonic in D, we have

$$\int_{\partial D} u(\zeta)P(\zeta,z)\, d\sigma(\zeta) = u(z), \qquad z \in D.$$

For a fixed z^0, we may take $G(\zeta, z^0)$ as a defining function ρ for the domain D.

In §1, we defined $\mathcal{H}^p(D)$ for $p > 0$ as the class of holomorphic functions f in D for which

$$\sup_{\epsilon>0} \int_{\partial D_\epsilon} |f|^p \, d\sigma < +\infty.$$

As remarked in [194], this definition does not depend on the choice of ρ.

If we set $\rho(\zeta) = G(\zeta, z^0)$, then it is easy to see that

$$\int_{\partial D_\epsilon} u(\zeta) P(\zeta, z^0) \, d\sigma = u(z^0) \tag{10.1}$$

for functions u that are harmonic in D (here $D_\epsilon = \{ z \in D : \rho(z) < -\epsilon \}$). Indeed, if $u \in \mathcal{C}(\overline{D})$, then

$$u(z^0) = \int_{\partial D} u(\zeta) P(\zeta, z^0).$$

Applying the classical Green formula to the domain $D \setminus \overline{D}_\epsilon$, we have

$$\int_{\partial(D\setminus D_\epsilon)} u(\zeta) \frac{\partial G(\zeta, z^0)}{\partial \nu(\zeta)} \, d\sigma = \int_{\partial(D\setminus D_\epsilon)} G(\zeta, z^0) \frac{\partial u(\zeta)}{\partial \nu(\zeta)} \, d\sigma = \epsilon \int_{\partial D_\epsilon} \frac{\partial u(\zeta)}{\partial \nu(\zeta)} \, d\sigma = 0,$$

that is,

$$\int_{\partial D} u(\zeta) P(\zeta, z^0) \, d\sigma = \int_{\partial D_\epsilon} u(\zeta) P(\zeta, z^0) \, d\sigma.$$

By passing to the limit, we obtain (10.1) for an arbitrary harmonic function u in D.

Suppose $f \in \mathcal{L}^p(\partial D)$, where $p > 1$. We write f^* for the maximal function

$$f^*(z) = \sup_{\epsilon>0} \frac{1}{\Lambda_{2n-1}(B(z,\epsilon) \cap \partial D)} \int_{B(z,\epsilon)\cap\partial D} |f(\zeta)| \, d\sigma.$$

As shown in [194, p. 10], for $p > 1$, the function f^* also lies in $\mathcal{L}^p(\partial D)$, and the Poisson integral u of f (that is, $u(z) = \int_{\partial D} f(\zeta) P(\zeta, z) \, d\sigma(\zeta)$ for $z \in D$) satisfies the inequality $|u(z - \epsilon\nu(z))| \le A f^*(z)$ for all $z \in \partial D$ and $\epsilon > 0$.

Lemma 10.4. *If $h \in \mathcal{O}(\partial D)$, and $\operatorname{Re} h > 0$ in D, then $h \in \mathcal{H}^p(D)$ for all p such that $0 < p < 1$.*

Proof. Let $h = |h| e^{i\alpha}$, where $|\alpha| < \pi/2$. Then h^p is holomorphic in D for every $p > 0$, since $\operatorname{Re} h > 0$, and $h^p = |h|^p (\cos p\alpha + i \sin p\alpha)$. But $p < 1$, so

$$|h|^p \le C_p \operatorname{Re} h^p, \qquad C_p = (\cos \pi p/2)^{-1}.$$

Choosing $G(\zeta, z^0)$ as the defining function $\rho(\zeta)$, and using (10.1), we have

$$\int_{\partial D_\epsilon} |h(\zeta)|^p P(\zeta, z^0) \, d\sigma \le C_p \int_{\partial D_\epsilon} \operatorname{Re} h^p P(\zeta, z^0) \, d\sigma = C_p \operatorname{Re} h^p(z^0) \le C_p |h^p(z^0)|,$$

that is, $h \in \mathcal{H}^p(D)$ for $p < 1$. □

Lemma 10.4 is proved the same way as for $n = 1$.

10.3 Analogue of Smirnov's theorem

Lemma 10.5. *If $f \in \mathcal{H}^p(D)$, and its normal boundary values lie in class $\mathcal{L}^q(\partial D)$, where $q > p$, then $f \in \mathcal{H}^q(D)$.*

Proof. Suppose $f \in \mathcal{H}^p(D)$, and $u(z) = |f(z)|^{p/2}$. Then

$$\sup_{\epsilon>0} \int_{\partial D_\epsilon} u^2 \, d\sigma < +\infty.$$

Consequently (see [194, pp. 8–9]), the subharmonic function u has a harmonic majorant v in D whose boundary values $\tilde{v}$ lie in class $\mathcal{L}^2(\partial D)$, and

$$v(z) = \int_{\partial D} \tilde{v}(\zeta) P(\zeta, z) \, d\sigma, \qquad z \in D$$

(see [75, p. 16]). If $\tilde{v}^*$ is the maximal function of $\tilde{v}$, then $\tilde{v}^* \in \mathcal{L}^2(\partial D)$, and

$$u(\zeta - \epsilon\nu(\zeta)) \leq v(\zeta - \epsilon\nu(\zeta)) \leq A\tilde{v}^*(\zeta) \tag{10.2}$$

for all $\zeta \in \partial D$ and $\epsilon > 0$.

If we set $\rho(\zeta) = G(\zeta, z)$, then

$$u(z) \leq \int_{\partial D_\epsilon} u(\zeta) P(\zeta, z) \, d\sigma$$

in view of (10.1). By (10.2) and Lebesgue's dominated convergence theorem,

$$\lim_{\epsilon \to 0+} \int_{\partial D_\epsilon} u(\zeta) P(\zeta, z) \, d\sigma = \int_{\partial D} u(\zeta) P(\zeta, z) \, d\sigma.$$

Consequently, $u(z) \leq \int_{\partial D} u(\zeta) P(\zeta, z) \, d\sigma$.

Since $P(\zeta, z)\, d\sigma$ is a probability measure on ∂D for each $z \in D$, and $2q/p > 1$, by using Jensen's inequality (see, for example, [71, §2.2]) we have

$$|f(z)|^q = u(z)^{2q/p} \leq \left(\int_{\partial D} |f|^{p/2} P(\zeta, z) \, d\sigma \right)^{2q/p} \leq \int_{\partial D} |f|^q P(\zeta, z) \, d\sigma. \tag{10.3}$$

Since $|f|^q \in \mathcal{L}^1(\partial D)$, the function $s(z) = \int_{\partial D} |f|^q P(\zeta, z) \, d\sigma$ satisfies the condition $\sup_{\epsilon>0} \int_{\partial D_\epsilon} s(\zeta) \, d\sigma < +\infty$ (see [194, p. 5]). Therefore we obtain from (10.3) that

$$\sup_{\epsilon>0} \int_{\partial D_\epsilon} |f|^q \, d\sigma \leq \sup_{\epsilon>0} \int_{\partial D_\epsilon} s(\zeta) \, d\sigma < +\infty,$$

that is, $f \in \mathcal{H}^q(D)$. □

10.4 Proof of the main result

Let $K = \{ z \in \overline{D} : \psi(z) = 1 \}$, $\psi \in \mathcal{C}^\alpha(\overline{D}) \cap \mathcal{O}(D)$, and $|\psi| < 1$ on $\overline{D} \setminus K$. We denote the function $1 - \psi$ by h. Then $h = 0$ on K and $h \neq 0$ on $\overline{D} \setminus K$, and moreover $\operatorname{Re} h = 1 - \operatorname{Re} \psi \geq 1 - |\psi| > 0$ in D. Now $fh^{n_0} \in \mathcal{L}^1(\partial D)$ is a CR-function on ∂D by Lemma 10.1, and fh^{n_0} is the boundary value of a function $g \in \mathcal{H}^1(D)$ by Theorem 7.1. Consider the function $F = gh^{-n_0}$, which is in $\mathcal{O}(D)$.

Lemma 10.6. *The function $F \in \mathcal{H}^1(D)$.*

Proof. We first prove that $F \in \mathcal{H}^p(D)$ for some $p > 0$. The function $g \in \mathcal{H}^1(D)$, and $\operatorname{Re} 1/h = \operatorname{Re} \bar{h}/|h|^2 > 0$, so $1/h \in \mathcal{H}^q(D)$ for all $q < 1$ (by Lemma 10.4). Then

$$\int_{\partial D_\epsilon} |F|^p \, d\sigma = \int_{\partial D_\epsilon} |gh^{-n_0}|^p \, d\sigma \leq \left(\int_{\partial D_\epsilon} |g|^{pp'} \, d\sigma \right)^{1/p'} \left(\int_{\partial D_\epsilon} |h|^{-pn_0 q'} \, d\sigma \right)^{1/q'},$$

where $1/p' + 1/q' = 1$. We choose $p' = 1/p > 1$, so that $q' = 1/(1-p)$, and moreover we choose p so that $pn_0 q' = pn_0/(1-p) < 1$, whence $p < 1/(n_0 + 1)$. Applying Lemma 10.5 (on ∂D we have $F = gh^{-n_0} = f \in \mathcal{L}^1(\partial D)$), we obtain that $F \in \mathcal{H}^1(D)$, that is, f is a CR-function on ∂D. □

Theorem 10.7 (Kytmanov). *Let D be a bounded domain in $\mathbf{C}^n$ such that ∂D is a surface of class $\mathcal{C}^\infty$. Suppose $K \subset \partial D$ is a compact set that is a peak set for functions in $\mathcal{O}(D) \cap \mathcal{C}^\alpha(\overline{D})$, where $\alpha > 0$. If $f \in \mathcal{L}^1_{\mathrm{loc}}(\partial D \setminus K)$ is a CR-function, and there is a constant $m > 0$ such that*

$$\int_{\partial D} |f(z)| d^m(z) \, d\sigma < +\infty,$$

where $d(z)$ is the distance from z to K, then there is a CR-distribution $\tilde{f}$ on ∂D such that $\tilde{f} \,|_{\partial D \setminus K} = f$ in $\mathcal{E}'(\partial D)$.

Proof. According to Lemma 10.1 and the remark following it, there is a constant $n_0 > 0$ such that $fh^{n_0} \in \mathcal{L}^1(\partial D)$, and $g = fh^{n_0}$ is a CR-function on ∂D ($h = 1 - \psi$, and ψ is a peak function for K). Then $g \in \mathcal{H}^1(D)$, and the function $F = gh^{-n_0}$ is holomorphic in D. The proof of Lemma 10.6 shows that $F \in \mathcal{H}^p(D)$ for some $p > 0$. Then the function $|F|^p$ is subharmonic and has a harmonic majorant u in D (see [75, §1]) of the form

$$u(z) = \int_{\partial D} P(\zeta, z) \, d\mu(\zeta), \qquad z \in D,$$

where μ is a positive Borel measure. Therefore $u(z) \leq C(-\rho(z))^{-k}$, where ρ is a defining function for D and $k > 0$. Consequently

$$|F(z)| \leq (u(z))^{1/p} \leq C^{1/p}(-\rho(z))^{-k/p},$$

that is, the holomorphic function F has finite order of growth at ∂D. In Chapter 3 we will show that there then exists a boundary value $\tilde{f}$ of F on ∂D in the sense of generalized functions. Since F is holomorphic in D, this $\tilde{f}$ will be a CR-distribution, and $\tilde{f} = f$ on $\partial D \setminus K$. □

10.5 Further results

In connection with Theorem 10.3, we mention a result of Lee and Wermer [134] on describing measures orthogonal to the space $\mathcal{R}(K)$. Let K be a rationally convex compact set in the boundary $S(0,1)$ of the ball $B(0,1)$ in $\mathbf{C}^2$. For every measure μ orthogonal to $\mathcal{R}(K)$, there exists a function $f \in \mathcal{L}^1(S(0,1))$ that extends holomorphically into $B(0,1)$ and has the property that for each $\varphi \in \mathcal{D}(\mathbf{C}^2)$,

$$\int \varphi\, d\mu = \int_{K^+} f\,\bar{\partial}\varphi \wedge d\zeta - \lim_{r\to 1^-} \int_{K_r^+} f\,\bar{\partial}\varphi \wedge d\zeta,$$

where K^+ is a neighborhood of K in $S(0,1)$ and $K_r^+ = \{\, z : z/r \in K^+ \,\}, 0 < r < 1$. From the construction of f, it is clear that f is a CR-function on $\partial D \setminus K$. Therefore if the singularity of f is removable, then $\mathcal{R}(K) = \mathcal{C}(K)$.

A local form of Theorem 10.3 is also given in [116, 118] for an arbitrary CR-manifold. We give it here in the case of a hypersurface.

Theorem 10.8 (Kytmanov). *Let Γ be a smooth oriented hypersurface in a domain $\Omega \subset \mathbf{C}^n$, and suppose that the Levi form of Γ is nondegenerate on $T_\zeta^c(\Gamma)$ for every $\zeta \in \Gamma$. Suppose $f \in \mathcal{L}^1(\Gamma)$ is a CR-function on $\Gamma \setminus K$, where $K = \{\, z \in \Gamma : \psi(z) = 1 \,\}$; ψ is a CR-function on Γ of class $\mathcal{C}^\alpha(\Gamma)$, where $0 < \alpha \le 1$; and $|\psi| < 1$ on $\Gamma \setminus K$. Then f is a CR-function on Γ.*

Proof. The outline of the proof is the same as in Theorem 10.3. The Levi form of Γ is nondegenerate, so for every point $\zeta \in \Gamma$ every CR-function on Γ given in a neighborhood of this point extends holomorphically into some domain D abutting Γ. In particular, the function $h = 1 - \psi$ extends holomorphically. In view of the local maximum principle, $h \ne 0$ in D. Moreover, fh^{n_0} is a CR-function on Γ by Lemma 10.1, that is, fh^{n_0} also extends holomorphically into D as a function g. Then the function $F = gh^{-n_0}$ is holomorphic in D, and its boundary value agrees with f. We need only show that the weak limit of F agrees with gh^{-n_0}, whence it will follow that $gh^{-n_0} = f$ is a CR-function on Γ. To find the weak limit of F, we construct a family of analytic discs Δ_ϵ shrinking toward ζ such that in each Δ_ϵ the function gh^{-n_0} extends holomorphically to a function of class $\mathcal{H}^1(\Delta_\epsilon)$ (here we apply Lemma 10.6 for $n = 1$). Applying Fubini's theorem, we obtain the required property. □

The hypothesis of nondegeneracy of the Levi form was replaced in [125] by the hypothesis of minimality of the CR-manifold Γ.

Chapter 3

Distributions Given on a Hypersurface

11 Harmonic representation of distributions

11.1 Statement of the problem

In this chapter, we will consider functions and distributions principally in $\mathbf{R}^n$. The problems to which the chapter is devoted are posed most naturally in $\mathbf{R}^n$, although the method of solution remains essentially the same as in the previous chapters.

We will denote points of $\mathbf{R}^n$ by the letters $x = (x_1, \ldots, x_n)$, $y = (y_1, \ldots, y_n)$, and so on. The scalar product of vectors x and y in $\mathbf{R}^n$ is $\langle x, y\rangle = x_1y_1 + \cdots + x_ny_n$, and $|x|^2 = \langle x, x\rangle$.

Let Ω be an arbitrary domain in $\mathbf{R}^n$, and let Γ be a connected, smooth (class $\mathcal{C}^\infty$), relatively closed, orientable hypersurface in Ω. If ρ is a defining function for Γ (that is, $\Gamma = \{x \in \Omega : \rho(x) = 0\}$, $\rho \in \mathcal{C}^\infty(\Omega)$, $d\rho \neq 0$ on Γ), then $\Omega^+ = \{x \in \Omega : \rho(x) > 0\}$, and $\Omega^- = \{x \in \Omega : \rho(x) < 0\}$. The unit normal vector $\nu(x) = \operatorname{grad}\rho/|\operatorname{grad}\rho|$, where, as usual, $\operatorname{grad}\rho = (\partial\rho/\partial x_1, \ldots, \partial\rho/\partial x_n)$. The orientation of Γ is chosen to be compatible with Ω^+.

We consider the space of distributions $\mathcal{D}'(\Gamma)$. The goal of this section is to solve the jump problem for distributions, that is, to represent each distribution $S \in \mathcal{D}'(\Gamma)$ as the difference of boundary values of functions that are harmonic in $\Omega^\pm$. (We will call such a representation a harmonic representation). In §3, we saw that the Bochner-Martinelli integral solves this problem in $\mathbf{C}^n$. Here we will first describe the class of harmonic functions in $\Omega^\pm$ whose boundary values are distributions, and we will somewhat modify the construction of a harmonic representation, having subsequent applications in view.

We denote by $\mathcal{G}(\Omega^+)$ the space of harmonic functions in Ω^+ of finite order of growth near Γ; that is, $f \in \mathcal{G}(\Omega^+)$ if $\Delta f = 0$ and for every ball $B(x^0, r)$.

where $x^0 \in \Gamma$, there exist $C > 0$ and $m > 0$ such that $|f(x)| \leq C\rho^{-m}(x)$ for all $x \in B(x^0, r) \cap \Omega^+$.

The definition does not depend on the choice of ρ, since if ρ_1 is also a defining function for Γ, then ρ/ρ_1 and ρ_1/ρ are class $\mathcal{C}^\infty$ functions in Ω, while $\rho/\rho_1 \neq 0$ and $\rho_1/\rho \neq 0$ (this is easy to show by passing to local coordinates). Therefore, if we take as $\rho(x)$ the distance function $d(x) = \inf_{y\in\Gamma} |x - y|$, then in some neighborhood W of Γ the function $d \in \mathcal{C}^\infty(W \cap \overline{\Omega^+})$, and it is a defining function. Consequently, the class $\mathcal{G}(\Omega^+)$ may be defined as follows: $f \in \mathcal{G}(\Omega^+)$ if for every ball $B(x^0, r)$, where $x^0 \in \Gamma$, there are constants $C > 0$ and $m > 0$ such that

$$|f(x)| \leq Cd^{-m}(x) \qquad \text{for } x \in B(x^0, r) \cap \Omega^+.$$

The class $\mathcal{G}(\Omega^-)$ is defined analogously.

If $f \in \mathcal{C}^\infty(\Omega^+)$, then we will say that the boundary value of f on Γ is a distribution f_0 if for every $\varphi \in \mathcal{D}(\Gamma)$ the limit

$$\lim_{\epsilon\to 0+} \int_\Gamma f(x + \epsilon\nu(x))\varphi(x)\, d\sigma \tag{11.1}$$

exists. If the limit (11.1) exists for each $\varphi \in \mathcal{D}(\Gamma)$, then in view of the weak completeness of $\mathcal{D}'(\Gamma)$, it does define a distribution in $\mathcal{D}'(\Gamma)$, which we denote f_0. We remark that the integral in (11.1) makes sense for sufficiently small positive ϵ since φ has compact support, and so the point $(x + \epsilon\nu(x)) \in \Omega^+$ if $x \in \operatorname{supp}\varphi$ and ϵ is sufficiently small.

If $\varphi \in \mathcal{D}(\Omega)$ and the boundary value of a function f in $\mathcal{C}^\infty(\Omega^+)$ is a distribution f_0, then

$$\lim_{\epsilon\to 0+} \int_\Gamma f(x + \epsilon\nu(x))\varphi(x + \epsilon\nu(x))\, d\sigma = f_0(\varphi) = \langle f_0, \varphi\rangle,$$

where $f_0(\varphi) = \langle f_0, \varphi\rangle$ is the value of f_0 on the restriction of φ to Γ. Indeed, passing to local coordinates, we obtain that $\varphi(x + \epsilon\nu(x))$ converges to $\varphi(x)$ on Γ in the topology of the space $\mathcal{D}(\Gamma)$, so our assertion follows from a well-known property of distributions (see, for example, [184, p. 70]).

Consider the domain $\Omega^+_\epsilon = \{\, x \in \Omega^+ : d(x) > \epsilon \,\}$. If the boundary value of $f \in \mathcal{C}^\infty(\Omega^+)$ is a distribution f_0, then in view of the previous argument,

$$\lim_{\epsilon\to 0+} \int_{\partial\Omega^+_\epsilon \setminus \partial\Omega} f(x)\varphi(x)\, d\sigma_\epsilon = f_0(\varphi) \qquad \text{for every } \varphi \in \mathcal{D}(\Omega)$$

(since $d\sigma_\epsilon/d\sigma$ is a function that converges to 1 in the topology of $\mathcal{D}(\Gamma)$).

Subsequently, we will use all of these definitions of the boundary values of $f \in \mathcal{C}^\infty(\Omega^+)$, and we will take as $\Omega^\pm_\epsilon$ the domain $\{\, x \in \Omega^\pm : d(x) > \epsilon \,\}$ for $\epsilon > 0$; that is, we will assume that the defining function $\rho(x)$ is

$$\begin{cases} d(x), & x \in \Omega^+, \\ -d(x), & x \in \Omega^-. \end{cases}$$

Then, as we mentioned, $\rho \in \mathcal{C}^\infty(W)$ in some neighborhood W of Γ.

11.2 Boundary values of harmonic functions of finite order of growth

Theorem 11.1. *If $f \in \mathcal{G}(\Omega^+)$, then the boundary value of f on Γ is a distribution f_0. If $f_0 = 0$ in some open set $U \subset \Gamma$, then f extends to U as a function of class $C^\infty(\Omega^+ \cup U)$, and $f = 0$ on U.*

Proof. Let $x^0 \in \Gamma$. We may assume that $x^0 = 0$ and that the normal $\nu(x^0) = (1, 0, \dots, 0)$. This can always be achieved by a unitary transformation and a translation (this transformation does not change the Laplace operator). In some ball $B(x^0, r)$, the surface Γ is the graph of a function $x_1 = h(x_2, \dots, x_n)$ with $h \in C^\infty(W_1)$, where W_1 is an open set in $\mathbf{R}^{n-1}$. Since $f \in \mathcal{G}(\Omega^+)$, there are constants $C > 0$ and $m > 0$ such that

$$|f(x)| \le C(x_1 - h(x_2, \dots, x_n))^{-m}, \qquad x \in B(x^0, r) \cap \Omega^+.$$

We choose $a \in \mathbf{R}$ such that the set $\{(a, x_2, \dots, x_n) : (x_2, \dots, x_n) \in W_1\}$ is relatively compact in $B(x^0, r) \cap \Omega^+$ (shrinking W_1 if necessary). Consider the function

$$F(x_1, \dots, x_n) = \psi(x_2, \dots, x_n) + \int_a^{x_1} f(\xi, x_2, \dots, x_n)\, d\xi.$$

We will select $\psi \in C^\infty(W_1)$ so that F will be harmonic on the set $\{(x_1, \dots, x_n) \in \Omega^+ : (x_2, \dots, x_n) \in W_1\}$. The Laplace operator $\Delta = \sum_{j=1}^n \partial^2/\partial x_j^2$ (in this chapter we use the classical definition of the Laplace operator), so

$$\Delta F = \widetilde{\Delta}\psi + \frac{\partial f}{\partial x_1}(x_1 \dots, x_n) + \int_a^{x_1} \widetilde{\Delta} f(\xi, x_2, \dots, x_n)\, d\xi,$$

where $\widetilde{\Delta} = \sum_{j=2}^n \partial^2/\partial x_j^2$. Since $\widetilde{\Delta} f(\xi, x_2, \dots, x_n) = -(\partial^2/\partial\xi^2) f(\xi, x_2, \dots, x_n)$, we have $\Delta F = \widetilde{\Delta}\psi + (\partial f/\partial x_1)(a, x_2, \dots, x_n)$. We obtain what was required by solving Poisson's equation $\widetilde{\Delta}\psi = -(\partial f/\partial x_1)(a, x_2, \dots, x_n)$ on W_1. We remark that the function $\psi(x_2, \dots, x_n)$ has no singularities on the set $\{(x_1, \dots, x_n) \in \Omega : (x_2, \dots, x_n) \in W_1\}$.

Thus, F is a harmonic function in a one-sided neighborhood of x^0. Hence

$$|F(x)| \le C_1 \int_a^{x_1} (\xi - h(x_2, \dots, x_n))^{-m}\, d\xi + C_2 \le C_3(x_1 - h(x_2, \dots, x_n))^{1-m}.$$

Applying this construction repeatedly, we find a harmonic function F_1 for which $\partial^s F_1/\partial x_1^s = f$, and $|F_1(x)| \le C\big|\ln|x_1 - h(x_2, \dots, x_n)|\big|$. Since $\ln t$ is integrable at $t = 0$, the subsequent primitive F_2 of F_1 will be bounded near Γ, and the next primitive will extend continuously to Γ. We have proved the following.

Lemma 11.2. *If $f \in \mathcal{G}(\Omega^+)$, then for every $x^0 \in \Gamma$ there is a harmonic function F in $B(x^0, r) \cap \Omega^+$ that extends continuously to $\Gamma \cap B(x^0, r)$, and there are constants $a_1, \dots, a_n$ such that $\big(a_1(\partial/\partial x_1) + \dots + a_n(\partial/\partial x_n)\big)^s F = f$.*

We now continue the proof of the theorem. Again suppose that $x^0 = 0$, and $\nu(x^0) = (1, 0, \dots, 0)$. Then $f = \partial^{2s}F/\partial x_1^{2s}$, where F is a harmonic function that extends continuously to $\Gamma \cap B(x^0, r)$ (we may always assume that the order of the derivative is even). Since F is harmonic,

$$\frac{\partial^{2s}F}{\partial x_1^{2s}} = (-\widetilde{\Delta})^s F = f.$$

For $\varphi \in \mathcal{D}(B(x^0, r))$, consider the integral

$$I = \int_\Gamma f(h(x_2, \dots, x_n) - \epsilon, x_2, \dots, x_n)\varphi(h(x_2, \dots, x_n), x_2, \dots, x_n)\, d\sigma,$$

in other words, the integral over the translate of Γ by the vector $-\epsilon\nu(x^0)$. The surface area element is

$$\begin{aligned} d\sigma &= \sqrt{1 + \left(\frac{\partial h}{\partial x_2}\right)^2 + \dots + \left(\frac{\partial h}{\partial x_n}\right)^2}\, dx_2 \wedge \dots \wedge dx_n \\ &= \psi_1(x_2, \dots, x_n)\, dx_2 \wedge \dots \wedge dx_n. \end{aligned}$$

Then

$$\begin{aligned} I &= \int_\Gamma (-\widetilde{\Delta})^s F(h - \epsilon, x_2, \dots, x_n)\varphi(x)\psi_1\, dx_2 \wedge \dots \wedge dx_n \\ &= \int_\Gamma (-\widetilde{\Delta})^s F(x_1 - \epsilon, x_2, \dots, x_n)\varphi(x)\psi_1\, dx_2 \wedge \dots \wedge dx_n \\ &= (-1)^s \int_\Gamma F(x_1 - \epsilon, x_2, \dots, x_n)\Delta^s(\varphi(x)\psi_1)\, dx_2 \wedge \dots \wedge dx_n. \end{aligned}$$

Since F extends continuously to Γ, the integral I has a limit as $\epsilon \to 0^+$, that is, f defines a distribution f_0 on Γ.

If the integral I tends to zero, then in local coordinates on Γ we obtain that the distribution f_0 satisfies the condition $Af_0 = 0$, where A is an elliptic operator with coefficients of class $\mathcal{C}^\infty$. Hence the restriction of F to Γ is a function of class $\mathcal{C}^\infty$, so F is a function of class $\mathcal{C}^\infty(\Omega^+ \cup (\Gamma \cap B(x^0, r)))$. This means that f too is a function of class $\mathcal{C}^\infty(\Omega^+ \cup (\Gamma \cap B(x^0, r)))$, and it equals zero on $\Gamma \cap B(x^0, r)$. □

Theorem 11.1 is a well-known statement. It may be found in the work of Roĭtberg [167] for all elliptic operators, in the work of Straube [199] for harmonic functions, and in [110] for a half-space. The proof we have given is close to the proof in [110, 199]. In the work of Straube, there is a more precise characterization of the space $\mathcal{G}(\Omega^+)$ in terms of the Sobolev spaces $\mathcal{W}_2^{-k}$, but we will not use this characterization here.

11.3 Corollaries

If $f \in \mathcal{G}(\Omega^+)$, then f determines a distribution in $\mathcal{D}'(\Omega)$.

Corollary 11.3. *If $f \in \mathcal{G}(\Omega^+)$ and $\varphi \in \mathcal{D}(\Omega)$, then the limit $\lim_{\epsilon\to 0^+} \int_{\Omega_\epsilon^+} f\varphi\, dv$ exists and determines a distribution in $\mathcal{D}'(\Omega)$.*

Proof. Recall that $\Omega_\epsilon^+ = \{x \in \Omega^+ : d(x) > \epsilon\}$. We may assume that $\operatorname{supp}\varphi$ lies in some ball $B(x^0, r)$, that $x^0 = 0 \in \Gamma$, and that $\nu(x^0) = (1, 0, \dots, 0)$. In Theorem 11.1, we proved that $f = \partial^s F/\partial x_1^s$ in $B(x^0, r) \cap \Omega^+$, and F extends continuously to $B(x^0, r) \cap \Gamma$. Therefore

$$\int_{\Omega_\epsilon^+} f\varphi\, dv = \int_{\Omega_\epsilon^+} \left(\frac{\partial^s F}{\partial x_1^s}\right) \varphi\, dv = \int_{\partial\Omega_\epsilon^+} \frac{\partial^{s-1} F}{\partial x_1^{s-1}} \varphi\, dx[1] - \int_{\Omega_\epsilon^+} \frac{\partial^{s-1} F}{\partial x_1^{s-1}} \frac{\partial\varphi}{\partial x_1}\, dv.$$

By Theorem 11.1, the first integral in this equation has a limit as $\epsilon \to 0^+$. We treat the second integral in exactly the same way by throwing the derivatives onto φ. We obtain the integral

$$\int_{\Omega_\epsilon^+} F \frac{\partial^s \varphi}{\partial x_1^s}\, dv,$$

which has a limit as $\epsilon \to 0^+$. □

Corollary 11.3 was given in [199].

The same assertion holds for $\mathcal{G}(\Omega^-)$.

We remark that if $f \in \mathcal{G}(\Omega^+)$ and $\psi \in C^\infty(\overline{\Omega^+})$, then the function $f\psi$ also defines a distribution on Γ, since for $\varphi \in \mathcal{D}(\Gamma)$,

$$\int_\Gamma f(x + \epsilon\nu(x))\psi(x + \epsilon\nu(x))\varphi(x)\, d\sigma = \int_\Gamma f(x + \epsilon\nu(x))[\psi(x + \epsilon\nu(x))\varphi(x)]\, d\sigma,$$

and $\psi(x + \epsilon\nu(x))\varphi(x)$ converges to $\psi(x)\varphi(x)$ in the topology of $\mathcal{D}(\Gamma)$ as $\epsilon \to 0^+$.

Moreover, we can differentiate a function $f \in \mathcal{G}(\Omega^+)$, and every derivative also lies in $\mathcal{G}(\Omega^+)$.

Corollary 11.4. *Suppose $f^+ \in \mathcal{G}(\Omega^+)$ and $f^- \in \mathcal{G}(\Omega^-)$, where $f_0^+ = f_0^-$, and $(\partial f^+/\partial\nu)_0 = (\partial f^-/\partial\nu)_0$ on $\Gamma \cap B(x^0, r)$. Then the functions f^+ and f^- are restrictions to Ω^+ and Ω^- of a function F that is harmonic in $\Omega^+ \cup \Omega^- \cup (\Gamma \cap B(x^0, r))$ (where $x^0 \in \Gamma$).*

Proof. Consider the distribution $S \in \mathcal{D}'(\Omega)$ given for $\varphi \in \mathcal{D}(\Omega)$ by

$$S(\varphi) = \lim_{\epsilon\to 0^+} \int_{\Omega_\epsilon^+} f^+\varphi\, dv + \lim_{\epsilon\to 0^+} \int_{\Omega_\epsilon^-} f^-\varphi\, dv.$$

We will show that $\Delta S = 0$ (in the sense of generalized functions) on the set $\Omega^+ \cup \Omega^- \cup (\Gamma \cap B(x^0, r))$. It suffices to prove this in the ball $B(x^0, r)$. If $\operatorname{supp}\varphi \subset B(x^0, r)$ and $x^0 \in \Gamma$, then

$$\Delta S(\varphi) = S(\Delta\varphi) = \lim_{\epsilon\to 0+} \int_{\Omega_\epsilon^+} f^+ \Delta\varphi\, dv + \lim_{\epsilon\to 0+} \int_{\Omega_\epsilon^-} f^- \Delta\varphi\, dv.$$

By Green's formula,

$$\int_{\Omega_\epsilon^+} f^+ \Delta\varphi\, dv = -\int_{\partial\Omega_\epsilon^+} f^+ \frac{\partial\varphi}{\partial\nu}\, d\sigma + \int_{\partial\Omega_\epsilon^+} \varphi \frac{\partial f^+}{\partial\nu}\, d\sigma,$$

$$\int_{\Omega_\epsilon^-} f^- \Delta\varphi\, dv = \int_{\partial\Omega_\epsilon^-} f^- \frac{\partial\varphi}{\partial\nu}\, d\sigma - \int_{\partial\Omega_\epsilon^-} \varphi \frac{\partial f^-}{\partial\nu}\, d\sigma.$$

Since $f_0^+ = f_0^-$ and $(\partial f^+/\partial\nu)_0 = (\partial f^-/\partial\nu)_0$ on $\Gamma \cap B(x^0, r)$, we obtain that $\Delta S = 0$. By Weyl's lemma, the distribution S is a harmonic function in $\Omega^+ \cup \Omega^- \cup (\Gamma \cap B(x^0, r))$, and $S = f^+$ in Ω^+, while $S = f^-$ in Ω^-. □

This corollary is a generalization of the well-known result that if two harmonic functions f^+ and f^- are smooth up to Γ, and their values and the values of their normal derivatives match on Γ, then they are harmonic extensions of each other.

For domains in $\mathbf{C}^n$, this result can be sharpened.

Corollary 11.5. *Suppose $\Omega \subset \mathbf{C}^n$, $f^+ \in \mathcal{G}(\Omega^+)$, and $f^- \in \mathcal{G}(\Omega^-)$. If $(f^+)_0 = (f^-)_0$ and $(\overline{\partial}_n f^+)_0 = (\overline{\partial}_n f^-)_0$ on $\Gamma \cap B(z^0, r)$, then there is a harmonic function F on $\Omega^+ \cup \Omega^- \cup (\Gamma \cap B(z^0, r))$ that agrees with f^+ on Ω^+ and with f^- on Ω^-.*

Proof. The derivative $\overline{\partial}_n f$, defined in §4, equals

$$\overline{\partial}_n f = \sum_{k=1}^{n} \frac{\partial f}{\partial \bar{z}_k} \frac{\partial\rho}{\partial z_k} \Big/ |\operatorname{grad}\rho|.$$

For the proof, repeat the arguments of Lemma 11.4, only in place of Green's formula, use its "complex" form

$$\int_{\partial\Omega_\epsilon^+} \varphi\, \overline{\partial}_n f^+\, d\sigma - \int_{\partial\Omega_\epsilon^+} f^+\, \partial_n\varphi\, d\sigma = \int_{\Omega_\epsilon^+} (f^+\Delta\varphi - \varphi\Delta f^+)\, dv, \quad \partial_n\varphi = \overline{\overline{\partial}_n\bar{\varphi}},$$

the proof of which is immediate from Stokes's formula and the fact that $\overline{\partial}_n f^+\, d\sigma$ is the restriction to $\partial\Omega_\epsilon^+$ of the form $(-1)^n 2^{1-n} i^{-n} \mu_{f^+}$ defined in §1. □

We obtain from Corollary 11.5 a version of the theorem on local holomorphic extension.

Corollary 11.6. *Suppose $\Omega \subset \mathbf{C}^n$, the function $f^+ \in \mathcal{G}(\Omega^+)$ is holomorphic in Ω^+, and $f^- \in \mathcal{G}(\Omega^-)$ has the property that $(f^+)_0 = (f^-)_0$ and $(\overline{\partial}_n f^-)_0 = 0$ on $\Gamma \cap B(z^0, r)$. Then $f^- \in \mathcal{O}(\Omega^-)$.*

Corollaries 11.4–11.6 are due to the author.

11.4 Theorems on harmonic extension

We denote by $\mathcal{G}(\Omega)$ the set of pairs $f = (f^+, f^-)$ for which $f^+ \in \mathcal{G}(\Omega^+)$, $f^- \in \mathcal{G}(\Omega^-)$, and $f_0^+ = f_0^-$ on Γ. We will consider the differential forms $\alpha(f^\pm) = \sum_{j=1}^n (-1)^{j-1}(\partial f^\pm/\partial x_j)\, dx[j]$. For $|\epsilon| > 0$, the restrictions of these forms to $\Gamma_\epsilon = \partial\Omega^\pm_{\pm\epsilon} \setminus \partial\Omega$ are

$$\alpha(f^+)\big|_{\Gamma_\epsilon} = \frac{\partial f^+}{\partial \nu}\, d\sigma, \qquad \text{if } \epsilon > 0,$$

$$\alpha(f^-)\big|_{\Gamma_\epsilon} = \frac{\partial f^-}{\partial \nu}\, d\sigma, \qquad \text{if } \epsilon < 0.$$

Indeed, we have already used the relation $dx[j]\big|_{\Gamma_\epsilon} = (-1)^{j-1}\cos\gamma_j\, d\sigma$ (see §3), where γ_j is the angle between the normal ν to Γ_ϵ and the axis Ox_j, so

$$\alpha(f^+)\big|_{\Gamma_\epsilon} = \sum_{j=1}^n \frac{\partial f^+}{\partial x_j} \cos\gamma_j\, d\sigma = \frac{\partial f^+}{\partial \nu}\, d\sigma$$

($d\sigma = d\sigma_\epsilon$ is the surface area element of Γ_ϵ; as a rule, we will omit the index ϵ on $d\sigma$).

By Theorem 11.1, the functions $\partial f^+/\partial\nu$ and $\partial f^-/\partial\nu$ define distributions on Γ, so their difference $(\partial f^+/\partial\nu)_0 - (\partial f^-/\partial\nu)_0$ is in $\mathcal{D}'(\Gamma)$. We will show that the converse is also true. Moreover, if $(\partial f^+/\partial\nu)_0 = (\partial f^-/\partial\nu)_0$, then by Corollary 11.5, the functions f^+ and f^- are the restrictions to Ω^+ and Ω^- of a function that is harmonic in Ω.

We denote the class of harmonic functions in Ω by $\mathcal{H}(\Omega)$. We may suppose that $\mathcal{H}(\Omega)$ consists of those pairs $(f^+, f^-) \in \mathcal{G}(\Omega)$ for which f^+ and f^- are harmonic extensions of each other into Ω.

Theorem 11.7 (Kytmanov). *For every distribution $S \in \mathcal{D}'(\Gamma)$, there exists a function $f = (f^+, f^-) \in \mathcal{G}(\Omega)$ (which we call a harmonic representation of S) such that $(\partial f^+/\partial\nu)_0 - (\partial f^-/\partial\nu)_0 = S$ on Γ, and f^+ and f^- are harmonic extensions of each other in $\Omega \setminus \operatorname{supp} S$. If $h \in \mathcal{G}(\Omega)$ is another harmonic representation of S, then $h - f = (h^+ - f^+, h^- - f^-) \in \mathcal{H}(\Omega)$.*

Proof. First consider a distribution S with compact support: $S \in \mathcal{E}'(\Gamma)$. We define a function $T_v S \in \mathcal{G}(\Omega)$ by

$$(T_v S)(x) = \frac{-1}{\sigma_n(n-2)} S_y\left(\frac{1}{|x-y|^{n-2}}\right), \qquad \text{if } n > 2,$$

$$(T_v S)(x) = \frac{1}{2\pi} S_y(\ln|x-y|), \qquad \text{if } n = 2,$$

where σ_n is the area of the unit sphere in $\mathbf{R}^n$; that is, T_v is a single-layer potential. If S is a continuous function with compact support in Γ, then it is well known that the single-layer potential $T_v S$ is continuous in $\mathbf{R}^n$, and $(\partial T_v^+ S/\partial\nu) - (\partial T_v^- S/\partial\nu) = S$ on Γ.

Lemma 11.8. *If $S \in \mathcal{E}'(\Gamma)$, then $T_v S$ is a harmonic function outside* $\operatorname{supp} S$, $T_v S = (T_v^+ S, T_v^- S) \in \mathcal{G}(\Omega)$, *and* $(\partial T_v^+ S/\partial\nu)_0 - (\partial T_v^- S/\partial\nu)_0 = S$.

Proof. The proof of the lemma is standard—it is a repetition of the proof of Theorem 3.6. The harmonicity of $T_v S$ outside $\operatorname{supp} S$ follows because $S(\varphi) = S(\psi\varphi)$, where $\psi \in \mathcal{D}(\Gamma)$ and $\psi = 1$ in a neighborhood of $\operatorname{supp} S$. Thus

$$(T_v S)(x) = \frac{-1}{\sigma_n(n-2)} S_y\left(\frac{\psi(y)}{|x-y|^{n-2}}\right) \qquad (n > 2),$$

and consequently $T_v S$ can be differentiated under the distribution sign everywhere outside $\operatorname{supp} S$.

Since S has compact support, S is a distribution with a finite order of singularity k, so $|S(\varphi)| \le C \sup_{\substack{\|\alpha\|\le k \\ \operatorname{supp} S}} |\partial^\alpha \varphi|$ for $\varphi \in \mathcal{D}(\Gamma)$. Then

$$|(T_v S)(x)| \le C_1 \sup_{\substack{\|\alpha\|\le k \\ y\in \operatorname{supp} S}} \partial_y^\alpha |x-y|^{2-n} \le C_2 d^{2-n-\|\alpha\|}(x),$$

that is, $T_v^+ S \in \mathcal{G}(\Omega^+)$ and $T_v^- S \in \mathcal{G}(\Omega^-)$. We will show that $(T_v^+ S)_0 = (T_v^- S)_0$ on Γ. The distribution $S \in \mathcal{E}'(\Gamma)$ extends to a current $S_\Gamma \in \mathcal{E}'_{n-1}(\Omega)$ in the following way: $S_\Gamma(\beta) = S(\beta/d\sigma)$ if $\beta \in \mathcal{D}^{n-1}(\Omega)$, since each form $\beta \in \mathcal{D}^{n-1}(\Omega)$, when restricted to Γ, takes the form $\psi\, d\sigma$, where $\psi \in \mathcal{D}(\Gamma)$.

On the other hand, each distribution $S = X_1 \dots X_j f$, where $f \in C(\Gamma)$, and $X_1, \dots, X_j$ are tangential vector fields on Γ, that is, $X_j = \sum_{m=1}^n a_m(x)(\partial/\partial x_m)$ and $X_j(\rho) = 0$. We will show that if $(T_v^+ S)_0 = (T_v^- S)_0$, then $(T_v^+ XS)_0 = (T_v^- XS)_0$, where X is a tangential vector field on Γ (see Lemma 3.7).

Let $\widetilde{X}$ be an arbitrary vector field in Ω, tangential to each Γ_ϵ, and agreeing with X on Γ. By hypothesis,

$$[X(T_v^+ S)_0 - X(T_v^- S)_0](\beta/d\sigma) = \lim_{\epsilon\to 0+}\left(\int_{\Gamma_\epsilon} \widetilde{X}(T_v^+ S)\beta - \int_{\Gamma_{-\epsilon}} \widetilde{X}(T_v^- S)\beta\right) = 0$$

for every form $\beta \in \mathcal{D}^{n-1}(\Omega)$. We must prove this equation for $T_v XS$. To do this, we choose $\widetilde{X}$ so that

$$\int_{\Gamma_\epsilon} (T_v^+ XS - \widetilde{X} T_v^+ S)\beta = \int_{\Gamma_{-\epsilon}} (T_v^- XS - \widetilde{X} T_v^- S)\beta$$

for a fixed form β. If $\widetilde{X} = \sum_{j=1}^n \tilde{a}_j(x)(\partial/\partial x_j)$, then

$$T_v XS - \widetilde{X} T_v S$$
$$= \frac{-1}{\sigma_n(n-2)} S_y\left(\sum_{j=1}^n a_j(y)\frac{\partial}{\partial x_j}|x-y|^{2-n} - \sum_{j=1}^n \tilde{a}_j(x)\frac{\partial}{\partial x_j}|x-y|^{2-n}\right),$$

so it suffices to show that

$$\int_{\Gamma_\epsilon}\Big(\sum_{j=1}^{n}\tilde a_j(x)\frac{\partial}{\partial x_j}T_v^+S\Big)\beta-\int_{\Gamma_{-\epsilon}}\Big(\sum_{j=1}^{n}\tilde a_j(x)\frac{\partial}{\partial x_j}T_v^-S\Big)\beta$$

$$=h(\epsilon)=\int_{\Gamma_\epsilon}(T_v^+XS)\beta-\int_{\Gamma_{-\epsilon}}(T_v^-XS)\beta.$$

We may assume that the functions $\partial T_v^+S/\partial x_j$ and $\partial T_v^-S/\partial x_j$ do not equal zero at the same time, for otherwise we can always transform S outside $\operatorname{supp}\beta$ so that this condition holds, while the behavior of the changed distribution $\widetilde{S}$ on Γ remains as before. Therefore the previous equation is solvable. Thus, $T_vS\in\mathcal{G}(\Omega)$.

The last assertion of the lemma is proved in the same way. □

Theorem 11.7 has now been proved for distributions with compact support. Suppose S is an arbitrary distribution in $\mathcal{D}'(\Gamma)$. We represent Ω as the union of an increasing sequence of domains Ω_k (that is, $\bigcup_{k=1}^{\infty}\Omega_k=\Omega$ and $\Omega_k\Subset\Omega_{k+1}$) such that the space $\mathcal{H}(\Omega)$ is dense in $\mathcal{H}(\Omega_k)$.

Consider a sequence of functions $\varphi_k\in\mathcal{D}(\Gamma)$ for which $\operatorname{supp}\varphi_k\subset\Gamma\cap\Omega_{k+1}$ and $\varphi_k=1$ on $\Gamma\cap\Omega_k$. We set $S_k=S\varphi_k$. Then $S_k\in\mathcal{E}'(\Gamma)$, and $\lim_{k\to\infty}S_k=S$. Therefore, by Lemma 11.8, there is a harmonic representation of the function S_k, which we denote by f_k. Generally speaking, f_k will not have a limit as $k\to\infty$, so it is necessary to correct it.

Since $S_{k+1}-S_k=0$ on $\Gamma\cap\Omega_k$, the function $f_{k+1}-f_k$ is harmonic in $(\Omega\setminus\Gamma)\cup\Omega_k$ (in view of Corollary 11.4). Consequently, there is a function $h_k\in\mathcal{H}(\Omega)$ such that $|h_k-(f_{k+1}-f_k)|<2^{-k}$ on Ω_k.

Consider the series $f=\sum_{k=0}^{\infty}(f_{k+1}-f_k-h_k)$ (the zeroth function, by definition, is zero). This series converges uniformly in $\Omega\setminus\Gamma$. Indeed, the remainder

$$\Big|\sum_{k=k_0}^{\infty}(f_{k+1}-f_k-h_k)\Big|\le\sum_{k=k_0}^{\infty}|f_{k+1}-f_k-h_k|<\sum_{k=k_0}^{\infty}2^{-k}=2^{1-k_0}$$

in Ω_{k_0}, that is, f is harmonic in $\Omega\setminus\Gamma$. On the other hand, the partial sum

$$t_{k_0}=\sum_{k=0}^{k_0-1}(f_{k+1}-f_k-h_k)=f_{k_0}-\sum_{k=1}^{k_0-1}h_k,$$

so $t_{k_0}\in\mathcal{G}(\Omega)$, and on $\Omega_{k_0}\cap\Gamma$ we have

$$\left(\frac{\partial t_{k_0}^+}{\partial\nu}\right)_0-\left(\frac{\partial t_{k_0}^-}{\partial\nu}\right)_0=S_{k_0}=\varphi_{k_0}S.$$

The remainder $f-t_{k_0}$ is a harmonic function in Ω_{k_0}, so $f\in\mathcal{G}(\Omega)$.

Moreover, it is clear that $\left(\frac{\partial f^+}{\partial\nu}\right)_0-\left(\frac{\partial f^-}{\partial\nu}\right)_0=S$ on Γ.

Thus, f is a harmonic representation of S. □

For the case of a half-space, Theorem 11.7 was proved in [110]. Tarkhanov [202] considered a similar construction, based on the double-layer potential.

We now give some corollaries of this theorem. Consider the space $\mathbf{R}^{n+1}$ with variables $(x, y) = (x_1, \ldots, x_n, y)$. Let $\mathbf{R}_+^{n+1} = \{ (x, y) \in \mathbf{R}^{n+1} : y > 0 \}$. Then $\Gamma = \partial \mathbf{R}_+^{n+1}$ may be identified with $\mathbf{R}^n$. A function f is in $\mathcal{G}(\mathbf{R}_+^{n+1})$ if f is harmonic in $\mathbf{R}_+^{n+1}$ and has finite-order growth under approach to $\mathbf{R}^n$; that is, if for every compact set $K \subset \mathbf{R}^n$ there are constants $C > 0$ and $m > 0$ for which

$$|f(x, y)| \le Cy^{-m}, \qquad 0 < y \le 1, \quad x \in K.$$

We denote by $\mathcal{H}_0(\mathbf{R}_+^{n+1})$ the class of harmonic functions f in $\mathbf{R}_+^{n+1}$ such that $f \in \mathcal{C}^\infty(\mathbf{R}_+^{n+1} \cup \mathbf{R}^n)$ and $f = 0$ on $\mathbf{R}^n$. Actually, this is the class of harmonic functions in $\mathbf{R}^{n+1}$ that are odd in y.

Corollary 11.9. *Every function $f \in \mathcal{G}(\mathbf{R}_+^{n+1})$ determines on $\mathbf{R}^n$ a distribution $f_0 \in \mathcal{D}'(\mathbf{R}^n)$ via*

$$\lim_{y \to 0^+} \int_{\mathbf{R}^n} f(x, y) \varphi(x) \, dx = f_0(\varphi), \qquad \varphi \in \mathcal{D}(\mathbf{R}^n). \tag{11.2}$$

For each distribution $S \in \mathcal{D}'(\mathbf{R}^n)$, there is a function $f \in \mathcal{G}(\mathbf{R}_+^{n+1})$ for which (11.2) holds (with S in place of f_0). Moreover, $f \in \mathcal{C}^\infty(\mathbf{R}_+^{n+1} \cup (\mathbf{R}^n \setminus \operatorname{supp} S))$, and $f = 0$ on $\mathbf{R}^n \setminus \operatorname{supp} S$. Thus, $\mathcal{D}'(\mathbf{R}^n) \cong \mathcal{G}(\mathbf{R}_+^{n+1})/\mathcal{H}_0(\mathbf{R}_+^{n+1})$.

Proof. The first part follows from Theorem 11.1. Consider the set $\mathbf{R}_-^{n+1} = \{ (x, y) : y < 0 \}$. By Theorem 11.7, we can find for each distribution $S \in \mathcal{D}'(\mathbf{R}^n)$ a function $h = (h^+, h^-) \in \mathcal{G}(\mathbf{R}^{n+1})$ for which $(\partial h^+/\partial \nu)_0 - (\partial h^-/\partial \nu)_0 = S$. But $\partial h^+/\partial \nu = \partial h^+/\partial y$ and $\partial h^-/\partial \nu = \partial h^-/\partial y$, and moreover $h^-(x, -y) \in \mathcal{G}(\mathbf{R}_+^{n+1})$, so we obtain $f = (\partial h^+/\partial y)(x, y) + (\partial h^-/\partial y)(x, -y)$. $\square$

The function f defined in the corollary will also be called a harmonic representation of the distribution S. Corollary 11.9 was given in [110].

In $\mathbf{C}^n$, we can replace the derivative $\partial f/\partial \nu$ by $\overline{\partial}_n f$.

Corollary 11.10. *If $\Omega \subset \mathbf{C}^n$, then for each distribution $S \in \mathcal{D}'(\Gamma)$ there is a function $f = (f^+, f^-) \in \mathcal{G}(\Omega)$ such that $(\overline{\partial}_n f^+)_0 - (\overline{\partial}_n f^-)_0 = S$ on Γ.*

Theorems 11.1 and 11.7 are analogues for harmonic functions of the theorem on analytic representation of distributions in $\mathcal{D}'(\mathbf{R}^1)$ (see [30]).

If Γ is the boundary of a bounded domain $\Omega^+ = D$ (and $\Gamma \in \mathcal{C}^\infty$), then we can take a harmonic function in D as a harmonic representation.

Corollary 11.11. *If $S \in \mathcal{D}'(\Gamma)$, and $P(x, y)$ is the Poisson kernel for D, then the function $f(y) = S_x(P(x, y))$ has the following properties: $f \in \mathcal{G}(D)$ and $(f)_0 = S$ on $\Gamma = \partial D$. If $(f)_0 = 0$ in $\mathcal{D}'(\Gamma)$, then $f \equiv 0$.*

Proof. The proof given in [199] consists in the following. First we prove in the standard way that the Poisson integral of a function $\varphi \in \mathcal{E}(\Gamma)$ converges to φ in the topology of $\mathcal{E}(\Gamma)$. Then for $S \in \mathcal{D}'(\Gamma)$ we have

$$\int_\Gamma f(y - \epsilon\nu(y))\varphi(y)\,d\sigma(y) = \int_\Gamma S_x(P(x, y - \epsilon\nu(y)))\varphi(y)\,d\sigma(y)$$
$$= S_x\Big(\int_\Gamma P(x, y - \epsilon\nu(y))\varphi(y)\,d\sigma(y)\Big) \to S(\varphi)$$

when $\epsilon \to 0^-$. From the definition of $P(x,y)$, it follows that $f \in \mathcal{G}(D) = \mathcal{G}(\Omega^+)$ for every distribution S. □

12 Multiplication of distributions

12.1 Different approaches to multiplication of distributions

The problem of multiplication of distributions arose contemporaneously with the theory of generalized functions. Schwartz [176] already encountered the fundamental difficulty: he gave an example showing that it is impossible to introduce an operation of multiplication in $\mathcal{D}'(\mathbf{R}^1)$ which is everywhere defined, associative, commutative, and compatible on multipliers with multiplication. Indeed, if we consider the product $\mathrm{P.V.}\,\frac{1}{x} \cdot x \cdot \delta$ (where δ is the delta function at 0), then $x \cdot \delta = 0$, so that $\mathrm{P.V.}\,\frac{1}{x}(x \cdot \delta) = 0$, while on the other hand $\mathrm{P.V.}\,\frac{1}{x} \cdot x = 1$, so that $(\mathrm{P.V.}\,\frac{1}{x} \cdot x)\delta = \delta$.

In view of this difficulty, the problem of multiplication of distributions can be approached in various ways. The attempt has been made to formulate a consistent system of axioms for the operation of multiplication which would include a wide circle of applications, and then to determine what multiplication it induces (see, for example, the work of Keller [93, 94] and Shirokov [185, 186]). However, to formulate a system of axioms that would include all applications is, of course, difficult.

Another method for introducing a multiplication is to use the Fourier transform. Hörmander introduced a multiplication of distributions in this way by using the wave front set (see [78, chap. 8]). It is now used most often in the theory of partial differential equations.

One of the most commonly used methods is the sequential approach of Mikusiński (see [15]). It consists in the following: let $f, g \in \mathcal{D}'(\mathbf{R}^n)$, and let ρ_k be an approximation to the identity; that is, $\rho_k \in \mathcal{D}(\mathbf{R}^n)$, $\operatorname{supp}\rho_k \subset B(0, r_k)$, $r_k \to 0$ when $k \to \infty$, and $\int \rho_k\,dv = 1$. Then $f * \rho_k \to f$ and $g * \rho_k \to g$ in the weak topology of $\mathcal{D}'(\mathbf{R}^n)$, and $f * \rho_k$ and $g * \rho_k$ are in $C^\infty(\mathbf{R}^n)$ (where $f * \rho_k$ is the convolution of the distribution f and the function ρ_k). A distribution $h \in \mathcal{D}(\mathbf{R}^n)$ is the product of the distributions f and g if for every approximation to the identity ρ_k, we have $\lim_{k\to\infty}(f * \rho_k)(g * \rho_k) = h$ in the weak topology of $\mathcal{D}'(\mathbf{R}^n)$. It is clear that if this limit exists for every sequence ρ_k, then it does not depend on the choice of ρ_k.

However, this limit exists rather infrequently, so we might select a sequence ρ_k by some method. We can obtain in this way a method of multiplication of distributions by using their analytic representations. For functions of one variable, this was considered by Tillmann [204] and Bremermann [30]. Then Ivanov [81]–[83] used it to define his own method of multiplication. In several variables, a direct use of the analytic representation leads to difficulties, as remarked by Itano [80]: products of distributions with disjoint supports and the product of a zero distribution with a nonzero one may be different from zero. These difficulties may be overcome by using the algebra of Vladimirov [210]. For distributions with point singular support, we can also apply the method of asymptotic decomposition (separation of the principal parts). The essence is this: suppose f and g are in $\mathcal{D}'(\mathbf{R}^n)$, and $f(x,\epsilon)$ and $g(x,\epsilon)$ are smooth functions for $\epsilon > 0$ such that $f(x,\epsilon) \to f(x)$ and $g(x,\epsilon) \to g(x)$ in the weak topology of $\mathcal{D}'(\mathbf{R}^n)$ when $\epsilon \to 0^+$. Assume that for every function $\varphi \in \mathcal{D}(\mathbf{R}^n)$ we have

$$\int f(x,\epsilon)g(x,\epsilon)\varphi(x)\,dx = \sum_{\substack{k,q\\ q\geq 0}} \epsilon^{\lambda_k} \ln^q \epsilon\, C_{q,k} + o(1) \qquad (\operatorname{Re}\lambda_k \leq 0)$$

as $\epsilon \to 0^+$. Then the product $f \cdot g$ is given by the functional

$$(fg)(\varphi) = \text{p. f.} \int f(x,\epsilon)g(x,\epsilon)\varphi(x)\,dv = C_{0,0}.$$

Various versions of this method were proposed by Ivanov [85]–[88], Khristov and Damyanov [97], Fisher [49], Li [137], and others.

In this section, we introduce a product of generalized functions by using their harmonic representations defined in §11 and the scheme of Ivanov [82, 83].

12.2 Definition of the product of distributions using harmonic representations

Let $\Omega \subset \mathbf{R}^n$, and suppose that $\Gamma \subset \Omega$ is a smooth, connected, oriented hypersurface. We denote by $\mathcal{G}^*(\Omega) = \mathcal{G}^*$ the algebra of functions generated by finite products and linear combinations of functions of the form

$$h(x,\epsilon) = \left(\frac{\partial f^+}{\partial \nu} - \frac{\partial f^-}{\partial \nu}\right), \qquad x \in \Gamma, \quad \epsilon > 0,$$

where $f = (f^+, f^-) \in \mathcal{G}(\Omega)$, and we denote by $\mathcal{H}^* \subset \mathcal{G}^*$ the set of those functions h in $\mathcal{G}^*$ for which

$$\lim_{\epsilon \to 0^+} \int_\Gamma h(x,\epsilon)\varphi(x)\,d\sigma = 0$$

for all $\varphi \in \mathcal{D}(\Gamma)$ (that is, functions in $\mathcal{H}^*$ represent the zero distribution). Let $\mathcal{D}^*(\Gamma) = \mathcal{D}^* = \mathcal{G}^*/\mathcal{H}^*$; we call $\mathcal{D}^*$ the space of hyperdistributions. If $h \in \mathcal{G}^*$, then we denote its class in $\mathcal{D}^*$ by h^*. If $f \in \mathcal{H}(\Omega)$, then it is clear that the function $h = ((\partial f^+/\partial\nu) - (\partial f^-/\partial\nu)) \in \mathcal{H}^*$. Thus, $\mathcal{D}' \subset \mathcal{D}^*$.

Suppose S_1 and S_2 are in $\mathcal{D}'(\Gamma)$, and f_1 and f_2 are their harmonic representations in $\mathcal{G}(\Omega)$. By the product $S_1 \circ S_2$ we mean the hyperdistribution

$$\left[\left(\frac{\partial f_1^+}{\partial\nu} - \frac{\partial f_1^-}{\partial\nu}\right)\left(\frac{\partial f_2^+}{\partial\nu} - \frac{\partial f_2^-}{\partial\nu}\right)\right]^* \in \mathcal{D}^*.$$

Theorem 12.1 (Kytmanov). *The product of distributions is well defined.*

Proof. It is enough to show that if $f \in \mathcal{G}(\Omega)$ and $h \in \mathcal{H}(\Omega)$, then $((\partial f^+/\partial\nu) - (\partial f^-/\partial\nu))((\partial h^+/\partial\nu) - (\partial h^-/\partial\nu)) \in \mathcal{H}^*$, since the difference of two harmonic representations of the same distribution lies in $\mathcal{H}(\Omega)$ by Theorem 11.7 (or by Corollary 11.4). Then the class

$$\left[\left(\frac{\partial f_1^+}{\partial\nu} - \frac{\partial f_1^-}{\partial\nu}\right)\left(\frac{\partial f_2^+}{\partial\nu} - \frac{\partial f_2^-}{\partial\nu}\right)\right]^*$$

will not depend on the choice of the harmonic representations f_1 and f_2 of the distributions S_1 and S_2. Suppose that φ is in $\mathcal{D}(\Gamma)$ and has support such that one of the derivatives $(\partial d/\partial x_j)(x) \neq 0$ on $\operatorname{supp}\varphi$. This means that in a neighborhood of $\operatorname{supp}\varphi$, we may choose a system of coordinates such that one of the coordinate functions is

$$\rho(x) = \begin{cases} d(x), & x \in \Omega^+ \cup \Gamma, \\ -d(x), & x \in \Omega^-, \end{cases}$$

that is, $\xi_1 = \rho = \epsilon$, and the remaining functions $\xi_2, \ldots, \xi_n$ are local coordinates on Γ. If $h \in \mathcal{H}(\Omega)$, then the function $h_1(\xi) = (\partial h^+/\partial\nu) - (\partial h^-/\partial\nu)$ vanishes for $\xi_1 = 0$, whence

$$h_1(\xi_1, \ldots, \xi_n) = \int_0^{\xi_1} \frac{\partial h}{\partial t}(t, \xi_2, \ldots, \xi_n)\, dt = \xi_1 \int_0^1 h_2(\xi_1\tau, \xi_2, \ldots, \xi_n)\, d\tau,$$

if we make the substitution $t = \xi_1\tau$. Therefore $h_1(\xi_1, \ldots, \xi_n) = \xi_1 h_3(\xi_1, \ldots, \xi_n)$ in a neighborhood of $\operatorname{supp}\varphi$, and $h_3 \in \mathcal{C}^\infty$ in this neighborhood. Then

$$\int_\Gamma \left[\frac{\partial f^+}{\partial\nu}(x + \epsilon\nu(x)) - \frac{\partial f^-}{\partial\nu}(x - \epsilon\nu(x))\right] \times$$
$$\times \left[\frac{\partial h^+}{\partial\nu}(x + \epsilon\nu(x)) - \frac{\partial h^-}{\partial\nu}(x - \epsilon\nu(x))\right] \varphi(x)\, d\sigma$$
$$= \int_\Gamma \left[\frac{\partial f^+}{\partial\nu}(x + \epsilon\nu(x)) - \frac{\partial f^-}{\partial\nu}(x - \epsilon\nu(x))\right] \epsilon\, \tilde{h}_3(\epsilon, x)\varphi(x)\, d\sigma.$$

However, $\epsilon h_3(\epsilon, x) \to 0$ in $\mathcal{D}(\Gamma)$ as $\epsilon \to 0^+$, so this integral also converges to zero as $\epsilon \to 0^+$ by a well-known property of distributions (see, for example, [184, §9]). □

Thus, Itano's counterexample [80] does not take place for the product we have introduced.

12.3 Properties of the product of distributions given on a hypersurface

We define the support of a hyperdistribution $h^* \in \mathcal{D}^*$ ($\operatorname{supp} h^* \subset \Gamma$) as follows: a point $x^0 \in \Gamma \setminus \operatorname{supp} h^*$ if there is a ball $B(x^0, r)$ such that for every $\varphi \in \mathcal{D}(\Gamma)$ with $\operatorname{supp} \varphi \subset B(x^0, r)$, we have $\lim_{\epsilon \to 0^+} \int_\Gamma h(x, \epsilon)\varphi(x)\, d\sigma = 0$, where $h(x, \epsilon) \in h^*$. It is clear that this definition does not depend on the choice of the representative $h \in h^*$.

We will say that the hyperdistribution h^* is a distribution on the open set $U \subset \Gamma$ if for all $\varphi \in \mathcal{D}(U)$ the limit $\lim_{\epsilon \to 0^+} \int_\Gamma h(x, \epsilon)\varphi(x)\, dx = S(\varphi)$ exists. If S is given by a function ψ of class $\mathcal{C}^\infty$, then we will say that the hyperdistribution h^* is a function of class $\mathcal{C}^\infty$ on U. We define $\operatorname{sing\,supp} h^*$ as the complement of the set of points $x^0 \in \Gamma$ in a neighborhood of which h^* is a function of class $\mathcal{C}^\infty$, and the supersingular support as the complement of the set of points $x^0 \in \Gamma$ in a neighborhood of which h^* is a distribution.

All of these supports are closed sets in Γ, and their definitions are compatible with the definitions of supports and singular supports of distributions.

Theorem 12.2 (Kytmanov). *If S_1 and S_2 are in $\mathcal{D}'$, then*

1. $\operatorname{supp}(S_1 \circ S_2) \subset \operatorname{supp} S_1 \cap \operatorname{supp} S_2$, *and in particular, if* $\operatorname{supp} S_1 \cap \operatorname{supp} S_2 = \varnothing$, *then* $S_1 \circ S_2 = 0$*;*

2. $\operatorname{sing\,supp}(S_1 \circ S_2) \subset \operatorname{sing\,supp} S_1 \cup \operatorname{sing\,supp} S_2$, *and in particular, if the singular supports of S_1 and S_2 are both empty, then so is the singular support of $S_1 \circ S_2$, that is, $S_1 \circ S_2$ is a function of class $\mathcal{C}^\infty$ on Γ;*

3. *the product of distributions agrees with multiplication on multipliers, that is, if $\psi \in \mathcal{C}^\infty(\Gamma)$ and $S \in \mathcal{D}'(\Gamma)$, then $\psi \circ T = \psi T$.*

Proof. (1) Suppose $x^0 \in \Gamma \setminus (\operatorname{supp} S_1 \cap \operatorname{supp} S_2)$, for example, $x^0 \in \Gamma \setminus \operatorname{supp} S_1$, and f_1 and f_2 are harmonic representations of S_1 and S_2 in $\mathcal{G}(\Omega)$. Since $x^0 \in \Gamma \setminus \operatorname{supp} S_1$, the function $f_1 = (f_1^+, f_1^-)$ is harmonic in a neighborhood $B(x^0, r)$ (see Theorem 11.7), so $(\partial f_1^+/\partial \nu) - (\partial f_1^-/\partial \nu) = \epsilon h(\epsilon, x)$ in $B(x^0, r)$, where $h \in \mathcal{C}^\infty(B(x^0, r))$ (see the proof of Theorem 12.1). Then, just as in Theorem 12.1, we obtain

$$\lim_{\epsilon\to 0+}\int_\Gamma \left(\frac{\partial f_1^+}{\partial \nu}-\frac{\partial f_1^-}{\partial \nu}\right)\left(\frac{\partial f_2^+}{\partial \nu}-\frac{\partial f_2^-}{\partial \nu}\right)\varphi(x)\,d\sigma$$
$$=\lim_{\epsilon\to 0+}\int_\Gamma \left(\frac{\partial f_2^+}{\partial \nu}-\frac{\partial f_2^-}{\partial \nu}\right)\epsilon h(\epsilon,x)\varphi(x)\,d\sigma = 0.$$

(2) Suppose $x^0 \notin (\operatorname{sing\,supp} S_1 \cup \operatorname{sing\,supp} S_2)$. Then in a neighborhood $B(x^0,r)$, the distributions S_1 and S_2 are functions of class $\mathcal{C}^\infty(B(x^0,r)\cap\Gamma)$, so if f_1 and f_2 are their harmonic representations, then we may assume that f_1 and f_2 are single-layer potentials of S_1 and S_2. Then $\partial f_j^+/\partial\nu$ and $\partial f_j^-/\partial\nu$ extend continuously to $\Gamma\cap B(x^0,r)$, and $(\partial f_j^+/\partial\nu)-(\partial f_j^-/\partial\nu) = S_j$ on $B(x^0,r)\cap\Gamma$ for $j=1, 2$.

The harmonic representation of the function S_1S_2 also has this property, that is, $(\partial f^+/\partial\nu)-(\partial f^-/\partial\nu) = S_1S_2$ on $\Gamma\cap B(x^0,r)$. Then the difference

$$\left(\frac{\partial f_1^+}{\partial \nu}-\frac{\partial f_1^-}{\partial \nu}\right)\left(\frac{\partial f_2^+}{\partial \nu}-\frac{\partial f_2^-}{\partial \nu}\right)-\left(\frac{\partial f^+}{\partial \nu}-\frac{\partial f^-}{\partial \nu}\right)$$

lies in $\mathcal{H}^*$. This proves property (2). We remark that it holds not only for functions S_1 and S_2 of class $\mathcal{C}^\infty$, but also for continuous functions S_1 and S_2.

(3) Suppose $\psi\in\mathcal{C}^\infty(\Gamma)$, $x^0\in\Gamma$, and $B(x^0,r)\Subset\Omega$ is some ball. Replacing ψ by the function $\alpha\psi$, where $\alpha=1$ in $B(x^0,r)$ and $\alpha\in\mathcal{D}(\Gamma)$, we consider the single-layer potential $T_v\alpha\psi$. Then a harmonic representation f of ψ and the function $T_v\alpha\psi$ give representations for ψ in $B(x^0,r)$, that is, their difference $f-T_v\alpha\psi$ is a harmonic function in $B(x^0,r)$. Therefore we may replace the function f in $B(x^0,r)$ by $T_v\alpha\psi$. The single-layer potential is a function of class $\mathcal{C}^\infty$ up to Γ since its derivatives satisfy the formula of Lemma 4.2. Consequently, the function $(\partial T_v^+\alpha\psi/\partial\nu)-(\partial T_v^-\alpha\psi/\partial\nu)$ is $\mathcal{C}^\infty$ up to Γ. Therefore this difference converges to $T_v\alpha\psi$ in the topology of $\mathcal{D}(\Gamma)$ as $\epsilon\to 0^+$. Again using properties of generalized functions from [184, §9], we obtain for $\varphi\in\mathcal{D}(B(x^0,r)\cap\Gamma)$ that

$$\lim_{\epsilon\to 0+}\int_\Gamma \left(\frac{\partial f^+}{\partial \nu}-\frac{\partial f^-}{\partial \nu}\right)\left(\frac{\partial f_1^+}{\partial \nu}-\frac{\partial f_1^-}{\partial \nu}\right)\varphi(x)\,d\sigma = (\psi S)(\varphi),$$

where f_1 is a harmonic representation of S. Theorems 12.1 and 12.2 were given in [110, 113] for distributions in $\mathcal{D}'(\mathbf{R}^n)$. □

12.4 Properties of products of distributions in $\mathcal{D}'(\mathbf{R}^n)$

In the second part of this section we will consider distributions in $\mathcal{D}'(\mathbf{R}^n)$, that is, we assume that $\Omega^+ = \mathbf{R}^{n+1}_+ = \{(x,y) : y>0\}$, where $x=(x_1,\dots,x_n)$, and $\Gamma=\partial\Omega^+=\mathbf{R}^n$. Harmonic representations of distributions in $\mathcal{D}'(\mathbf{R}^n)$ are functions $f\in\mathcal{G}(\mathbf{R}^{n+1}_+)$, and $\mathcal{G}^*$ is the algebra generated by finite products and linear combinations of functions in $\mathcal{G}(\mathbf{R}^{n+1}_+)$. The space $\mathcal{H}_0(\Omega)=\mathcal{H}_0$ consists of the harmonic functions in $\mathbf{R}^{n+1}$ that are equal to zero on $\mathbf{R}^n$, and $\mathcal{H}^* \subset \mathcal{G}^*$ contains the functions represented by the zero distribution.

Besides the properties enumerated in Theorem 12.2, we note the following properties of multiplication in $\mathcal{D}'(\mathbf{R}^n)$.

Theorem 12.3 (Kytmanov). *The following hold:*

1. *Leibniz's rule holds for distributions S_1 and S_2 in $\mathcal{D}'(\mathbf{R}^n)$:*

$$\frac{\partial}{\partial x_k}(S_1 \circ S_2) = \frac{\partial S_1}{\partial x_k} \circ S_2 + S_1 \circ \frac{\partial S_2}{\partial x_k}, \qquad k = 1, \dots, n.$$

2. *Suppose S_1, S_2, and $S_1 \circ S_2$ are in $\mathcal{D}'(\mathbf{R}^n)$. If S_1 and S_2 are homogeneous distributions of degrees p and q, then $S_1 \circ S_2$ is a homogeneous distribution of degree $p + q$.*

3. *Suppose $x' = (x_1, \dots, x_m)$ and $x'' = (x_{m+1}, \dots, x_n)$, while $S_1 \in \mathcal{D}'(\mathbf{R}^m)$ and $S_2 \in \mathcal{D}'(\mathbf{R}^{n-m})$, and f_1 and f_2 are their harmonic representations in $\mathbf{R}^{m+1}_+$ and $\mathbf{R}^{n-m+1}_+$. Then $(f_1 + \mathcal{H}_0^{x'}) \times (f_2 + \mathcal{H}_0^{x''}) \subset f + \mathcal{H}^*$, where f is a harmonic representation of the direct product $S_1 \otimes S_2 \in \mathcal{D}'(\mathbf{R}^n)$. Multiplication of distributions agrees in this way with the direct product ($S_1 \circ S_2 = S_1 \otimes S_2$).*

Proof. (1) Suppose f_1 and f_2 are harmonic representations of S_1 and S_2. Since $\partial f_j/\partial x_k$ is a harmonic representation of $\partial S_j/\partial x_k$ for $j = 1, 2$, we have

$$\frac{\partial}{\partial x_k} f_j^* = \left(\frac{\partial f_j}{\partial x_k}\right)^* = \frac{\partial f_j}{\partial x_k} + \mathcal{H}^*, \qquad j = 1, 2.$$

But Leibniz's rule holds for $f_1 f_2$:

$$\frac{\partial}{\partial x_k}(f_1 f_2) = \frac{\partial f_1}{\partial x_k} f_2 + f_1 \frac{\partial f_2}{\partial x_k}, \qquad k = 1, \dots, n.$$

Hence property (1) holds.

(2) The distribution S_1 is homogeneous of degree p if $t^{p+n} S_1(\varphi(tx)) = S_1(\varphi)$ for all $\varphi \in \mathcal{D}(\mathbf{R}^n)$ for $t > 0$. If the harmonic representation f_1 of the distribution S_1 is a homogeneous function of degree p (in $\mathbf{R}^{n+1}$), then S_1 is a homogeneous distribution of degree p. Indeed

$$\begin{aligned} t^{p+n} S_1(\varphi(tx)) &= t^{p+n} \lim_{\epsilon \to 0+} \int_{y=\epsilon} f_1(x, y)\varphi(tx)\, dx \\ &= t^p \lim_{\epsilon \to 0+} \int_{y=\epsilon t} f_1(x/t, y/t)\varphi(x)\, dx \\ &= \lim_{\epsilon \to 0+} \int_{y=\epsilon t} f_1(x, y)\varphi(x)\, dx = S_1(\varphi). \end{aligned}$$

Conversely, for any homogeneous distribution S_1 of degree p, we can find a harmonic representation which will be a homogeneous function in $\mathbf{R}^{n+1}$ of degree p. Let

$$P(x,y) = \frac{\Gamma((n+1)/2)}{\pi^{(n+1)/2}} \frac{y}{(|x|^2 + y^2)^{(n+1)/2}}$$

be the Poisson kernel for the half-space. If $S \in \mathcal{E}'(\mathbf{R}^n)$, then a harmonic representation f of the distribution S is given by the formula

$$f(x,y) = S_\xi(P(x-\xi,y)), \qquad y \neq 0. \tag{12.1}$$

Since the derivative with respect to y of the fundamental solution of Laplace's equation is

$$\begin{aligned}\frac{-1}{(n-1)\sigma_{n+1}} \frac{\partial}{\partial y}(|x-\xi|^2 + y^2)^{1-n/2} &= \frac{y}{\sigma_{n+1}}(|x-\xi|^2+y^2)^{-(n+1)/2} \\ &= \frac{1}{2}P(x-\xi,y),\end{aligned}$$

it follows (see Theorem 11.7) that the difference

$$\frac{\partial T_v^+ S}{\partial \nu} - \frac{\partial T_v^- S}{\partial \nu} = \frac{\partial T_v^+ S}{\partial y} - \frac{\partial T_v^- S}{\partial y} = S_\xi(P(x-\xi,y)).$$

Moreover, (12.1) gives a harmonic representation of distributions in $\mathcal{O}'_{-n}$, that is, distributions that extend to functions $\varphi \in C^\infty(\mathbf{R}^n)$ decreasing at infinity faster than $(1+|x|)^{-n}$ (together with all derivatives).

Now if $S_1 \in \mathcal{O}'_{-n}$ is homogeneous of degree p, then the function f_1 defined by (12.1) is also homogeneous of degree p, since

$$\begin{aligned}f_1(tx,ty) &= S_\xi(P(tx-\xi,ty)) = t^{-n}S_\xi(P(x-\xi/t,y)) \\ &= t^p S_\xi(P(x-\xi,y)) = t^p f_1(x,y).\end{aligned}$$

If S_1 is an arbitrary homogeneous distribution in $\mathcal{D}'(\mathbf{R}^n)$, then we see from the general scheme of constructing a harmonic representation (see Theorem 11.7) that we can always choose a harmonic representation that is a homogeneous function. Hence property (2) holds.

(3) It is enough to show that the boundary value of $f_1(x',y) \times f_2(x'',y)$ is $S_1 \otimes S_2$ in $\mathcal{D}'(\mathbf{R}^n)$. Consider the integral

$$\int_{\mathbf{R}^{n-m}} f_2(x'',\epsilon)\varphi(x)\,dx'' = \psi(x',\epsilon), \qquad \varphi \in \mathcal{D}(\mathbf{R}^n).$$

Then $\psi(x',\epsilon) \in \mathcal{D}(\mathbf{R}^m)$ and $\psi(x',\epsilon) \to S_2(\varphi(x)) = \psi(x')$ in $\mathcal{D}(\mathbf{R}^m)$ as $\epsilon \to 0^+$, since the operation $\varphi \to S_1(\varphi)$ is continuous from $\mathcal{D}(\mathbf{R}^n)$ into $\mathcal{D}(\mathbf{R}^m)$. Hence

$$\begin{aligned}\int_{\mathbf{R}^n} f_1(x',\epsilon)f_2(x'',\epsilon)\varphi(x)\,dx &= \int_{\mathbf{R}^m} f_1(x',\epsilon)\psi(x',\epsilon)\,dx' \to \langle S_1, \psi(x')\rangle \\ &= \langle S_1, \langle S_2, \varphi(x)\rangle\rangle = (S_1 \otimes S_2)(\varphi)\end{aligned}$$

as $\epsilon \to 0^+$. □

12.5 Multiplication of hyperfunctions with compact support

If we do not put any growth condition near $\mathbf{R}^n$ on harmonic functions in $\mathbf{R}^{n+1}_+$, then their boundary values will be hyperfunctions (see [78, chap. 9]). We can define, for example, the product of hyperfunctions with compact support in $\mathbf{R}^n$ (that is, analytic functionals). We will assume that $\mathbf{R}^n$ is embedded into $\mathbf{C}^n$ as follows: if $z \in \mathbf{C}^n$, then $\operatorname{Re} z = x \in \mathbf{R}^n$. If $K \subset \mathbf{C}^n$ is a compact set, then $\mathcal{O}'(K)$ consists of the linear forms u on the space of entire functions $\mathcal{O}(\mathbf{C}^n)$ for which the condition

$$|u(\varphi)| \leq C_\omega \sup_\omega |\varphi|, \qquad \varphi \in \mathcal{O}(\mathbf{C}^n),$$

holds in every neighborhood ω of K. The elements of $\mathcal{O}'(K)$ are called analytic functionals. Each distribution $u \in \mathcal{E}'$ is an analytic functional, but the converse is false.

Since there do not exist holomorphic functions with compact support (other than the zero function), the problem of localization is considerably more difficult for analytic functionals than for distributions.

If $K \subset \mathbf{R}^n$, then every analytic functional $u \in \mathcal{O}'(K)$ can be defined on functions holomorphic in a "complex" neighborhood of K (see, for example, [78, chap. 9]). Thus, by using the Poisson kernel $P(x,y)$ for a half-space, we can introduce the idea of a harmonic representation of $u \in \mathcal{O}'(K)$.

Theorem 12.4 (Hörmander). *Suppose $u \in \mathcal{O}'(K)$, $K \subset \mathbf{R}^n$, and $f(x,y) = u_\xi(P(x-\xi,y))$. Then f is a harmonic function in $\mathbf{R}^{n+1} \setminus K$, odd in y. If $\Phi \in \mathcal{H}(\mathbf{R}^{n+1})$, then*

$$u\left(\frac{\partial\Phi}{\partial y}(x,0)\right) = -\frac{1}{2}\int_{\mathbf{R}^{n+1}} f(x,y)\Delta(\chi\Phi)\,dx\,dy, \tag{12.2}$$

where $\chi \in \mathcal{D}(\mathbf{R}^{n+1})$ and $\chi = 1$ in a neighborhood of K. Every function $f(x,y)$ that is harmonic in $\mathbf{R}^{n+1} \setminus K$ and odd in y determines some functional $u \in \mathcal{O}'(K)$ by (12.2), and if $h(x,y)$ determines the same functional u, then $(h-f) \in \mathcal{H}(\mathbf{R}^{n+1})$.

Proof. The proof may be found in [78, chap. 9]; we just make some remarks about the idea. It is clear that $f(x,y) = u_\xi(P(x-\xi,y))$ is harmonic outside K, since the functional u is defined on functions that are analytic in a neighborhood of K, and in particular on the kernels $P(x-\xi,y)$ for $(x,y) \notin K$. Since $\Delta(\chi\Phi) = 0$ in a neighborhood of K, the integral in (12.2) is defined. Since $P(x,y) = 2(\partial/\partial y)g(x,y)$, where $g(x,y)$ is the fundamental solution of Laplace's equation in $\mathbf{R}^n$, we have

$$\Delta P(x,y) = 2\frac{\partial}{\partial y}\Delta(g(x,y)) = 2\frac{\partial}{\partial y}\delta,$$

where δ is the delta function at 0. Therefore, if $x^0 \in \mathbf{R}^n$ and $\chi = 1$ in a neighborhood of $(x^0, 0)$, where $\chi \in \mathcal{D}(\mathbf{R}^{n+1})$, then

$$\int_{\mathbf{R}^{n-1}} P(x^0 - \xi, y)\Delta(\chi\Phi)\, d\xi\, dy = \Delta_{(\xi,y)} P(x^0 - \xi, y)(\chi\Phi)$$
$$= 2\frac{\partial}{\partial y}\delta_{(x^0,0)}(\chi\Phi) = -2\frac{\partial}{\partial y}\Phi(x^0, 0).$$

Since the right-hand side of (12.2) is the limit of Riemann sums, and (12.2) is already proved for each term in this sum, we obtain (12.2) in the general case. □

As Hörmander remarks, (12.2) completely determines the functional u, since if $\varphi \in \mathcal{O}(\mathbf{C}^n)$, then there always exists $\Phi \in \mathcal{O}(\mathbf{C}^{n+1})$ such that $\Phi(z, 0) = 0$, $\partial\Phi/\partial w = \varphi$ for $w = 0$, and

$$\sum_{j=1}^{n} \frac{\partial^2\Phi}{\partial x_j^2} + \frac{\partial^2\Phi}{\partial w^2} = 0, \qquad (z, w) = (z_1, \dots, z_n, w) \in \mathbf{C}^{n+1}$$

(see [78, chap. 9]).

Since the left-hand side of (12.2) does not depend on χ, we may take χ to be even in y, while it suffices to take Φ to be odd in w. Then (12.2) takes the form

$$u\left(\frac{\partial\Phi}{\partial y}(x, 0)\right) = -\int_{\mathbf{R}_+^{n+1}} f(x, y)\Delta(\chi\Phi)\, dx\, dy. \tag{12.3}$$

Using Green's formula, we can transform the right-hand side of (12.3) (the derivative $\partial/\partial y$ is the derivative along the inner normal to $\partial\mathbf{R}_+^{n+1} = \mathbf{R}^n$):

$$-\int_{\mathbf{R}_+^{n+1}} f(x, y)\Delta(\chi\Phi)\, dx\, dy = -\lim_{\epsilon\to 0+}\int_{\{y>\epsilon\}} f(x, y)\Delta(\chi\Phi)\, dx\, dy$$
$$= \lim_{\epsilon\to 0+}\int_{\{y=\epsilon\}} \left(f(x, y)\frac{\partial}{\partial y}(\chi\Phi) - \frac{\partial f}{\partial y}\chi\Phi\right) dx.$$

Consequently, (12.3) can be rewritten in the form

$$u\left(\frac{\partial\Phi}{\partial y}(x, 0)\right) = \lim_{\epsilon\to 0+}\int_{\{y=\epsilon\}} \left(f\frac{\partial}{\partial y}(\chi\Phi) - \frac{\partial f}{\partial y}\chi\Phi\right) dx. \tag{12.4}$$

When u is a distribution with compact support, f has finite order of growth, so

$$\lim_{\epsilon\to 0+}\int_{\{y=\epsilon\}} \frac{\partial f}{\partial y}\chi\Phi\, dx = \lim_{\epsilon\to 0+}\int_{\mathbf{R}^n} \frac{\partial f}{\partial y}(x, \epsilon)\epsilon\chi\Phi_1\, dx = 0.$$

Consequently, we obtain from (12.4) the usual formula

$$u(\varphi(x, 0)) = \lim_{\epsilon\to 0+}\int_{\mathbf{R}^n} f(x, \epsilon)\varphi(x, \epsilon)\, dx,$$

where $\varphi(x, y) = \partial(\chi\Phi)/\partial y$.

We remark that $f(x,y) = u_\xi(P(x-\xi,y))$ converges to zero when $|x|+|y| \to \infty$, and there is no other harmonic representation for u with a zero at infinity by Liouville's theorem for harmonic functions. So we can introduce a multiplication for analytic functionals by using the scheme of Ivanov.

Let $\mathcal{H}_0^0(\mathbf{R}^n)$ be the space of harmonic functions in $\mathbf{R}^{n+1}\setminus K$ that tend to zero at infinity, where K is a compact set in $\mathbf{R}^n$ (not assumed to be fixed). Then $\mathcal{H}_0^0$ is isomorphic to the space of hyperfunctions with compact support. Consider the algebra $\mathcal{H}_0^{0^*}$ consisting of finite products and linear combinations of functions in $\mathcal{H}_0^0$. We call the element $f_1 f_2$ of the algebra $\mathcal{H}_0^{0^*}$ the product of the analytic functionals $u_1 \in \mathcal{O}'(K_1)$ and $u_2 \in \mathcal{O}'(K_2)$ when the f_j are harmonic representations of the functionals u_j.

It is clear that the properties of multiplication listed in Theorem 12.3 carry over to this multiplication.

12.6 Multiplication in the sense of Mikusiński

Theorem 12.5 (Zaslavskiĭ). *Suppose the distributions S_1 and S_2 in $\mathcal{D}'(\mathbf{R}^n)$ are such that the product $S = S_1 \circ S_2 \in \mathcal{D}'(\mathbf{R}^n)$ exists in the sense of Mikusiński. Then harmonic representations f_1 and f_2 of S_1 and S_2 satisfy the condition*

$$\lim_{\epsilon\to 0+} \int_{\mathbf{R}^n} f_1(x,\epsilon) f_2(x,\epsilon)\varphi(x)\,dx = S(\varphi)$$

for all $\varphi \in \mathcal{D}(\mathbf{R}^n)$, that is, the hyperdistribution $(f_1 f_2)^*$ agrees with the distribution S.

Proof. Suppose $\varphi \in \mathcal{D}(\mathbf{R}^n)$, and $\chi \in \mathcal{D}(\mathbf{R}^n)$ is equal to 1 in a neighborhood of $\operatorname{supp}\varphi$. We set $R_j = \chi S_j$; then $R_j \in \mathcal{E}'(\mathbf{R}^n)$. We construct a harmonic representation g_j for R_j by using the Poisson kernel for a half-space:

$$g_j(x,y) = R_{j\xi}(P(x-\xi,y)), \qquad j = 1,2.$$

Then $\int_{\{y=\epsilon\}}(f_1 f_2 - g_1 g_2)\varphi\,dx = \int_{\{y=\epsilon\}}[(f_1-g_1)f_2 + g_1(f_2-g_2)]\,dx$, and $f_j - g_j = y h_j$ in a neighborhood of $\operatorname{supp}\varphi$, since in this neighborhood the functions f_j and g_j represent the same distribution Q_j (see Theorem 11.7). Hence

$$\lim_{\epsilon\to 0+} \int_{\{y=\epsilon\}} (f_1 f_2 - g_1 g_2)\varphi\,dx = 0$$

(also see the proof of Theorems 12.1 and 12.2). Consequently, it suffices to show that

$$\lim_{\epsilon\to 0+} \int_{\{y=\epsilon\}} g_1 g_2 \varphi\,dx = S(\varphi).$$

Consider a function $\psi \in \mathcal{D}(\mathbf{R}^1)$ such that $\operatorname{supp}\psi \subset [-1,1]$, and $\psi = 1$ on $[-1/2, 1/2]$. We set

$$\Psi(x,y) = \begin{cases} 1, & |x| < 1/|\ln y|, \\ |x \ln|\ln y|| - |\ln|\ln y|/\ln y|, & |x| \geq 1/|\ln y|, \end{cases}$$

and $Q(x,y) = P(x,y)\Psi(x,y)$.

Lemma 12.6. *$P(x,y) - Q(x,y) \to 0$ in the topology of $\mathcal{E}(\mathbf{R}^n)$ as $y \to 0^+$.*

Proof. We write $P(x,y) - Q(x,y) = P(x,y)(1 - \Psi(x,y))$ and use the homogeneity of $P(x,y)$ in (x,y) when $y > 0$:

$$\begin{aligned} |\partial_x^\alpha (P(x,y) - Q(x,y))| &\leq \sum_{\|\beta\| \leq \|\alpha\|} C_\beta^\alpha |\partial_x^\beta P(x,y) \partial_x^{\alpha-\beta}(1 - \Psi(x,y))| \\ &\leq \sum_{\|\beta\| \leq \|\alpha\|} C_\beta^\alpha y^{-n-\|\beta\|} \left|\partial_x^\beta P(x/y,1) \partial_x^{\alpha-\beta}(1 - \Psi(x,y))\right| ; \end{aligned}$$

here $\alpha = (\alpha_1, \dots, \alpha_n)$ and β are multi-indices, $\|\alpha\| = \alpha_1 + \dots + \alpha_n$, while $\partial_x^\alpha f = \partial^{\|\alpha\|} f / \partial_{x_1}^{\alpha_1} \dots \partial_{x_n}^{\alpha_n}$, and the C_β^α are polynomial coefficients.

Furthermore,

$$\partial_x^{\alpha-\beta}(1 - \Psi(x,y)) = \begin{cases} 0, & \text{if } |x| > 1/|\ln y| + 1/|\ln|\ln y||, \quad \|\alpha - \beta\| > 0, \\ 1, & \text{if } |x| > 1/|\ln y| + 1/|\ln|\ln y||, \quad \alpha = \beta, \\ 0, & \text{if } |x| < 1/|\ln y|, \\ -|\ln|\ln y||^{\|\alpha-\beta\|} \partial^{\|\alpha-\beta\|}\psi(|x\ln|\ln y|| - |\ln|\ln y|/\ln y|), \\ & \text{if } 1/|\ln y| \leq |x| \leq 1/|\ln y| + 1/|\ln|\ln y||. \end{cases}$$

Therefore, denoting $\xi = x/y$, we have

$$\begin{aligned} &\sup_{x \in \mathbf{R}^n} |\partial_x^\alpha (P(x,y) - Q(x,y))| \\ &\qquad \leq c_\alpha \max_{\|\beta\| \leq \|\alpha\|} y^{-n-\|\beta\|} |\ln|\ln y||^{\|\alpha-\beta\|} \max_{|\xi| \geq 1/y|\ln|y||} |\partial_\xi^\beta P(\xi,1)|. \end{aligned}$$

It is easy to show that

$$|\partial_\xi^\beta P(\xi,1)| \leq C'_\beta |\xi|^{-n-1-\|\beta\|}, \qquad |\xi| > 1/y|\ln y|,$$

that is,

$$|\partial_\xi^\beta P(\xi,1)| \leq C'_\beta (y|\ln y|)^{n+1+\|\beta\|}, \qquad |\xi| > 1/y|\ln y|.$$

Hence

$$\begin{aligned} \sup_{x \in \mathbf{R}^n} |\partial_x^\alpha (P(x,y) - Q(x,y))| &\leq b_\alpha \max_{\|\beta \leq \|\alpha\|} y^{-n-\|\beta\|} |\ln|\ln y||^{\|\alpha-\beta\|} (y|\ln y|)^{n+1+\|\beta\|} \\ &= b_\alpha \max_{\|\beta \leq \|\alpha\|} y|\ln y|^{n+1+\|\beta\|} |\ln|\ln y||^{\|\alpha\|-\|\beta\|} \longrightarrow 0 \end{aligned}$$

as $y \to 0$. □

We now continue with the proof of the theorem. We denote $h_j(x,y) = R_{j\xi}(Q(x-\xi,y))$.

In view of Lemma 12.6, we have $h_j - g_j \to 0$ ($j = 1, 2$) as $y \to 0^+$, uniformly on each compact set in $\mathbf{R}^n$, and in particular on $\operatorname{supp}\varphi$. Therefore it suffices to show that

$$\lim_{\epsilon\to 0+} \int_{\mathbf{R}^n} h_1(x,\epsilon)h_2(x,\epsilon)\varphi(x)\,dx = S(\varphi).$$

This follows at once from the following two assertions:

1. $Q(x,y)C(y)$ is an approximation to the identity if $C(y) \to 1$ as $y \to 0^+$;

2. h_j agrees with $S_{j\xi}(Q(x-\xi,y))$ for all $x \in \operatorname{supp}\varphi$ for sufficiently small positive y.

To verify assertion (1), observe that $\operatorname{supp} Q(x,y)C(y)$ is contained in the ball $B(0, 1/|\ln y| + 1/|\ln|\ln y||)$, since this ball contains the support of $\Psi(x,y)$. By Lebesgue's theorem,

$$1/C(y) = \int_{\mathbf{R}^n} Q(x,y)\,dx = \int_{\mathbf{R}^n} P(\xi,1)\Psi(\xi y,y)\,d\xi \to \int_{\mathbf{R}^n} P(\xi,1)\,d\xi = 1$$

as $y \to 0^+$.

To prove assertion (2), it is enough to remark that the value of the convolution $S_{j\xi}(Q(x-\xi,y))$ depends only on the value of the distribution S_j on the set $x + \operatorname{supp} Q(x,y)$, and $\operatorname{diam}\operatorname{supp} Q(x,y) \le 1/|\ln y| + 1/|\ln|\ln y|| \to 0$ as $y \to 0^+$. In a neighborhood of the set $\operatorname{supp}\varphi$, the distributions S_j and R_j agree, $j = 1, 2$. □

Theorem 12.5 was given in [226].

12.7 Multipliable distributions

The theorems given in this section show that multiplication of distributions by using harmonic functions satisfies a number of natural properties. We will be interested in the question of when the product of distributions (or hyperfunctions) is again a distribution (or hyperfunction). Since the product of S_1 and S_2 does not depend on the choice of representation, it suffices to find only one pair of harmonic representations f_1 and f_2 for which the product $f_1 f_2$ defines a distribution.

The first class of multipliable distributions consists of distributions S_1 and S_2 for which the product $f_1 f_2$ is again a harmonic function. Since in this case $f_1 f_2$ has finite order of growth near $\mathbf{R}^n$, the function $f_1 f_2$ determines some distribution in $\mathcal{D}'(\mathbf{R}^n)$ (this argument is valid also for analytic functionals).Since

$$\Delta(f_1 f_2) = f_2\Delta f_1 + f_1\Delta f_2 + 2\left(\sum_{j=1}^{n} \frac{\partial f_1}{\partial x_j}\frac{\partial f_2}{\partial x_j} + \frac{\partial f_1}{\partial y}\frac{\partial f_2}{\partial y}\right),$$

we have that $\Delta(f_1 f_2) = 0$ if and only if the gradients of f_1 and f_2 are orthogonal. For example, this will be the case if f_1 and f_2 depend on different groups of variables.

When $n = 1$, we can give a complete answer to the question of when $f_1 f_2$ is a harmonic function. We suppose that $\mathbf{R}^2 = \mathbf{C}$ and $z = x + iy$.

Theorem 12.7 (Zaslavskiĭ). *Suppose f_1 and f_2 are harmonic in $\mathbf{C}_+ = \mathbf{R}^2_+$. The product $f_1 f_2$ is harmonic in $\mathbf{C}_+$ if and only if one of the following four conditions holds:*

(a) $\partial f_j/\partial \bar{z} = 0$ for $j = 1, 2$;

(b) $\partial f_j/\partial z = 0$ for $j = 1, 2$;

(c) either f_1 or f_2 is a constant;

(d) there is a constant μ such that f_1 and μf_2 are conjugate harmonic functions.

Proof. Condition (a) means that f_j is holomorphic for $j = 1, 2$, so $f_1 f_2$ is also. Condition (b) means that f_j is antiholomorphic for $j = 1, 2$, so $f_1 f_2$ is also.

Suppose (d) holds, that is, $\partial f_1/\partial x = -\mu \partial f_2/\partial y$ and $\partial f_1/\partial y = \mu \partial f_2/\partial x$. Then the gradients of f_1 and f_2 are orthogonal, so $\Delta(f_1 f_2) = 0$.

We now prove the necessity. Suppose $\Delta(f_1 f_2) = 0$. Then $(\partial f_1/\partial x)(\partial f_2/\partial x) + (\partial f_1/\partial y)(\partial f_2/\partial y) = 0$. If $(\partial f_2/\partial x, \partial f_2/\partial y) = (0, 0)$ in $\mathbf{C}_+$, then we have condition (c). If $(\partial f_2/\partial x, \partial f_2/\partial y) \not\equiv 0$, then there is a function $\lambda = \lambda(x, y)$ such that $(\partial f_1/\partial x, \partial f_1/\partial y) = \lambda(\partial f_2/\partial y, -\partial f_2/\partial x)$ where $\operatorname{grad} f_2 \neq 0$. Thus, λ is defined on a dense open set $U \subset \mathbf{C}_+$. On U, we have

$$\begin{aligned} 0 = \Delta f_1 = \frac{\partial^2 f_1}{\partial x^2} + \frac{\partial^2 f_1}{\partial y^2} &= \lambda \frac{\partial^2 f_2}{\partial x \partial y} + \frac{\partial \lambda}{\partial x}\frac{\partial f_2}{\partial y} - \lambda \frac{\partial^2 f_2}{\partial x \partial y} - \frac{\partial \lambda}{\partial y}\frac{\partial f_2}{\partial x} \\ &= \frac{\partial \lambda}{\partial x}\frac{\partial f_2}{\partial y} - \frac{\partial \lambda}{\partial y}\frac{\partial f_2}{\partial x}. \end{aligned}$$

On the other hand, $(\partial/\partial y)(\partial f_1/\partial x) = (\partial/\partial x)(\partial f_1/\partial y)$, whence

$$\lambda \frac{\partial^2 f_2}{\partial y^2} + \frac{\partial \lambda}{\partial y}\frac{\partial f_2}{\partial y} = -\lambda \frac{\partial^2 f_2}{\partial x^2} - \frac{\partial \lambda}{\partial x}\frac{\partial f_2}{\partial x},$$

that is,

$$\frac{\partial \lambda}{\partial y}\frac{\partial f_2}{\partial y} + \frac{\partial \lambda}{\partial x}\frac{\partial f_2}{\partial x} = 0.$$

Thus, we have the system

$$\begin{cases} \dfrac{\partial \lambda}{\partial x}\dfrac{\partial f_2}{\partial y} - \dfrac{\partial \lambda}{\partial y}\dfrac{\partial f_2}{\partial x} = 0, \\ \dfrac{\partial \lambda}{\partial x}\dfrac{\partial f_2}{\partial x} + \dfrac{\partial \lambda}{\partial y}\dfrac{\partial f_2}{\partial y} = 0. \end{cases}$$

Its determinant is $(\partial f_2/\partial y)^2+(\partial f_2/\partial x)^2 = (\partial f_2/\partial x+i\partial f_2/\partial y)(\partial f_2/\partial x-i\partial f_2/\partial y)$. If this is equal to zero, then we obtain that either $\partial f_2/\partial \bar{z} = 0$ or $\partial f_2/\partial z = 0$, which leads to (a) or (b). If the determinant is not identically equal to zero, then $\partial\lambda/\partial x = \partial\lambda/\partial y = 0$ on an open set, so $\lambda \equiv \text{const}$. Setting $\mu = -\lambda$, we find that

$$\frac{\partial f_1}{\partial x} + \frac{\partial}{\partial y}(\mu f_2) = 0 \quad \text{and} \quad \frac{\partial f_1}{\partial y} = \frac{\partial}{\partial x}(\mu f_2).$$

Thus, for $n = 1$, the hypothesis $\Delta(f_1 f_2) = 0$ implies a well-known condition for the existence of a product (see, for example, [30]). When $n > 1$, the class of harmonic functions f_1 and f_2 for which $\Delta(f_1 f_2) = 0$ is considerably wider. Theorem 12.7 was given in [225]. □

12.8 Boundary values of polyharmonic functions of finite order of growth

Before we consider another class of multipliable distributions, we study the boundary values of polyharmonic functions.

Lemma 12.8. *Let $f(x,y)$ be a polyharmonic function in $\mathbf{R}^{n+1}_+$, that is, $\Delta^m f = 0$. Then*

$$f(x,y) = f_0(x,y) + y f_1(x,y) + \cdots + y^{m-1} f_{m-1}(x,y), \tag{12.5}$$

where the $f_j(x,y)$ are harmonic functions in $\mathbf{R}^{n+1}_+$. If f has finite order of growth as $y \to 0^+$, then all the f_j also have finite order of growth, so $f_j \in \mathcal{G}(\mathbf{R}^{n+1}_+)$.

Proof. The proof is by induction on m. When $m = 1$, equation (12.5) is obvious. Suppose $\Delta^m f = 0$ (where $m > 1$). Then we must have $f(x,y) = \sum_{j=0}^{m-1} y^j f_j(x,y)$, whence

$$\Delta f = \sum_{j=0}^{m-2} y^j \left[(j+2)(j+1) f_{j+2} + 2(j+1)\frac{\partial f_{j+1}}{\partial y}\right].$$

Since $\Delta^{m-1}(\Delta f) = 0$, we have $\Delta f = \sum_{j=0}^{m-2} y^j \varphi_j(x,y)$ by the induction hypothesis. To determine the functions f_j, it suffices to solve the system of equations

$$\varphi_j = (j+2)(j+1) f_{j+2} + 2(j+1)\frac{\partial f_{j+1}}{\partial y}, \qquad j = 0, 1, \ldots, m-2.$$

In particular, $\varphi_{m-2} = 2(m-1)\partial f_{m-1}/\partial y$. We seek a solution of this equation in the form

$$f_{m-1}(x,y) = \frac{1}{2(m-1)} \int_1^y \varphi_{m-2}(x,\xi)\, d\xi + \psi(x).$$

Then

$$\begin{aligned}\Delta f_{m-1} &= \frac{1}{2(m-1)}\int_1^y \widetilde{\Delta}\varphi_{m-2}\,d\xi + \frac{1}{2(m-1)}\frac{\partial\varphi_{m-2}}{\partial y} + \widetilde{\Delta}\psi \\ &= -\frac{1}{2(m-1)}\int_1^y \frac{\partial^2\varphi_{m-2}}{\partial\xi^2}\,d\xi + \frac{1}{2(m-1)}\frac{\partial\varphi_{m-2}}{\partial y} + \widetilde{\Delta}\psi \\ &= \frac{1}{2(m-1)}\frac{\partial\varphi_{m-2}}{\partial y}(x,1) + \widetilde{\Delta}\psi = 0\end{aligned}$$

(we recall that $\widetilde{\Delta}$ is the Laplace operator in $\mathbf{R}^n$). Hence we can find ψ. It is clear that f_{m-1} has finite order of growth as $y \to 0^+$ if the function φ_{m-2} has this property.

Furthermore, we find f_{m-2} in the same way from the equation $\varphi_{m-3} = (m-1)(m-2)f_{m-1} + 2(m-2)\partial f_{m-2}/\partial y$, etc. □

We also obtain from the proof of Lemma 12.8 that if $f \in C^\infty(U \cup \mathbf{R}_+^{n+1})$, then $f_j \in C^\infty(U \cup \mathbf{R}_+^{n+1})$, where U is an open set in $\mathbf{R}^n$.

Corollary 12.9. *Suppose f is a polyharmonic function in $\mathbf{R}_+^{n+1}$ having finite order of growth as $y \to 0^+$. Then $f(x,y) \to S \in \mathcal{D}'(\mathbf{R}^n)$ in the weak topology of $\mathcal{D}'(\mathbf{R}^n)$ as $y \to 0^+$, and $S = \lim_{y\to 0^+} f_0(x,y)$.*

Lemma 12.8 and Corollary 12.9 were proved in [112].

Corollary 12.9 shows that a sufficient condition for the existence of the product of the distributions S_1 and S_2 is the polyharmonicity of the product of their harmonic representations f_1 and f_2, and

$$\Delta^m(f_1 f_2) = 2\Delta^{m-1}\left(\sum_{j=1}^{n+1}\frac{\partial f_1}{\partial x_j}\frac{\partial f_2}{\partial x_j}\right) = 2^m \sum_{j_1,\ldots,j_m=1}^{n+1} \frac{\partial^m f_1}{\partial x_{j_1}\ldots\partial x_{j_m}}\frac{\partial^m f_2}{\partial x_{j_1}\ldots\partial x_{j_m}}$$

(where $x_{n+1} = y$).

12.9 The class of homogeneous multipliable distributions

We consider the class $\mathfrak{M}$ of polyharmonic functions f of the form

$$f(x,y) = P_k(x,y)(|x|^2 + y^2)^{m-k-(n+1)/2}, \qquad k \geq 0, \quad m \geq 1, \tag{12.6}$$

where the $P_k(x,y)$ are homogeneous polynomials of degree k (in x and y), and $\Delta^m P_k = 0$. These functions are polyharmonic functions of order m (that is, $\Delta^m f = 0$) since they are Kelvin transforms (of polyharmonic functions) for the polynomials $P_k(x,y)$. Indeed, if $\Delta^m\varphi = 0$, then the Kelvin transform of φ is the function $\varphi^* = R^{2m-n-1}\varphi(x/R^2, y/R^2)$, where $R = \sqrt{|x|^2+y^2}$. Then $\Delta^m\varphi^* = 0$ (see, for example, [189, p. 535]).

We now clarify which distributions define functions of class $\mathfrak{M}$.

Lemma 12.10. *If a function f of the form (12.6) belongs to $\mathfrak{M}$, then*

$$\int_{\mathbf{R}^n} P_k(x,y)x^\alpha(|x|^2+y^2)^{m-k-(n+1)/2}\,dx=0$$

for all monomials $x^\alpha = x_1^{\alpha_1}\dots x_n^{\alpha_n}$ for which $\|\alpha\| < k+1-2m$, if $k+1-2m>0$.

Proof. Consider the integral

$$I(y)=\int_{B(0,1)} P_k(x,y)x^\alpha(|x|^2+y^2)^{m-k-(n+1)/2}\,dx,\qquad \|\alpha\|<k+1-2m.$$

We remark that a finite limit $\lim_{y\to 0+} I(y) = C$ exists. Indeed, by Lemma 12.8 we can represent $f(x,y)$ in the form $f = f_0 + yf_1 + \dots + y^{m-1}f_{m-1}$, so

$$\lim_{y\to 0+} I(y) = \sum_{j=0}^{m-1} \lim_{y\to 0+} y^j \int_{B(0,1)} f_j(x,y)x^\alpha\,dx.$$

The functions f_j are class $\mathcal{C}^\infty$ on the set $(\mathbf{R}^n\setminus\{0\})\cup\mathbf{R}^{n+1}_+$ since f has this property. We will show the existence of the limit

$$\lim_{y\to 0+}\int_{B(0,1)} f_j(x,y)x^\alpha\,dx,\qquad j=0,1,\dots,m-1.$$

By Lemma 11.2, there are functions $F_j\in\mathcal{G}(\mathbf{R}^{n+1}_+)$ that extend continuously to $\mathbf{R}^{n+1}_+\cup\{x\in\mathbf{R}^n : |x|<2\}$ such that $\partial^{2p}F_j/\partial y^{2p} = f_j$. Then $f_j(x,y) = (-1)^p\widetilde{\Delta}^pF_j(x,y)$, and we have

$$\int_{B(0,1)}\frac{\partial F_j}{\partial x_k}(x,y)\varphi(x)\,dx$$
$$=-\int_{B(0,1)} F_j\frac{\partial\varphi}{\partial x_k}\,dx + (-1)^{k-1}\int_{S(0,1)} F_j(x,y)\varphi(x)\,dx[k].$$

Thus

$$\int_{S(0,1)} f_j(x,y)x^\alpha\,dx = (-1)^p\int_{B(0,1)} F_j(x,y)\widetilde{\Delta}^p x^\alpha\,dx + \int_{S(0,1)} L_1F_jL_2x^\alpha\,d\sigma,$$

where L_1 and L_2 are certain differential operators. It remains to observe that F_j is class $\mathcal{C}^\infty$ on the set $\mathbf{R}^{n+1}_+\cup\{x : 0<|x|<2\}$, since f_j has this property. Therefore the limit $\lim_{y\to 0+} I(y) = C$ exists.

Making in $I(y)$ the change of variables $x = r\xi$, $r>0$, $|\xi|=1$, we have

$$I(y)=\sum_{s\ge 0} y^s\int_0^1\frac{r^{\|\alpha\|+n-1+k-s}\,dr}{(r^2+y^2)^{k-m+(n+1)/2}}\int_{S(0,1)} P_k^s(\zeta)\zeta^\alpha\,d\sigma,$$

where $P_k(x,y) = \sum_{s\ge 0} y^sP_k^s(x)$, and $d\sigma$ is the surface area element for the sphere.

Making the substitution $r = yt$, we obtain

$$I(y) = y^{\|\alpha\|-k-1+2m} \int_0^{1/y} \sum_{s\geq 0} \frac{t^{\|\alpha\|+n-1+k-s}\,dt}{(1+t^2)^{k-m+(n+1)/2}} \int_{S(0,1)} P_k^s(\zeta)\zeta^\alpha\,d\sigma. \tag{12.7}$$

Since $I(y)$ has a limit as $y \to 0^+$, we have from (12.7) that

$$J = \int_0^\infty \sum_{s\geq 0} \frac{t^{\|\alpha\|+n-1+k-s}\,dt}{(1+t^2)^{k-m+(n+1)/2}} \int_{S(0,1)} P_k^s(\zeta)\zeta^\alpha\,d\sigma = 0,$$

since this improper integral converges. It is easy to see that

$$\int_{\mathbf{R}^n} P_k(x,y)(|x|^2+y^2)^{m-k-(n+1)/2} x^\alpha\,dx = J\cdot y^{\|\alpha\|-k-1+2m} = 0.$$

□

Lemma 12.11. *If $f(x,y)$ of the form (12.6) lies in $\mathfrak{M}$, and $k+1-2m \geq 0$, then*

$$I = \mathrm{P.V.}\int_{\mathbf{R}^n} P_k(x,y)x^\alpha(|x|^2+y^2)^{m-k-(n+1)/2}\,dx = C_\alpha, \qquad \|\alpha\| = k+1-2m,$$

where

$$C_\alpha = \sum_{s\geq 1} \frac{\pi^{n/2}\Gamma(s-1/2)\widetilde{\Delta}^{k+1-s-m}(P_k^{2s-1}(\xi)\xi^\alpha)}{2^{2(k+1-s-m)}(k+1-s-m)!\,\Gamma(k-m+(n+1)/2)} \tag{12.8}$$

($\widetilde{\Delta}$ is the Laplace operator on $\mathbf{R}^n$).

Proof. Consider the integral $I(y) = \int_{B(0,1)} P_k(x,y)x^\alpha(|x|^2+y^2)^{m-k-(n+1)/2}\,dx$. Passing to polar coordinates, as in Lemma 12.10, we obtain

$$I(y) = \sum_{s\geq 0} \int_0^{1/y} \frac{t^{2k+n-s-2m}\,dt}{(1+t^2)^{k-m+(n+1)/2}} \int_{S(0,1)} P_k^s(\xi)\xi^\alpha\,d\sigma.$$

The polynomials P_k^s have degree $k-s$, so $P_k^s\xi^\alpha$ has degree $2k+1-2m-s$. Thus, if s is even, the degree of the homogeneous polynomial $P_k^s\xi^\alpha$ is odd, so $\int_{S(0,1)} P_k^s(\xi)\xi^\alpha\,d\sigma = 0$. Consequently

$$I(y) = \sum_{s\geq 1} \int_0^{1/y} \frac{t^{2k+n+1-2s-2m}\,dt}{(1+t^2)^{k-m+(n+1)/2}} \int_{S(0,1)} P_k^{2s-1}(\xi)\xi^\alpha\,d\sigma.$$

It is clear that $I(y) \to I$ as $y \to 0^+$, where

$$I = \sum_{s\geq 1} \int_0^\infty \frac{t^{2k+n+1-2s-2m}\,dt}{(1+t^2)^{k-m+(n+1)/2}} \int_{S(0,1)} P_k^{2s-1}(\xi)\xi^\alpha\,d\sigma.$$

However,

$$\int_0^\infty \frac{t^{2k+n+1-2s-2m}\,dt}{(1+t^2)^{k-m+(n+1)/2}} = \frac{1}{2}\mathrm{B}(k-s-m+1+n/2, s-1/2)$$

and

$$\int_{S(0,1)} P_k^{2s-1}(\xi)\xi^\alpha\,d\sigma = \frac{2\pi^{n/2}\widetilde{\Delta}^{k+1-s-m}(P_k^{2s-1}(\xi)\xi^\alpha)}{2^{2(k+1-s-m)}(k+1-s-m)!\,\Gamma(k+1-s-m+n/2)}.$$

The last equality follows from Gauss's formula for homogeneous polynomials (see, for example, [189, chap. 11]). Indeed, if P is a harmonic polynomial, then by the mean-value property, $\int_{S(0,1)} P\,d\sigma = P(0)\sigma_n$. Gauss's representation for $P_k^{2s-1}\xi^\alpha$ has the form

$$P_k^{2s-1}\xi^\alpha = \sum_{l\geq 0}(|\xi|/2)^{2l} Z_{2(k+1-m-s-l)}(\xi),$$

where the homogeneous harmonic polynomials

$$Z_{2(k+1-m-s-l)} = \frac{(2k-2m-2s-2l+n/2+1)}{l!\,\Gamma(2k-2m-2s-l+2+n/2)}\times$$
$$\times\sum_{j\geq 0}(-1)^j\frac{\Gamma(2k-2m-2s-2l-j+n/2+1)}{\Gamma(j+1)}(|\xi|/2)^{2j}\widetilde{\Delta}^{j+2}(P_k^{2s-1}\xi^\alpha).$$

Thus $\int_{S(0,1)} P_k^{2s-1}\xi^\alpha\,d\sigma = \sigma_n Z_0$, while $k+1-m-s-l=0$ when $l = k+1-m-s$, and

$$\sigma_n Z_0 = \frac{(n/2-1)\Gamma(n/2-1)\sigma_n\widetilde{\Delta}^{k+1-m-s}(P_k^{2s-1}\xi^\alpha)}{(k+1-n-s)!\,\Gamma(k+1-m-s+n/2)2^{2(k+1-m-s)}}$$
$$= \frac{2\pi^{n/2}\widetilde{\Delta}^{k+1-s-m}(P_k^{2s-1}\xi^\alpha)}{2^{2(k+1-m-s)}(k+1-m-s)!\,\Gamma(k+1-m-s+n/2)}.$$

Hence we obtain that $I = C_\alpha$. □

We remark that if P_k is expanded in even powers of y, then $C_\alpha = 0$ for all α such that $\|\alpha\| = k+1-2m$.

Lemmas 12.10 and 12.11 were given in [112].

Let $h(x) = Q_k(x)|x|^{2m-2k-n-1}$, where Q_k is an arbitrary homogeneous polynomial in $\mathbf{R}^n$ of degree k. We denote by $\widehat{h}$ the distribution obtained by regularizing h, that is,

$$\widehat{h}(\varphi) = \text{p.f.}\int_{\mathbf{R}^n} h(x)\varphi(x)\,dx$$
$$= \text{P.V.}\int_{\mathbf{R}^n} h(x)\Big(\varphi(x) - \sum_{\|\alpha\|\leq k+1-2m}\frac{\partial^\alpha\varphi(0)}{\alpha!}x^\alpha\Big)\,dx, \qquad \varphi\in\mathcal{D}(\mathbf{R}^n),$$

if $k+1-2m \geq 0$. If $k+1-2m < 0$, we have $\widehat{h}(\varphi) = \int_{\mathbf{R}^n} h(x)\varphi(x)\,dx$, since in this case the integral is absolutely convergent. In other words, $\widehat{h}$ is the Hadamard finite part of the integral.

Lemma 12.12. *Let $Q(x)$ be an arbitrary polynomial in $\mathbf{R}^n$. Then*

$$\widetilde{Q}(x,y) = \sum_{l\geq 0} \frac{(-1)^l}{(2l)!} y^{2l} \widetilde{\Delta}(Q(x))$$

is a harmonic extension of the polynomial Q to the whole space $\mathbf{R}^n$.

The lemma is proved by a direct calculation.

If $h(x) = Q_k(x)|x|^{2m-2k-n-1}$, then we denote by $\widetilde{h}(x,y)$ the function in $\mathfrak{M}$ of the form

$$\widetilde{h}(x,y) = \widetilde{Q}_k(x,y)(|x|^2+y^2)^{m-k-(n+1)/2},$$

where $\widetilde{Q}_k$ is defined in Lemma 12.12.

Theorem 12.13 (Kytmanov). *The distribution $\widehat{h}$ coincides with the boundary value of the function $\widetilde{h}(x,y)$, that is,*

$$\lim_{\epsilon\to 0^+} \int_{\mathbf{R}^n} \widetilde{h}(x,\epsilon)\varphi(x)\,dx = \widehat{h}(\varphi), \qquad \varphi \in \mathcal{D}(\mathbf{R}^n).$$

Proof. If $k+1-2m < 0$, then the equality follows because we can pass to the limit under the integral sign by the Lebesgue dominated convergence theorem. If $k+1-2m \geq 0$, then according to Lemmas 12.11 and 12.10, we have

$$\text{P.V.} \int_{\mathbf{R}^n} \widehat{h}(x,y)x^\alpha\,dx = 0,$$

if $\|\alpha\| \leq k+1-2m$. Therefore

$$\int_{\mathbf{R}^n} \widehat{h}(x,\epsilon)\varphi(x)\,dx = \text{P.V.} \int_{\mathbf{R}^n} \widehat{h}(x,\epsilon)\Big(\varphi(x) - \sum_{\|\alpha\|\leq k+1-2m} \frac{\partial^\alpha\varphi(0)}{\alpha!}x^\alpha\Big)\,dx.$$

In the last integral we may pass to the limit as $y \to 0^+$ since the function $h(x)\big(\varphi(x) - \sum_{\|\alpha\|\leq k+1-m}(1/\alpha!)\partial^\alpha\varphi(0)x^\alpha\big)$ has an integrable singularity at 0. □

Theorem 12.13 was given in [112].

We denote by $\mathfrak{M}_h$ the functions in $\mathfrak{M}$ of the form (12.6) whose numerator $P_k(x,y)$ expands in even powers of y, and by $\mathfrak{M}_\delta$ the set of functions in $\mathfrak{M}$ whose numerator expands in odd powers of y. It is clear that every function $f \in \mathfrak{M}$ is a linear combination of functions in $\mathfrak{M}_h$ and $\mathfrak{M}_\delta$.

It follows from Theorem 12.13 that the boundary values of functions $f \in \mathfrak{M}_h$ are the distributions $\widehat{f}(x,0)$.

Theorem 12.14 (Kytmanov). *Suppose $f(x, y)$ of the form (12.6) lies in $\mathfrak{M}_\delta$. Then its boundary value is the distribution*

$$f_0 = \begin{cases} (-1)^{k+1-2m} \sum_{\|\alpha\|=k+1-2m} \dfrac{C_\alpha}{\alpha!}\partial^\alpha \delta, & k+1-2m \geq 0, \\ 0, & k+1-2m < 0, \end{cases}$$

where C_α is defined by (12.8).

The proof of the theorem follows from Lemmas 12.10 and 12.11.

Thus, we have completely described the class of distributions represented by functions of $\mathfrak{M}$.

Example. Suppose $f(x, y) = y(|x|^2 + y^2)^{-(n+1)/2}$. Then

$$f_0 = \frac{\pi^{(n+1)/2}}{\Gamma((n+1)/2)}\delta,$$

since $C_\alpha = C_{(0,\dots,0)} = \pi^{n/2}\Gamma(1/2)/\Gamma((n+1)/2)$ by formula (12.8).

We now consider the question of when the product of two functions in $\mathfrak{M}$ is again in $\mathfrak{M}$.

Suppose

$$\begin{aligned} f(x, y) &= P_k(x, y)(|x|^2 + y^2)^{m-k-(n+1)/2}, \qquad \text{and} \\ g(x, y) &= Q_l(x, y)(|x|^2 + y^2)^{s-l-(n+1)/2}; \qquad \text{then} \\ fg &= P_k Q_l(|x|^2 + y^2)^{(m+s-(n+1)/2)-l-k-(n+1)/2}. \end{aligned}$$

Hence in order that $fg \in \mathfrak{M}$, it is necessary and sufficient that n be odd ($n = 2p-1$), $m + s \geq p$, and $\Delta^{m+s-p}(P_k Q_l) = 0$.

Theorem 12.15 (Kytmanov). *Suppose f, g, and fg are in $\mathfrak{M}$.*

1. *If f and g are in $\mathfrak{M}_h$, then $fg \in \mathfrak{M}_h$, and $\widehat{f}(x, 0) \circ \widehat{g}(x, 0) = \widehat{fg}(x, 0)$.*
2. *If f and g are in $\mathfrak{M}_\delta$, then $f_0 \circ g_0 = 0$.*
3. *If $f \in \mathfrak{M}_h$ and $g \in \mathfrak{M}_\delta$, then $fg \in \mathfrak{M}_\delta$, and*

$$(-1)^{k-2p-m}\widehat{f}(x, 0) \circ \sum_{\|\alpha\|=l+1-2s} \frac{C_\alpha}{\alpha!}\partial^\alpha \delta = \sum_{\|\beta\|=l+k+2p-2m-2s-1} \frac{C_\beta}{\beta!}\partial^\beta \delta.$$

Theorems 12.14 and 12.15 were obtained in [112]. For $n = 1$, identities of this form were given in [85].

Example. If $f = (p-1)!\,\pi^{-p}y(|x|^2+y^2)^{-p}$, then $f_0 = \delta$. If $h(x) = Q_k(x)|x|^{-2k}$, where $\Delta Q_k = 0$, then

$$\widehat{h}\circ\delta = (-1)^k \sum_{\|\alpha\|=k} \frac{(p-1)!\,\widetilde{\Delta}^k(Q_k x^\alpha)}{k!\,(k+p-1)!}\delta^{(\alpha)}.$$

In particular, if $h = x_j|x|^{-2}$, then $\widehat{x_j|x|^{-2}}\circ\delta = (-1/2p)\partial\delta/\partial x_j$. When $p=1$, this identity reduces to the classical one: $\mathrm{P.V.}\,x^{-1}\circ\delta = -\delta'/2$ (see [15]).

We remark that the wave front set of the delta function is $\{0\}\times(\mathbf{R}^n\setminus\{0\})$, so Hörmander's method for multiplication of distributions is not applicable to the product of the delta function with $x_j|x|^{-2}$.

12.10 Further results

Ivanov [88] considers a more general class of functions. Let $f_0(x), \ldots, f_m(x)$ be an associated system of infinitely differentiable functions on $\mathbf{R}^n\setminus\{0\}$ that are formally homogeneous of degree $-\nu$ (where $\nu > 0$); that is, for all $k = 1, \ldots, m$, $x \neq 0$, and $t > 0$, we have

$$f_0(tx) = t^{-\nu}f_0(x) \quad\text{and}\quad f_k(tx) = t^{-\nu}f_k(x) + t^{-\nu}(\log t)f_{k-1}(x).$$

If $\nu < n$, then $f_k(x)$ can be identified with a regular distribution, since the integral $\int_{\mathbf{R}^n} f_k(x)\varphi(x)\,dx$ converges. If $\nu \geq n$, then as the distribution generated by this element we consider the Hadamard finite part, that is,

$$f_k(\varphi) = \mathrm{p.\,f.}\int_{\mathbf{R}^n} f_k\varphi\,dx.$$

We denote the class of distributions generated in this way by $\mathcal{P}$. If $P(x,y)$ is the Poisson kernel for the half-space, then we can define the Poisson representation for $f \in \mathcal{P}$:

$$f(x,y) = f_\xi(P(x-\xi,y)) = f(x) * P(x,y).$$

Then $f(x,y)$ is a harmonic representation of the distribution f.

If is clear that if $f(x)$ is an associated formally homogeneous function of degree $-\nu$, then $f(x,y)$ is also an associated homogeneous function of the same degree. We denote the class of such representations by $\mathcal{P}_y$.

Consider the algebra $\mathcal{P}_y^*$ generated by the elements of the space $\mathcal{P}_y$ and consisting of functions of the form

$$f(x,y) = \prod_{j=1}^{k} f_j(x,y), \qquad f_j \in \mathcal{P}_y. \tag{12.9}$$

Functions of the form (12.9) are homogeneous, and for $y=0$ they are infinitely differentiable in $\mathbf{R}^n$ away from 0.

Sometimes the boundary value of $f(x, y)$ at $y = 0$ determines a distribution, which we may suppose to be the product of the functions $f_j(x, 0)$. But usually this boundary value does not exist, and we consider instead the asymptotic expansion of $f(x, y)$ as $y \to 0^+$.

Making the change of variables $x = y\xi$, where $\xi \in \mathbf{R}^n$, we obtain from (12.9) the representation

$$f(x, y) = y^{-\nu} \sum_{j=0}^{s} g_j(\xi, 1) \log^j y.$$

Suppose the degree of homogeneity of the function $f_j(x, y)$ in (12.9) is ν_j, and $\nu = \sum_{j=1}^{k} \nu_j$. If $\nu < n$, then $f(x, 0)$ is an integrable function in a neighborhood of the origin in $\mathbf{R}^n$, and by Lebesgue's dominated convergence theorem we can pass to the limit as $y \to 0^+$ in the integral

$$F(y) = \int_{\mathbf{R}^n} f(x, y)\varphi(x)\, dx, \qquad \varphi \in \mathcal{D}(\mathbf{R}^n), \tag{12.10}$$

to obtain

$$\lim_{y \to 0^+} F(y) = \int_{\mathbf{R}^n} f(x, 0)\varphi(x)\, dx.$$

Consequently, in this case the product $f(x, 0) = \prod_{j=1}^{k} f_j(x, 0)$ is defined.

If $\nu \geq n$, then we have the following.

Theorem 12.16 (Ivanov). *If $\nu \geq n$ and ν is not an integer, say $[\nu] = m$, then as $y \to 0^+$ the integral (12.10) has an asymptotic expansion of the form*

$$\begin{aligned} F(y) &= \sum_{\|\alpha\|=0}^{m-n} \frac{\partial^\alpha \varphi(0)}{\alpha!} y^{-\nu+\|\alpha\|+n} \sum_{j=0}^{s} \log^j y \int_{\mathbf{R}^n} g_j(\xi, 1)\xi^\alpha\, d\xi \\ &\qquad + \text{p.f.} \int_{\mathbf{R}^n} f(x)\varphi(x)\, dx + o(1) \\ &= F_\sigma(y) + F_r(0) + F_i(y). \end{aligned}$$

Proof. We write (12.10) in the form

$$F(y) = \int_{\mathbf{R}^n} f(x, y) T_{m-n}(x, \varphi)\, dx + \int_{\mathbf{R}^n} f(x, y)(\varphi - T_{m-n}(x, \varphi))\, dx,$$

where $T_j(x, \varphi) = \sum_{\|\alpha\| \leq j} \partial^\alpha \varphi(0) x^\alpha / \alpha!$ is the Taylor polynomial of φ.

In the first integral, we make the change of variable $x = y\xi$, which gives the singular part $F_\sigma(y)$; in the second integral we may pass to the limit as $y \to 0^+$, which lets us represent the integral as the sum $F_r(0) + F_i(y)$ of a regular part and an infinitesimal part. $\square$

Theorem 12.16 was given in [88]. If ν is an integer, $F(y)$ has a similar expansion (see [88]).

Thus, each function $f \in \mathcal{P}_y^*$ is represented in the form

$$f = f_\sigma(x, y) + \text{p.f.}\, f(x, 0) + f_i(x, y).$$

Consequently, the following theorem holds.

Theorem 12.17 (Ivanov). *The algebra $\mathcal{P}_y^*$ is the direct sum of three linear subspaces:*

$$\mathcal{P}_y^* = \mathcal{P}_\sigma^*(y) \oplus \mathcal{P}_r^*(y) \oplus \mathcal{P}_i^*(y).$$

Here $\mathcal{P}_r^*(y) \subset \mathcal{D}'(\mathbf{R}^n)$, but none of these subspaces is an ideal or even a subalgebra of $\mathcal{P}_y^*$.

When we constructed the space of hyperdistributions, it was in fact an isomorphism: $\mathcal{P}_y^*/\mathcal{P}_i^*(y) = \mathcal{P}_\sigma^*(y) \oplus \mathcal{P}_r^*(y)$.

We remark that Ivanov's work [88] considers, in place of the Poisson kernel, any δ-kernel with certain properties.

13 The generalized Fourier transform

13.1 Functions of slow growth

Let $\mathcal{S} = \mathcal{S}(\mathbf{R}^n)$ be the space of rapidly decreasing functions, that is, $\varphi \in \mathcal{S}(\mathbf{R}^n)$ if $\varphi \in C^\infty(\mathbf{R}^n)$, and for every derivative $\partial^\alpha \varphi$ and every $k \geq 0$,

$$|\partial^\alpha \varphi(x)| \leq C_{\alpha,k}(1 + |x|)^{-k}, \qquad x \in \mathbf{R}^n.$$

We introduce a topology in $\mathcal{S}$ by the system of seminorms

$$\varphi \mapsto \sup_{x \in \mathbf{R}^n} |\partial^\alpha \varphi(x)|(1 + |x|)^k.$$

The space $\mathcal{S}'$ is the set of continuous linear functionals on $\mathcal{S}$. Elements of $\mathcal{S}'$ are called distributions of slow growth.

The Fourier transform $F[\varphi]$ of a function $\varphi \in \mathcal{S}$ is defined as usual:

$$F[\varphi](t) = \int_{\mathbf{R}^n} \varphi(x) e^{i\langle t, x\rangle}\, dx.$$

Then $F[\varphi] \in \mathcal{S}$. The Fourier transform of a distribution T in $\mathcal{S}'$ is the functional

$$F[T](\varphi) = T(F[\varphi]), \qquad \varphi \in \mathcal{S}.$$

Then $F[T] \in \mathcal{S}'$.

It is well known what a large role these ideas play in mathematical physics (see, for example, [30, 78, 212]). Closely connected with the Fourier transform is the Laplace transform, which allows us to view the Fourier transform as the boundary value of a holomorphic function and to bring to bear the powerful methods of complex analysis [30, 212]. At the same time, we showed in §11 that distributions can be viewed as boundary values of harmonic functions. Such a representation turned out to be useful in the question of multiplication of distributions (see §12). Here we will show that the harmonic representation can be useful in studying the Fourier transform of distributions in $\mathcal{S}'$. We will basically follow the results of [120]. A similar construction for distributions with compact support was considered by Zaslavskiĭ in [226].

A Borel measure μ on $\mathbf{R}^n$ is a measure of slow growth if there exists a number $m > 0$ for which

$$\int_{\mathbf{R}^n} (1+|x|)^{-m} |d\mu| < +\infty. \tag{13.1}$$

If (13.1) holds for a measure μ of the form $d\mu = f\,dx$, where $f \in \mathcal{L}^1_{\mathrm{loc}}(\mathbf{R}^n)$, then f is a function of slow growth in $\mathbf{R}^n$.

Measures and functions of slow growth determine distributions in $\mathcal{S}'$. Moreover, every distribution T in $\mathcal{S}'$ is a derivative of finite order of a continuous function f of slow growth (see, for example, [212, p. 94]), that is, $T = \partial^\alpha f$ in $\mathcal{S}'$, where $\alpha = (\alpha_1, \dots, \alpha_n)$ is a multi-index.

As in §12, we will consider $\mathbf{R}^n$ as the boundary of the half-space $\mathbf{R}^{n+1}_+ = \{\,(x,y) : x \in \mathbf{R}^n,\ y > 0\,\}$. We recall that the space $\mathcal{G}(\mathbf{R}^{n+1}_+)$ consists of harmonic functions f in $\mathbf{R}^{n+1}_+$ having finite order of growth as $y \to 0^+$. In other words, for every compact set $K \subset \mathbf{R}^n$, there are constants $m \geq 0$ and $C > 0$ such that

$$|f(x,y)| \leq C y^{-m}, \qquad x \in K, \quad 0 < y \leq 1.$$

If μ is a measure of slow growth, we define the generalized Fourier transform $\widetilde{F}[\mu]$ of μ in the following way:

$$\widetilde{F}[\mu](x,y) = F[\mu(t)e^{-|t|y}] = \int_{\mathbf{R}^n} e^{i\langle t,x\rangle - |t|y}\, d\mu(t), \qquad y > 0. \tag{13.2}$$

Our construction (13.2) is reminiscent of the construction of Bros and Iagolnitzer [31] (and also [78, chap. 9]), who introduced a nonlinear Fourier transform. Ours differs in that the function $|t|$ is not analytic in a neighborhood of the origin, so that the generalized transform $\widetilde{F}$ is not connected with the analytic wave front set, which is the case for the transform of Bros and Iagolnitzer (see [78, chap. 9]).

When $f \in \mathcal{L}^1(\mathbf{R}^n)$, the function $\widetilde{F}[f]$ is, in the terminology of the book [195], the Abel mean of the function $f(t)e^{i\langle t,x\rangle}$.

Theorem 13.1 (Kytmanov). *Let μ be a measure of slow growth. Then $\widetilde{F}[\mu] \in \mathcal{G}(\mathbf{R}^{n+1}_+)$, and for every $\varphi \in \mathcal{S}$,*

$$\lim_{y\to 0^+} \int_{\mathbf{R}^n} \widetilde{F}[\mu](x,y)\varphi(x)\,dx = F[\mu](\varphi) = \langle F[\mu], \varphi\rangle.$$

Proof. It is clear that $\widetilde{F}[\mu]$ is a harmonic function in $\mathbf{R}^{n+1}_+$. In view of (13.1), we have

$$\begin{aligned}\left|\widetilde{F}[\mu]\right| &\le \int_{\mathbf{R}^n} e^{-|t|y}\,|d\mu| = \int_{\mathbf{R}^n} e^{-|t|y}(1+|t|)^m(1+|t|)^{-m}\,|d\mu| \\ &\le Cy^{-m}\int_{\mathbf{R}^n}(1+|t|)^{-m}\,|d\mu| = C_1 y^{-m}.\end{aligned}$$

We remark that this estimate holds for every $x \in \mathbf{R}^n$; in particular, the function $\widetilde{F}$ is bounded for fixed y, and consequently is a function of slow growth in $\mathbf{R}^n$.

For every $\varphi \in \mathcal{S}$, we have

$$\begin{aligned}\int_{\mathbf{R}^n} \widetilde{F}[\mu](x,y)\varphi(x)\,dx &= \int_{\mathbf{R}^n}\varphi(x)\,dx \int_{\mathbf{R}^n} e^{i\langle t,x\rangle - |t|y}\,d\mu(t) \\ &= \int_{\mathbf{R}^n} e^{-|t|y}\,d\mu(t)\int_{\mathbf{R}^n}\varphi(x)e^{i\langle t,x\rangle}\,dx \\ &= \int_{\mathbf{R}^n} F[\varphi](t)e^{-|t|y}\,d\mu(t).\end{aligned}$$

On every compact set $K \subset \mathbf{R}^n$, the function $e^{-|t|y}F[\varphi]$ converges uniformly to $F[\varphi]$ as $y \to 0^+$, while $\left|e^{-|t|y}F[\varphi]\right| \le |F[\varphi]|$ in $\mathbf{R}^n$, and $F[\varphi] \in \mathcal{S}$, so in the last integral we can pass to the limit as $y \to 0^+$ under the integral sign. □

A number of properties of the Fourier transform F carry over to $\widetilde{F}$:

1. $\partial_x^\alpha \widetilde{F}[\mu] = \widetilde{F}[(it)^\alpha \mu(t)]$, where $(it)^\alpha = i^{\|\alpha\|} t_1^{\alpha_1} \dots t_n^{\alpha_n}$;
2. $\widetilde{F}[\mu](x + x^0, y) = \widetilde{F}[e^{i\langle x^0, t\rangle}\mu(t)](x,y)$;
3. if f is a function of slow growth, and A is an orthogonal transformation in $\mathbf{R}^n$, then

$$\widetilde{F}[f(At)](x,y) = \widetilde{F}[f(t)](Ax,y).$$

These properties follow at once from the definition of $\widetilde{F}$.

We now consider some **examples**.

I. If δ_{t^0} is the delta function at t^0, then $\widetilde{F}[\delta_{t^0}] = e^{i\langle t^0, x\rangle - |t^0|y}$, and in particular $\widetilde{F}[\delta] = 1$.

II. $\widetilde{F}[1] = (2\pi)^n P(x, y)$, where $P(x, y)$ is the Poisson kernel for the half-space. Indeed, consider

$$\widetilde{F}[1] = \int_{\mathbf{R}^n} e^{i\langle t, x \rangle - |t|y}\, dt.$$

We find this expression by following [193, chap. 3, §2]. We first compute the Fourier transform of $e^{-|t|^2/2}$. Let

$$h(x) = \int_{\mathbf{R}^n} e^{-|t|^2/2} e^{i\langle t, x \rangle}\, dt.$$

Since the right-hand side is invariant with respect to orthogonal transformations, $h(x) = h(|x|)$. Then

$$\frac{\partial h}{\partial x_j} = i \int_{\mathbf{R}^n} e^{-|t|^2/2} t_j e^{i\langle t, x \rangle}\, dt = -i \int_{\mathbf{R}^n} e^{i\langle t, x \rangle} \frac{\partial}{\partial t_j} e^{-|t|^2/2}\, dt$$

$$= i \int_{\mathbf{R}^n} e^{-|t|^2/2} \frac{\partial}{\partial t_j} e^{i\langle t, x \rangle} = -x_j h(x),$$

that is, $\partial h/\partial |x| = -|x|h$, whence $h = Ce^{-|x|^2/2}$.

To find C, we set $x = 0$. Then (putting $x = r\xi$, $|\xi| = 1$, $r > 0$)

$$h(0) = \int_{\mathbf{R}^n} e^{-|t|^2/2}\, dt = \int_0^\infty e^{-r^2/2} r^{n-1}\, dr \int_{S(0,1)} d\sigma$$

$$= \frac{\pi^{n/2} 2^{n/2}}{\Gamma(n/2)} \int_0^\infty e^{-\tau} \tau^{(n/2)-1}\, d\tau = (2\pi)^{n/2}.$$

Therefore $h(x) = (2\pi)^{n/2} e^{-|x|^2/2}$. Hence for $u > 0$,

$$\int_{\mathbf{R}^n} e^{-u|t|^2} e^{i\langle x, t \rangle}\, dt = (\pi/u)^{n/2} e^{-|x|^2/4u}. \tag{13.3}$$

On the other hand,

$$e^{-\gamma} = \frac{1}{\sqrt{\pi}} \int_0^\infty \frac{e^{-u}}{\sqrt{u}} e^{-\gamma^2/4u}\, du, \qquad \gamma > 0. \tag{13.4}$$

To prove (13.4), we compute (by the residue theorem)

$$\frac{1}{2\pi i} \int_{-\infty}^{+\infty} \frac{e^{i\gamma x}}{1 + x^2}\, dx = \frac{e^{i\gamma x}}{i + x}\bigg|_{x=i} = e^{-\gamma/2i},$$

so $e^{-\gamma} = (1/\pi) \int_{-\infty}^{+\infty} e^{i\gamma x}/(1 + x^2)\, dx$. We represent $(1 + x^2)$ as $\int_0^\infty e^{-(1+x^2)u}\, du$. Then (using (13.3)),

$$e^{-\gamma} = \frac{1}{\pi} \int_0^\infty e^{-u}\, du \int_{-\infty}^{+\infty} e^{i\gamma x} e^{-ux^2}\, dx = \int_0^\infty \frac{e^{-u}}{\sqrt{\pi u}} e^{-\gamma^2/4u}\, du.$$

We now return to $\widetilde{F}[1]$. From (13.4), we obtain

$$\begin{aligned}\widetilde{F}[1] &= \int_{\mathbf{R}^n} e^{i\langle t,x\rangle - |t|y}\,dt = \frac{1}{\sqrt{\pi}}\int_{\mathbf{R}^n}\left(\int_0^\infty \frac{e^{-u}}{\sqrt{u}}\, e^{-|t|^2y^2/4u}\,du\right)e^{i\langle t,x\rangle}\,dt \\ &= \frac{1}{\sqrt{\pi}}\int_0^\infty \frac{e^{-u}}{\sqrt{u}}\,du \int_{\mathbf{R}^n} e^{-|t|^2y^2/4u} e^{i\langle t,x\rangle}\,dt.\end{aligned}$$

We now apply (13.3), changing u to $y^2/4u$, to get

$$\begin{aligned}\widetilde{F}[1] &= \frac{1}{\sqrt{\pi}}\int_0^\infty \frac{e^{-u}}{\sqrt{u}}\,(4\pi u y^{-2})^{n/2} e^{-u|x|^2/y^2}\,du \\ &= \frac{(2\pi)^n}{\pi^{(n+1)/2}y^n}\int_0^\infty e^{-u(1+|x|^2/y^2)}u^{(n-1)/2}\,du \\ &= \frac{(2\pi)^n y}{\pi^{(n+1)/2}(y^2+|x|^2)^{(n+1)/2}}\int_0^\infty e^{-u}u^{(n+1)/2-1}\,du \\ &= \frac{(2\pi)^n\Gamma\left(\frac{n+1}{2}\right)y}{\pi^{(n+1)/2}(y^2+|x|^2)^{(n+1)/2}} = (2\pi)^n P(x,y).\end{aligned}$$

III. $\widetilde{F}[t^\alpha] = (-i)^{\|\alpha\|}\partial_x^\alpha \widetilde{F}[1] = (-i)^{\|\alpha\|}(2\pi)^n \partial_x^\alpha P(x,y)$.

13.2 Distributions of slow growth

When $T \in \mathcal{S}'$, the generalized Fourier transform of T cannot be defined by (13.2), because the function $e^{i\langle t,x\rangle-|t|y}$ is not smooth in t in $\mathbf{R}^n$. Instead, we proceed in the following way. Since $T \in \mathcal{S}'$, there is a continuous function f of slow growth such that $\partial^\alpha f = T$ in $\mathcal{S}'$. Consider the function $\Phi(x,y) = (-ix)^\alpha \widetilde{F}[f](x,y)$. It is polyharmonic in $\mathbf{R}^{n+1}_+$, so by Lemma 12.8 it can be represented in the form

$$\Phi(x,y) = \Phi_0(x,y) + y\Phi_1(x,y) + \cdots + y^k\Phi_k(x,y),$$

where the $\Phi_j \in \mathcal{G}(\mathbf{R}^{n+1}_+)$ for $j = 0, \dots, k$, are functions of slow growth for fixed y (see the proof of Lemma 12.8).

We define the generalized Fourier transform of T via $\widetilde{F}[T](x,y) = \Phi_0(x,y)$.

Corollary 13.2. *For $T \in \mathcal{S}'$ and $\varphi \in \mathcal{S}$, we have*

$$\lim_{y\to 0+}\int_{\mathbf{R}^n}\widetilde{F}[T](x,y)\varphi(x)\,dx = F[T](\varphi).$$

Proof. Since $\Phi_j \in \mathcal{G}(\mathbf{R}^{n+1}_+)$, the function $\Phi_j \to T_j$ in $\mathcal{S}'$ as $y \to 0^+$. Therefore $y^j\Phi_j(x,y) \to 0$ as $y \to 0^+$ for $j = 1, \dots, k$. Using Theorem 13.1, we obtain

$$\begin{aligned}\lim_{y\to 0+}\int_{\mathbf{R}^n}\widetilde{F}[T](x,y)\varphi(x)\,dx &= \lim_{y\to 0+}\int_{\mathbf{R}^n}(-ix)^\alpha\widetilde{F}[f](x,y)\varphi(x)\,dx \\ &= \langle F[f],(-ix)^\alpha\varphi(x)\rangle = \langle F[T],\varphi\rangle.\end{aligned}$$

□

It is clear that the generalized Fourier transform is not uniquely defined, but the difference of two such transforms produces the zero distribution on $\mathbf{R}^n$, so by Corollary 11.9 this difference is a harmonic function in $\mathbf{R}^{n+1}$ that equals zero on $\mathbf{R}^n$.

It is easy to see, using Theorem 13.1 and Corollaries 13.2 and 11.9, that a harmonic representation of a distribution $T \in \mathcal{S}'$ is a function f in $\mathcal{G}(\mathbf{R}^{n+1}_+)$ for which there are positive constants C, m, and k such that

$$|f(x,y)| \le C(1+|x|)^m y^{-k}, \qquad x \in \mathbf{R}^n, \quad 0 < y \le 1.$$

Example. Let us find $\widetilde{F}[\partial^\alpha \delta]$. Since $(-ix)^\alpha \widetilde{F}[\delta] = (-ix)^\alpha$,

$$\widetilde{F}[\partial^\alpha \delta](x,y) = (-i)^{\|\alpha\|} \sum_{j\ge 0} (-i)^j y^{2j} \widetilde{\Delta}^j x^\alpha/(2j)!$$

(see Lemma 12.12).

13.3 The inversion formula

We now consider the question of the inverse of the generalized Fourier transform. First suppose that f is a continuous function of slow growth. Then $f(t)e^{-|t|y}$ is a function of class $\mathcal{L}^2(\mathbf{R}^n) \cap \mathcal{L}^1(\mathbf{R}^n)$ for $y > 0$. Consequently $\widetilde{F}[f] \in \mathcal{L}^2(\mathbf{R}^n)$, and $\widetilde{F}[f] \to 0$ when $|x| \to \infty$ (for y fixed). By Plancherel's theorem,

$$f(t) = (2\pi)^{-n} e^{|t|y} \int_{\mathbf{R}^n} \widetilde{F}[f](x,y) e^{-i\langle t,x\rangle}\, dx \tag{13.5}$$

(where the integral is understood in the sense of a principal value). It turns out, as in the formula for the Bros-Iagolnitzer transform [31], that we can transform (13.5) into an integral of a certain differential form.

We represent the function $|\xi| - |t|$ in the form

$$|\xi| - |t| = \sum_{k=1}^n (\xi_k - t_k)\varphi_k(t,\xi), \quad \text{where} \quad \varphi_k = \frac{\xi_k + t_k}{|\xi| + |t|}.$$

Then $|\varphi_k| \le 1$ for all points ξ and t in $\mathbf{R}^n$. We define

$$F_k(x,y,t) = \int_{\mathbf{R}^n} \varphi_k(\xi,t) f(\xi) e^{i\langle \xi,x\rangle - |\xi|y}\, d\xi.$$

Consider the differential form

$$\omega(x,y,t) = \frac{-i}{(2\pi)^n}\, e^{-i\langle t,x\rangle + |t|y} \left(\sum_{k=1}^n (-1)^k F_k\, dy \wedge dx[k] + i\widetilde{F}[f]\, dx \right).$$

For fixed t and $y > 0$, the coefficients of the form ω belong to $\mathcal{L}^2(\mathbf{R}^n)$ and tend to zero as $|x| \to \infty$. We find $d\omega$:

$$\begin{aligned}
i(2\pi)^n\, d_{x,y}\omega &= \sum_{k=1}^{n}(-1)^k d_{x,y}(e^{-i\langle t,x\rangle+|t|y}F_k)\wedge dy\wedge dx[k] \\
&\quad + ie^{-i\langle t,x\rangle+|t|y}\left(\frac{\partial\widetilde{F}}{\partial y}+|t|\widetilde{F}\right)dy\wedge dx \\
&= \sum_{k=1}^{n} e^{-i\langle t,x\rangle+|t|y}\left(\frac{\partial F_k}{\partial x_k}-it_kF_k\right)dy\wedge dx \\
&\quad + ie^{-i\langle t,x\rangle+|t|y}\left(\frac{\partial\widetilde{F}}{\partial y}+|t|y\right)dy\wedge dx \\
&= e^{-i\langle t,x\rangle+|t|y}\left(\sum_{k=1}^{n}\int_{\mathbf{R}^n} i\varphi_k(\xi,t)f(\xi)(\xi_k-t_k)e^{i\langle\xi,x\rangle-\,\xi|y}\,d\xi\right. \\
&\quad \left.+\,i\int_{\mathbf{R}^n}(|t|-|\xi|)f(\xi)e^{i\langle\xi,x\rangle-|\xi|y}\,d\xi\right)dy\wedge dx = 0.
\end{aligned}$$

Using Stokes's formula, we obtain $\int_{\Gamma_x}\omega = \int_{\{y=\epsilon\}}\omega$, where Γ is a hypersurface in $\mathbf{R}^{n+1}_+$ of the form

$$\Gamma = \{\,(x,y) : y = \psi(x)\,\}, \tag{13.6}$$

where $\psi > 0$, and the functions ψ and $|\operatorname{grad}\psi|$ are bounded in $\mathbf{R}^n$.

Theorem 13.3 (Kytmanov). *Suppose f is a continuous function of slow growth in $\mathbf{R}^n$. Then*

$$f(t) = \int_{\Gamma_x}\omega(x,y,t),$$

where Γ is a hypersurface of the form (13.6).

Suppose $T \in \mathcal{S}'$, and f is a continuous function of slow growth for which $\partial^\alpha f = T$ in $\mathcal{S}'$. We consider the form ω for f, and we define a form ω_φ for $\varphi \in \mathcal{D}(\mathbf{R}^n)$ via

$$\omega_\varphi(x,y) = (-1)^{\|\alpha\|}\int_{\mathbf{R}^n}\omega(x,y,t)\partial^\alpha\varphi(t)\,dt.$$

Then the coefficients of ω_φ are in $\mathcal{L}^2(\mathbf{R}^n)$ for fixed $y > 0$, and they tend to zero as $|x| \to \infty$.

Corollary 13.4. *If $T \in \mathcal{S}'$, then $T(\varphi) = \int_\Gamma \omega_\varphi$ for every $\varphi \in \mathcal{D}$, where Γ is a hypersurface of the form (13.6).*

Proof. Indeed,

$$\begin{aligned}\int_\Gamma \omega_\varphi &= \int_{\{y=\epsilon\}} \omega_\varphi = (-1)^{\|\alpha\|} \int_{\{y=\epsilon\}} \int_{\mathbf{R}^n} \omega(x,y,t)\partial^\alpha \varphi(t)\,dt \\ &= \frac{(-1)^{\|\alpha\|}}{(2\pi)^n} \int_{\mathbf{R}^n} dx \int_{\mathbf{R}^n} e^{-i\langle t,x\rangle + |t|\epsilon} \widetilde{F}(x,\epsilon)\partial^\alpha \varphi(t)\,dt \\ &= (-1)^{\|\alpha\|} \int_{\mathbf{R}^n} f(t)\partial^\alpha \varphi(t)\,dt = T(\varphi).\end{aligned}$$

□

We remark that the inversion formula holds for a wider class of hypersurfaces Γ than (13.6).

13.4 Analogue of Vladimirov's theorem

We denote by $\mathcal{W}_2^s$ the Hilbert space of functions g of slow growth in $\mathbf{R}^n$ with finite norm

$$\|g\|_{(s)}^2 = \int_{\mathbf{R}^n} |g(t)|^2 (1+|t|^2)^s\,dt,$$

and by $\mathcal{H}_s$ the space of distributions $T \in \mathcal{S}'$ that are Fourier transforms of functions in $\mathcal{W}_2^s$ (that is, $T = F[g]$) with norm $\|T\|_s = \|g\|_{(s)}$. For the properties of these spaces, see the book [212, §10]. The space $\mathcal{H}_s$ is dual to $\mathcal{H}_{-s}$. If $T \in \mathcal{H}_s$, and $Q \in \mathcal{H}_p$, then we can define in $\mathcal{S}'$ the convolution

$$T * Q = F^{-1}\big[F[T]\cdot F[Q]\big] \tag{13.7}$$

(where F^{-1} is the inverse Fourier transform).

Let K be a closed set in $\mathbf{R}^n$ with nonempty interior. We define a function $P_K(x,y)$ by $P_K(x,y) = \widetilde{F}[\chi_K(t)]$, where χ_K is the characteristic function of the set K.

Here are some properties of the function P_K.

1. *The function $P_K(x,y) \in \mathcal{H}_s$ for every s for fixed $y > 0$. In particular, $P_K \in \mathcal{L}^2(\mathbf{R}^n)$.*

   Indeed, $P_K(x,y) = F[\chi_K e^{-|t|y}]$, and $\chi_K(t)e^{-|t|y} \in \mathcal{W}_2^s$ for every s.

2. *$P_K(x,y) \to F[\chi_K]$ as $y \to 0^+$ in the norm of the space $\mathcal{H}_s$ for $s < -n/2$.* Since $\chi_K \in \mathcal{W}_2^s$ for $s < -n/2$, we have

$$\begin{aligned}\|P_K(x,y) - F[\chi_K]\|_s^2 &= \|(e^{-|t|y}-1)\chi_K(t)\|_{(s)}^2 \\ &\le \int_{\mathbf{R}^n} (1-e^{-|t|y})^2(1+|t|^2)^s\,dt \to 0\end{aligned}$$

   as $y \to 0^+$.

3. $P_K(x, y) \to 0$ *as* $|x| \to \infty$ *for fixed* $y > 0$.

If $T \in \mathcal{H}_s$, we define a function $\widehat{T}_K(x, y)$ by

$$\widehat{T}_K(x,y) = (2\pi)^{-n} T(t) * P_K(t,y) = (2\pi)^{-n} T_t(P_K(x-t,y)).$$

By (13.7) and the definition of $\mathcal{H}_s$, we obtain that $\widehat{T}_K \in \mathcal{H}_s$ (for fixed $y > 0$), and moreover (for $T = F[g]$ and $g \in \mathcal{W}_s^2$)

$$\|\widehat{T}_K(x,y)\|_s = \|g(t)e^{-|t|y}\chi_K(t)\|_{(s)} \le \|g\|_{(s)} = \|T\|_s, \qquad y > 0.$$

Theorem 13.5 (Kytmanov). *Suppose* $T = F[g]$ *and* $g \in \mathcal{W}_2^s$. *In order that* $g = 0$ *almost everywhere outside* K, *it is necessary and sufficient that*

$$\widehat{T}_K(x,y) = \widetilde{F}[g](x,y).$$

In other words, for a function g *with support in* K *we have the integral representation*

$$\widetilde{F}[g](x,y) = (2\pi)^{-n} \langle F[g]_t, P_K(x-t,y)\rangle. \tag{13.8}$$

In particular, if $g \in \mathcal{L}^2(\mathbf{R}^n)$, *then* $T \in \mathcal{L}^2(\mathbf{R}^n)$, *and*

$$\widetilde{F}[g](x,y) = \frac{1}{(2\pi)^n} \int_{\mathbf{R}^n} F[g](t) P_K(x-t,y)\, dt.$$

This theorem is an analogue of Vladimirov's theorem for the generalized Cauchy-Bochner transform for cones (see [212, p. 152]).

Proof. If $g = 0$ almost everywhere outside K, then

$$\begin{aligned}\widetilde{F}[g] &= F[g(t)e^{-|t|y}] = F[g(t)\chi_K(t)e^{-|t|y}] \\ &= F\big[F^{-1}[F[g]] \cdot F^{-1}[F[\chi_K(t)e^{-|t|y}]]\big] \\ &= F\big[F^{-1}[T] \cdot F^{-1}[P_K]\big] = \widehat{T}_K(x,y).\end{aligned}$$

Conversely, suppose $\widetilde{F}[g] = \widehat{T}_K(x,y)$. This means that $\widehat{T}_K(x,y) \to T = F[g]$ when $y \to 0^+$ in $\mathcal{S}'$ (see Theorem 13.1). On the other hand,

$$\begin{aligned}\widehat{T}_K(x,y) &= (2\pi)^{-n} F^{-1}\big[F[T] \cdot F[P_K]\big] = F\big[F^{-1}[T] \cdot F^{-1}[P_k]\big] \\ &= F[g(t) \cdot \chi_K(t)e^{-|t|y}] = \widetilde{F}[g(t)\chi_K(t)](x,y),\end{aligned}$$

so $\widehat{T}_K(x,y) \to F[g\chi_K]$ in $\mathcal{S}'$ as $y \to 0^+$. Then we have $g(t) = g(t)\chi_K(t)$ in $\mathcal{S}'$, and so $g = 0$ almost everywhere outside K. □

We now give some corollaries. Suppose K is a closed cone in $\mathbf{R}^n$ with vertex at the origin and nonempty interior. We have

$$P_K(x,y) = i(-1)^{n-1}\frac{\partial^{n-1}}{\partial y^{n-1}}\int_{K\cap S(0,1)} \frac{d\sigma(\xi)}{\langle \xi, x\rangle + iy}. \tag{13.9}$$

For the proof, it suffices to pass to polar coordinates $t = r\xi$ in (13.2), where $|\xi| = 1$ and $r > 0$, and to integrate in r.

We obtain from (13.9) that when $y \to 0^+$, the function P_K extends to the interior points of the dual cone K^* as an analytic function.

For cones K, we can also represent P_K in the form

$$P_K(x,y) = i(-1)^n y\frac{\partial^n}{\partial y^n}\int_{K\cap B(0,1)} \frac{dt}{\langle t, x\rangle + iy}.$$

Indeed, suppose $K = \{t : h(t) > 0\}$, where h is a homogeneous function of degree p. The measure $d\sigma(t)$ is given by the differential form $\sum_{k=1}^n (-1)^{k-1} t_k\, dt[k]$. Therefore, if h is smooth and $\partial h/\partial t_1 \neq 0$, we have

$$d\sigma = \frac{1}{\partial h/\partial t_1}\sum_{k=1}^n t_k \frac{\partial h}{\partial t_k}\, dt[1]\big|_{\partial K} = \frac{1}{\partial h/\partial t_1} ph\, dt[1]\big|_{\partial K} = 0.$$

Hence, by Stokes's formula,

$$P_K(x,y) = (n-1)!\int_{K\cap S(0,1)} \frac{d\sigma(t)}{(-i\langle t, x\rangle + y)^n} = n!\int_{K\cap B(0,1)} \frac{y\, dt}{(-i\langle t, x\rangle + y)^{n+1}}.$$

For certain cones K, these formulas let us carry the computation of P_K to the end. We give some examples.

(1) Suppose $n = 1$ and $K = \{t : t \geq 0\}$. Then $P_K(x,y) = i(x+iy)^{-1}$, and for $g \in \mathcal{L}^2(\mathbf{R}^1)$ with support in K, formula (13.8) reduces to Cauchy's formula

$$\widetilde{F}[g](z) = \frac{1}{2\pi i}\int_{-\infty}^{+\infty} \frac{F[g]}{t - z}\, dt, \qquad z = x + iy,$$

for functions $\widetilde{F}[g] \in \mathcal{H}^2(\mathbf{R}^2_+)$.

If $K = \mathbf{R}^1$, then

$$\frac{1}{2\pi}P_K(x,y) = \frac{y}{\pi(x^2+y^2)}$$

is the Poisson kernel for the upper half-plane.

(2) For $n = 2$, consider a cone K in $\mathbf{R}^2$ bounded by the lines $\varphi = \alpha$ and $\varphi = \beta$, where $\alpha < \beta$ and $\varphi = \arctan \xi_2/\xi_1$. Then by (13.9),

$$P_K(x,y) = -i\frac{\partial}{\partial y}\int_\alpha^\beta \frac{d\varphi}{|x|\cos(\varphi - \psi) + iy},$$

where $\cos\psi = x_1/|x|$ and $\sin\psi = x_2/|x|$. We have

$$J = \int_\alpha^\beta \frac{d\varphi}{|x|\cos(\varphi-\psi)+iy} = \int_\alpha^\beta \frac{|x|\cos(\varphi-\psi)-iy}{|x|^2\cos^2(\varphi-\psi)+y^2}\,d\varphi.$$

Computations show that

$$\int_\alpha^\beta \frac{|x|\cos(\varphi-\psi)\,d\varphi}{|x|^2\cos^2(\varphi-\psi)+y^2} = |x|\int_\alpha^\beta \frac{d\sin(\varphi-\psi)}{|x|^2+y^2-|x|^2\sin^2(\varphi-\psi)}$$

$$= \frac{1}{2\sqrt{|x|^2+y^2}}\ln\left|\frac{\sqrt{|x|^2+y^2}+|x|\sin(\varphi-\psi)}{\sqrt{|x|^2+y^2}-|x|\sin(\varphi-\psi)}\right|\Bigg|_\alpha^\beta$$

$$= \frac{1}{2\sqrt{|x|^2+y^2}}\ln\left|\frac{(\sqrt{|x|^2+y^2}+|x|\sin(\beta-\psi))(\sqrt{|x|^2+y^2}-|x|\sin(\alpha-\psi))}{(\sqrt{|x|^2+y^2}-|x|\sin(\beta-\psi))(\sqrt{|x|^2+y^2}+|x|\sin(\alpha-\psi))}\right|$$

$$= \frac{1}{2\sqrt{|x|^2+y^2}}\ln\frac{|x|^2+y^2-x_1^2\sin\alpha\sin\beta-x_2^2\cos\alpha\cos\beta+x_1x_2\sin(\alpha+\beta)+}{|x|^2+y^2-x_1^2\sin\alpha\sin\beta-x_2^2\cos\alpha\cos\beta+x_1x_2\sin(\alpha+\beta)-}$$

$$\xrightarrow{\textit{fraction continues}} \frac{+\sqrt{|x|^2+y^2}\,(x_1(\sin\beta-\sin\alpha)-x_2(\cos\beta-\cos\alpha))}{-\sqrt{|x|^2+y^2}\,(x_1(\sin\beta-\sin\alpha)-x_2(\cos\beta-\cos\alpha))}.$$

Now

$$iy\int_\alpha^\beta \frac{d\varphi}{|x|^2\cos^2(\varphi-\psi)+y^2} = \frac{i}{\sqrt{|x|^2+y^2}}\arctan\frac{y\tan(\varphi-\psi)}{\sqrt{|x|^2+y^2}}\Bigg|_\alpha^\beta$$

$$= \frac{i}{\sqrt{|x|^2+y^2}}\times$$

$$\times\arctan\frac{y\sqrt{|x|^2+y^2}\sin(\beta+\alpha)}{y^2\cos(\beta-\alpha)+x_1^2\cos\alpha\cos\beta+x_2^2\sin\alpha\sin\beta+x_1x_2\sin(\alpha+\beta)}$$

if x_1 and x_2 are positive. When $\alpha = 0$ and $\beta = \pi/2$, the integral J equals

$$\frac{1}{2\sqrt{|x|^2+y^2}}\times$$

$$\times\left[\ln\frac{|x|^2+y^2+x_1x_2+(x_1+x_2)\sqrt{|x|^2+y^2}}{|x|^2+y^2+x_1x_2-(x_1+x_2)\sqrt{|x|^2+y^2}} - 2i\arctan\frac{y\sqrt{|x|^2+y^2}}{x_1x_2}\right]$$

if x_1 and x_2 are positive. When $\alpha = -\pi/4$ and $\beta = \pi/4$, the integral J equals

$$\frac{1}{2\sqrt{|x|^2+y^2}}\left[\ln\frac{2y^2+3x_1^2+x_2^2+2\sqrt{2}\,x_1\sqrt{|x|^2+y^2}}{2y^2+3x_1^2+x_2^2-2\sqrt{2}\,x_1\sqrt{|x|^2+y^2}} - 2i\arctan\frac{2y\sqrt{|x|^2+y^2}}{x_1^2-x_2^2}\right]$$

if $x_1 > 0$ and $x_1^2 - x_2^2 > 0$.

(3) Suppose $n = 4$, and K is the future light cone $\{x \in \mathbf{R}^4 : x_1 \geq 0, x_1^2 - |x'|^2 \geq 0\}$, where $|x'|^2 = x_2^2 + x_3^2 + x_4^2$. Then $P_K(x, y) = -i\partial^3 I/\partial y^3$ by (13.9), where $I = \int_{K \cap S(0,1)} (\langle t, x\rangle + iy)^{-1}\, d\sigma(t)$. Computing this integral gives

$$I = \pi\left[\frac{\pi}{\sqrt{|x|^2+y^2}+y} + \frac{\sqrt{2}}{|x'|}\arctan\frac{2\sqrt{2}\,x_1 y}{2y^2+x_1^2-|x'|^2}\right.$$
$$-\frac{yx_1}{|x'|\,|x|^2}\ln\frac{2y^2+(x_1+|x'|)^2}{2y^2+(x_1-|x'|)^2} + \frac{2\sqrt{|x|^2+y^2}}{|x|^2}\arctan\frac{x_1^2-|x'|^2}{2y\sqrt{|x|^2+y^2}}$$
$$-\frac{\sqrt{2}i}{2|x'|}\ln\frac{2y^2+(x_1-|x'|)^2}{2y^2+(x_1+|x'|)^2} + \frac{2iyx_1}{|x'|\,|x|^2}\arctan\frac{x_1^2-|x'|^2}{2y^2+x_1^2-|x'|^2}$$
$$\left.-\frac{i\sqrt{|x|^2+y^2}}{|x|^2}\ln\frac{2y^2+3x_1^2+|x'|^2+2\sqrt{2}\,x_1\sqrt{|x|^2+y^2}}{2y^2+3x_1^2+|x'|^2-2\sqrt{2}\,x_1\sqrt{|x|^2+y^2}}\right]$$

if $x \in K$ (this function has no singularity when $x' = 0$).

Now consider as the closed set K the ball $B(0, a)$. Then

$$P_B(x, y) = \int_{B(0,a)} e^{i\langle t,x\rangle - |t|y}\, dt.$$

We use a formula from [195, Chap. IV, Theorem 3.3]: if $f \in \mathcal{L}^1(\mathbf{R}^n)$ and $f = f(|x|)$, then

$$F[f](x) = (2\pi)^{n/2}|x|^{1-n/2}\int_0^\infty f(r)I_{(n-2)/2}(|x|r)r^{n/2}\, dr, \tag{13.10}$$

where $I_{(n-2)/2}$ is a Bessel function. We obtain

$$P_B(x, y) = \int_{B(0,a)} e^{-|t|y}e^{i\langle t,x\rangle}\, dt = (2\pi)^{n/2}|x|^{1-n/2}\int_0^a I_{(n-2)/2}(|x|r)e^{-yr}r^{n/2}\, dr.$$

For $n = 2s + 1$, this integral can be fully computed. Recall the well-known formula for Bessel functions

$$\frac{I_{\nu+1}}{x^\nu} = -\frac{d}{dx}\left(\frac{I_\nu}{x^\nu}\right).$$

Hence

$$I_\nu = (-1)^k x^\nu \left(\frac{1}{x}\frac{d}{dx}\right)^k \left(\frac{I_{\nu-k}}{x^{\nu-k}}\right).$$

Since $I_{-1/2} = (\cos x)\sqrt{2/(\pi x)}$, we have for $\nu = s - 1/2$ that

$$I_{s-1/2} = (-1)^s\sqrt{\frac{2}{\pi}}\, x^{s-1/2}\left(\frac{1}{x}\frac{d}{dx}\right)^s \cos x, \tag{13.11}$$

and for $\nu = s$ that

$$I_s = (-1)^s x^s \left(\frac{1}{x}\frac{d}{dx}\right)^s I_0. \qquad (13.12)$$

Then (for $n = 2s+1$)

$$\begin{aligned}
P_B(x,y) &= (2\pi)^{n/2}|x|^{1-n/2}\sqrt{\frac{2}{\pi}}\int_0^a (-1)^s|x|^{s-1/2}r^{s-1/2}\left(\frac{1}{|x|r}\frac{d}{d|x|r}\right)^s \times \\
&\quad \times (\cos|x|r)e^{-yr}r^{n/2}\,dr \\
&= 2^{s+1}\pi^s(-1)^s\left(\frac{1}{|x|}\frac{d}{d|x|}\right)^s \int_0^a e^{-yr}\cos|x|r\,dr \\
&= 2^{s+1}\pi^s(-1)^s\left(\frac{1}{|x|}\frac{d}{d|x|}\right)^s \left[\frac{e^{-ay}(|x|\sin|x|a - y\cos|x|a)}{|x|^2+y^2} - \frac{y}{|x|^2+y^2}\right] \\
&= (2\pi)^n\widetilde{P}(x,y) + e^{-ay}(-1)^s\pi^s 2^{s+1}\left(\frac{1}{|x|}\frac{d}{d|x|}\right)^s \frac{|x|\sin|x|a - y\cos|x|a}{|x|^2+y^2},
\end{aligned}$$

where $\widetilde{P}(x,y)$ is the Poisson kernel for the ball.

13.5 Determination of the Fourier transform of some distributions

The Fourier transform of the function $|t|^\lambda$ is given by the formula

$$F\left[|t|^\lambda\right] = \frac{2^{k+n}\pi^{n/2}\Gamma((\lambda+n)/2)}{|x|^{\lambda+n}\Gamma(-\lambda/2)}$$

(see, for example, [32, p. 244]), but this is not applicable for $\lambda = -n, -n-2, \ldots$. Let us find $F[\text{p.f.}\,|t|^{-n-2k}]$. The Hadamard finite part (p.f.) of the integral has the form

$$\langle \text{p.f.}\,|t|^{-n-2k}, \varphi\rangle = \int_{B(0,1)} \frac{\varphi(t) - T_{2k}(t,\varphi)}{|t|^{n+2k}}\,dt + \int_{\mathbf{R}^n\setminus B(0,1)} \frac{\varphi(t) - T_{2k-1}(t,\varphi)}{|t|^{n+2k}}\,dt,$$

where $T_m(t,\varphi) = \sum_{\|\alpha\|\le m} t^\alpha \partial^\alpha\varphi(0)/\alpha!$ is the Taylor polynomial of φ (this definition is somewhat different from the definition given in §12).

Lemma 13.6. *Let $f = f(|x|)$ be a function of slow growth in $\mathbf{R}^n$.*

1. *If $n = 2s+1$ and $s \ge 0$, then*

$$\widetilde{F}[f](x,y) = (-1)^s 2^{s+1}\pi^s\left(\frac{1}{|x|}\frac{d}{d|x|}\right)^s \int_0^\infty e^{-yr}f(r)\cos(|x|r)\,dr.$$

2. *If $n = 2s+2$ and $s \ge 0$, then*

$$\widetilde{F}[f](x,y) = (-1)^{s+1}(2\pi)^{s+1}\left(\frac{1}{|x|}\frac{d}{d|x|}\right)^s \frac{\partial}{\partial y}\int_0^\infty e^{-yr}f(r)I_0(|x|r)\,dr.$$

The proof follows from formulas (13.10)–(13.12).

We now find $\widetilde{F}[\ln|t|]$. To do this we apply formulas from [165, p. 268] and [164, p. 550]. We obtain for $n = 2s+1$ that

$$\begin{aligned}&\widetilde{F}[\ln|t|](x,y)\\&= (-1)^{s+1}2^{s+1}\pi^s \left(\frac{1}{|x|}\frac{d}{d|x|}\right)^s \left(\frac{Cy + |x|\arctan(|x|/y) + (1/2)y\ln(|x|^2+y^2)}{|x|^2+y^2}\right);\end{aligned}$$

and for $n = 2s+2$ that

$$\begin{aligned}&\widetilde{F}[\ln|t|](x,y) = (-1)^s(2\pi)^{s+1} \times\\&\times \left(\frac{1}{|x|}\frac{d}{d|x|}\right)^s \left[\frac{y\left(2-\ln 2 - C + \ln(y+\sqrt{|x|^2+y^2})/(|x|^2+y^2)\right)}{(|x|^2+y^2)^{3/2}} - \frac{1}{|x|^2+y^2}\right],\end{aligned}$$

where C is Euler's constant.

Now we compute the boundary values of these functions as $y \to 0$. We have (for $n = 2s+1$)

$$\begin{aligned}&\int_{\mathbf{R}^n} \widetilde{F}[\ln|t|](x,y)\varphi(x)\,dx\\&\qquad = \int_{B(0,1)} \widetilde{F}(\varphi(x)-\varphi(0))\,dx + \int_{\mathbf{R}^n\setminus B(0,1)} \widetilde{F}\varphi\,dx + \varphi(0)\int_{B(0,1)} \widetilde{F}\,dx.\end{aligned}$$

Since the product $|\widetilde{F}(x,0)|\,|x|^{2s+2-\epsilon}$ is bounded as $|x| \to \infty$, we may pass to the limit under the integral sign in the integral over $\mathbf{R}^n \setminus B(0,1)$. Then

$$\begin{aligned}\lim_{y\to 0+}\int_{\mathbf{R}^n\setminus B(0,1)} \widetilde{F}\varphi\,dx &= (-1)^{s+1}2^s\pi^{s+1}\int_{\mathbf{R}^n\setminus B(0,1)} \varphi\left(\frac{1}{|x|}\frac{d}{d|x|}\right)^s \frac{1}{|x|}\,dx\\&= -2^s\pi^{s+1}(2s-1)!!\int_{\mathbf{R}^n\setminus B(0,1)} \varphi(x)|x|^{-2s-1}\,dx.\end{aligned}$$

We may also pass to the limit under the integral sign in the $\varphi(x)-\varphi(0)$ integral, so

$$\begin{aligned}&\lim_{y\to 0+}\int_{B(0,1)} \widetilde{F}(\varphi(x)-\varphi(0))\,dx\\&\qquad = (-1)^{s+1}2^s\pi^{s+1}\int_{B(0,1)} (\varphi(x)-\varphi(0))\left(\frac{1}{|x|}\frac{d}{d|x|}\right)^s \frac{1}{|x|}\,dx\\&\qquad = -2^s\pi^{s+1}(2s-1)!!\int_{B(0,1)} (\varphi(x)-\varphi(0))|x|^{-2s-1}\,dx.\end{aligned}$$

By passing to polar coordinates, we compute

$$\int_{B(0,1)} \widetilde{F}\,dx = \frac{(-1)^{s+1}2^{2s+2}\pi^{2s}}{(2s-1)!!}\int_0^1 \left(\frac{1}{r}\frac{d}{dr}\right)^s \psi(r,y)r^{2s}\,dr,$$

where $r = |x|$, and

$$\psi(r,y) = \frac{r\arctan(r/y) + Cy + (y/2)\ln(r^2+y^2)}{(r^2+y^2)} = \frac{Cy}{r^2+y^2} + \frac{1}{2}\left(\ln(r^2+y^2)\arctan\frac{r}{y}\right)'_r.$$

Computations give

$$\begin{aligned}
&\int_0^1 \left(\frac{1}{r}\frac{d}{dr}\right)^s \psi(r,y) r^{2s}\,dr \\
&= r^{2s-1}\left(\frac{1}{r}\frac{d}{dr}\right)^{s-1}\psi(r,y)\Big|_0^1 - (2s-1)\int_0^1 \left(\frac{1}{r}\frac{d}{dr}\right)^{s-1}\psi(r,y) r^{2s-2}\,dr \\
&= \Bigg[r^{2s-1}\left(\frac{1}{r}\frac{d}{dr}\right)^{s-1}\psi - (2s-1)r^{2s-3}\left(\frac{1}{r}\frac{d}{dr}\right)^{s-2}\psi \\
&\quad + \cdots + (-1)^{s-1}(2s-1)!!\,r\psi\Bigg]\Bigg|_0^1 + (2s-1)!!\,(-1)^s\int_0^1 \psi\,dr.
\end{aligned}$$

But $\int_0^1 \psi\,dr = C\arctan\frac{1}{y} + \frac{1}{2}\ln(1+y^2)\arctan\frac{1}{y}$, so $\lim_{y\to 0+}\int_0^1 \psi\,dr = C\pi/2$, and

$$\begin{aligned}
\lim_{y\to 0+} r^{2s-j}\left(\frac{1}{r}\frac{d}{dr}\right)^{s-j}\psi(r,y)\Big|_0^1 &= r^{2s-j}\left(\frac{1}{r}\frac{d}{dr}\right)^{s-j}\psi(r,0)\Big|_0^1 \\
&= \frac{\pi}{2}r^{2s-j}\left(\frac{1}{r}\frac{d}{dr}\right)^{s-j}\left(\frac{1}{r}\right)\Big|_0^1 \\
&= \frac{\pi}{2}(-1)^{s-j}(2s-2j-1)!!.
\end{aligned}$$

Then

$$\begin{aligned}
\lim_{y\to 0+}\int_0^1 \left(\frac{1}{r}\frac{d}{dr}\right)^s \psi(r,y) r^{2s}\,dr = \frac{\pi}{2}(-1)^{s-1}(2s-1)!!\left(1+\frac{1}{3}+\cdots+\frac{1}{2s-1}\right) \\
+ \frac{\pi C}{2}(-1)^s(2s-1)!!.
\end{aligned}$$

Consequently,

$$\lim_{y\to 0+}\int_{B(0,1)} \widetilde{F}\,dx = (2\pi)^{s+1}\left(1+\frac{1}{3}+\cdots+\frac{1}{2s-1} - C\right).$$

(When $s = 0$ we take the sum $1 + (1/3) + \cdots + 1/(2s-1)$ to be zero.)

Finally, we have

$$F[\ln|t|](x) = -2^s\pi^{s+1}(2s-1)!!\ \mathrm{p.\,f.}\,|x|^{-2s-1} + (2\pi)^{2s+1}\left(1+\frac{1}{3}+\cdots+\frac{1}{(2s-1)}-C\right)\delta(x).$$

In precisely the same way, we obtain for $n = 2s+2$ that

$$F[\ln|t|](x) = -(2\pi)^{s+1}(2s)!!\ \mathrm{p.\,f.}\,|x|^{-2s-2} + (2\pi)^{2s+2}\left(\frac{1}{2}+\cdots+\frac{1}{2s}+\ln 2-C\right)\delta(x).$$

(When $s=0$, we take the sum $(1/2)+\cdots+1/(2s)$ to be zero.)

Using the inverse Fourier transform, we find $F[\mathrm{p.\,f.}\,|x|^{-n}]$ for $n = 2s+1$:

$$F[\mathrm{p.\,f.}\,|x|^{-2s-1}] = \frac{2^{s+1}\pi^s}{(2s-1)!!}\left[-\ln|t|+1+\frac{1}{3}+\cdots+\frac{1}{2s-1}-C\right]; \tag{13.13}$$

and for $n = 2s+2$:

$$F[\mathrm{p.\,f.}\,|x|^{-2s-2}] = \frac{2\pi^{s+1}}{s!}\left[-\ln|t|+\frac{1}{2}+\cdots+\frac{1}{2s}+\ln 2-C\right]. \tag{13.14}$$

When $s=0$ (that is, $n=2$), formula (13.14) agrees with formula (68) in [212, p. 110], since the constant c_0 in that formula will be

$$c_0 = \int_0^1 \frac{1-I_0(u)}{u}\,du - \int_1^\infty \frac{I_0(u)}{u}\,du.$$

Using the formula $\int_0^\infty (e^{-x}-I_0(x))/x\,dx = -\ln 2$ (see [165, p. 185]) and the well-known identity $C = \int_0^1 (1-e^{-x})/x\,dx - \int_1^\infty e^{-x}/x\,dx$, we obtain

$$\begin{aligned} c_0+\ln 2 &= \int_0^1 \frac{1-I_0}{x}\,dx - \int_1^\infty \frac{I_0}{x}\,dx - \int_0^\infty \frac{e^{-x}-I_0}{x}\,dx \\ &= \int_0^1 \frac{1-e^{-x}}{x}\,dx - \int_1^\infty \frac{e^{-x}}{x}\,dx = C, \end{aligned}$$

that is, $c_0 = C - \ln 2$.

We can now accomplish the determination of $F[\mathrm{p.\,f.}\,|x|^{-n-2k}]$ as follows. First we compute $\Delta\,\mathrm{p.\,f.}\,|x|^{-n-2k}$. By definition (with $\varphi\in\mathcal{D}(\mathbf{R}^n)$),

$$\begin{aligned} &\langle \Delta\,\mathrm{p.\,f.}\,|x|^{-n-2k},\varphi\rangle = \langle \mathrm{p.\,f.}\,|x|^{-n-2k},\Delta\varphi\rangle \\ &= \int_{B(0,1)} (\Delta\varphi - T_{2k}(t,\Delta\varphi))|t|^{-n-2k}\,dt + \int_{\mathbf{R}^n\setminus B(0,1)} (\Delta\varphi - T_{2k-1}(t,\Delta\varphi))|t|^{-n-2k}\,dt \end{aligned}$$

(where $T_m(t,\varphi)$ is the Taylor polynomial of φ of order m). Then

$$\int_{B(0,1)} \Delta(\varphi - T_{2k+2}(t,\varphi))|t|^{-n-2k}\,dt = \int_{B(0,1)} (\Delta\varphi - T_{2k}(t,\Delta\varphi))|t|^{-n-2k}\,dt.$$

Using Green's formula

$$\int_{B(0,1)} (\psi\Delta\varphi - \varphi\Delta\psi)\,dt = \int_{S(0,1)} \left(\psi\frac{\partial\varphi}{\partial|t|} - \varphi\frac{\partial\psi}{\partial|t|}\right) d\sigma(t),$$

we obtain

$$\begin{aligned}\langle \text{p.f.}\,|x|^{-n-2k}, \Delta\varphi\rangle &= \langle \text{p.f.}\,\Delta|x|^{-n-2k}, \varphi\rangle \\ &\quad - \int_{S(0,1)} (n+4k+2) \sum_{\|\alpha\|=2k+2} \frac{t^\alpha\partial^\alpha\varphi(0)}{\alpha!}\,d\sigma(t).\end{aligned}$$

Indeed,

$$\begin{aligned}\int_{B(0,1)} (\Delta\varphi - T_{2k}(t,\Delta\varphi))|t|^{-n-2k}\,dt &= \int_{B(0,1)} \Delta(\varphi - T_{2k+2})|t|^{-n-2k}\,dt \\ = \int_{B(0,1)} (\varphi - T_{2k+2}(t,\varphi))\Delta|t|^{-n-2k} &+ \int_{S(0,1)} \frac{\partial}{\partial|t|}(\varphi - T_{2k+2}(t,\varphi))\,d\sigma \\ &+ \int_{S(0,1)} (\varphi - T_{2k+2})(n+2k)\,d\sigma.\end{aligned}$$

Here we may apply Green's formula because there is an integrable singularity at zero. This means that

$$\begin{aligned}&\int_{B(0,1)} (\Delta\varphi - T_{2k}(t,\Delta\varphi))|t|^{-n-2k}\,dt + \int_{\mathbf{R}^n\setminus B(0,1)} (\Delta\varphi - T_{2k-1}(t,\Delta\varphi))|t|^{-n-2k}\,dt \\ &\quad = \langle \text{p.f.}\,\Delta|x|^{-n-2k}, \varphi\rangle - \int_{S(0,1)} \frac{\partial}{\partial|t|}(T_{2k+2}(t,\varphi) - T_{2k+1}(t,\varphi))\,d\sigma \\ &\quad\quad - (n+2k)\int_{S(0,1)} (T_{2k+2}(t,\varphi) - T_{2k+1}(t,\varphi))\,d\sigma \\ &\quad = \langle \text{p.f.}\,\Delta|x|^{-n-2k}, \varphi\rangle - (n+4k+2)\int_{S(0,1)} (T_{2k+2}(t,\varphi) - T_{2k+1}(t,\varphi))\,d\sigma.\end{aligned}$$

Also

$$\int_{S(0,1)} \sum_{\|\alpha\|=2k+2} \frac{t^\alpha\partial^\alpha\varphi(0)}{\alpha!}\,d\sigma(t) = \int_{S(0,1)} P_{2k+2}(t)\,d\sigma(t).$$

Extending P_{2k+2} into the ball $B(0,1)$ by Gauss's formula (see [189, p. 476]):

$$P_{2k+2}(t) = \sum_{s=0}^{k+1} \left(\frac{|t|}{2}\right)^{2s} Z_{k-2s},$$

we have $\int_{S(0,1)} P_{2k+2}(t)\,d\sigma(t) = \sigma_n Z_0 2^{-2k-2}$, where

$$Z_0 = \frac{((n/2)-1)\Gamma((n/2)-1)\Delta^{k+1}P_{2k+2}}{(k+1)!\,\Gamma((n/2)+k+1)} = \frac{\Gamma(n/2)(\Delta^{k+1}\varphi)(0)}{(k+1)!\,\Gamma((n/2)+k+1)}.$$

Thus

$$\int_{S(0,1)} P_{2k+2}(t)\,d\sigma(t) = \frac{\pi^{n/2}(\Delta^{k+1}\varphi)(0)(n+4k+2)}{(k+1)!\,2^{2k+1}\Gamma((n/2)+k+1)}.$$

Then

$$\begin{aligned}&\Delta\,\text{p. f.}\,|x|^{-n-2k}\\ &\quad= (n+2k)(2k+2)\,\text{p. f.}\,|x|^{-n-2k-2} - \frac{(n+4k+2)\pi^{n/2}\Delta^{k+1}\delta(x)}{2^{2k+1}\Gamma((n/2)+k+1)(k+1)!}.\end{aligned}$$

Hence

$$\begin{aligned}F[\text{p. f.}\,|t|^{-n-2k}] &= \frac{(-1)^k|x|^{2k}}{2^k k!\,n(n+2)\dots(n+2k-2)}\times\\ &\times\left\{F[\text{p. f.}\,|t|^{-n}] + \frac{2\pi^{n/2}}{\Gamma(n/2)}\left(\frac{1}{2}+\dots+\frac{1}{2k}+\frac{1}{n}+\dots+\frac{1}{n+2k-2}\right)\right\}.\end{aligned}$$

By applying (13.13) and (13.14), we have the following assertion.

Theorem 13.7 (Kytmanov). *When $n = 2s+1$,*

$$\begin{aligned}F[\text{p. f.}\,|x|^{-2(s+k)-1}] &= \frac{(-1)^k|x|^{2k}\pi^s 2^{s-k+1}}{k!\,(2(s+k)-1)!!}\times\\ &\times\left[-\ln|x| - C + \frac{1}{2}+\dots+\frac{1}{2k}+1+\frac{1}{3}+\dots+\frac{1}{2s+2k-1}\right],\end{aligned} \tag{13.15}$$

and when $n = 2s+2$,

$$\begin{aligned}F[\text{p. f.}\,|t|^{-2(s+k+1)}] &= \frac{(-1)^k|x|^{2k}\pi^{s+1}}{2^{2k-1}k!\,(s+k)!}\times\\ &\times\left[-\ln|x| - C + \ln 2 + \frac{1}{2}+\dots+\frac{1}{2k}+\frac{1}{2}+\dots+\frac{1}{2k+2s}\right].\end{aligned}$$

When $s = 0$ (that is, $n = 1$), formula (13.15) agrees with a formula from [32, p. 104].

Chapter 4

The $\overline{\partial}$-Neumann Problem for Smooth Functions and Distributions

14 Statement of the $\overline{\partial}$-Neumann problem

14.1 The Hodge operator

Subsequently we shall need some properties of the Hodge operator. We first consider the space $\mathbf{R}^m$ with the usual Euclidean metric. Let $I = (i_1, \ldots, i_p)$ be an increasing multi-index, $dx_I = dx_{i_1} \wedge \cdots \wedge dx_{i_p}$, and $dx[I] = dx_{j_1} \wedge \cdots \wedge dx_{j_{m-p}}$, where $j_1 < \cdots < j_{m-p}$, and $j_k \neq i_l$ for $k = 1, \ldots, m-p$ and $l = 1, \ldots, p$. We recall that the symbol $\sigma(I)$ is defined by $dx_I \wedge dx[I] = \sigma(I)\, dx$ (see §1).

The Hodge star operator $\star$ acts on the form dx_I in the following way: $\star dx_I = \sigma(I)\, dx[I]$. We extend it to an arbitrary form $\varphi = \sum_I' \varphi_I\, dx_I$ by linearity:

$$\star\varphi = \sum_I{}' \varphi_I \sigma(I)\, dx[I].$$

We now give the main properties of the Hodge operator.
(1) $dx_I \wedge \star dx_I = dx = dx_1 \wedge \cdots \wedge dx_m$.
(2) $\star\star dx_I = (-1)^{mp+p} dx_I$.
We obtain from (1) and (2) that if φ and ψ are two p-forms, then

$$\varphi \wedge \star\bar{\psi} = \sum_I{}' \varphi_I\, dx_I \bigwedge \star \sum_J{}' \bar{\psi}_J\, dx_J = \Big(\sum_I{}' \varphi_I \bar{\psi}_I\Big)\, dx.$$

Consequently, a scalar product (φ, ψ) may be defined for p-forms φ and ψ with coefficients of class $\mathcal{L}^2$ in a domain $D \subset \mathbf{R}^m$ by

$$(\varphi, \psi) = \int_D \varphi \wedge \star\bar{\psi}.$$

Then

$$(\varphi, \varphi) = \|\varphi\|^2 = \int_D \sum_I{}' |\varphi_I|^2 \, dx.$$

This scalar product is called the Hodge product.

Now consider $\mathbf{C}^n \cong \mathbf{R}^n$ with coordinates $z = (z_1, \dots, z_n)$ and $z_j = x_j + iy_j$ for $j = 1, \dots, n$. We have defined the volume form dv in $\mathbf{C}^n$ as (see §1)

$$dv = dx \wedge dy = (i/2)^n \, dz \wedge d\bar{z}.$$

If $I = (i_1, \dots, i_p)$ and $J = (j_1, \dots j_q)$ are increasing multi-indices, and we write the (p, q)-form $dz_I \wedge d\bar{z}_J$ in terms of the forms dx_j and dy_k, apply the Hodge operator, and then express dx_j and dy_k in terms of dz_j and $d\bar{z}_k$, we obtain the following equality:

(3) $\star(dz_I \wedge d\bar{z}_J) = 2^{p+q-n}(-1)^{pn} i^n \, \sigma(I)\sigma(J) \, dz[J] \wedge d\bar{z}[I]$ (a detailed calculation may be found in [217, Chap. V, Lemma 1.2]).

The Hodge operator extends to (p, q)-forms $\varphi = \sum'_{I,J} \varphi_{I,J} \, dz_I \wedge d\bar{z}_J$ by linearity. Thus, $\star\varphi$ is a form of type $(n - p, n - q)$. Properties (1) and (2) carry over to the following:

(4) $dz_I \wedge d\bar{z}_J \wedge \star\overline{dz_I \wedge d\bar{z}_J} = 2^{p+q} \, dv$;

(5) $\star\star(dz_I \wedge d\bar{z}_J) = (-1)^{p+q} \, dz_I \wedge d\bar{z}_J$.

By using the Hodge operator, it is easy to find the operators formally dual to d, $\overline{\partial}$, and ∂. For example, we find $\overline{\partial}^*$. If φ is a $(p, q-1)$-form and ψ is a (p, q)-form, φ and ψ have smooth coefficients of class $\mathcal{L}^2(D)$, and ψ has compact support in D, then $(\overline{\partial}\varphi, \psi) = (\varphi, \overline{\partial}^*\psi)$, and

$$\begin{aligned}(\overline{\partial}\varphi, \psi) &= \int_D \overline{\partial}\varphi \wedge \star\bar{\psi} = \int_D d\varphi \wedge \star\bar{\psi} = \int_D d(\varphi \wedge \star\bar{\psi}) + (-1)^{p+q} \int_D \varphi \wedge d\star\bar{\psi} \\ &= (-1)^{p+q} \int_D \varphi \wedge \overline{\partial}\star\bar{\psi} = -\int_D \varphi \wedge \star(\overline{\star\partial\star\psi}),\end{aligned}$$

so $\overline{\partial}^* = -\star\partial\star$.

In just the same way, we see that $-\partial^* = -\star\overline{\partial}\star$. The operator $\overline{\partial}^*$ carries forms of type (p, q) into forms of type $(p, q-1)$. By definition, $\overline{\partial}^* = 0$ for forms of type $(p, 0)$.

We consider the operator $\square = \overline{\partial}^*\overline{\partial} + \overline{\partial}\,\overline{\partial}^*$, which is known as the complex

Laplacian. If φ is a function, then

$$\Box\varphi = \overline{\partial}^*\overline{\partial}\varphi = \overline{\partial}^* \sum_{k=1}^{n} \frac{\partial\varphi}{\partial\bar{z}_k}\, d\bar{z}_k = -\star\partial \sum_{k=1}^{n} 2^{1-n} i^n (-1)^{k-1} \frac{\partial\varphi}{\partial\bar{z}_k}\, dz[k] \wedge d\bar{z}$$

$$= -\star 2^{1-n} i^n \sum_{k=1}^{n} \frac{\partial^2\varphi}{\partial\bar{z}_k\partial z_k}\, dz \wedge d\bar{z} = -2\sum_{k=1}^{n} \frac{\partial^2\varphi}{\partial\bar{z}_k\partial z_k} = -2\Delta\varphi,$$

that is, $\Box = -2\Delta$ for functions, and this identity continues to hold for forms (see, for example, [62, p. 106]). Thus, in $\mathbf{C}^n$, harmonic forms in the sense of $\Box$ are forms with harmonic coefficients. It is also easy to show that $\Box = \partial\partial^* + \partial^*\partial$.

Finally, if a (p, q)-form φ is given in a neighborhood of a smooth hypersurface Γ, then a simple but important property of the Hodge operator is that it takes the normal part of φ into the tangential part of $\star\varphi$, and conversely.

14.2 Statement of the problem

Suppose $n > 1$, and $D = \{\, z : \rho(z) < 0 \,\}$ is a bounded domain in $\mathbf{C}^n$ with boundary of class $\mathcal{C}^1$, where $\rho \in \mathcal{C}^1(\overline{D})$ and $d\rho \neq 0$ on ∂D. If $F \in \mathcal{C}^1(\overline{D})$, then we recall that

$$\overline{\partial}_n F = \sum_{k=1}^{n} \frac{\partial F}{\partial\bar{z}_k} \rho_k.$$

In subsection 6.2, we showed that if we write the form $\overline{\partial}F = \overline{\partial}_\tau F + \lambda\overline{\partial}\rho/|\overline{\partial}\rho|$, then $\lambda = \overline{\partial}_n F$, that is, $\overline{\partial}_n F$ is the coefficient of the normal part of the form $\overline{\partial}F$. If we denote the outer unit normal to ∂D at z by $\nu(z)$, and $s(z) = i\nu(z)$, then

$$\overline{\partial}_n F = \frac{1}{2}\left(\frac{\partial F}{\partial\nu} + i\frac{\partial F}{\partial s}\right).$$

On the other hand,

$$\star\overline{\partial}F = \star\sum_{k=1}^{n} \frac{\partial F}{\partial\bar{z}_k}\, d\bar{z}_k = 2^{1-n} i^n \sum_{k=1}^{n} (-1)^{k-1} \frac{\partial F}{\partial\bar{z}_k}\, dz[k] \wedge d\bar{z} = 2^{1-n} i^n (-1)^n \mu_F,$$

where μ_F is defined in §1. The restriction of $\star\overline{\partial}F$ to ∂D equals $\overline{\partial}_n F\, d\sigma$ (by Lemma 3.5). The normal part of $\overline{\partial}F$ is carried into the tangential part of $\star\overline{\partial}F$.

We consider the following problem (the $\overline{\partial}$-Neumann problem): for a given function φ on ∂D, find a harmonic function F in D such that

$$\overline{\partial}_n F = \varphi \quad \text{on} \quad \partial D. \tag{14.1}$$

This problem is an exact analogue of the usual Neumann problem for harmonic functions.

Just as for the usual Neumann problem, the problem (14.1) is not always solvable. There is a necessary orthogonality condition. Indeed, if F is a harmonic function of class $\mathcal{C}^1(\overline{D})$, then $\star\overline{\partial}F$ is a ∂-closed form in D, since $0 = \Box F = \overline{\partial}^*\overline{\partial}F = -\star\overline{\partial}(\star\overline{\partial}F)$, that is, $\partial(\star\overline{\partial}F) = 0$. Hence, if $\varphi = \overline{\partial}_n F$ on ∂D, and f is a holomorphic function on $\overline{D}$, then

$$\int_{\partial D} \varphi \bar{f}\, d\sigma = \int_{\partial D} \bar{f} \star\overline{\partial}F = \int_D d(\bar{f}\star\overline{\partial}F) = \int_D \partial(\bar{f}\star\overline{\partial}F) = \int_D \bar{f}\,\partial\star\overline{\partial}F = 0.$$

Thus, a necessary condition for solvability of (14.1) is the orthogonality condition

$$\int_{\partial D} \varphi \bar{f}\, d\sigma = 0 \tag{14.2}$$

for all $f \in \mathcal{O}(\overline{D})$. (We may replace the class $\mathcal{O}(\overline{D})$ by $\mathcal{O}(D)\cap\mathcal{C}^\infty(\overline{D})$ or by $\mathcal{A}(D)$, but we will usually consider the class $\mathcal{O}(\overline{D})$.)

Condition (14.2) agrees with the necessary condition for solvability of the ∂_τ-problem for the tangential ∂ operator (see, for example, the survey [74]). To solve (14.1), we will use solvability results for the ∂_τ-problem (see §18).

It is also natural to pose problem (14.1) for distributions. Suppose $\partial D \in \mathcal{C}^\infty$. As proved in Chapter 3, each distribution $S \in \mathcal{E}'(\partial D)$ is the boundary value of a harmonic function F that has finite order of growth at ∂D, that is,

$$\lim_{\epsilon\to 0+} \int_{\partial D} F(z - \epsilon\nu(z))\psi(z)\, d\sigma = S(\psi)$$

for every function $\psi \in \mathcal{C}^\infty(\partial D)$. Therefore, the $\overline{\partial}$-Neumann problem for distributions can be stated as follows: for a given distribution $S \in \mathcal{E}'(\partial D)$, find a harmonic function F of finite order of growth at ∂D such that weak limit of $\overline{\partial}_n F$ equals S on ∂D.

This weak limit exists, since $\partial F/\partial\bar{z}_k$ has finite order of growth and so defines some distribution on ∂D, while $\rho_k \in \mathcal{C}^\infty(\partial D)$. The distribution S also satisfies the necessary condition for solvability:

$$S(\bar{f}) = 0 \qquad \text{for all } f \in \mathcal{O}(\overline{D}).$$

We now compare (14.1) with the $\overline{\partial}$-Neumann problem for forms (see [50]): given a form ψ of type (p,q) in D, find a form F for which $\Box F = \psi$ in D, and the normal parts of the forms F and $\overline{\partial}^* F$ are zero on ∂D. If F and ψ are functions, then we need to find a function F such that

$$\begin{aligned} \Box F &= \psi \qquad \text{in } D, \\ \overline{\partial}_n F &= 0 \qquad \text{on } \partial D. \end{aligned} \tag{14.3}$$

If we do not pay attention to the smoothness of the functions, then (14.1) and (14.3) are equivalent. Indeed, one solution of (14.3) is the volume potential F_ψ.

Subtracting it from (14.3), we obtain $\square(F - F_\psi) = 0$, and $\overline{\partial}_n(F - F_\psi) = \varphi$ on ∂D, that is, we have (14.1). Conversely, given (14.1), we take the single-layer potential $F_\varphi^\pm$ for φ and extend F_φ^- into D as a smooth function to obtain that $\overline{\partial}_n(F - F_\varphi^+ + F_\varphi^-) = 0$, and $\square(F - F_\varphi^+ + F_\varphi^-) = \psi$, that is, we have (14.3).

Problem (14.1) is more natural for studying the boundary properties of holomorphic functions.

The $\overline{\partial}$-Neumann problem for forms arose in the works of Spencer and then was studied by many authors. An especially large role was played here by Kohn. Currently, interest in this problem has revived in connection with finding explicit integral representations of solutions for forms of type (p, q), $q > 0$. The solution in strongly pseudoconvex domains, based on a priori estimates, was given in [50]. The case $q = 0$ is exceptional, but in [50] (see Theorem 3.1.19) it was shown that it reduces to the case of forms of type $(p, 1)$. We will consider a direct solution of (14.1), and besides for a wider class of functions than in [50], in particular, for distributions.

14.3 The homogeneous $\overline{\partial}$-Neumann problem

We first consider the homogeneous $\overline{\partial}$-Neumann problem

$$\square F = 0 \quad \text{in } D \qquad \text{and} \qquad \overline{\partial}_n F = 0 \quad \text{on } \partial D. \tag{14.4}$$

It is clear that holomorphic functions F satisfy (14.4). Later we will show that the converse is also true. First we reformulate the problem. Let MF be the Bochner-Martinelli operator, that is,

$$(MF)(z) = \int_{\partial D} F(\zeta)\, U(\zeta, z), \qquad z \notin \partial D.$$

We will write M^+F or M^-F depending on whether we are considering the operator MF in D or outside $\overline{D}$.

Theorem 14.1. *Let F be a harmonic function in D. In order that $\overline{\partial}_n F = 0$ on ∂D, it is necessary and sufficient that $M^+F = F$ in D (that is, F is represented by the Bochner-Martinelli integral).*

Proof. First consider the case that $F \in \mathcal{C}^1(\overline{D})$ and ∂D is piecewise smooth. Then by the Bochner-Martinelli formula for smooth functions,

$$\int_{\partial D} F(\zeta)\, U(\zeta, z) - \int_{\partial D} \mu_F(\zeta) g(\zeta, z) = \begin{cases} F(z), & z \in D, \\ 0, & z \notin \overline{D}; \end{cases} \tag{14.5}$$

$$\begin{aligned} \mu_F\big|_{\partial D} g(\zeta, z) &= -2^{n-1} i^{-n} (-1)^n \overline{\partial}_n F (n-2)!\, (2\pi i)^{-n} |\zeta - z|^{2-2n}\, d\sigma \\ &= -(n-2)!\, (\pi^{-n}/2) \overline{\partial}_n F |\zeta - z|^{2-2n}\, d\sigma. \end{aligned}$$

If $\overline{\partial}_n F = 0$, then it follows from (14.5) that $M^+F = F$ in D. If $M^+F = F$ in D, then $M^-F = 0$ on ∂D by Theorem 2.5 on the jump of the Bochner-Martinelli integral; then $M^-F \equiv 0$ since M^-F is a harmonic function and $M^-F(z) \to 0$ when $|z| \to \infty$. Thus, we obtain from (14.5) that

$$\int_{\partial D} \frac{\bar{\partial}_n F(\zeta)}{|\zeta - z|^{2n-2}} \, d\sigma(\zeta) = 0$$

for all $z \notin \partial D$. Now applying the theorem of Keldysh-Lavrent'ev (see, for example, [127, p. 418]) on the density of fractions of the form $1/|\zeta - z|^{2n-2}$ in the space of functions $\mathcal{C}(K)$ (where $\Lambda^{2n}(K) = 0$), we obtain that $\bar{\partial}_n F = 0$ on ∂D. $\qquad \square$

We need the Keldysh-Lavrent'ev theorem for smooth functions.

Lemma 14.2. *Suppose $\partial D \in \mathcal{C}^k$. Then linear combinations of fractions of the form $1/|\zeta - z|^{2n-2}$ for $\zeta \in \partial D$ and $z \notin \partial D$ are dense in the space $\mathcal{C}^k(\partial D)$.*

Proof. Let V be a neighborhood of ∂D with smooth boundary ∂V and such that a given function $f \in \mathcal{C}^k(\partial D)$ lies in $\mathcal{C}^k(\overline{V})$. We write for f in V the Bochner-Martinelli formula (1.4) for smooth functions: for $z \in \partial D$, we have

$$f(z) = \int_{\partial V} f(\zeta) \, U(\zeta, z) - \int_V \bar{\partial} f \wedge U(\zeta, z)$$
$$= \int_{\partial V} f(\zeta) \, U(\zeta, z) - \int_{V \setminus \partial D} \bar{\partial} f \wedge U(\zeta, z).$$

We obtain the result by replacing the integral with an appropriate Riemann sum and $\partial g / \partial \zeta_k$ with a difference quotient.

Now suppose F is a harmonic function in D having finite order of growth at ∂D. Then (14.5) will hold for F. Indeed, if we write (14.5) for the domain $D_\epsilon = \{ z : \rho(z) < -\epsilon \}$ and pass to the limit as $\epsilon \to 0^+$, then

$$MF_0 + \frac{(n-2)!}{2\pi^n} (\bar{\partial}_n F_\zeta)_0 \left(\frac{1}{\zeta - z|^{2n-2}} \right) = \begin{cases} F(z), & z \in D, \\ 0, & z \notin \overline{D}, \end{cases}$$

and $(MF_0)(z) = (F_\zeta)_0(U(\zeta, z)/d\sigma(\zeta))$ (here F_0 and $(\bar{\partial}_n F)_0$ are the boundary values of F and $\bar{\partial}_n F$, that is, distributions in $\mathcal{E}'(\partial D)$).

If $(\bar{\partial}_n F)_0 = 0$ on ∂D, then it is clear that $M^+F_0 = F$ in D. If $M^+F_0 = F$ in D, then $M^-F_0 = 0$ on ∂D (in the sense of $\mathcal{E}'(\partial D)$) by Theorem 3.6, so (see Theorem 11.1) the function M^-F_0 extends to ∂D as a function of class $\mathcal{C}^\infty(\mathbf{C}^n \setminus D)$, and consequently $M^-F_0 \equiv 0$, that is,

$$(\bar{\partial}_n F)_0 \left(\frac{1}{|\zeta - z|^{2n-2}} \right) = 0, \qquad z \notin \partial D.$$

By Lemma 14.2, linear combinations of the fractions $1/|\zeta - z|^{2n-2}$ are dense in $\mathcal{C}^\infty(\partial D)$, so $(\bar{\partial}_n F)_0 = 0$ on ∂D. $\qquad \square$

Theorem 14.1 was noted by Aronov [16] for smooth functions, and by the author [117] for distributions.

15 Holomorphic functions represented by the Bochner-Martinelli integral

15.1 Smooth functions

Theorem 15.1. *Let D be a bounded domain with piecewise-smooth boundary ∂D, and suppose $F \in \mathcal{C}^1(\overline{D})$. In order that $F \in \mathcal{O}(D)$, it is necessary and sufficient that $M^+F = F$ in D.*

Proof. The necessity follows from (1.5), the Bochner-Martinelli integral representation. We prove the sufficiency. From Theorem 14.1, we obtain that $\overline{\partial}_n F = 0$ on ∂D, and F is harmonic in D, that is, the form $\star\overline{\partial}F$ is ∂-closed. Then

$$0 = \int_{\partial D} \overline{F}(\star\overline{\partial}F) = \int_D \partial\overline{F} \wedge \star\overline{\partial}F$$
$$= 2^{1-n} i^n \int_D \sum_{k=1}^n \left|\frac{\partial F}{\partial \bar{z}_k}\right|^2 dz \wedge d\bar{z} = 2\int_D \sum_{k=1}^n \left|\frac{\partial F}{\partial \bar{z}_k}\right|^2 dv.$$

Hence $\partial F/\partial \bar{z}_k = 0$ in D for all $k = 1, \dots, n$, so $F \in \mathcal{O}(D)$. □

This theorem was given in the book of Folland and Kohn [50] for functions $F \in \mathcal{C}^\infty(\overline{D})$ and $\partial D \in \mathcal{C}^\infty$. Aronov and Kytmanov [20] remarked that it extends to functions of class $\mathcal{C}^1$.

Corollary 15.2. *If $f \in \mathcal{C}^1(\partial D)$ and $\partial D \in \mathcal{C}^1$, then a necessary and sufficient condition for the existence of a function $F \in \mathcal{O}(D) \cap \mathcal{C}^1(\overline{D})$ such that $F(z) = f(z)$ for $z \in \partial D$ is that $M^- f = 0$ for $z \notin \overline{D}$.*

The proof follows from Theorem 15.1 and Theorem 2.5 on the jump of the Bochner-Martinelli integral. Corollary 15.2 was given in [20, 21].

We remark that it is not assumed in these results that the domain D has connected boundary.

Since $U(\zeta, z)$ is a $\overline{\partial}$-closed form of class $\mathcal{C}^\infty(\overline{D})$ when $z \notin \overline{D}$, Theorem 15.1 sharpens Theorem 7.1 for smooth functions.

When $n > 1$, Theorem 15.1 admits a converse.

Theorem 15.3 (Kytmanov). *If $M^+ f$ is holomorphic in D, $f \in \mathcal{C}^1(\partial D)$, and $\partial D \in \mathcal{C}^1$ is connected, then the boundary value of $M^+ f$ coincides with f.*

Proof. Since $M^+ f$ is holomorphic in D, while $\partial D \in \mathcal{C}^1$ is connected, we have $\overline{\partial}_n M^+ f = 0$ in D. Using Corollary 4.9, we obtain that $\overline{\partial}_n M^- f = 0$ on ∂D. Since $M^- f = O(|z|^{1-2n})$, and $\partial M^- f/\partial \bar{z}_k = O(|z|^{-2n})$ as $|z| \to \infty$, we find by applying Stokes's formula that

$$0 = \int_{\partial D} \overline{M^- f} \star\overline{\partial} M^- f = -2^{1-n} i^n \int_{\mathbb{C}^n \setminus D} \sum_{k=1}^n \left|\frac{\partial M^- f}{\partial \bar{z}_k}\right|^2 dz \wedge d\bar{z}.$$

Consequently, $M^- f$ is holomorphic in $\mathbf{C}^n \setminus \overline{D}$, and by Hartogs's theorem it extends holomorphically into $\mathbf{C}^n$. But then $M^- f \equiv 0$, since $M^- f \to 0$ as $|z| \to \infty$ (here we use that $n > 1$ and that ∂D is connected). Now applying Corollary 15.2, we obtain the required assertion $M^+ f = f$ on ∂D. □

This result was given in [107].

It is clear that Theorem 15.3 is not true when $n = 1$. Also, it is not true if ∂D is not connected: it suffices to set $f = 1$ on one connected component of ∂D and $f = 0$ on the remaining components.

The question of what functions are represented by the Bochner-Martinelli integral was also studied by Serbin [177, 179]. One of his results states that the Bochner-Martinelli integral $M^+ f$ is represented by the Bochner-Martinelli formula, that is, $M^+ M^+ f = M^+ f$. Theorem 15.1 shows that this result is not true. In the ball $B(0, 1)$, for example, we have $M^+ \bar{z}_k = (n-1)\bar{z}_k/n$ (see Lemma 5.2), while $M^+ M^+ \bar{z}_k = (n-1)^2 n^{-2} \bar{z}_k \neq M^+ \bar{z}_k$.

15.2 Continuous functions

Theorem 15.4 (Kytmanov, Aĭzenberg). *Let D be a bounded domain with boundary of class $\mathcal{C}^2$. A necessary and sufficient condition for a function $F \in \mathcal{C}(\overline{D})$ to be holomorphic in D is that $M^+ F = F$ in D.*

Proof. If $M^+ F = F$ in D, then $M^- F$ extends continuously to $\mathbf{C}^n \setminus D$ by Theorem 3.1, and $M^- F = 0$ on ∂D. By the uniqueness theorem for harmonic functions, $M^- F \equiv 0$. Then $\overline{\partial}_n M^- F \equiv 0$, so (by Theorem 4.10), $\overline{\partial}_n M^+ F = \overline{\partial}_n F$ extends continuously to $\overline{D}$, and $\overline{\partial}_n F = 0$ on ∂D. Let $\rho(z)$ be a defining function for D, and $D_\epsilon = \{ z \in D : \rho(z) < -\epsilon \}$. Then

$$\int_{\partial D_\epsilon} \overline{F}(\star \overline{\partial} F) = 2^{1-n} i^n \int_{D_\epsilon} \sum_{k=1}^{n} \left| \frac{\partial F}{\partial \bar{z}_k} \right|^2 dz \wedge d\bar{z} \to 0$$

when $\epsilon \to 0^+$. Consequently $\partial F / \partial \bar{z}_k = 0$ in D for all $k = 1, \dots, n$. □

Theorem 15.4 was proved in [123]. We get corollaries from this theorem just as we did from Theorem 15.1.

Corollary 15.5. *If $f \in \mathcal{C}(\partial D)$ and $\partial D \in \mathcal{C}^2$, then a necessary and sufficient condition for f to extend into D as a function F of class $\mathcal{A}(D)$ is that $M^- f = 0$ outside $\overline{D}$.*

Corollary 15.6. *Suppose $n > 1$, $\partial D \in \mathcal{C}^2$ is connected, and $f \in \mathcal{C}(\partial D)$. If $M^+ f$ is holomorphic in D, then $M^+ f \in \mathcal{C}(\overline{D})$, and $M^+ f = f$ on ∂D.*

As in §5, we consider an orthonormal basis $\{P_{k,s,t}\}$ in $\mathcal{L}^2(S(0, 1))$ consisting of homogeneous harmonic polynomials of degree k in z and degree s in $\bar{z}$, where $k, s = 0, 1, \dots$, and $t \geq 1$.

Corollary 15.7. *Suppose $\partial D \in C^2$ is connected, and $f \in C(\partial D)$. A necessary and sufficient condition for f to extend into D as a function $F \in \mathcal{A}(D)$ is that*

$$\int_{\partial D} f(\star\partial P_{k,s,t}) = 0$$

for all k, s, and t.

Proof. The function $M^- f$ is harmonic outside $\overline{D}$, so to prove $M^- f = 0$ outside $\overline{D}$ it suffices to prove that $M^- f = 0$ outside some ball $B(0,R) \supset \overline{D}$. When $z \notin B(0,R)$, the function $|\zeta - z|^{2-2n}$ is harmonic in ζ in $\overline{B}(0,R)$, and this means that it can be represented by a series in the harmonic polynomials $P_{k,s,t}$ that converges uniformly on $\overline{B}(0,R)$. Since $U(\zeta,z) = 2^{n-1} i^n \star\partial_\zeta g(\zeta,z)$, we obtain the required equality $\int_{\partial D} f(\zeta)U(\zeta,z) = 0$ for $z \notin \overline{B}(0,R)$. □

Corollaries 15.5 and 15.7 were given in [123], and Corollary 15.6 in [107].

When $n = 1$, Corollary 15.7 reduces to the classical criterion for the existence of a holomorphic extension, which consists in the orthogonality of f to the monomials z^k, $k = 0, 1, \dots$, since in this case $P_{k,s,t} = az^k + b\bar{z}^s$, and so $\star\partial P_{k,s,t} = cz^{k-1}\,dz$, $k \geq 1$.

15.3 Functions with the one-dimensional holomorphic extension property

Suppose $\partial D \in C^1$ and $f \in C(\partial D)$. We will say that f *has the one-dimensional holomorphic extension property if, for every complex line l, there is a function $F_l \in C(\overline{D} \cap l)$, holomorphic at the interior points (with respect to l) of $\overline{D} \cap l$, such that $F_l = f_l$ on $\partial D \cap l$, where f_l is the restriction $f\big|_{\partial D \cap l}$.*

Lemma 15.8. *If f has the one-dimensional holomorphic extension property, then $M^- f = 0$ in $\mathbf{C}^n \setminus \overline{D}$.*

Proof. Suppose z lies in the unbounded component of $\mathbf{C}^n \setminus \overline{D}$. We make an orthogonal linear transformation and a translation in $\mathbf{C}^n$ so that z is taken to 0 and the plane $\{ z : z_1 = 0 \}$ does not intersect D. This can be done for points z sufficiently far from $\overline{D}$. Then the kernel $U(\zeta,z)$ goes into $U(\zeta,0)$. Moreover, making the change of variables $\zeta_1 = 1/v_1$ and $\zeta_j = v_j/v_1$ for $j = 2, \dots, n$, we obtain

$$d\zeta = d\zeta_1 \wedge \dots \wedge d\zeta_n = \begin{vmatrix} -1/v_1^2 & 0 & \dots & 0 \\ -v_2/v_1^2 & 1/v_1 & \dots & 0 \\ & & \vdots & \\ -v_n/v_1^2 & 0 & \dots & 1/v_1 \end{vmatrix} dv = -dv/v_1^{n+1}.$$

Here $\mathbf{C}^n_z \setminus \{ z : z_1 = 0 \}$ is mapped biholomorphically onto $\mathbf{C}^n_v \setminus \{ v : v_1 = 0 \}$. We remark that

$$\sum_{k=1}^{n} (-1)^{k-1} \bar{\zeta}_k \, d\bar{\zeta}[k] = \sum_{k=1}^{n} \overline{\Delta}_k \, d\bar{v}[k]$$

where

$$\Delta_k = \begin{vmatrix} \zeta_1 & \zeta_2 & \dots & \zeta_n \\ \zeta'_{1v_1} & \zeta'_{2v_1} & \dots & \zeta'_{nv_1} \\ [k] & & \dots & \\ \zeta'_{1v_n} & \zeta'_{2v_n} & \dots & \zeta'_{nv_n} \end{vmatrix} = \begin{vmatrix} 1/v_1 & v_2/v_1 & \dots & v_n/v_1 \\ -1/v_1^2 & -v_2/v_1^2 & \dots & -v_n/v_1^2 \\ 0 & 1/v_1 & \dots & 0 \\ [k] & & \dots & \\ 0 & 0 & \dots & 1/v_1 \end{vmatrix}$$
$$= \begin{cases} 1/v_1^n, & k=1, \\ 0, & k \neq 1. \end{cases}$$

Therefore

$$U(\zeta,0) = \frac{(n-1)!\,(-1)^n}{(2\pi i)^n}\,\frac{dv_1}{v_1} \wedge \frac{d\bar{v}[1] \wedge dv[1]}{(1+|v_2|^2+\dots+|v_n|^2)^n}$$
$$= \frac{(n-1)!\,(-1)^n}{(2\pi i)^n}\,\frac{dv_1}{v_1} \wedge \lambda(v_2,\dots,v_n).$$

An analogous representation for the kernel $U(\zeta,z)$ is given in the book of Griffiths and Harris (see [62, chap. 3]). It shows that the Bochner-Martinelli integral is the average of the Cauchy kernel over the set of lines passing through z (see also the survey [74, §1]).

Under the change of variables being considered, the lines $\{\,\zeta : \zeta_1 = t,\ \zeta_2 = c_2 t,\ \dots,\ \zeta_n = c_n t,\ t \in \mathbf{C}\}$ go over to the lines $\{\,v : v_2 = c_2,\ \dots,\ v_n = c_n\,\}$, and the boundary ∂D goes into the boundary ∂G of some domain. If f has the one-dimensional holomorphic extension property, then after the change of variables it will have the one-dimensional holomorphic extension property along the lines $l_{c_2,\dots,c_n} = \{\,v : v_2 = c_2,\ \dots,\ v_n = c_n\,\}$. Moreover,

$$\int_{\partial D} f(\zeta)\,U(\zeta,0) = \frac{(-1)^n (n-1)!}{(2\pi i)^n} \int_{\partial G} f^*(v)\,\frac{dv_1}{v_1} \wedge \lambda(v_2,\dots,v_n)$$
$$= \frac{(-1)^n (n-1)!}{(2\pi i)^n} \int_{G'} \lambda(v_2,\dots,v_n) \int_{\partial G \cap l_{v_2,\dots,v_n}} f^*(v)\,\frac{dv_1}{v_1},$$

where G' is the projection of ∂G under the mapping $v \mapsto (v_2,\dots,v_n)$. We remark that $\partial G \cap l_{v_2,\dots,v_n}$ is a smooth curve for almost all $(v_2,\dots,v_n) \in G'$ by Sard's theorem. The point 0 lies outside the domain bounded by ∂G, so the inner integral equals zero for almost all $(v_2,\dots,v_n) \in G'$. Therefore $M^- f = 0$ for points z sufficiently far from $\overline{D}$, that is, $M^- f = 0$ in the unbounded component of $\mathbf{C}^n \setminus \overline{D}$.

Suppose z lies in a bounded component $Q \subset \mathbf{C}^n \setminus \overline{D}$; we suppose that $z = 0$. Consider a cone Π_ϵ with center at the origin formed by complex lines passing through 0, where ϵ is the size of the area of the part of the hypersurface ∂Q lying inside Π_ϵ. After the same change of variables, the cone goes into a cylinder formed by the lines $\{\,v : v_2 = c_2,\ \dots,\ v_n = c_n\,\}$, where Π_ϵ may be chosen so that (1) the area of $\partial(Q \cap \Pi_\epsilon)$ converges to zero as $\epsilon \to 0^+$, whence

$$\int_{\partial D} f(\zeta)\, U(\zeta, 0) = \lim_{\epsilon \to 0^+} \int_{\partial D \setminus \partial(Q \cap \Pi_\epsilon)} f(\zeta)\, U(\zeta, 0);$$

(2) under the change of variables, $\overline{\partial D \setminus \partial(Q \cap \Pi_\epsilon)}$ goes into the compact set K_ϵ. Then just as above,

$$\int_{K_\epsilon} f^*(v)\, \frac{dv_1}{v_1} \wedge \lambda(v_2, \dots, v_n) = 0,$$

so $M^- f = 0$ outside $\overline{D}$. □

From this lemma and Corollary 15.5, we obtain the following.

Theorem 15.9 (Agranovskiĭ, Val'skiĭ, Stout). *If D is bounded with $\partial D \in \mathcal{C}^2$, and $f \in \mathcal{C}(\partial D)$ has the one-dimensional holomorphic extension property, then f extends into D as a function F in $\mathcal{A}(D)$.*

Theorem 15.9 was proved by Agranovskiĭ and Val'skiĭ [1] for the ball (see also [157]), and in the general case by Stout [196]. Lemma 15.8 is due to the author (see [10, §24]).

Other applications of Theorem 15.4 may be found in [103].

15.4 Generalizations for differential forms

When $n = 2$, another proof can be given for Theorem 15.5 (see [123]). Shaimkulov used it [181] to prove the following analogue of the theorem of the brothers Riesz.

Theorem 15.10 (Shaimkulov). *If $n = 2$, $\partial D \in \mathcal{C}^k$ is connected, $k \geq 9$, and μ is a measure on ∂D for which*

$$\int_{\partial D} (\star \partial P_{m,s,t} / d\sigma)\, d\mu = 0$$

for all m, s, and t, then $\mu = f\, d\sigma$, where $f \in \mathcal{H}^1(D)$.

The notation $\alpha/d\sigma$ for a form of type $(2n-1)$ means that we restrict α to ∂D and then take the coefficient of $d\sigma$ in this restriction.

In a subsequent section, we will generalize this result to all $n \geq 2$ when $\partial D \in \mathcal{C}^\infty$. Theorem 15.10 generalizes an analogue of the theorem of the brothers Riesz obtained by Aĭzenberg (see, for example, [7]).

We now consider a generalization of Theorem 15.1 to the case of differential forms. Let

$$\gamma = \sum_{I,J}{}' \gamma_{I,J}(z)\, dz_I \wedge d\bar{z}_J, \tag{15.1}$$

where $I = (i_1, \dots, i_p)$ and $J = (j_1, \dots, j_q)$ are increasing multi-indices. If the coefficients $\gamma_{I,J} \in \mathcal{C}^1(\overline{D})$ and $\overline{\partial}\gamma = 0$, then the second term in Koppelman's formula vanishes:

$$\int_D \overline{\partial}\gamma(\zeta) \wedge U_{p,q}(\zeta, z) \equiv 0. \tag{15.2}$$

(If $p = q = 0$, then this means that $M^+\gamma = \gamma$ in D.) It turns out that the converse is also true.

Theorem 15.11 (Kytmanov). *If ∂D is piecewise smooth, the coefficients $\gamma_{I,J}$ in (15.1) are in $\mathcal{C}^1(\overline{D})$ and are harmonic in D, and the integral in (15.2) is zero for $z \in D$, then $\overline{\partial}\gamma = 0$ in D.*

Proof. Weinstock's theorem [216] shows that every function $f \in \mathcal{C}^1(\overline{D})$ that is harmonic in D can be approximated in the metric of $\mathcal{C}^1(\overline{D})$ by functions harmonic on $\overline{D}$. Hence such a function can be approximated in the metric of $\mathcal{C}^1(\overline{D})$ by linear combinations of fractions $g(\zeta, z^k)$, where $\zeta \in \overline{D}$ and $z^k \notin \overline{D}$ (here $g(\zeta, z)$ is the fundamental solution of Laplace's equation).

It is clear that if γ satisfies (15.2), then each of the forms $\gamma_I = \sum'_J \gamma_{I,J}\, d\bar{z}_J$ also satisfies this equation, that is, we may assume that γ is a form of type $(0, q)$:

$$\gamma = \sum_J{}' \gamma_J(z)\, d\bar{z}_J. \tag{15.3}$$

If $\int_D \overline{\partial}\gamma(\zeta) \wedge U_{0,q}(\zeta, z) = 0$ for $z \in D$, then, since this function is continuous in $\mathbf{C}^n$, is harmonic in $\mathbf{C}^n \setminus \overline{D}$, and tends to zero as $|z| \to \infty$, we have

$$\int_D \overline{\partial}\gamma(\zeta) \wedge U_{0,q}(\zeta, z) \equiv 0.$$

Writing this identity for each increasing multi-index J, we obtain

$$\int_D \sum_K{}' \sum_{k \notin K} \frac{\partial \gamma_K}{\partial \bar{\zeta}_k}\, d\bar{\zeta}_k \wedge d\bar{\zeta}_K \wedge \sum_{m \notin J} \sigma(J, m) \frac{\partial g}{\partial \zeta_m}\, d\bar{\zeta}[J, m] \wedge d\zeta = 0, \tag{15.4}$$

where $K = (k_1, \dots, k_q)$. Approximating $\bar{\gamma}_J$ on $\overline{D}$ in the metric of $\mathcal{C}^1(\overline{D})$ by linear combinations of fractions $g(\zeta, z^s)$ for $\zeta \in \overline{D}$ and $z^s \notin \overline{D}$, we have from (15.4) that

$$\int_D \sum_K{}' \sum_{k \notin K} \frac{\partial \gamma_K}{\partial \bar{\zeta}_k}\, d\bar{\zeta}_k \wedge d\bar{\zeta}_K \wedge \sum_{m \notin J} \sigma(J, m) \frac{\partial \bar{\gamma}_J}{\partial \zeta_m}\, d\bar{\zeta}[J, m] \wedge d\zeta = 0. \tag{15.5}$$

Summing (15.5) on J, we obtain $\int_D |\overline{\partial}\gamma|^2\, d\bar{\zeta} \wedge d\zeta = 0$, that is, $\overline{\partial}\gamma = 0$. □

Theorem 15.11 was given in [108, 109].

Theorem 15.12 (Kytmanov). *If $\partial D \in \mathcal{C}^2$, and the form γ has coefficients of class $\mathcal{C}^2(\overline{D})$, then a necessary and sufficient condition for $\overline{\partial}\gamma$ to be 0 is that (15.2) holds for $z \in D$.*

Proof. Consider the form

$$\gamma_1(z) = \int_{\partial D} \gamma(\zeta) \wedge U_{p,q}(\zeta, z), \qquad z \in D.$$

The coefficients of γ_1 are harmonic in D and lie in class $\mathcal{C}^1(\overline{D})$ (by Theorem 4.3 and Lemma 4.2). Formula (1.9) shows that $\overline{\partial}\gamma_1 = \overline{\partial}\gamma$ in D, so (15.2) holds for γ_1. Consequently $\overline{\partial}\gamma_1 = 0$ by Theorem 15.11, that is, $\overline{\partial}\gamma = 0$. □

Theorem 15.12 was proved in [108, 109].

Condition (15.2) may be replaced by a differential condition.

Theorem 15.13 (Kytmanov). *If $\partial D \in \mathcal{C}^1$, and the form γ of type (p, q) has coefficients of class $\mathcal{C}^2(\overline{D})$, then $\overline{\partial}\gamma = 0$ in D if and only if $\overline{\partial}^*\overline{\partial}\gamma = 0$ in D and $(\overline{\partial}\gamma)_n = 0$ on ∂D (where α_n denotes the normal part of the form α).*

Proof. In view of (15.3), we may assume that γ is a form of type $(0, q)$. Then $\partial(\star\overline{\partial}\gamma) = 0$ in D, and the tangential part of the form $\star\overline{\partial}\gamma$ is zero on ∂D. Then the restriction of the form $(\star\overline{\partial}\gamma) \wedge \bar{\gamma}_J\, d\zeta_J = 0$ on ∂D. Therefore

$$
\begin{aligned}
0 &= \sum_J{}' \int_{\partial D} (\star\overline{\partial}\gamma) \wedge \bar{\gamma}_J\, d\zeta_J = \sum_J{}' \int_D d((\star\overline{\partial}\gamma) \wedge \bar{\gamma}_J\, d\zeta_J) \\
&= (-1)^{q+1} \sum_J{}' \int_D \star\overline{\partial}\gamma \wedge \partial\bar{\gamma}_J \wedge d\zeta_J \\
&= (-1)^{q+1} \sum_J{}' \int_D \star \sum_K{}' \sum_{m \notin K} \frac{\partial \gamma_K}{\partial \bar{\zeta}_m}\, d\bar{\zeta}_m \wedge d\bar{\zeta}_K \wedge \sum_{k \notin J} \frac{\partial \bar{\gamma}_J}{\partial \zeta_k}\, d\zeta_k \wedge d\zeta_J \\
&= (-1)^{q+1} 2^{q+1-n} i^n \times \\
&\quad \times \sum_J{}' \int_D \sum_K{}' \sum_{m \notin k} \sigma(K, m) \frac{\partial \gamma_K}{\partial \bar{\zeta}_m}\, d\zeta[K, m] \wedge d\bar{\zeta} \wedge \sum_{k \notin J} \frac{\partial \bar{\gamma}_J}{\partial \zeta_k}\, d\zeta_k \wedge d\zeta_J \\
&= (-1)^{(q+1)(n+1)} 2^{q+1-n} i^n \sum_I{}' \int_D \sum_{K \cup m = I} \sum_{J \cup k = I} \sigma(K, m) \frac{\partial \gamma_K}{\partial \bar{\zeta}_m} \sigma(J, k) \frac{\partial \bar{\gamma}_J}{\partial \zeta_k}\, d\zeta \wedge d\bar{\zeta} \\
&= (-1)^{(q+1)(n+1)} 2^{q+1-n} i^n \sum_I{}' \int_D \sum_{K \cup m = I} \left| \sigma(K, m)\sigma(I) \frac{\partial \gamma_K}{\partial \bar{\zeta}_m} \right|^2 d\zeta \wedge d\bar{\zeta} \\
&= (-1)^{(q+1)(n+1)} 2^{q+1-n} i^n \int_D |\overline{\partial}\gamma|^2\, d\zeta \wedge d\bar{\zeta},
\end{aligned}
$$

that is, $\overline{\partial}\gamma = 0$. □

We remark that we can deduce from the conditions $\overline{\partial}^*\overline{\partial}\gamma = 0$ and $(\overline{\partial}\gamma)_n = 0$ on ∂D that γ satisfies (15.2) (the same as for functions). The converse is also true, since $U_{p,q}$ satisfies the relation $\overline{\partial}^*\overline{\partial}U_{p,q} = 0$.

The forms γ representable in D by the first term in Koppelman's formula were studied by Tarkhanov [201] (see also [203]).

Theorem 15.9 was generalized by Agranovskiĭ and Semenov [2].

Theorem 15.14 (Agranovskiĭ, Semenov). *Suppose $\partial D \in \mathcal{C}^2$ is connected, V is an arbitrary open set in D, and Λ_V is the family of complex lines intersecting V. If $f \in \mathcal{C}(\partial D)$ satisfies the one-dimensional holomorphic extension property along the lines of Λ_V, then f extends holomorphically into D.*

The proof of this theorem also uses Corollary 15.5 and Theorem 15.4.

Other families of complex lines that suffice for holomorphic extension were found in [57].

16 Iterates of the Bochner-Martinelli integral

16.1 The theorem on iterates

Let D be a bounded domain in $\mathbf{C}^n$ with connected, class $\mathcal{C}^\infty$ boundary, and suppose $n > 1$. We consider the Sobolev space $\mathcal{W}_2^s = \mathcal{W}_2^s(D)$, where s is a natural number. This space consists of the functions $f \in \mathcal{L}^2(D)$ such that all derivatives $\partial^\alpha f$ through order s lie in $\mathcal{L}^2(D)$. If $(f, g) = \int_D f\bar{g}\, dv$ is the scalar product in $\mathcal{L}^2(D)$, then the scalar product in $\mathcal{W}_2^s(D)$ is given by

$$(f, g)_s = \sum_{\|\alpha\| \le s} (\partial^\alpha f, \partial^\alpha g), \qquad \alpha = (\alpha_1, \dots, \alpha_{2n}).$$

We will also need the space $\mathcal{W}_2^{s+\lambda}(\partial D)$ for $0 < \lambda < 1$. It consists of the functions $f \in \mathcal{W}_2^s(\partial D)$ for which the integral

$$\int_{\partial D}\int_{\partial D} \sum_{\|\alpha\|=s} \frac{|\partial^\alpha f(z) - \partial^\alpha f(\zeta)|^2}{|\zeta - z|^{2n+2\lambda-1}}\, dv_\zeta dv_z$$

converges.

Properties of these spaces may be found, for example, in the survey [45]. We will need the following properties.

1. When $s \ge 1$, the restriction of a function $f \in \mathcal{W}_2^s(D)$ to ∂D lies in the space $\mathcal{W}_2^{s-1/2}(\partial D)$, and the restriction operator is continuous.

2. If we denote the subspace of harmonic functions in $\mathcal{W}_2^s(D)$ by $\mathcal{G}_2^s = \mathcal{G}_2^s(D)$, then the restriction operator from $\mathcal{G}_2^s$ to $\mathcal{W}_2^{s-1/2}(\partial D)$ is a linear topological isomorphism (see [45]). The same may be said for the space $\mathcal{G}_2^s(\mathbf{C}^n \setminus \overline{D})$ and

$\mathcal{W}_2^{s-1/2}(\partial D)$ (except that we must put in $\mathcal{G}_2^s(\mathbf{C}^n \setminus \overline{D})$ only those harmonic functions that tend to zero as $|z| \to \infty$). Then $\mathcal{W}_2^s = \mathcal{G}_2^s \oplus \mathcal{N}_2^s$, where $\mathcal{N}_2^s$ consists of the functions in $\mathcal{W}_2^s$ that are equal to zero on ∂D.

If $s \geq 1$ and $f \in \mathcal{W}_2^s(D)$, then we have by the Bochner-Martinelli formula (1.4) for smooth functions that

$$f(z) = \int_{\partial D} f(\zeta)\, U(\zeta, z) - \int_D \overline{\partial} f \wedge U(\zeta, z), \qquad z \in D.$$

(It is clear that this formula holds for functions in $\mathcal{W}_2^s$ when $s \geq 1$.) We denote the first integral by Mf and the second by Tf. (If we need to indicate that we are considering these operators in D, then we will put a plus sign (M^+ or T^+), and outside $\overline{D}$ we will use a minus sign.) Then $I = M + T$ in D, where I is the identity operator in $\mathcal{W}_2^s$. The operator $\overline{\partial}$ is a bounded operator from $\mathcal{W}_2^s$ into $\mathcal{W}_2^{s-1}$ (see, for example, [45]), so the operator T is a bounded operator from $\mathcal{W}_2^s$ into $\mathcal{W}_2^s$, and consequently M too is bounded in $\mathcal{W}_2^s$.

Let $\mathcal{A}_2^s(D)$ be the space of holomorphic functions in $\mathcal{W}_2^s(D)$.

Theorem 16.1 (Romanov). *We have*

$$\lim_{m\to\infty} M^m = P_A \quad \text{and} \quad \lim_{m\to\infty} T^m = P_N$$

in the strong operator topology of $\mathcal{W}_2^1(D)$, where P_A is a projection from $\mathcal{W}_2^1$ onto $\mathcal{A}_2^1$, and P_N is a projection from $\mathcal{W}_2^1$ onto $\mathcal{N}_2^1$.

Theorem 16.1 was proved by Romanov [169].

We will assume subsequently that $f \in \mathcal{G}_2^1$. Since M takes $\mathcal{G}_2^s$ into $\mathcal{G}_2^s$, it is enough to prove that $T^m f \to 0$ in the topology of $\mathcal{W}_2^1$ as $m \to \infty$.

For $f \in \mathcal{G}_2^1$, the operator T can be described in another way. Let $\nu(z)$ be the outer unit normal to ∂D at z, and let $\tau = i\nu$. We saw in §13 that $\overline{\partial}_n f = \frac{1}{2}\left(\frac{\partial f}{\partial \nu} + i\frac{\partial f}{\partial \tau}\right)$. We write $\partial_n f = \frac{1}{2}\left(\frac{\partial f}{\partial \nu} - i\frac{\partial f}{\partial \tau}\right)$, $\overline{\partial}_{-n} f = -\frac{1}{2}\left(\frac{\partial f'}{\partial \nu} + i\frac{\partial f'}{\partial \tau}\right)$, and $\partial_{-n} f = -\frac{1}{2}\left(\frac{\partial f'}{\partial \nu} - i\frac{\partial f'}{\partial \tau}\right)$, where f' is the harmonic extension of f from ∂D to $\mathbf{C}^n \setminus \overline{D}$ that is equal to zero at infinity. Then

$$\begin{aligned}
\int_{\partial D} \overline{\partial}_n f(\zeta) g(\zeta, z)\, d\sigma &= \int_{\partial D} \star\overline{\partial} f(\zeta) g(\zeta, z) = -\int_D (\star\overline{\partial} f) \wedge \partial_\zeta g \\
&= -2^{1-n} i^n \int_D \sum_{k=1}^n \frac{\partial f}{\partial \bar{\zeta}_k} \frac{\partial g}{\partial \zeta_k}\, d\zeta \wedge d\bar{\zeta} \\
&= (-1)^{n+1} i^n 2^{1-n} \int_D \overline{\partial} f \wedge U(\zeta, z),
\end{aligned}$$

that is, $Tf = -2^{n-1} i^n \int_{\partial D} \overline{\partial}_n f\, g(\zeta, z)\, d\sigma$.

We write

$$T_\nu f = -2^{n-1} i^n \int_{\partial D} f(\zeta) g(\zeta, z)\, d\sigma = \frac{(n-2)!}{2\pi^n} \int_{\partial D} f(\zeta) |\zeta - z|^{2-2n}\, d\sigma(\zeta)$$

for the operator given by the single-layer potential. Then $T = T_\nu \overline{\partial}_n$. Moreover, we have by Lemma 4.2 that $\overline{\partial}_n T_\nu f + \overline{\partial}_{-n} T_\nu f = f$ on ∂D, so $(\overline{\partial}_n + \overline{\partial}_{-n}) T_\nu = I$, and consequently $\overline{\partial}_n + \overline{\partial}_{-n} = T_\nu^{-1}$. Hence $M = I - T = T_\nu (T_\nu^{-1} - \overline{\partial}_n) = T_\nu \overline{\partial}_{-n}$ in D.

16.2 Auxiliary results

We consider the restrictions to ∂D of the derivatives $\partial^\alpha f$ (in z and $\bar{z}$) through order $s-1$ for $f \in \mathcal{G}_2^s(D)$, where $\alpha = (\alpha_1, \ldots, \alpha_{2n})$. We denote these derivatives by f_α, and we extend them into $\mathbf{C}^n \setminus \overline{D}$ and into D so that we can talk about $\overline{\partial}_n f_\alpha$ and $\overline{\partial}_{-n} f_\alpha$. We consider the bilinear form B_s given on f and g in $\mathcal{G}_2^s$ as follows:

$$B_s(f,g) = \sum_{\|\alpha\| \leq s-1} \langle T_\nu^{-1} f_\alpha, g_\alpha \rangle = \sum_{\|\alpha\| \leq s-1} \int_{\partial D} (T_\nu^{-1} f_\alpha) \bar{g}_\alpha\, d\sigma.$$

The form B_s defines a scalar product on $\mathcal{G}_2^s(D)$; indeed

$$\begin{aligned} B_s(f,f) &= \sum_{\|\alpha\| \leq s-1} \langle T_\nu^{-1} f_\alpha, f_\alpha \rangle = \sum_{\|\alpha\| \leq s-1} \langle \overline{\partial}_n f_\alpha + \overline{\partial}_{-n} f_\alpha, f_\alpha \rangle \\ &= \sum_{\|\alpha\| \leq s-1} \left(\int_{\partial D} (\overline{\partial}_n f_\alpha) \bar{f}_\alpha\, d\sigma + \int_{\partial D} (\overline{\partial}_{-n} f_\alpha) \bar{f}_\alpha\, d\sigma \right) \\ &= \sum_{\|\alpha\| \leq s-1} \left(\int_{\partial D} \star \overline{\partial} f_\alpha \cdot \bar{f}_\alpha + \int_{\partial D} \star \overline{\partial}'_\alpha \cdot \bar{f}'_\alpha \right), \end{aligned}$$

where f'_α is the harmonic extension of f_α to $\mathbf{C}^n \setminus \overline{D}$ that is equal to zero at infinity. Applying Stokes's formula, we obtain

$$\begin{aligned} B_s(f,f) &= 2^{1-n} i^n \sum_{\|\alpha\| \leq s-1} \left(\int_D \sum_{k=1}^n \left| \frac{\partial f_\alpha}{\partial \bar{z}_k} \right|^2 dz \wedge d\bar{z} + \int_{\mathbf{C}^n \setminus D} \sum_{k=1}^n \left| \frac{\partial f'_\alpha}{\partial \bar{z}_k} \right|^2 dz \wedge d\bar{z} \right) \\ &= 2 \sum_{\|\alpha\| \leq s-1} \left(\int_D \sum_{k=1}^n \left| \frac{\partial f_\alpha}{\partial \bar{z}_k} \right|^2 dv + \int_{\mathbf{C}^n \setminus D} \sum_{k=1}^n \left| \frac{\partial f'_\alpha}{\partial \bar{z}_k} \right|^2 dv \right) \geq 0. \end{aligned}$$

If $B_s(f,f) = 0$, then, in particular,

$$\int_{\mathbf{C}^n \setminus D} \sum_{k=1}^n \left| \frac{\partial f'}{\partial \bar{z}_k} \right|^2 dv = 0,$$

that is, f' is holomorphic outside $\overline{D}$. Therefore $f' \equiv 0$, so the restriction of f to ∂D is zero (in view of the connectedness of ∂D), and consequently $f \equiv 0$.

Moreover,

$$\begin{aligned}
B_s(f,g) &= \sum_{\|\alpha\|\le s-1}\left(\int_{\partial D}\overline{\partial}_n f_\alpha\cdot\bar g_\alpha\,d\sigma+\int_{\partial D}\overline{\partial}_{-n} f'_\alpha\cdot g'_\alpha\,d\sigma\right)\\
&= \sum_{\|\alpha\|\le s-1}\left(\int_{\partial D} f_\alpha\,\partial_n\bar g_\alpha\,d\sigma+\int_{\partial D} f'_\alpha\,\partial_{-n}\bar g'_\alpha\,d\sigma\right)\\
&= \sum_{\|\alpha\|\le s-1}\left(\int_{\partial D}\overline{\overline{\partial}_n g_\alpha\cdot\bar f_\alpha}\,d\sigma+\int_{\partial D}\overline{\overline{\partial}_n g'_\alpha\cdot\bar f'_\alpha}\,d\sigma\right)=\overline{B_s(g,f)}
\end{aligned}$$

by Stokes's formula and the harmonicity of f and g.

Lemma 16.2. *The norm determined by the form B_s is equivalent to the norm of $\mathcal{G}_2^s$.*

Proof. The proof follows from the topological isomorphism of $\mathcal{W}_2^{s-1/2}(\partial D)$ and $\mathcal{G}_2^s(D)$ (or $\mathcal{G}_2^s(\mathbf{C}^n\setminus\overline{D})$), the fact that $\overline{\partial}_n$ is a bounded operator from $\mathcal{G}_2^s$ into $\mathcal{G}_2^{s-1}$, and the equation

$$B_s(f,g)=\sum_{\|\alpha\|\le s-1}\left(\int_D \star\overline{\partial} f_\alpha\wedge\partial\bar g_\alpha+\int_{\mathbf{C}^n\setminus D}\star\overline{\partial} f'_\alpha\wedge\partial\bar g'_\alpha\right).$$

□

Lemma 16.3. *In the space $\mathcal{G}_2^1$ with the bilinear form B_1, the operator M is positive, and T is nonnegative.*

Proof. Consider

$$B_1(Tf,f)=B_1(T_\nu\overline{\partial}_n f,f)=\langle\overline{\partial}_n f,f\rangle=2\int_D\sum_{k=1}^n\left|\frac{\partial f}{\partial\bar z_k}\right|^2 dv\ge 0. \tag{16.1}$$

If $B_1(Tf,f)=0$, then f is a holomorphic function in $\mathcal{A}_2^1$, that is, $\operatorname{Ker}T=\mathcal{A}_2^1$.

Moreover, $B_1(Mf,f)\ge 0$, and if $B_1(Mf,f)=0$, then we obtain as above that f extends holomorphically into $\mathbf{C}^n\setminus\overline{D}$ and equals zero at infinity, so $f\equiv 0$ since ∂D is connected. □

Corollary 16.4. *The operators M and T are self-adjoint in $\mathcal{G}_2^1$, $\operatorname{Ker}T=\mathcal{A}_2^1$, $\operatorname{Ker}M=\{0\}$, and $0<M\le I$.*

The proof follows immediately from the preceding lemma and the representation $I=M+T$. The self-adjointness of T follows from (16.1).

Lemmas 16.2 and 16.3 and Corollary 16.4 were proved by Romanov [169].

16.3 Proof of the theorem on iterates and some corollaries

The spectra of the operators M and T are concentrated on the segment $[0,1]$, so the spectral resolutions of M and T have the form

$$M = \int_0^1 \lambda E(d\lambda), \qquad T = \int_0^1 (1-\lambda)E(d\lambda)$$

(see, for example, [11, §75]), where $E(d\lambda)$ is the operator measure for the operator M. Hence

$$M^k = \int_0^1 \lambda^k E(d\lambda), \qquad T^k = \int_0^1 (1-\lambda)^k E(d\lambda).$$

Passing to the limit as $k \to \infty$, we obtain

$$\lim_{k\to\infty} M^k = E(1), \qquad \lim_{k\to\infty} T^k = E(0).$$

Here $E(1)$ is the projection operator onto the subspace of $\mathcal{G}_2^{\frac{1}{2}}$ corresponding to the eigenvalue 1 of M, which equals $\mathcal{A}_2^{\frac{1}{2}} = \operatorname{Ker} T$, and $E(0)$ is the operator of projection onto the subspace corresponding to the eigenvalue 0 of M, that is, onto $\operatorname{Ker} M = \{0\}$ (see Corollary 16.4).

Corollary 16.5. *If $f \in \mathcal{G}_2^{\frac{1}{2}}(D)$, and $Mf = f$ in D, then f is holomorphic in D (that is, $f \in \mathcal{A}_2^{\frac{1}{2}}$).*

Proof. The proof follows from Corollary 16.4, since the equality $Mf = f$ means that $f \in \operatorname{Ker} T = \mathcal{A}_2^{\frac{1}{2}}$. □

Thus, Theorem 15.4 on the holomorphicity of continuous functions represented by the Bochner-Martinelli integral carries over to functions in $\mathcal{G}_2^{\frac{1}{2}}$.

Remark. In the author's paper [117], Theorem 16.1 was stated for the spaces $\mathcal{W}_2^s$ for $s \geq 1$. Recently, Professor Straube gave an example showing that Theorem 16.1 cannot be true for all domains D and all spaces $\mathcal{W}_2^s(D)$ for $s \geq 1$.

Example 16.6. Suppose $\lim_{m\to\infty} M^m = P_A$ in the strong operator topology of $\mathcal{W}_2^s(D)$ for all $s \geq 1$. Then $P_A : W_2^s \to \mathcal{A}_2^s$. Consequently, P_A is a projection operator from $\mathcal{W}_2^s$ into $\mathcal{A}_2^s$ for all $s \geq 1$. This implies that the space $\mathcal{C}^\infty(\overline{D}) \cap \mathcal{O}(D)$ is dense in $\mathcal{A}_2^{\frac{1}{2}}$ (since $\mathcal{C}^\infty(\overline{D})$ is dense in $\mathcal{W}_2^s$, and this property is preserved under application of the operator P_A). But such a density does not hold for every domain D according to an example in [25].

Thus, the question of which domains D satisfy Theorem 16.1 for $W_2^s(D)$ for all $s \geq 1$ remains open. Lemma 5.2 and Theorem 5.6 show that it has a positive answer in the ball in $\mathbf{C}^n$. We also note that the question of convergence of the iterates of the operator M remains open for distributions. Theorem 17.1 thus has a conditional character. Nonetheless, Theorem 17.2 is true for all domains D.

Corollary 16.7. *If $f \in \mathcal{G}_2^1$, then*

$$f = \lim_{k\to\infty} M^k f + \sum_{l=0}^{\infty} M^l T_\nu \overline{\partial}_n f,$$

where the series converges in the metric of $\mathcal{W}_2^1$.

Proof. If $f \in \mathcal{G}_2^1$, then $I = M + T = M + T_\nu\overline{\partial}_n = M^k + \sum_{l=0}^{k-1} M^l T_\nu \overline{\partial}_n$, and we obtain the result by passing to the limit as $k \to \infty$. □

Corollary 16.8. *If $f \in \mathcal{G}_2^1$, and $M^k f = M^m f$ for some k and m, with $k > m$, then $f \in \mathcal{A}_2^1$.*

Proof. Since $M^k f = M^{k-m}(M^m f) = M^m f$, we have $M^{l(k-m)}(M^m f) = M^m f$, that is,

$$\lim_{l\to\infty} M^{l(k-m)}(M^m f) = M^m f,$$

so $M^m f \in \mathcal{A}_2^1$ by Theorem 16.1. Then $\overline{\partial}_n M^m f = 0$ in D. Let $g = M^{m-1} f$; then $\overline{\partial}_n M^+ g = 0$. By Theorem 4.10 on the jump of the "normal" derivative, $\overline{\partial}_{-n} M^- g = 0$ on ∂D, so $M^- g$ is holomorphic in $\mathbf{C}^n \setminus \overline{D}$ just as in Theorem 15.3, that is, $M^- g \equiv 0$. Consequently $M^+ g = g$, and so g is holomorphic in D. Since $g = M^{m-1} f$, we conclude by induction that $f \in \mathcal{A}_2^1$. □

Corollary 16.9. *If $f \in \mathcal{G}_2^1$, and there is a k such that $(M^k)^- f = 0$ outside $\overline{D}$, then $f \in \mathcal{A}_2^1(D)$.*

Consider twice the Bochner-Martinelli singular integral $M_\sigma f$ for a function $f \in \mathcal{W}_2^{s-1/2}(\partial D)$, where $s \geq 1$. Theorem 2.6 shows that $M_\sigma f \in \mathcal{W}_2^{s-1/2}(\partial D)$. By the Sokhotskiĭ-Plemelj formula (2.6),

$$M_\sigma f = 2M^+ f - f = (2M^+ - I) f = (M^+ - T^+) f.$$

Corollary 16.10. *If there is some k for which $M_\sigma^k f = f$, then f extends holomorphically into D.*

Proof. The operator $M_\sigma^k - I = (2M^+ - I)^k - I = \sum_{s=0}^{k-1} (2M^+ - I)^s (2M^+ - 2I)$, and $(2M^+ - I) + I = 2M^+ > 0$. Therefore $g = 0$ if $\sum_{s=0}^{k-1} (2M^+ - I)^s g = 0$, hence $g = 2(M^+ - I) f = 0$, that is, $M^+ f = f$, and consequently f extends holomorphically into D. □

If f does not extend holomorphically into D, then the equality $M_\sigma^k f = f$ cannot hold for any k, and moreover, $M_\sigma^k f \to 0$ in the metric of $\mathcal{W}_2^{1/2}(\partial D)$ when $k \to \infty$, since $\|(M^+ - T^+) f\| < \|f\|$.

17 Uniqueness theorem for the $\overline{\partial}$-Neumann problem

17.1 Proof of the theorem

In §14, we showed that the condition $\overline{\partial}_n F = 0$ on ∂D is equivalent to the condition $MF = F$ in D. In §15, we proved a theorem on the holomorphicity of continuous functions F for which $MF = F$, that is, we established a uniqueness theorem for the $\overline{\partial}$-Neumann problem for such functions. Here we will prove a uniqueness theorem for distributions. We will carry over the results of the preceding section on iterates of the Bochner-Martinelli integral to distributions, and as a corollary we will obtain the uniqueness theorem.

We consider the space $\mathcal{G}(D)$ of harmonic functions F in D having finite order of growth at ∂D (see Chap. 3). Each function F in $\mathcal{G}(D)$ determines a distribution $F_0 \in \mathcal{E}'(\partial D)$ via

$$F_0(\varphi) = \lim_{\epsilon\to 0+} \int_{\partial D} F(z - \epsilon\nu(z))\varphi(z)\,d\sigma(z),$$

where $\varphi \in \mathcal{C}^\infty(\partial D)$. Subsequently we will suppose that $\varphi \in \mathcal{C}^\infty(\partial D)$ has been extended into D as a harmonic function in $\mathcal{C}^\infty(\overline{D})$, and we will identify it with this extension. In particular, $\overline{\partial}_n\varphi$ is defined.

If $S \in \mathcal{E}'(\partial D)$, we may define the operators $(T_v S)(z) = -2^{n-1} i^n S_\zeta(g(\zeta, z))$ and $(MS)(z) = S_\zeta(U(\zeta, z)/d\sigma(\zeta))$. It is clear that $T_v S$ and MS are in $\mathcal{G}(D)$ for $z \in D$, and $T_v S$ and MS are in $\mathcal{G}(\mathbf{C}^n \setminus \overline{D})$ for $z \notin \overline{D}$. If $F \in \mathcal{G}(D)$, then $(\partial F/\partial \bar{z}_k) \in \mathcal{G}(D)$, so the distribution $(\overline{\partial}_n F)_0 = \sum_{k=1}^{n} \rho_k (\partial F/\partial \bar{z}_k)_0$ is defined.

Green's formula (1.3) for D and $\mathbf{C}^n \setminus \overline{D}$ shows that the identity operator on $\mathcal{E}'(\partial D)$ decomposes as the sum $I = M^+ + T^+ = M^+ + T_v\overline{\partial}_n$ in D and $I = -M^- + T_v\overline{\partial}_{-n}$ outside $\overline{D}$.

Theorem 3.6 on the jump of the Bochner-Martinelli integral for distributions says that $M^+ - M^- = I$ on ∂D. We have $M^+ = I + M^- = I + T_v\overline{\partial}_{-n} - I = T_v\overline{\partial}_{-n}$. Recall that to find $\overline{\partial}_{-n} f$, we need to extend f to $\mathbf{C}^n \setminus \overline{D}$ as a harmonic function F that is equal to zero at infinity (then automatically $|F(z)| \le C|z|^{2-2n}$), and take the normal part of the form $\overline{\partial} F$ with the opposite sign. Hence $I = T_\nu(\overline{\partial}_{-n} + \overline{\partial}_n)$, that is, $T_v^{-1} = \overline{\partial}_{-n} + \overline{\partial}_n$, the same as in §15 for functions in $\mathcal{G}_2^s$.

Theorem 17.1 (Kytmanov). *Suppose D is a domain such that $M^k \to P_A$ as $k \to \infty$ in the strong operator topology of $\mathcal{W}_2^s(D)$ for all s, and let $F \in \mathcal{G}(D)$. Then $(M^k F)_0$ converges in $\mathcal{E}'(\partial D)$ as $k \to \infty$ to some CR-distribution S.*

Proof. Consider the bilinear form $B(F, \varphi) = T_v^{-1} F_0(\bar{\varphi}) = F_0(T_v^{-1}\bar{\varphi})$ for $F \in \mathcal{G}(D)$ and $\varphi \in \mathcal{C}^\infty(\partial D)$. (The single-layer operator T_v is self-adjoint and real, $(T_v F)(\varphi) = F(T_v\varphi)$, since $g(\zeta, z) = g(z, \zeta)$ and $i^n g$ is a real function; consequently T_v^{-1} too is self-adjoint and real.) Moreover,

$$B(TF, \varphi) = B(T_v\overline{\partial}_n F, \varphi) = (\overline{\partial}_n F)_0(\bar{\varphi}) = F_0\left(\overline{\overline{\partial}_n \varphi}\right),$$

and by Stokes's formula

$$\int_{\partial D} G\,\overline{\partial}_n F\,d\sigma = \int_{\partial D} F\,\partial_n G\,d\sigma.$$

Therefore $B(TF,\varphi) = (T_v^{-1}F)_0\left(\overline{T_v\overline{\partial}_n\varphi}\right) = B(F,T\varphi)$. In exactly the same way, we see that $B(MF,\varphi) = B(F,M\varphi)$, that is, the operators T_v, T, and M are self-adjoint with respect to the bilinear form B.

If we consider the sequence M^kF, then

$$(M^kF)_0(\bar{\varphi}) = B(T_vM^kF,\varphi) = B(F,M^kT_v\varphi) = F_0\left(T_v^{-1}\overline{M^kT_v\varphi}\right).$$

But $M^kT_v\varphi$ converges in the metric of $\mathcal{C}^\infty(\overline{D})$ to some holomorphic function, in view of the hypothesis of the theorem and the embedding theorem for $\mathcal{W}_2^s$. Therefore the sequence $(M^kF)_0(\bar{\varphi})$ has a limit for every function $\varphi \in \mathcal{C}^\infty(\partial D)$, and this means that it determines some distribution $S \in \mathcal{E}'(\partial D)$.

We will show that S is a CR-distribution. By the Hartogs-Bochner Theorem 7.1, we need to show that $\overline{\partial}_\tau S = 0$, where $\overline{\partial}_\tau S$ is the value of S on functions $\overline{\partial_\tau\varphi}$ of the form $\partial_\tau\varphi\,d\sigma = \partial\varphi \wedge d\zeta[k,m] \wedge d\bar{\zeta}\big|_{\partial D}$ (each ∂-exact form is the sum of such forms). If we consider the restriction of the form B to $\mathcal{G}_2^s(D)$, then $B(F,\varphi) = B_1(F,\varphi)$ (see §16). Suppose $\mathcal{Y}_2^s \subset \mathcal{G}_2^s$ consists of the functions h such that $B(f,h) = 0$ for all $f \in \mathcal{A}_2^s(D)$. It follows from the hypothesis of the theorem that the projection operator $P_A : \mathcal{G}_2^s \to \mathcal{A}_2^s$ is bounded, so $\mathcal{G}_2^s = \mathcal{A}_2^s \oplus \mathcal{Y}_2^s$. Indeed, if $f \in \mathcal{G}_2^s$, then

$$B(P_Af,P_Af) = \lim_{k\to\infty} B(M^kf,P_Af) = \lim_{k\to\infty} B(f,M^kP_Af) = B(f,P_Af),$$

so $(f - P_Af) \in \mathcal{Y}_2^s$. If $f+h = 0$ and $f \in \mathcal{A}_2^s$ and $h \in \mathcal{Y}_2^s$, then $0 = B(f+h,f+h) = B(f,f) + B(h,h)$, that is, $f = 0$ and $h = 0$.

For every function $f \in \mathcal{G}_2^s$, the function $T_v\partial_\tau f \in \mathcal{Y}_2^s$, since $B(h,T_v\partial_\tau f) = h(\overline{\partial_\tau f}) = -\overline{\partial}_\tau h(\bar{f}) = 0$ by Stokes's formula, if $h \in \mathcal{A}_2^s$. Consider

$$\begin{aligned}\overline{\partial}_\tau S(\bar{\varphi}) &= \lim_{k\to\infty} \overline{\partial}_\tau(M^kF)_0(\bar{\varphi}) = -\lim_{k\to\infty}(M^kF)_0(\overline{\partial_\tau\varphi})\\ &= -\lim_{k\to\infty} B(T_vM^kF,\partial_\tau\varphi) = -\lim_{k\to\infty} B(M^kF,T_v\partial_\tau\varphi)\\ &= -\lim_{k\to\infty} B(F,M^kT_v\partial_\tau\varphi).\end{aligned}$$

Since $T_v\partial_\tau\varphi \in \mathcal{Y}_2^s$, we obtain from the self-adjointness of the operator M and the fact that $Mg = g$ for functions in $\mathcal{A}_2^s$ that $M^kT_v\partial_\tau\varphi \in \mathcal{Y}_2^s$. By the hypothesis of the theorem, $M^kT_v\partial_\tau\varphi \to P_AT_v\partial_\tau\varphi$ as $k \to \infty$ in $\mathcal{G}_2^s$ (for all s), and $P_AT_v\partial_\tau\varphi = 0$ in view of the decomposition of $\mathcal{G}_2^s$ as the direct sum of $\mathcal{A}_2^s$ and $\mathcal{Y}_2^s$. Therefore $M^kT_v\partial_\tau\varphi \to 0$ as $k \to \infty$ in the metric of $\mathcal{C}^\infty(\partial D)$, that is, $\overline{\partial}_\tau S(\bar{\varphi}) = 0$. □

Theorem 17.1 was given in [117] without the supplementary hypothesis of convergence of the iterates of the operator M.

Theorem 17.2 (Kytmanov). *If $F \in \mathcal{G}(D)$ and $(\overline{\partial}_n F)_0 = 0$ in $\mathcal{E}'(\partial D)$, then $F \in \mathcal{O}(D)$.*

Proof. By Theorem 14.1, the condition $(\overline{\partial}_n F)_0 = 0$ is equivalent to the condition $(MF)_0 = F_0$. Theorem 11.1 shows that if for a function $f \in \mathcal{G}(D)$ the distribution $f_0 = 0$ on some part Γ of the boundary ∂D, then f extends to this part of the boundary as a function of class $\mathcal{C}^\infty(D \cup \Gamma)$ that is equal to zero on Γ. The same property holds for the function $\overline{\partial}_n F$ (with the same proof). Therefore, from the fact that $(\overline{\partial}_n F)_0 = 0$ on ∂D, we obtain that $\overline{\partial}_n F$ extends to ∂D as a function of class $\mathcal{C}^\infty(\overline{D})$, and $\overline{\partial}_n F = 0$ on ∂D. Now consider the integral $\int_{\partial D_\epsilon} \overline{F}\, \overline{\partial}_n F \, d\sigma$, where D_ϵ is, as usual, $\{ z \in D : \rho(z) < -\epsilon \}$, $\epsilon > 0$.

On the one hand,

$$\int_{\partial D_\epsilon} \overline{F}\, \overline{\partial}_n F \, d\sigma = \int_{\partial D_\epsilon} \overline{F}(\star \overline{\partial} F) = 2^{1-n} i^n \int_{D_\epsilon} \sum_{k=1}^n \left| \frac{\partial F}{\partial \bar{z}_k} \right|^2 dz \wedge d\bar{z},$$

and on the other hand, the integral $\int_{\partial D_\epsilon} \overline{F}\, \overline{\partial}_n F \, d\sigma$ has limit zero as $\epsilon \to 0^+$, since $\overline{\partial}_n F$ is in $\mathcal{C}^\infty(\overline{D})$ and is equal to zero on ∂D (see p. 107 or the proof of Theorem 12.1). Hence we conclude that

$$\lim_{\epsilon \to 0+} \int_{D_\epsilon} \sum_{k=1}^n \left| \frac{\partial F}{\partial \bar{z}_k} \right|^2 dv = 0,$$

that is, F is holomorphic in D. □

Since the extension of the CR-distribution F_0 into D (by Theorem 7.1) is given by the Bochner-Martinelli integral MF_0 (or by the Poisson integral), the behavior of F near ∂D is determined by the behavior of these integrals.

17.2 Corollaries of the uniqueness theorem

Corollary 17.3. *If μ is a Borel measure on ∂D, where $\partial D \in \mathcal{C}^\infty$ is connected, and*

$$\int_{\partial D} d\mu_\zeta (U(\zeta, z)/d\sigma(\zeta)) = 0, \qquad z \notin \overline{D},$$

then $\mu = f\, d\sigma$, where f is the boundary value of a function in the Hardy space $\mathcal{H}^1(D)$.

Proof. By Theorem 3.6 on the jump of the Bochner-Martinelli integral for distributions, $M\mu$ agrees with the extension of μ into D as a function in $\mathcal{G}(D)$, and therefore Theorems 17.2 and 14.1 are applicable. Then $M\mu \in \mathcal{O}(D)$, and since the extension of μ into D is given by the Poisson integral, $M\mu \in \mathcal{H}^1(D)$. □

Corollary 17.3 was given in [117]. It generalizes Theorem 15.10 and an analogue of the theorem of the brothers Riesz obtained by Aĭzenberg (see, for example, [7]) for the given class of domains.

Corollary 17.4. *If for $S \in \mathcal{E}'(\partial D)$ the condition $S(\partial_n P_{k,s,t}) = 0$ holds for all k, s, and t, then S is a CR-distribution (here $\{P_{k,s,t}\}$ is a basis for $\mathcal{L}^2(S(0,1))$ of homogeneous harmonic polynomials of degree k in z and s in $\bar{z}$).*

Proof. The proof goes through the same as the proof of Corollary 15.7. Corollary 17.4 was given in [117]. □

18 Solvability of the $\overline{\partial}$-Neumann problem

18.1 The tangential $\overline{\partial}_\tau$-equation

We now turn to solvability of the $\overline{\partial}$-Neumann problem (14.1). We first recall the definition of the tangential $\overline{\partial}$-operator. Let D be a bounded domain in $\mathbf{C}^n$ with $\partial D \in \mathcal{C}^\infty$. Suppose the form ψ of type (p,q) has smooth coefficients on ∂D. Then by definition we set $\overline{\partial}_\tau \psi = (\overline{\partial}\tilde{\psi})_\tau$, where $\tilde{\psi}$ is a smooth extension of ψ into a neighborhood of the boundary of the domain (the tangential part of a form ψ is defined in §3). If this condition is written in integral form, we obtain

$$\int_{\partial D} \overline{\partial}_\tau \psi \wedge \varphi = \int_{\partial D} \overline{\partial}\tilde{\psi} \wedge \varphi = (-1)^{p+q} \int_{\partial D} \psi \wedge \overline{\partial}\varphi,$$

where φ is a form of type $(n-p, n-q-2)$ with coefficients of class $\mathcal{C}^\infty(\partial D)$.

The tangential $\overline{\partial}_\tau$-equation is the following: given a form u of type (p,q) on ∂D, find a form ψ on ∂D such that $\overline{\partial}_\tau \psi = u$, that is,

$$\int_{\partial D} u \wedge \varphi = (-1)^{p+q+1} \int_{\partial D} \psi \wedge \overline{\partial}\varphi$$

for every form φ of type $(n-p, n-q-1)$ with smooth coefficients.

A necessary condition for solvability of the $\overline{\partial}_\tau$-equation is $\overline{\partial}_\tau u = 0$, or

$$\int_{\partial D} u \wedge \overline{\partial}\varphi = 0$$

for forms φ of type $(n-p, n-q-2)$. The Lewy example shows that this is not a sufficient condition (when $q = n-1$ it is in general degenerate), so one replaces it by another necessary condition

$$\int_{\partial D} u \wedge \varphi = 0$$

for all forms φ of type $(n-p, n-q-1)$ with smooth coefficients on $\overline{D}$ such that $\overline{\partial}\varphi = 0$ on $\overline{D}$.

Many works have been devoted to the study of the $\overline{\partial}_\tau$-equation (see the survey [74]). We remark that when $p = n$ and $q = n - 1$, we obtain

$$\int_{\partial D} u\varphi = 0$$

for all functions φ that are holomorphic on $\overline{D}$. In other words, this condition agrees with (14.2) up to conjugation. Therefore we may speak of the ∂_τ-equation instead of the $\overline{\partial}_\tau$-equation. It is clear that all the results valid for the $\overline{\partial}_\tau$-equation remain true for the ∂_τ-equation.

Thus, we consider the necessary condition for solvability of the ∂-problem (14.2):

$$\int_{\partial D} \varphi \bar{f}\, d\sigma = 0 \qquad \text{for all } f \in \mathcal{O}(\overline{D}). \tag{18.1}$$

This condition was first investigated by Dautov in [40]. He showed that if D is a strongly pseudoconvex domain, and φ satisfies (18.1), then under certain smoothness conditions on φ and ∂D there is a ∂-closed form γ in D of type $(n-1, n)$ such that the restriction of γ to ∂D coincides with $\varphi\, d\sigma$, and he gave an integral representation for the form γ. Subsequently Dautov repeatedly improved this result (in terms of the smoothness of φ, γ, and ∂D) (see [7, 10]), and he also showed that in $\mathbf{C}^2$, if each function φ orthogonal to $\bar{f}$ is the restriction of a ∂-closed form, then D is a domain of holomorphy (see §23).

A result of Henkin is given in [10, Theorem 26.9] stating that for $\varphi \in \mathcal{C}^\infty(\partial D)$ and $\partial D \in \mathcal{C}^\infty$, condition (18.1) is sufficient for the extendibility of φ as a ∂-closed form γ in (weakly) pseudoconvex domains D (see also [173]). These results were extended to the case $\varphi \in \mathcal{W}_2^s(\partial D)$ in [27].

Theorem 18.1 (Boas, Shaw). *Suppose $n > 1$ and D is a bounded (weakly) pseudoconvex domain with boundary of class $\mathcal{C}^\infty$. If $\varphi \in \mathcal{W}_2^s(\partial D)$, where $s \geq 1$, satisfies (18.1), then there is a form α of type $(n-2, n)$ with coefficients in $\mathcal{W}_2^s(\partial D)$ such that the restriction of $\partial\alpha$ to ∂D agrees with $\varphi\, d\sigma$. Moreover, the operator sending φ to α is bounded in $\mathcal{W}_2^s(\partial D)$.*

The proof of this theorem uses the solvability of the $\overline{\partial}$-Neumann problem for forms obtained by Kohn in [50, 99] and the jump Theorem 3.8 for differential forms.

We will also need the following result of Kohn [50, 99].

Theorem 18.2 (Kohn). *If D is a strongly pseudoconvex domain with boundary of class $\mathcal{C}^\infty$, then for every form ψ of type (p, q), where $q > 0$, with coefficients of class $\mathcal{W}_2^s(D)$, the $\overline{\partial}$-Neumann problem (14.3) has a unique solution $\varphi = N\psi$, which belongs to the class $\mathcal{W}_2^{s+1}(\overline{D})$, and the Neumann operator N is bounded.*

We obtain as a corollary of this theorem that if α is a form of type (p, q), where $q > 0$, with coefficients of class $\mathcal{W}_2^s(D)$ such that $\overline{\partial}\alpha = 0$, then we may take the form $\beta = \overline{\partial}^* N\alpha$ as a solution of the equation $\alpha = \overline{\partial}\beta$. Here the coefficients of β lie in the class $\mathcal{W}_2^s(D)$, and the operator $\overline{\partial}^* N$ is bounded in $\mathcal{W}_2^s(D)$. Indeed, if we apply Theorem 18.2 to α, then $\overline{\partial}\,\overline{\partial}^* N\alpha + \overline{\partial}^*\overline{\partial}N\alpha = \alpha$; since $\overline{\partial}\alpha = 0$, we have $\overline{\partial}\,\overline{\partial}^*\overline{\partial}N\alpha = 0$. Consider the Hodge scalar product

$$0 = (\overline{\partial}\,\overline{\partial}^*\overline{\partial}N\alpha, \overline{\partial}N\alpha) = (\overline{\partial}^*\overline{\partial}N\alpha, \overline{\partial}^*\overline{\partial}N\alpha).$$

Here we applied Stokes's formula and used the condition that the normal part of the form $\overline{\partial}N\alpha$ equals zero. Hence $\overline{\partial}^*\overline{\partial}N\alpha = 0$, that is, $\alpha = \overline{\partial}\,\overline{\partial}^* N\alpha$.

18.2 The $\overline{\partial}$-Neumann problem for smooth functions

Theorem 18.3 (Kytmanov). *Suppose D is a bounded strongly pseudoconvex domain with boundary $\partial D \in \mathcal{C}^\infty$, and $n > 1$. If the function $\varphi \in \mathcal{G}_2^s(D)$, where $s \geq 2$, satisfies (18.1), then there exists a harmonic function $F \in \mathcal{W}_2^{s-1}(D)$ (that is, $F \in \mathcal{G}_2^{s-1}(D)$) such that $\overline{\partial}_n F = \varphi$ on ∂D, and F may be chosen so that it also satisfies (18.1). The function F defined in this way is unique; we denote it by $N\varphi$; and the Neumann operator N is bounded.*

Proof. Suppose that we have found a form α of type $(n-2, n)$ with coefficients of class $\mathcal{W}_2^{s-1}(D)$ such that $\partial^*\partial\alpha = 0$ in D, and the restriction of $\partial\alpha$ to ∂D agrees with $\varphi\, d\sigma$, that is, the form $\partial\alpha$ is ∂^*-closed. Then it will be ∂^*-exact in D: $\partial\alpha = \partial^*(F dz \wedge d\bar{z})$, where $F \in \mathcal{W}_2^{s-1}(D)$. But $\partial\partial^*(F dz \wedge d\bar{z}) = \partial\partial\alpha = 0$, so F is a harmonic function (since $\Box = \partial\partial^*$ for forms of type (n, n)). Moreover, $\partial^*(F dz \wedge d\bar{z}) = -\star\overline{\partial}\star(F dz \wedge d\bar{z}) = 2^n(-1)^{n+1} i^n \star\overline{\partial}F$, and since the restriction of $\partial\alpha$ to ∂D equals $\varphi\, d\sigma$, we have $\star\overline{\partial}F\big|_{\partial D} = \overline{\partial}_n F\, d\sigma = -2^{-n} i^n \varphi\, d\sigma$.

Using Theorem 18.1, we can find a form β of type $(n-2, n)$ with coefficients in $\mathcal{W}_2^s(D)$ such that the restriction of $\partial\beta$ to ∂D agrees with $\varphi\, d\sigma$. Let $\gamma = \partial^*\partial\beta$, and choose a form α in $\mathcal{W}_2^{s-1}(D)$ of type $(n-2, n)$ such that $\partial^*\partial\alpha = \gamma$, and the restriction of $\partial\alpha$ to ∂D equals zero. Then the form $\beta - \alpha$ will satisfy the condition $\partial^*\partial(\beta - \alpha) = 0$, and the restriction of $\partial(\beta - \alpha)$ to ∂D will equal $\varphi\, d\sigma$. The form $\star\gamma$ has type $(0, 2)$, so the $\overline{\partial}$-Neumann problem for forms is solvable for it. The form $N\star\gamma$ has the properties that the coefficients of $N\star\gamma$ lie in $\mathcal{W}_2^{s-1}(D)$, that $\overline{\partial}\,\overline{\partial}^* N\star\gamma + \overline{\partial}^*\overline{\partial}N\star\gamma = \star\gamma$, and moreover that the normal parts of the forms $N\star\gamma$ and $\overline{\partial}N\star\gamma$ equal zero on ∂D. Hence $\star\gamma = \star\partial^*\overline{\partial}\beta = \overline{\partial}\,\overline{\partial}^* N\star\gamma + \overline{\partial}^*\overline{\partial}N\star\gamma$. Using the equalities $\partial^* = -\star\overline{\partial}\star$, $\overline{\partial}^* = -\star\partial\star$, and $\star\star\psi = (-1)^{p+q}\psi$ (where ψ is a form of type (p, q)), we obtain $\partial^*\partial\beta = \partial\partial^*\star N\star\gamma + \partial^*\partial\star N\star\gamma$. Consequently, the form $\overline{\partial}^*\overline{\partial}N\star\gamma$ is $\overline{\partial}$-closed. Therefore, using that the normal part of $\overline{\partial}N\star\gamma$ equals zero on ∂D, we have

$$0 = (\overline{\partial}\,\overline{\partial}^*\overline{\partial}N\star\gamma, \overline{\partial}N\star\gamma) = (\overline{\partial}^*\overline{\partial}N\star\gamma, \overline{\partial}^*\overline{\partial}N\star\gamma),$$

that is, $\overline{\partial}^*\overline{\partial}N\star\gamma = 0$. Thus, $\partial^*\partial\beta = \partial^*\partial\star N\star\gamma$. Since the normal part of the form $N\star\gamma$ equals zero on ∂D, the tangential part of the form $\star N\star\gamma$ also equals zero on ∂D. Hence the restriction of the form $\partial \star N \star \gamma$ to ∂D equals zero. Consequently, we may take the form $\star N \star \gamma$ as α.

Now suppose $F \in \mathcal{G}_2^{s-1}(D)$ is a solution of the $\overline{\partial}$-Neumann problem (14.1) for the function $\varphi \in \mathcal{G}_2^s(D)$. Let P be the Szegő projection from $\mathcal{L}^2(\partial D)$ onto the subspace of holomorphic functions. For strongly pseudoconvex domains, the Szegő projection preserves smoothness (see, for example, the survey [74]). Then $F - PF$ is also a solution of (14.1) in $\mathcal{G}_2^{s-1}(D)$, and $F - PF$ satisfies (18.1) by the definition of the Szegő projection P. We denote this solution by $N\varphi$. Since the difference of two solutions of the $\overline{\partial}$-Neumann problem is a holomorphic function (see Theorem 17.2), the Neumann operator N is well defined, and so all the operators appearing in the proof will be bounded in the corresponding spaces. Then N too is a bounded operator from $\mathcal{G}_2^s$ into $\mathcal{G}_2^{s-1}$.

If (18.1) holds for a function $\varphi \in \mathcal{G}_2^s$, then $\overline{\partial}_n N\varphi = \varphi$, and if φ is an arbitrary function in $\mathcal{G}_2^s$, then

$$N\overline{\partial}_n \varphi = \varphi - P\varphi. \tag{18.2}$$

□

Theorem 18.3 was given in [117] for $\varphi \in \mathcal{C}^\infty(\partial D)$.

Corollary 18.4. *Suppose (18.1) holds for a function $\varphi \in \mathcal{G}_2^2(D)$. A solution to the $\overline{\partial}$-Neumann problem is given by the series*

$$F = \sum_{k=0}^{\infty} M^k T_v \varphi, \tag{18.3}$$

which converges to F in the metric of $\mathcal{G}_2^1(D)$.

Proof. If $\varphi \in \mathcal{G}_2^2(D)$, then for some solution F_1 of (14.1) we have

$$F_1 = MF_1 + TF_1 = MF_1 + T_v \overline{\partial}_n F_1 = MF_1 + T_v \varphi.$$

Hence $F_1 = M^k F_1 + \sum_{l=0}^{k-1} M^l T_v \varphi$. Since $M^k F_1$ converges in $\mathcal{G}_2^1$ to some function in $\mathcal{A}_2^1$ (see Theorem 16.1), the series (18.3) converges to a solution of the $\overline{\partial}$-Neumann problem. Generally speaking, the solution F given by the series (18.3) is not the canonical one, since it is orthogonal to holomorphic functions in the sense of the bilinear form B_1, and not in the sense of (18.1). □

Corollary 18.4 was given in [117].

Corollary 18.5. *If D is an arbitrary domain with boundary of class $\mathcal{C}^\infty$, and the series (18.3) converges in $\mathcal{G}_2^1$ for a function $\varphi \in \mathcal{G}_2^2(D)$, then it determines a solution F of the $\overline{\partial}$-Neumann problem (14.1).*

Proof. Indeed, for $\psi \in \mathcal{G}_2^2$ we have

$$\begin{aligned}\langle \overline{\partial}_n F, \psi\rangle &= B_1(T_v\overline{\partial}_n F, \psi) = B_1(TF, \psi) = B_1(F, T\psi)\\ &= B_1\left(\sum_{k=0}^{\infty} M^k T_v\varphi, T\psi\right) = B_1\left(T_v\varphi, \sum_{k=0}^{\infty} M^k T\psi\right)\\ &= B_1\left(T_v\varphi, \sum_{k=0}^{\infty} M^k T_v\overline{\partial}_n\psi\right) = \left\langle \varphi, \sum_{k=0}^{\infty} M^k T_v\overline{\partial}_n\psi\right\rangle\\ &= \langle \varphi, \psi - \lim_{k\to\infty} M^k\psi\rangle = \langle\varphi,\psi\rangle,\end{aligned}$$

since φ is orthogonal to holomorphic functions, and $\lim_{k\to\infty} M^k\psi \in \mathcal{A}_2^1$ (see §16). Thus $\overline{\partial}_n F = \varphi$. Hence if D is an arbitrary domain, then the series (18.3) gives a solution to the $\overline{\partial}$-Neumann problem when this series converges (at least weakly). □

Remark. We have proved the solvability of the $\overline{\partial}$-Neumann problem in strongly pseudoconvex domains. To do this we needed to solve the $\overline{\partial}$-problem, the $\overline{\partial}_\tau$-problem, and the $\overline{\partial}$-Neumann problem for forms. But the $\overline{\partial}$-problem and $\overline{\partial}_\tau$-problem are solvable in weakly pseudoconvex domains. Therefore the $\overline{\partial}$-Neumann problem (14.1) for functions is solvable in those weakly pseudoconvex domains for which the $\overline{\partial}$-Neumann problem is solvable for forms of type (p,q) with $q>0$ (in any case for $\varphi \in \mathcal{C}^\infty$). In particular, (14.1) is solvable for weakly pseudoconvex domains with real-analytic boundary (see [99]).

18.3 The $\overline{\partial}$-Neumann problem for distributions

Theorem 18.6 (Kytmanov). *Suppose that D satisfies the hypotheses of Theorem 18.3, $S \in \mathcal{E}'(\partial D)$, and $S(\bar f) = 0$ for all functions $f \in \mathcal{O}(\overline{D})$. Then there is a function $F \in \mathcal{G}(D)$ such that $(\overline{\partial}_n F)_0 = S$ on ∂D.*

Proof. If $\varphi \in \mathcal{C}^\infty(\partial D)$, then $Q\varphi = \varphi - P\varphi$ satisfies (18.1). Here P is the Szegő operator; we know that $P\varphi \in \mathcal{C}^\infty(\partial D)\cap\mathcal{A}(D)$ in strongly pseudoconvex domains, and P is a bounded operator on $\mathcal{C}^\infty(\partial D)$ (see the survey [74]). It follows from Theorem 18.3 and the embedding theorem for $\mathcal{W}_2^s$ that $Q\varphi = \overline{\partial}_n NQ\varphi$ and $NQ\varphi \in \mathcal{C}^\infty(\partial D)$.

If $F \in \mathcal{G}(D)$ is the solution of the $\overline{\partial}$-Neumann problem orthogonal to holomorphic functions, then $(F)_0(\bar\varphi) = F_0(\overline{P\varphi} + \overline{Q\varphi}) = F_0(\overline{Q\varphi}) = F_0(\partial_n \overline{NQ\varphi}) = (\overline{\partial}_n F)_0(\overline{NQ\varphi}) = S(\overline{NQ\varphi})$ (by Stokes's formula).

If we define a distribution F_0 via $F_0(\bar\varphi) = S(\overline{NQ\varphi})$, then it is clear that $F_0 \in \mathcal{E}'(\partial D)$, since P and N are bounded in $\mathcal{C}^\infty(\partial D)$. Moreover, $(\overline{\partial}_n F)_0(\bar\varphi) = F_0(\partial_n\bar\varphi) = F_0\left(\overline{\overline{\partial}_n\varphi}\right) = S\left(\overline{NQ\overline{\partial}_n\varphi}\right)$. But $Q\overline{\partial}_n\varphi = \overline{\partial}_n\varphi$, so by (18.2) we have $(\partial_n F)_0(\bar\varphi) = S\left(\overline{N\overline{\partial}_n\varphi}\right) = S(\bar\varphi - P\varphi) = S(\varphi)$. Since our solution F satisfies the condition $F_0(\bar f) = S(\overline{NQf}) = 0$ (if $f \in \mathcal{O}(\overline{D})$), we may denote it by NS. □

Theorem 18.6 was given in [117].

Remark. The distribution S satisfies Corollaries 18.4 and 18.5. In other words, the series (18.3) converges to a solution of the $\overline{\partial}$-Neumann problem in $\mathcal{E}'(\partial D)$ if D is a strongly pseudoconvex domain, and in the case of an arbitrary domain, the series (18.3) is a solution of the $\overline{\partial}$-Neumann problem if it converges in $\mathcal{E}'(\partial D)$. The proof follows from Theorem 17.2.

18.4 Generalization to differential forms

The following problem is a generalization of (14.1) to differential forms. Suppose D is a bounded domain in $\mathbf{C}^n$ with $\partial D \in \mathcal{C}^\infty$, and φ is a form of type $(p, q+1)$ given on ∂D, where $0 \le q \le n-1$ and $0 \le p \le n$. We wish to find a form α of type (p,q) in D such that

$$(\overline{\partial}\alpha)_n = \varphi_n \quad \text{on } \partial D \qquad \text{and} \qquad \overline{\partial}^*\overline{\partial}\alpha = 0 \quad \text{in } D. \tag{18.4}$$

When $p = q = 0$, we obtain (14.1), since if α is a function, then $(\overline{\partial}\alpha)_n = \overline{\partial}_n\alpha \wedge \overline{\partial}\rho/|\overline{\partial}\rho|$.

We first find a necessary condition for solvability of (18.4). We consider the form $\star\varphi$, which has type $(n-q-1, n-p)$, and we integrate it against a form $\bar{\beta}$, where β is a form of type (p,q) such that $\overline{\partial}\beta = 0$ in $\overline{D}$:

$$\int_{\partial D} (\star\varphi) \wedge \bar{\beta} = \int_{\partial D} (\star\overline{\partial}\alpha) \wedge \bar{\beta} = \int_D d(\star\overline{\partial}\alpha \wedge \bar{\beta}) = \int_D \partial(\star\overline{\partial}\alpha \wedge \bar{\beta}) = 0$$

in view of (18.4) and the fact that $\overline{\partial}\beta = 0$. Thus, we find that a necessary condition for solvability of (18.4) is that

$$\int_{\partial D} (\star\varphi) \wedge \bar{\beta} = 0 \tag{18.5}$$

for all forms β of type (p,q) with $\partial\bar{\beta} = 0$ on $\overline{D}$ and coefficients of class $\mathcal{C}^\infty(\overline{D})$.

Condition (18.5) is none other than the necessary condition for solvability of the tangential ∂_τ-equation for the form $\star\varphi$. Later on we will use that this condition is sufficient (in strongly pseudoconvex domains D) for the existence of a form γ of class $\mathcal{C}^\infty(\overline{D})$ of type $(n-q-1, n-p)$ such that the tangential part of $\partial\gamma$ agrees with $(\star\varphi)_\tau$ on ∂D (if the coefficients of φ are class $\mathcal{C}^\infty(\partial D)$) (see the survey [74]).

We showed in Theorem 15.12 that if $\varphi = 0$, then every form satisfying (18.4) is $\overline{\partial}$-closed, that is, Theorem 15.12 is a uniqueness theorem for the problem (18.4).

Theorem 18.7 (Kytmanov). *If D is a bounded strongly pseudoconvex domain in $\mathbf{C}^n$ (where $n > 1$), $\partial D \in \mathcal{C}^\infty$, and (18.5) holds for the form φ of type $(p, q+1)$ with coefficients of class $\mathcal{C}^\infty(\partial D)$, then (18.4) is solvable, and the solution may be taken of class $\mathcal{C}^\infty(\overline{D})$.*

Proof. If $q = n - 1$ and $p = n$, then (18.5) says that $\star\varphi$ extends antiholomorphically into D (by the Hartogs-Bochner theorem). A form α of type $(n, n-1)$ can be written $\alpha = \sum_{k=1}^{n}(-1)^{k-1}\alpha_k\, dz \wedge d\bar{z}[k]$, while $(\bar{\partial}\alpha)_n = \bar{\partial}\alpha$, and the condition $\bar{\partial}^*\bar{\partial}\alpha = 0$ indicates antiholomorphicity of the function $\star\bar{\partial}\alpha$. Therefore the problem consists of finding antiholomorphic functions α_k of class $\mathcal{C}^\infty(\overline{D})$ such that $\sum_{k=1}^{n} \partial\alpha_k/\partial\bar{z}_k = \varphi$ in D. This problem is always solvable, because we can take these functions all equal, $\alpha_k = \alpha$, to obtain the equation $\sum_{k=1}^{n} \partial\alpha/\partial\bar{z}_k = \varphi$, which is solvable in strongly pseudoconvex domains. For arbitrary p, the same argument applies to each term of φ containing a fixed set of differentials $dz_{i_1} \wedge \cdots \wedge dz_{i_p}$.

Suppose $0 \le q \le n-2$. We assume that we have found a form γ of type $(n-q-2, n-p)$ such that $\partial^*\partial\gamma = 0$ in D, and the tangential part of $\partial\gamma$ on ∂D agrees with the tangential part of $\star\varphi$. Then the form $\partial\gamma$ is ∂^*-closed in D, so it will be ∂^*-exact, that is, $\partial\gamma = \partial^*\alpha$, where α is a form of type $(n-q, n-p)$. But $\partial\partial^*\alpha = \partial\partial\gamma = 0$, so $\bar{\partial}^*\bar{\partial}\star\alpha = \star\partial\partial^*\alpha = 0$, and since the tangential part of $\partial\gamma$ agrees with the tangential part of $\star\varphi$, the normal part of $-\bar{\partial}\star\alpha$ agrees with φ_n.

We first find a form μ of type $(n-q-2, n-p)$ such that the tangential part of $\partial\mu$ agrees on ∂D with the tangential part of $\star\varphi$. Consider the form $\gamma = \partial^*\partial\mu$, and choose a form α such that $\partial^*\partial\alpha = \gamma$, and the tangential part of $\partial\alpha$ on ∂D equals zero. Then the form $\mu - \alpha$ will satisfy the condition $\partial^*\partial(\mu - \alpha) = 0$ in D, and the tangential part of $\partial(\mu - \alpha)$ on ∂D agrees with the tangential part of the form $\star\varphi$. The form $\star\gamma$ has type $(p, q+2)$, so the $\bar{\partial}$-Neumann problem is solvable for it (see Theorem 18.2), that is, the form $N\star\gamma$ will satisfy the conditions $\bar{\partial}\bar{\partial}^* N\star\gamma + \bar{\partial}^*\bar{\partial}N\star\gamma = \star\gamma$, and $(\bar{\partial}N\star\gamma)_n = 0$ and $(N\star\gamma)_n = 0$ on ∂D. Hence $\partial\partial^*\alpha = \gamma = \partial\partial^*\star N\star\gamma + \partial^*\partial\star N\star\gamma$.

Consequently, the form $\bar{\partial}^*\bar{\partial}N\star\gamma$ is $\bar{\partial}$-closed, so (since $(\bar{\partial}N\star\gamma)_n = 0$)

$$0 = (\bar{\partial}\bar{\partial}^*\bar{\partial}N\star\gamma, \bar{\partial}N\star\gamma) = (\bar{\partial}^*\bar{\partial}N\star\gamma, \bar{\partial}^*\bar{\partial}N\star\gamma),$$

that is, $\bar{\partial}^*\bar{\partial}N\star\gamma = 0$. Then $\partial^*\partial\alpha = \partial^*\partial\star N\star\gamma$. Since the normal part of the form $N\star\gamma$ equals zero on ∂D, the tangential part of the form $\star N\star\gamma$ also is zero on ∂D. Therefore the tangential part of the form $\partial\star N\star\gamma$ is zero on ∂D. Consequently, we may take $\star N\star\gamma$ as α. □

19 Integral representation for the solution of the $\bar{\partial}$-Neumann problem in the ball

19.1 The $\bar{\partial}$-Neumann problem in the ball

Let $B(0,1)$ be the unit ball with center at the origin in $\mathbf{C}^n$, and let $S(0,1)$ be its boundary. An integral representation for the operator $\bar{\partial}^* N$ in the ball $B(0,1)$ for forms of type (p,q), where $q > 0$, was found in [68, 69, 166]. When $q = 0$, we have $\bar{\partial}^* N \equiv 0$. An integral representation for the Neumann operator N in the ball in $\mathbf{C}^2$ for forms of type $(0,1)$ was found in [98]. Here we will give an integral representation for solutions of (14.1) in the ball in $\mathbf{C}^n$.

In the case of the ball $B(0,1)$, the operator $\overline{\partial}_n F$ has the form $\overline{\partial}_n F = (\partial F/\partial \bar{z}_1)\bar{z}_1 + \cdots + (\partial F/\partial \bar{z}_n)\bar{z}_n$. If F is a harmonic function, then $\overline{\partial}_n F$ is also a harmonic function, which makes it easier to find the kernel of the integral representation.

Lemma 19.1. *If $\varphi \in \mathcal{W}_2^s(S(0,1))$ satisfies (18.1), then $N\varphi \in \mathcal{W}_2^s(S(0,1))$.*

Proof. Let $\{P_{k,s,t}\}$ be an orthonormal basis in $\mathcal{L}^2(S(0,1))$ consisting of homogeneous harmonic polynomials of degree k in z and s in $\bar{z}$. Then $\overline{\partial}_n P_{s,k,t} = sP_{k,s,t}$. Suppose $\varphi = \sum_{k,s,t} a_{k,s,t} P_{k,s,t}$, and the series $\sum_{k,s,t} |a_{k,s,t}|^2$ converges.

Since φ is orthogonal to holomorphic functions, $a_{k,0,t} = 0$, so

$$N\varphi = \sum_{k,s,t} \frac{a_{k,s,t}}{s} P_{k,s,t}.$$

The series

$$\sum_{k,s,t} \frac{|a_{k,s,t}|^2}{s^2} \leq \sum_{k,s,t} |a_{k,s,t}|^2,$$

that is, $N\varphi \in \mathcal{L}^2(S(0,1))$. It is clear that we can never expect better behavior of $N\varphi$ in terms of the space $\mathcal{L}^2$. Thus, the operator N is bounded in $\mathcal{L}^2$, and its norm $\|N\| = 1$. Since the same argument holds for derivatives, N is bounded also in $\mathcal{W}_2^s(S(0,1))$ for $s \geq 0$. □

Consider for $n > 1$ the function

$$\begin{aligned} K(\zeta, z) = \frac{(n-1)!}{2\pi^n} \Bigg[&\frac{1}{(n-1)} \frac{1}{|\zeta - z|^{2n-2}} - \frac{1}{(n-1)} \frac{1}{(1 - \langle \bar{\zeta}, z \rangle)^{n-1}} \\ &+ \frac{1}{(n-1)} \sum_{j=0}^{n-2} \frac{n-1-(j+1)\langle \bar{z}, \zeta \rangle}{(j+1)|\zeta - z|^{2(j+1)}(1 - \langle \bar{\zeta}, z \rangle)^{n-j-1}} \\ &- \frac{1}{(1 - \langle \bar{\zeta}, z \rangle)^n} \sum_{j=0}^{n-2} \frac{1}{j+1} - \frac{1}{(1 - \langle \bar{\zeta}, z \rangle)^n} \ln \frac{|\zeta - z|^2}{1 - \langle \bar{\zeta}, z \rangle} \Bigg], \end{aligned}$$

and for $n = 1$,

$$K(\zeta, z) = -\ln \frac{|\zeta - z|^2}{1 - \bar{\zeta} z}$$

($|\zeta| \leq 1$ and $|z| < 1$, so $\ln(1 - \langle \bar{\zeta}, z \rangle)$ is a holomorphic function, and we suppose that $\ln 1 = 0$).

Theorem 19.2 (Kytmanov). *Suppose $\varphi \in \mathcal{W}_2^s(S(0,1))$ satisfies (18.1). Then $N\varphi \in \mathcal{W}_2^s(S(0,1))$, and*

$$(N\varphi)(z) = \int_{S(0,1)} \varphi(\zeta) K(\zeta, z)\, d\sigma(\zeta), \qquad z \in B(0,1); \tag{19.1}$$

if $\varphi \in \mathcal{E}'(S(0,1))$, then $N\varphi = \varphi_\zeta(K(\zeta, z))$.

Theorem 19.2 was given in [117].

19.2 Auxiliary results

We recall that the Poisson kernel for the ball is

$$P(\zeta, z) = \frac{(n-1)!}{2\pi^n} \frac{1 - |z|^2}{|\zeta - z|^{2n}},$$

and the Szegő kernel in the ball has the form

$$H(\bar{\zeta}, z) = \frac{(n-1)!}{2\pi^n} \frac{1}{(1 - \langle \bar{\zeta}, z \rangle)^n}.$$

Lemma 19.3. *We have $\overline{\partial}_n(K(\zeta, z)) = P(\zeta, z) - H(\bar{\zeta}, z)$, where the operator $\overline{\partial}_n$ is taken in the z variable.*

Proof. Indeed,

$$\overline{\partial}_n \left[\frac{1}{(1 - \langle \bar{\zeta}, z \rangle)^n} \ln \frac{|\zeta - z|^2}{1 - \langle \bar{\zeta}, z \rangle} \right] = \frac{1}{(1 - \langle \bar{\zeta}, z \rangle)^n} \frac{|z|^2 - \langle \zeta, \bar{z} \rangle}{|\zeta - z|^2},$$

$$\text{and} \quad \overline{\partial}_n \left[\frac{1}{|\zeta - z|^{2n-2}} \right] = (n-1) \frac{\langle \zeta, \bar{z} \rangle - |z|^2}{|\zeta - z|^{2n}}.$$

Moreover,

$$\begin{aligned}
&\overline{\partial}_n \frac{n - 1 - (j+1)\langle \zeta, \bar{z} \rangle}{(n-1)(j+1)(1 - \langle \bar{\zeta}, z \rangle)^{n-j-1} |\zeta - z|^{2(j+1)}} \\
&= \frac{(\langle \zeta, \bar{z} \rangle - |z|^2)(n - 1 - (j+1)\langle \zeta, \bar{z} \rangle)}{(n-1)|\zeta - z|^{2(j+2)}(1 - \langle \bar{\zeta}, z \rangle)^{n-j-1}} - \frac{\langle \zeta, \bar{z} \rangle}{(n-1)|\zeta - z|^{2(j+1)}(1 - \langle \bar{\zeta}, z \rangle)^{n-j-1}} \\
&= \frac{j\langle \zeta, \bar{z} \rangle - n + 1}{(n-1)(1 - \langle \bar{\zeta}, z \rangle)^{n-j-1} |\zeta - z|^{2(j+1)}} + \frac{n - 1 - (j+1)\langle \zeta, \bar{z} \rangle}{(n-1)(1 - \langle \bar{\zeta}, z \rangle)^{n-j-2} |\zeta - z|^{2(j+2)}}.
\end{aligned}$$

Hence

$$\begin{aligned}
&\frac{2\pi^n}{(n-1)!}\overline{\partial}_n K(\zeta,z)\\
&= \frac{\langle\zeta,\bar z\rangle - |z|^2}{|\zeta - z|^{2n}} + \frac{1}{n-1}\sum_{j=0}^{n-2}\Bigg[\frac{j\langle\zeta,\bar z\rangle - n + 1}{(1-\langle\bar\zeta,z\rangle)^{n-j-1}|\zeta - z|^{2(j+1)}}\\
&\quad - \frac{(j+1)\langle\zeta,\bar z\rangle - n + 1}{(1-\langle\bar\zeta,z\rangle)^{n-j-2}|\zeta - z|^{2(j+2)}}\Bigg] - \frac{|z|^2 - \langle z,\bar\zeta\rangle}{(1-\langle\bar\zeta,z\rangle)^{n}|\zeta - z|^{2}}\\
&= \frac{\langle\zeta,\bar z\rangle - |z|^2}{|\zeta - z|^{2n}} - \frac{\langle\zeta,\bar z\rangle - 1}{|\zeta - z|^{2n}} - \frac{1}{(1-\langle\bar\zeta,z\rangle)^{n-1}|\zeta - z|^{2}} - \frac{|z|^2 - \langle\zeta,\bar z\rangle}{(1-\langle\bar\zeta,z\rangle)^{n}|\zeta - z|^{2}}\\
&= \frac{1-|z|^2}{|\zeta - z|^{2n}} - \frac{1}{(1-\langle\bar\zeta,z\rangle)^{n}}.
\end{aligned}$$

□

Lemma 19.4. *The kernel $K(\zeta,z)$ is harmonic in z (and in ζ).*

Proof. We use the formula

$$\Delta(uv) = u\Delta v + v\Delta u + \sum_{k=1}^{n}\left(\frac{\partial u}{\partial z_k}\frac{\partial v}{\partial \bar z_k} + \frac{\partial u}{\partial \bar z_k}\frac{\partial v}{\partial z_k}\right).$$

Now

$$\begin{aligned}
\Delta\frac{\ln|\zeta - z|^2}{(1-\langle\bar\zeta,z\rangle)^n} &= \frac{1}{(1-\langle\bar\zeta,z\rangle)^n}\Delta\ln|\zeta - z|^2 + \sum_{k=1}^{n}\frac{\partial}{\partial z_k}\frac{1}{(1-\langle\bar\zeta,z\rangle)^n}\frac{\partial}{\partial \bar z_k}\ln|\zeta - z|^2\\
&= \frac{n-1}{(1-\langle\bar\zeta,z\rangle)^n|\zeta - z|^2} - \frac{n(|\zeta|^2 - \langle\bar\zeta,z\rangle)}{(1-\langle\bar\zeta,z\rangle)^{n+1}|\zeta - z|^2}\\
&= -\frac{1}{|\zeta - z|^2(1-\langle\bar\zeta,z\rangle)^n}.
\end{aligned}$$

(We used here that $|\zeta| = 1$.) Moreover

$$\Delta\frac{1}{|\zeta - z|^{2j}} = \frac{j(j+1-n)}{|\zeta - z|^{2(j+1)}},$$

$$\Delta\frac{n-1-(j+1)\langle\zeta,\bar z\rangle}{(1-\langle\bar\zeta,z\rangle)^{n-j-1}} = \frac{(j+1)(j+1-n)}{(1-\langle\bar\zeta,z\rangle)^{n-j}}, \qquad \text{and}$$

$$\begin{aligned}
\sum_{k=1}^{n}\frac{1}{(j+1)(1-\langle\bar\zeta,z\rangle)^{n-j-1}}\frac{\partial}{\partial \bar z_k}(n-1-(j+1)\langle\zeta,\bar z\rangle)\frac{\partial}{\partial z_k}\frac{1}{|\zeta - z|^{2(j+1)}}&\\
= -\frac{(j+1)(1-\langle\zeta,\bar z\rangle)}{(1-\langle\bar\zeta,z\rangle)^{n-j-1}|\zeta - z|^{2(j+2)}}&.
\end{aligned}$$

However,

$$\sum_{k=1}^{n} \frac{n-1-(j+1)\langle\zeta,\bar{z}\rangle}{j+1} \frac{\partial}{\partial z_k}(1-\langle\bar{\zeta},z\rangle)^{j+1+n} \frac{\partial}{\partial \bar{z}_k} \frac{1}{|\zeta-z|^{2(j+1)}}$$
$$= \frac{(n-j-1)(n-1-(j+1)\langle\zeta,\bar{z}\rangle)}{(1-\langle\bar{\zeta},z\rangle)^{n-j-1}|\zeta-z|^{2(j+2)}}.$$

Therefore

$$\begin{aligned}
&\frac{2\pi^n}{(n-1)!}\Delta K(\zeta,z)\\
&= \frac{1}{n-1}\sum_{j=0}^{n-2}\Bigg[\frac{(n-1-(j+1)\langle\zeta,\bar{z}\rangle)(j+2-n)}{(1-\langle\bar{\zeta},z\rangle)^{n-j-1}|\zeta-z|^{2(j+2)}}\\
&\quad + \frac{j+1-n}{(1-\langle\bar{\zeta},z\rangle)^{n-j}|\zeta-z|^{2(j+1)}} + \frac{(n-j-1)(n-1-(j+1)\langle\zeta,\bar{z}\rangle)}{(1-\langle\bar{\zeta},z\rangle)^{n-j-1}|\zeta-z|^{2(j+2)}}\\
&\quad - \frac{(j+1)(1-\langle\zeta,\bar{z}\rangle)}{(1-\langle\bar{\zeta},z\rangle)^{n-j-1}|\zeta-z|^{2(j+2)}}\Bigg] + \frac{1}{(1-\langle\bar{\zeta},z\rangle)^{n}|\zeta-z|^{2}}\\
&= \frac{1}{n-1}\sum_{j=0}^{n-2}\left[\frac{n-2-j}{|\zeta-z|^{2(j+2)}(1-\langle\bar{\zeta},z\rangle)^{n-j-1}} - \frac{n-j-1}{|\zeta-z|^{2(j+1)}(1-\langle\bar{\zeta},z\rangle)^{n-j}}\right]\\
&\quad + \frac{1}{(1-\langle\bar{\zeta},z\rangle)^{n}|\zeta-z|^{2}}\\
&= -\frac{1}{(1-\langle\bar{\zeta},z\rangle)^{n}|\zeta-z|^{2}} + \frac{1}{(1-\langle\bar{\zeta},z\rangle)^{n}|\zeta-z|^{2}} = 0.
\end{aligned}$$

□

19.3 Proof of the main theorem

Suppose $\varphi \in \mathcal{W}_2^s(S(0,1))$. We consider

$$F(z) = \int_{S(0,1)} \varphi(\zeta)K(\zeta,z)\,d\sigma(\zeta).$$

In view of Lemma 19.4, this function is harmonic in $B(0,1)$. Then

$$\overline{\partial}_n F = \int_{S(0,1)} \varphi(\zeta)\,\overline{\partial}_n K(\zeta,z)\,d\sigma = \int_{S(0,1)} \varphi(\zeta)(P(\zeta,z) - H(\bar{\zeta},z))\,d\sigma.$$

If φ satisfies (18.1), then $\int_{S(0,1)} \varphi(\zeta)H(\bar{\zeta},z)\,d\sigma(\zeta) = 0$, and so $\overline{\partial}_n F = \varphi$ on $S(0,1)$.

Thus, formula (19.1) gives a solution to the $\overline{\partial}$-Neumann problem (14.1). We will show that $F = N\varphi$, in other words, that (19.1) is the canonical solution. It is enough to show that the power series expansion of F does not contain "pure" monomials z^α (that is, F is orthogonal to holomorphic functions). Indeed, since $K(\zeta, z) = \overline{K(z,\zeta)}$, the expansion of $K(\zeta, z)$ in the system $\{P_{k,s,t}\}$ has the form

$$K(\zeta, z) = \sum_{k,s,t} C_{k,s,t} \bar{P}_{k,s,t}(\zeta) P_{k,s,t}(z).$$

But since $|\zeta - z|^{2-2n} - (1 - \langle \bar{\zeta}, z\rangle)^{1-n} = \sum_{l=1}^n \bar{z}_l \varphi_l(\zeta, z)$, while

$$|\zeta - z|^{-2(j+1)}(1 - \langle \bar{\zeta}, z\rangle)^{1+j-n} - (1 - \langle \bar{\zeta}, z\rangle)^{-n} = \sum_{l=1}^n \bar{z}_l f_l(\zeta, z)$$

and $\ln \dfrac{|\zeta - z|^2}{1 - \langle \bar{\zeta}, z\rangle} = \sum_{l=1}^n \bar{z}_l \psi_l(\zeta, z)$, we obtain that $K(\zeta, z) = \sum_{l=1}^n \bar{z}_l g_l(\zeta, z)$.

Consequently, the expansion of K does not contain polynomials $P_{k,0,t}$, that is,

$$\int_{S(0,1)} K(\zeta, z) P_{k,0,t}(\zeta)\, d\sigma(\zeta) = 0$$

for all s and t. We also consider

$$\int_{S(0,1)} K(\zeta, z) P_{k,s,t}(\zeta)\, d\sigma(\zeta).$$

If $s > 0$, we have

$$\begin{aligned}
&\int_{S(0,1)} K(\zeta, z) P_{k,s,t}(\zeta)\, d\sigma \\
&\quad = \frac{1}{s}\int_{S(0,1)} \bar{\partial}_n P_{k,s,t} K(\zeta, z)\, d\sigma = \frac{1}{s}\int_{S(0,1)} P_{k,s,t}\, \partial_n^\zeta K(\zeta, z)\, d\sigma \\
&\quad = \frac{1}{s}\int_{S(0,1)} P_{k,s,t}\, \bar{\partial}_n^z K(\zeta, z)\, d\sigma = \frac{1}{s}\int_{S(0,1)} P_{k,s,t}(P(\zeta, z) - H(\bar{\zeta}, z))\, d\sigma \\
&\quad = \frac{1}{s} P_{k,s,t}(z).
\end{aligned}$$

Thus, if $\varphi = \sum_{s>0} a_{k,s,t} P_{k,s,t}$, then

$$\int_{S(0,1)} \varphi(\zeta) K(\zeta, z)\, d\sigma = \sum_{s>0} \frac{a_{k,s,t}}{s} P_{k,s,t}(z) = (N\varphi)(z).$$

The proof for distributions φ is analogous.

Chapter 5

Some Applications and Open Problems

20 Multidimensional logarithmic residues

20.1 The residue formula for smooth functions

Suppose Ω is a domain in $\mathbf{C}^n$, and $f_1, \dots, f_n \in \mathcal{O}(\Omega)$. Consider the zero set of the mapping $f = (f_1, \dots, f_n)$: $E_f = \{ a \in \Omega : f_1(a) = \cdots = f_n(a) = 0 \}$. We suppose that E_f is discrete in Ω. We denote by $\omega(f)$ the form $U(f, 0)$, that is,

$$\omega(f) = \frac{(n-1)!}{(2\pi i)^n} \sum_{k=1}^{n} (-1)^{k-1} \frac{\bar{f}_k}{|f|^{2n}} \, d\bar{f}[k] \wedge df,$$

where, as usual, $df = df_1 \wedge \cdots \wedge df_n$ and $|f|^2 = |f_1|^2 + \cdots + |f_n|^2$.

Lemma 20.1. *The current* $[\omega(f)] \in \mathcal{D}'_{0,1}(\Omega)$, *and*

$$\bar{\partial}[\omega(f)] = \sum_{a \in E_f} \mu_a \delta_a \, dv, \tag{20.1}$$

where μ_a *is the multiplicity of the zero* $a \in E_f$, δ_a *is the delta function at* a, *and* dv *is the volume element in* $\mathbf{C}^n$.

Proof. First we prove that $\omega(f)$ defines a current in Ω, that is, the integral $\int \varphi \wedge \omega(f)$ converges absolutely for every form $\varphi \in \mathcal{D}^{0,1}(\Omega)$.

It is enough to show that $\int_K d\bar{z}_j \wedge \omega(f)$ converges when K is a compact set in Ω. Consider a compact set K that is the closure of a neighborhood of a point $a \in E_f$ and contains no other points of E_f. The surface $S_f(r) = \{ z \in K :$

$|f_1|^2+\cdots+|f_n|^2 = r^2\}$ is smooth and compact for almost all $r_0 \geq r \geq 0$ by Sard's theorem. Let $B_f(r) = \{z : |f|^2 \leq r^2\} \subset K$. Then by Fubini's theorem,

$$\int_{B_f(r_0)} d\bar{z}_j \wedge \omega(f) = \int_0^{r_0} \int_{S_f(r)} d\bar{z}_j \wedge \omega(f)$$
$$= C\int_0^{r_0} r^{-2n} \int_{S_f(r)} d\bar{z}_j \wedge \sum_{k=1}^{n} (-1)^{k-1} \bar{f}_k \, d\bar{f}[k] \wedge df$$
$$= nC\int_0^{r_0} r^{-2n} \int_{B_f(r)} d\bar{z}_j \wedge d\bar{f} \wedge df.$$

But

$$\int_{B_f(r)} |d\bar{z}_j \wedge d\bar{f} \wedge df| \leq C_1\, dr \int_{B_f(r)} |d\bar{f} \wedge df| = C_1 dr \int_{B(0,r)} |d\bar{z} \wedge dz| = C_2 r^{2n}\, dr,$$

so the original integral converges.

Since the form $\omega(f)$ has no singularities besides points $a \in E_f$, it defines a current $[\omega(f)] \in \mathcal{D}'_{0,1}$. We now find $\overline{\partial}[\omega(f)]$. Since $\omega(f)$ is $\overline{\partial}$-closed, $\overline{\partial}\omega(f) = \overline{\partial}[\omega(f)] = 0$ outside points $a \in E_f$. Consider a neighborhood of some point $a \in E_f$ with closure K and a function $\varphi \in \mathcal{D}(\Omega)$ with support in K. Then

$$\overline{\partial}[\omega(f)](\varphi) = [\omega(f)](\overline{\partial}\varphi) = \int_K \omega(f) \wedge \overline{\partial}\varphi.$$

Suppose the zero a is simple, that is, the Jacobian $I_f = df/dz$ is different from zero at a. Then we may assume that the mapping $w = f(z)$ is biholomorphic on K, and so

$$\int_K \omega(f) \wedge \overline{\partial}\varphi = \int_{f(K)} U(w,0) \wedge \overline{\partial}\varphi(f^{-1}(w))$$

(the operator $\overline{\partial}$ being invariant under biholomorphic mappings). Since $\overline{\partial}[U(z,0)] = \delta\, dv$ by (6.8), we obtain

$$\int_K \omega(f) \wedge \overline{\partial}\varphi = \varphi(f^{-1}(0)) = \varphi(a).$$

Thus (20.1) is proved for a mapping with simple zeroes. If a is a multiple zero (and its multiplicity equals μ_a), then, by considering the mapping $f_\zeta = f(z) - \zeta$ in a neighborhood K of a, we obtain that for almost all ζ the mapping f_ζ has μ_a simple zeroes in K (this is the dynamic definition of the multiplicity of a zero—see [10, §2]). Applying (20.1) in K to the mapping f_ζ and passing to the limit as $\zeta \to 0$, we obtain $\overline{\partial}[\omega(f)] = \mu_a \delta_a\, dv$ in K, that is, (20.1) is proved. □

Lemma 20.1 generalizes formula (6.8).

20.2 The formula for logarithmic residues

Theorem 20.2 (Roos, Yuzhakov). *Suppose $f = (f_1, \dots, f_n)$ is a holomorphic mapping given in a neighborhood of the closure of a bounded domain $D \subset \mathbf{C}^n$ with piecewise-smooth boundary. Suppose f has no zeroes on ∂D, and $\varphi \in \mathcal{A}(D)$. Then*

$$\int_{\partial D} \varphi(\zeta)\,\omega(f) = \sum_{a \in E_f \cap D} \mu_a \varphi(a). \tag{20.2}$$

Formula (20.2) was obtained by Yuzhakov [223] and Roos [170].

Proof. Since ∂D is compact and does not contain zeroes of the vector-valued function f, the set E_f is finite in D. We consider the current $[\chi_D \omega(f)]$, where χ_D is the characteristic function of D, and find $\bar\partial[\chi_D \omega(f)]$. If $\varphi \in \mathcal{C}^\infty$ has compact support lying in a neighborhood V of $\overline{D}$ (we may assume that $\operatorname{supp}\varphi$ contains only those zeroes of f which lie in D), then

$$\begin{aligned}\bar\partial[\chi_D \omega(f)](\varphi) &= \int_D \omega(f) \wedge \bar\partial\varphi = \int_V \omega(f) \wedge \bar\partial\varphi - \int_{V\setminus\overline{D}} \omega(f) \wedge \bar\partial\varphi \\ &= \sum_{a \in E_f} \mu_a \varphi(a) + \int_{V\setminus\overline{D}} d(\varphi\,\omega(f)) = \sum_{a \in E_f} \mu_a \varphi(a) - \int_{\partial D} \varphi\,\omega(f),\end{aligned}$$

that is,

$$\int_{\partial D} \varphi\omega(f) = \sum_{a \in E_f} \mu_a \varphi(a) - \int_D \omega(f) \wedge \bar\partial\varphi. \tag{20.3}$$

Taking $\varphi \in \mathcal{A}(D)$, we obtain (20.2). Although $\varphi \in \mathcal{A}(D)$ in the theorem, that is, $\varphi \in \mathcal{O}(D) \cap \mathcal{C}(\overline{D})$, we can approximate D from inside in the usual way by domains D_ϵ with smooth boundary, so that $\varphi \in \mathcal{C}^\infty(\overline{D}_\epsilon)$, and outside $\overline{D}_\epsilon$ we can extend φ to be a $\mathcal{C}^\infty$ function with compact support. □

We remark that (20.3) is a generalization of the Bochner-Martinelli formula for smooth functions.

When $n > 1$, we consider the form

$$\omega_1(f) = \frac{(n-2)!}{(2\pi i)^n f_1 |f|^{2n-2}} \sum_{k=2}^{n} (-1)^k \bar f_k\, d\bar f[1,k] \wedge df.$$

We can verify in the obvious way that $d\omega_1 = \bar\partial\omega_1 = \omega(f)$ outside the surface $\{f_1 = 0\}$.

We represent ∂D as a union $S_1 \cup S_2$ such that the surfaces S_1 and S_2 have piecewise-smooth boundary ∂S_1, and the zero set $\{\, z : f_1(z) = 0 \,\}$ does not intersect S_2.

Corollary 20.3. *Under the hypotheses of Theorem 20.2,*

$$\int_{S_1} \varphi\omega(f) - \int_{\partial S_1} \varphi\omega_1(f) = \sum_{a\in E_f\cap D} \mu_a\varphi(a)$$

(a local version of the theorem on logarithmic residues).

Proof. The proof follows from the equalities

$$\int_{\partial D} \varphi\omega(f) = \int_{S_1} \varphi\omega(f) + \int_{S_2} \varphi\omega(f) = \int_{S_1} \varphi\omega(f) - \int_{\partial S_1} \varphi\omega_1(f).$$

□

We will not discuss the numerous applications of (20.2) in multidimensional complex analysis, but refer the reader to the monographs [10, 62] and the survey [5]. Here we will consider a generalization of (20.2) to the case when the mapping f has zeroes on ∂D.

20.3 The singular Bochner-Martinelli integral

In §2, we defined the singular Bochner-Martinelli integral as the Cauchy principal value of the improper integral, that is,

$$\int_{\partial D} f(\zeta)\, U(\zeta, z) = \lim_{r\to 0+} \int_{\partial D\setminus B(z,r)} f(\zeta)\, U(\zeta, z), \qquad z \in \partial D.$$

We took balls $B(z, r)$ as the family of domains shrinking to z. Generally speaking, if we take a different family of domains, then the value of the singular integral will change. For example, in $\mathbf{C}^1$ the Bochner-Martinelli integral reduces to the Cauchy integral, and if we take

$$\text{P.V.} \int_{-1}^{1} \frac{f(t)}{t}\, dt = \lim_{\epsilon\to 0+} \left[\int_{-1}^{-\epsilon b} \frac{f(t)}{t}\, dt + \int_{\epsilon a}^{1} \frac{f(t)}{t}\, dt \right], \qquad a, b > 0,$$

then this integral will depend on a and b. It is easy to extend this example to the case $n > 1$.

It turns out that if we delete domains defined by using holomorphic mappings, then the principal value of the Bochner-Martinelli integral does not change.

Theorem 20.4 (Prenov, Tarkhanov). *Suppose D is a bounded domain in $\mathbf{C}^n$ with piecewise-smooth boundary, and $f \in \mathcal{L}^1(\partial D)$ satisfies the Dini condition at $z \in \partial D$. Then*

$$\begin{aligned}\lim_{r\to 0+} \int_{\partial D\setminus B(z,r)} f(\zeta)\, U(\zeta, z) &= \text{P.V.} \int_{\partial D} f(\zeta)\, U(\zeta, z) \\ &= \lim_{r\to 0+} \int_{\partial D\setminus B_h(r)} f(\zeta)\, U(\zeta, z),\end{aligned}$$

where $h = (h_1, \dots, h_n)$ *is a holomorphic mapping in a neighborhood of the point* $z \in \partial D$ *having a single simple zero at* z.

We recall that the Dini condition consists in the following (see Corollary 2.7): let $\omega_f(\delta, z) = \sup_{|\zeta - z| \le \delta} |f(\zeta) - f(z)|$ be the modulus of continuity of f at z; then $\int_0^1 \omega_f(\delta, z)/\delta \, d\delta < \infty$.

Proof. We write $B_h(r) = \{\zeta : |h(\zeta)| < r\}$, and we represent the Bochner-Martinelli integral in the form

$$\int_{\partial D \setminus B_h(r)} f(\zeta)\, U(\zeta, z) = \int_{\partial D \setminus B_h(r)} (f(\zeta) - f(z))\, U(\zeta, z) + f(z) \int_{\partial D \setminus B_h(r)} U(\zeta, z).$$

We showed in Corollary 2.7 that if f satisfies the Dini condition at z, then the integral $\int_{\partial D} (f(\zeta) - f(z))\, U(\zeta, z)$ converges absolutely, so it does not depend on the form of the domain $B_h(r)$, as long as this shrinks to z. To prove the theorem it remains to prove the equality

$$\text{P.V.} \int_{\partial D} U(\zeta, z) = \lim_{r \to 0+} \int_{\partial D \setminus B_h(r)} U(\zeta, z).$$

We transform this integral:

$$\int_{\partial D \setminus B_h(r)} U(\zeta, z) = \int_{\partial (D \setminus B_h(r))} U(\zeta, z) + \int_{D \cap \partial B_h(r)} U(\zeta, z) = \int_{D \cap S_h(r)} U(\zeta, z).$$

It remains to prove that

$$\lim_{r \to 0+} \int_{D \cap S(z, r)} U(\zeta, z) = \lim_{r \to 0+} \int_{D \cap S_h(r)} U(\zeta, z).$$

We may assume that $z = 0$, $h(0) = 0$, and the Jacobian $I_h(0) \neq 0$. First of all, $|h(\zeta)| = |\zeta (\partial h / \partial \zeta)(0)| + o(|\zeta|)$ as $\zeta \to 0$ (where $\partial h / \partial \zeta$ is the Jacobian matrix of h), so

$$\lim_{r \to 0+} \int_{D \cap S_h(r)} U(\zeta, z) = \lim_{r \to 0+} \int_{D \cap \{\zeta : |\zeta (\partial h / \partial \zeta)(0)| = r\}} U(\zeta, z).$$

Since $I_h(0) \neq 0$, the Hermitian form $\left| \dfrac{\partial h}{\partial \zeta}(0) \zeta \right|^2$ can be reduced by a unitary transformation to a sum of squares of variables, that is, to the form $\sum_{j=1}^n a_j^2 |\zeta_j|^2$, where $a_j > 0$ for $j = 1, \dots, n$. Under a unitary transformation, the form of $U(\zeta, z)$ does not change, and D is taken into some domain that we again denote by D. Thus, we need to prove the equality

$$\lim_{r \to 0+} \int_{D \cap S(0, r)} U(\zeta, 0) = \lim_{r \to 0+} \int_{D \cap \left\{\zeta : \sum_{j=1}^n a_j^2 |\zeta_j|^2 = r^2\right\}} U(\zeta, 0).$$

We may replace the domain D in these integrals by the tangent cone Π to ∂D at 0 (the boundary of this cone is formed by all the rays tangent to ∂D at 0; since ∂D is piecewise smooth, $\partial\Pi$ also is piecewise smooth). Finally, using that the form $U(\zeta,0)$ is homogeneous of degree zero, we make the change of variables $\zeta = rw$ in the first integral, and $\zeta = crw$ in the second, so that the ellipsoid $\{\,\zeta : \sum_{j=1}^{n} a_j^2|\zeta_j|^2 = r^2\,\}$ goes into the ellipsoid $\Gamma = \{\,w : |w_1|^2/c_1^2 + \cdots + |w_n|^2/c_n^2 = 1\,\}$, where $0 < c_j < 1$ for each $j = 1, \ldots, n$.

Now we prove that $\int_{\Gamma\cap\Pi} U(w,0) = \int_{S(0,1)\cap\Pi} U(w,0)$. Consider the domain G bounded by the surfaces $\Gamma \cap \Pi$ and $S(0,1)\cap\Pi$ and the part M of the conical surface $\partial\Pi$ lying between Γ and $S(0,1)$. By the Bochner-Martinelli formula,

$$\begin{aligned}\int_{S(0,1)\cap\Pi} U(w,0) - \int_{\Gamma\cap\Pi} U(w,0) &= \int_{\partial G} U(w,0) - \int_M U(w,0)\\ &= -\int_M U(w,0).\end{aligned}$$

It remains to show that $\int_M U(w,0) = 0$. We pass from w to the real coordinates $w_j = \xi_j + i\xi_{n+j}$ for $j = 1, \ldots, n$; then

$$\begin{aligned}\operatorname{Re} U(w,0) &= \frac{(n-1)!}{2\pi^n}\sum_{k=1}^{2n}(-1)^{k-1}\frac{\xi_k}{|\xi|^{2n}}\,d\xi[k],\\ \operatorname{Im} U(w,0) &= -\frac{(n-2)!}{4\pi^n}d\biggl(\sum_{k=1}^{n}\frac{1}{|\xi|^{2n-2}}\,d\xi[k,n+k]\biggr), \qquad n>1\end{aligned}$$

(when $n = 1$, we have $\operatorname{Im} U(w,0) = \operatorname{Im}(2\pi i w)^{-1}\,dw = -(4\pi)^{-1}\,d\ln|\xi|^2)$.

We carry out the integration over M as follows: first over the line segments $l_b = \{\,\xi : \xi_j = b_j t,\ j = 1,\ldots,2n\,\}$ lying in M, where $b_1^2 + \cdots + b_{2n}^2 = 1$ and $t > 0$, and then over the set of all lines intersecting M (that is, over the cycle $S(0,1)\cap\partial\Pi$). We find the limits of the integration parameter t as follows: $S(0,1) = \{\,\xi : |\xi| = 1\,\}$, so $S(0,1)\cap l_b$ is the point $t = 1$, and $\Gamma\cap l_b$ is the point $t_0 = ((b_1^2 + b_{n+1})^2c_1^{-2} + \cdots + (b_n^2 + b_{2n})c_n^{-2})^{-1/2}$.

The restriction of the form $\operatorname{Re} U(w,0)$ to M is zero. Indeed, if the surface $\partial\Pi$ is given by a homogeneous function ψ, that is, $\partial\Pi = \{\,\xi : \psi(\xi) = 0\,\}$ and $\psi(t\xi) = t^l\psi(\xi)$ for $t \in \mathbf{R}^1$, then at the smooth points, $d\psi = \sum_{j=1}^{2n}(\partial\psi/\partial\xi_j)\,d\xi_j = 0$ on $\partial\Pi$. If $\partial\psi/\partial\xi_1 \neq 0$, then $d\xi_1 = -(\partial\psi/\partial\xi_1)^{-1}\sum_{j=2}^{2n}(\partial\psi/\partial\xi_j)\,d\xi_j$, so on $\partial\Pi$ we have

$$\begin{aligned}\sum_{k=1}^{2n}(-1)^{k-1}\xi_k\,d\xi[k] &= -\left(\frac{\partial\psi}{\partial\xi_1}\right)^{-1}\left(\sum_{k=1}^{2n}\xi_k\frac{\partial\psi}{\partial\xi_k}\right)d\xi[1]\\ &= -l\psi(\xi)\left(\frac{\partial\psi}{\partial\xi_1}\right)^{-1} d\xi[1] = 0.\end{aligned}$$

We consider

$$\int_M \operatorname{Im} U(w,0)$$
$$= -\frac{(n-2)!}{4\pi^n}\int_{S(0,1)\cap\partial\Pi}\sum_{k=1}^n db[k,n+k]\int_{t_0}^1 \frac{dt}{t}$$
$$= -\frac{(n-2)!}{8\pi^n}\int_{S(0,1)\cap\partial\Pi}\sum_{k=1}^n db[k,n+k]\ln\left(\frac{b_1^2+b_{n+1}^2}{c_1^2}+\cdots+\frac{b_n^2+b_{2n}^2}{c_n^2}\right).$$

On $S(0,1)$, the expression $b_k^2+b_{n+k}^2 = 1-(b_1^2+b_{n+1}^2)-\dots[k]\dots-(b_n^2+b_{2n}^2)$, so

$$\operatorname{Im}\int_M U(w,0)$$
$$= -\frac{(n-2)!}{8\pi^n}\int_{S(0,1)\cap\partial\Pi}\sum_{k=1}^n \ln\Phi_k(b_1,\dots,[k],\dots,[n+k],\dots,b_{2n})\,db[k,n+k],$$

where $\Phi_k > 0$ on S does not depend on b_k and b_{n+k}, so the form in the integrand is closed. Hence, by Stokes's formula,

$$\operatorname{Im}\int_M U(w,0) = -\frac{(n-2)!}{8\pi^n}\int_{S(0,1)\cap\Pi} d\sum_{k=1}^n \ln\Phi_k\,db[k,n+k] = 0.$$

□

Theorem 20.4 was proved in [162].

20.4 The formula for logarithmic residues with singularities on the boundary

Suppose D is a bounded domain in $\mathbf{C}^n$ with piecewise-smooth boundary, and $f=(f_1,\dots,f_n)$ is a holomorphic mapping on $\overline{D}$ having a finite number of zeroes on $\overline{D}$, all zeroes that lie on ∂D being simple. For $z\in\partial D\cap E_f$, we denote the expression

$$\lim_{\epsilon\to 0+}\frac{\operatorname{vol}\{f[S(z,\epsilon)\cap D]\}}{\operatorname{vol} S(z,\epsilon)}$$

by $\tau_f(z)$. In other words, we consider the tangent cone not to the surface ∂D (see §2), but to the image $f(\partial D)$ at the point $z\in E_f$. The mapping f is biholomorphic in a neighborhood of z, so $f(\partial D)$ is also a piecewise-smooth surface in a neighborhood of the origin.

Theorem 20.5 (Prenov, Tarkhanov). *Suppose $\varphi\in\mathcal{A}(D)$ satisfies the Dini condition on ∂D. Then*

$$\text{P V.}\int_{\partial D}\varphi\omega(f) = \sum_{a\in E_f\cap D}\mu_a\varphi(a) + \sum_{a\in E_f\cap\partial D}\tau_f(a)\varphi(a).$$

Proof. Consider the domain $D_\epsilon = D \setminus \bigcup_{a\in E_f\cap\partial D} B(a,\epsilon)$. By the Roos-Yuzhakov formula (20.2),

$$\int_{\partial D_\epsilon} \varphi\,\omega(f) = \sum_{a\in E_f\cap D} \mu_a \varphi(a), \qquad \text{and}$$

$$\text{P.V.}\int_{\partial D} \varphi\,\omega(f) = \lim_{\epsilon\to 0+} \int_{\partial D\setminus\bigcup_{a\in E_f\cap\partial D} B(a,\epsilon)} \varphi\,\omega(f), \qquad \text{so}$$

$$\int_{\partial D_\epsilon} \varphi\,\omega(f) = \int_{\partial D\setminus\bigcup_a B(a,\epsilon)} \varphi\,\omega(f) - \sum_a \int_{S(a,\epsilon)\cap D} \varphi\,\omega(f).$$

Consider $\int_{D\cap S(a,\epsilon)} \varphi\,\omega(f)$. Since f is biholomorphic in a neighborhood of a, we can make the change of variables $w = f(z)$. If we denote the mapping f^{-1} by h, then

$$\int_{D\cap S(a,\epsilon)} \varphi\,\omega(f) = \int_{f(D)\cap S_h(\epsilon)} \varphi(h)\,U(w,0).$$

By Theorem 20.4,

$$\begin{aligned}\lim_{\epsilon\to 0+} &\int_{f(D)\cap S_h(\epsilon)} \varphi(h)\,U(w,0)\\ &= \lim_{\epsilon\to 0+} \int_{f(D)\cap S(0,\epsilon)} \varphi(h)\,U(w,0)\\ &= \lim_{\epsilon\to 0+} \int_{f(D)\cap S(0,\epsilon)} (\varphi(h)-\varphi(a))\,U(w,0) + \varphi(a)\tau_f(a).\end{aligned}$$

Now $\varphi(h)$ satisfies the Dini condition at 0 since φ satisfies the Dini condition at a. Consider

$$\begin{aligned}&\int_{f(D)\cap S(0,\epsilon)} (\varphi(h)-\varphi(a))\,U(w,0)\\ &\qquad = C\epsilon^{-2n} \int_{f(D)\cap S(0,\epsilon)} (\varphi(h)-\varphi(a)) \sum_{k=1}^{n} (-1)^{k-1}\bar{w}_k\, d\bar{w}[k]\wedge dw.\end{aligned}$$

Since $\varphi \in \mathcal{A}(D)$, we have that $|\varphi(h)-\varphi(a)| \to 0$ as $\epsilon \to 0^+$, and

$$\int_{S(0,\epsilon)} \left|\sum_{k=1}^{n} (-1)^{k-1}\bar{w}_k\, d\bar{w}[k]\wedge dw\right| = C_1\epsilon^{2n},$$

that is,

$$\int_{f(D)\cap S(0,\epsilon)} (\varphi(h)-\varphi(a))\,U(w,0) \to 0$$

as $\epsilon \to 0^+$. This concludes the proof. □

Theorem 20.5 was given in [162].

Corollary 20.6. *If ∂D is smooth, then under the hypotheses of Theorem 20.5,*

$$\text{P.V.}\int_{\partial D}\varphi\,\omega(f)=\sum_{a\in E_f\cap D}\mu_a\varphi(a)+\frac{1}{2}\sum_{a\in E_f\cap\partial D}\varphi(a).$$

Proof. The proof follows from Theorem 20.5 and the fact that under a biholomorphic mapping, a smooth surface goes into a smooth surface, and for a smooth ∂D the solid angle $\tau_f(a)=1/2$. □

In Theorem 20.5 we obtained that on ∂D, the value $\varphi(a)$ is multiplied not by the solid angle $\tau(a)$ of the tangent cone to ∂D at a but by the solid angle $\tau_f(a)$ in the image. In general, these quantities do not coincide. Here is an example: let $w_1=z_1+z_2$ and $w_2=2(1-i)z_1+z_2$. If $z_j=\xi_j+i\xi_{2+j}$ and $w_j=\eta_j+i\eta_{2+j}$ for $j=1, 2$, then

$$\begin{aligned}\eta_1&=\xi_1+\xi_2,\\ \eta_2&=2\xi_1+\xi_2+2\xi_3,\\ \eta_3&=\xi_3+\xi_4,\\ \eta_4&=-2\xi_1+\xi_4+2\xi_3.\end{aligned}\tag{20.4}$$

This mapping is nondegenerate and smooth, and this means that it takes the tangent cone to a surface into the tangent cone to the image of this surface. We take two orthogonal planes: $\Pi_1=\{\xi:\xi_1=0\}$ and $\Pi_2=\{\xi:\xi_2=0\}$. The mapping (20.4) takes them into the plane $f(\Pi_1)$:

$$\begin{aligned}\eta_1&=\xi_2, & \eta_2&=\xi_2+2\xi_3,\\ \eta_3&=\xi_3+\xi_4, & \eta_4&=2\xi_3+\xi_4,\end{aligned}$$

or $\eta_1-\eta_2-2\eta_3+2\eta_4=0$, and into the plane $f(\Pi_2)$:

$$\begin{aligned}\eta_1&=\xi_1, & \eta_2&=2\xi_1+2\xi_3,\\ \eta_3&=\xi_3+\xi_4, & \eta_4&=-2\xi_1+2\xi_3+\xi_4,\end{aligned}$$

or $6\eta_1-\eta_2-2\eta_3+2\eta_4=0$. The cosine of the angle between these planes equals $\sqrt{2}/2$, that is, the angle equals $\pi/4$, while the angle between the planes Π_1 and Π_2 is $\pi/2$.

This example is due to Prenov.

Corollary 20.6 can be carried over to a wider class of mappings f. Let f be such that at each point $a\in E_f\cap\partial D$ the lowest homogeneous parts P_j of the Taylor expansions of the f_j at a also have an isolated zero at a.

Theorem 20.7 (Kytmanov). *If D is a bounded domain, and ∂D is smooth, then for $\varphi\in\mathcal{A}(D)$ satisfying the Dini condition on ∂D we have*

$$\text{P.V.}\int_{\partial D}\varphi\,\omega(f)=\sum_{a\in E_f\cap D}\mu_a\varphi(a)+\frac{1}{2}\sum_{a\in E_f\cap\partial D}\mu_a\varphi(a).$$

Proof. We consider the domain $D_\epsilon = D \setminus \bigcup_{a\in E_f\cap\partial D} B(a,\epsilon)$, just as in Theorem 20.5. By the Roos-Yuzhakov formula,

$$\int_{\partial D_\epsilon} \varphi\,\omega(f) = \sum_{a\in E_f\cap D} \mu_a\varphi(a).$$

Consider $\int_{S(a,\epsilon)\cap D}\varphi\,\omega(f)$; smoothness considerations show that

$$\lim_{\epsilon\to 0+}\int_{S(a,\epsilon)\cap D}\varphi\,\omega(f) = \lim_{\epsilon\to 0+}\int_{S(a,\epsilon)\cap D}\varphi\,\omega(P),$$

where $P = (P_1,\dots,P_n)$ is the holomorphic mapping consisting of the lowest homogeneous parts of the Taylor expansions of the f_j at a. Furthermore,

$$\int_{S(a,\epsilon)\cap D}\varphi\,\omega(P) = \int_{S(a,\epsilon)\cap D}(\varphi(\zeta)-\varphi(a))\,\omega(P) + \varphi(a)\int_{S(a,\epsilon)\cap D}\omega(P).$$

The integral of the function $\varphi(\zeta)-\varphi(a)$ over the set $S(a,\epsilon)\cap D$ may be replaced by the integral over the set $S_P(\delta)\cap D$ and the part $M_{\epsilon,\delta}$ of the set ∂D lying between $S(a,\epsilon)$ and $S_P(\delta)$. Since φ satisfies the Dini condition, it is easy to show that the improper integral $\int_{\partial D\cap B(a,\epsilon)}(\varphi(\zeta)-\varphi(a))\,\omega(P)$ is absolutely convergent, so $\int_{M_{\epsilon,\delta}}(\varphi(\zeta)-\varphi(a))\,\omega(P)\to 0$ as $\epsilon,\delta\to 0$, while

$$\begin{aligned}\left|\int_{S_P(\delta)\cap D}(\varphi(\zeta)-\varphi(a))\,\omega(P)\right| &\le C\omega_\varphi(\delta,a)\delta^{-2n}\int_{B_P(\delta)}|d\bar P\wedge dP|\\ &\le C_1\omega_\varphi(\delta,a)\to 0\end{aligned}$$

as $\delta\to 0^+$.

Therefore $\int_{S(a,\epsilon)\cap D}(\varphi(\zeta)-\varphi(a))\,\omega(P)\to 0$ as $\epsilon\to 0^+$. We are interested in the integral $\int_{S(a,\epsilon)\cap D}\omega(P)$. The form $\omega(P)$ equals

$$\omega(P) = \frac{(n-1)!}{|P|^{2n}(2\pi i)^n}\sum_{k=1}^{n}(-1)^{k-1}\bar P_k\,d\bar P[k]\wedge dP.$$

Consequently, $\omega(P)$ is invariant with respect to rotations $z\mapsto ze^{i\varphi}$. The set $S(a,\epsilon)$ also is invariant with respect to rotations, and

$$\lim_{\epsilon\to 0+}\int_{S(a,\epsilon)\cap D}\omega(P) = \lim_{\epsilon\to 0+}\int_{S(a,\epsilon)\cap\Pi}\omega(P),$$

where Π is the tangent cone to D at a. In this case it reduces to a half-space, and we may suppose that $a=0$ and $\Pi = \{\,z : \operatorname{Im} z_1 \le 0\,\}$. Making the change of variables $z\mapsto ze^{i\pi}$, we obtain that

$$\int_{S(0,\epsilon)\cap\Pi}\omega(P) = \int_{S(0,\epsilon)\cap(\mathbf{C}^n\setminus\Pi)}\omega(P), \qquad \text{that is,}$$

$$\int_{S(0,\epsilon)}\omega(P) = \frac{1}{2}\int_{S(0,\epsilon)\cap\Pi}\omega(P).$$

By the Roos-Yuzhakov formula, $\int_{S(a,\epsilon)} \omega(P) = \mu(a)$, since the multiplicity of the zero a of the mapping f agrees with the multiplicity of the zero a of the system P (see [10, §22]). □

Apparently Theorem 20.7 holds for arbitrary mappings f having isolated zeroes on $\overline{D}$.

As one example of an application of Theorem 20.5 (or of Corollary 20.6), we give for $n = 1$ a proof due to Aĭzenberg of the Poisson summation formula for real-analytic functions.

Theorem 20.8. *If φ is a real-analytic function on $[a,b]$, where $a < b$ and a and b are not integers, then*

$$\sum_{k\in[a,b]} \varphi(k) = \sum_{m=-\infty}^{+\infty} \int_a^b \varphi(x)e^{2\pi imx}\,dx$$

(where k and m are integers).

Proof. Consider the entire function $f(z) = e^{2\pi iz} - 1$ in $\mathbf{C}^1$, which vanishes only at the integer points lying on the real axis, and has only simple zeroes. Suppose the contour γ_1 lies in the upper half-plane and in the domain of analyticity of φ, with initial point b and final point a. Then by Theorem 20.5,

$$\text{P.V.} \int_a^b \varphi(x)\frac{e^{2\pi ix}}{e^{2\pi ix}-1}\,dx + \int_{\gamma_1} \varphi(z)\frac{e^{2\pi iz}\,dz}{e^{2\pi iz}-1} = \frac{1}{2}\sum_{k\in[a,b]} \varphi(k).$$

Since $\operatorname{Im} z > 0$ on γ_1, we have $|e^{2\pi iz}| < 1$, so

$$\frac{e^{2\pi iz}}{e^{2\pi iz}-1} = -\sum_{m=1}^{\infty} e^{2\pi imz}.$$

Moreover, the partial sums of this series equal

$$s_p = \frac{e^{2\pi iz}(1-e^{2\pi i(p+1)z})}{e^{2\pi iz}-1};$$

the modulus of the denominator is bounded away from zero on γ_1, since $|e^{2\pi iz}| < 1$ on γ_1 and the numbers a and b are not integers; and

$$|1-e^{2\pi i(p+1)z}| \le 1 + e^{2\pi(p+1)\operatorname{Re} iz} = 1 + e^{-2\pi(p+1)y} \le 2.$$

Therefore the series obtained by expanding $f(z)$ can be integrated term-by-term on γ_1. Then

$$\int_{\gamma_1} \varphi(z)\frac{e^{2\pi iz}\,dz}{e^{2\pi iz}-1} = -\sum_{m=1}^{\infty}\int_{\gamma_1} \varphi(z)e^{2\pi imz}\,dz.$$

Since $\varphi(z)$ is holomorphic, $-\int_{\gamma_1}\varphi(z)e^{2\pi imz}\,dz = \int_a^b \varphi(x)e^{2\pi imx}\,dx$. Hence

$$\frac{1}{2}\sum_{k\in[a,b]}\varphi(k) = \mathrm{P.V.}\int_a^b \varphi(x)\frac{e^{2\pi ix}}{e^{2\pi ix}-1}\,dx + \sum_{m=1}^{\infty}\int_a^b \varphi(x)e^{2\pi imx}\,dx.$$

Now, by choosing a contour γ_2 lying in the lower half-plane, starting at a and ending at b, and applying the same argument to the domain bounded by $[a,b]$ and γ_2 (except that $|e^{2\pi iz}|>1$ in the lower half-plane), we obtain

$$\frac{1}{2}\sum_{k\in[a,b]}\varphi(k) = -\mathrm{P.V.}\int_a^b \varphi(x)\frac{e^{2\pi ix}}{e^{2\pi ix}-1}\,dx + \sum_{m=0}^{-\infty}\int_a^b \varphi(x)e^{2\pi imx}\,dx.$$

We get the result by combining these two formulas. □

By induction, we can prove the following.

Theorem 20.9 (Aĭzenberg). *Let $G\subset \mathbf{R}^n$ be a polytope with sides parallel to the coordinate planes. Suppose ∂G does not contain points with integer coordinates, and let φ be a real-analytic function on $\overline{G}$. Then*

$$\sum_{k\in G}\varphi(k) = \sum_k \int_G \varphi(x)e^{2\pi i\langle k,x\rangle}\,dx,$$

where $k=(k_1,\dots,k_n)$ is an integer-valued vector, and the series on the right-hand side converges in the sense of Pringsheim (that is, according to parallelograms).

By using Theorem 20.5, it is possible to eliminate the hypothesis that ∂G does not contain integer points.

21 Multidimensional analogues of Carleman's formula

21.1 The classical Carleman-Goluzin-Krylov formula

Let D be a simply connected domain in $\mathbf{C}^1$ with piecewise-smooth boundary. If a set $M\subset\partial D$ has positive linear Lebesgue measure, then this set is a uniqueness set for the class $\mathcal{H}^p(D)$.

Theorem 21.1 (Carleman, Goluzin, Krylov). *If $f\in\mathcal{H}^p(D)$, where $p\geq 1$, then*

$$f(z) = \lim_{k\to\infty}\frac{1}{2\pi i}\int_M \frac{f(\zeta)}{\zeta - z}e^{k(\varphi(\zeta)-\varphi(z))}\,d\zeta, \qquad z\in D, \tag{21.1}$$

where $\varphi\in\mathcal{O}(D)$, and $\operatorname{Re}\varphi(z)\to 1$ if $z\to z^0\in M$ along a nontangential path, while $\operatorname{Re}\varphi(z)\to 0$ if $z\to z^0\in\partial D\setminus M$.

Proof. We solve the Dirichlet problem for the characteristic function χ of the set M in ∂D, and then we complete the resulting function to a holomorphic function (using that D is simply connected). This will be the function $\varphi(z)$. The function $e^{m\varphi(z)}$ has the following properties:

(a) $|e^{m\varphi(z)}| \leq e^{m \operatorname{Re} \varphi(z)} \leq e^m$, that is, $e^{m\varphi} \in \mathcal{H}^\infty(D)$,

(b) the boundary value of $|e^{m\varphi(z)}|$ equals e^m on M and equals 1 on $\partial D \setminus M$.

We apply Cauchy's formula to $fe^{m\varphi}$:

$$2\pi i f(z)e^{m\varphi(z)} = \int_M \frac{f(\zeta)}{\zeta - z} e^{m\varphi(\zeta)}\, d\zeta + \int_{\partial D\setminus M} \frac{f(\zeta)}{\zeta - z} e^{m\varphi(\zeta)}\, d\zeta.$$

Now

$$\left| \int_{\partial D\setminus M} \frac{f(\zeta)}{\zeta - z} e^{m(\varphi(\zeta)-\varphi(z))}\, d\zeta \right| \leq \int_{\partial D\setminus M} \frac{|f(\zeta)|\, e^{-m \operatorname{Re} \varphi(z)}}{|\zeta - z|}\, d|\zeta|.$$

Since $e^{-m \operatorname{Re} \varphi(z)} \to 0$ as $m \to \infty$, the integral

$$\int_{\partial D\setminus M} \frac{|f(\zeta)|\, e^{-m \operatorname{Re} \varphi(z)}}{|\zeta - z|}\, d|\zeta| \to 0 \qquad \text{as } m \to \infty.$$

□

Theorem 21.1 may be found in [163].

Thus, (21.1) gives a solution to the problem of extending a function given on a uniqueness set M holomorphically into the domain D. In the multidimensional case, many analogues of Carleman's formula are known. Various generalizations of this formula and applications are given in the book of Aĭzenberg [6]. Here we will discuss those formulas that are connected with the Bochner-Martinelli integral, but first we consider the question of what hypotheses must be imposed on a function f given on a uniqueness set M so that it will extend holomorphically into D. The Hartogs-Bochner theorem and the theorems of §8 give an extension of f into the envelope of holomorphy of the set on which it is considered. Here we are interested in conditions for f to extend into a given domain D, generally speaking not the envelope of holomorphy of M.

21.2 Holomorphic extension from a part of the boundary

Suppose D is a bounded domain in $\mathbf{C}^n$, and ∂D is a Lyapunov surface. We denote by $\mathcal{O}_p^\perp(\partial D)$ the subspace of functions in $\mathcal{L}^q(\partial D)$, $q > 1$, satisfying the following property: if $f \in \mathcal{L}^p(\partial D)$, $p > 1$, where $(1/p) + (1/q) = 1$, then f is the boundary value of a function $F \in \mathcal{H}^p(D)$ if and only if $\int_{\partial D} f\psi\, d\sigma = 0$ for all $\psi \in \mathcal{O}_p^\perp(\partial D)$. For example, if $n > 1$ we may take as $\mathcal{O}_p^\perp$ the functions of the form $\overline{\partial}\varphi/d\sigma$, if $\varphi \in \mathcal{D}^{n,n-2}(\overline{D})$ (see the Hartogs-Bochner Theorem 7.1), or if $\partial D \in \mathcal{C}^\infty$, the functions of the form $\partial_n F$, where $F \in \mathcal{C}^1(\overline{D})$ is harmonic in D (see Theorem 17.2), etc.

Theorem 21.2 (Tarkhanov). *Let M be a set of positive Lebesgue surface measure in ∂D, and $f \in \mathcal{L}^p(M)$, where $p > 1$. A necessary and sufficient condition for the existence of a function F in $\mathcal{H}^p(D)$ whose nontangential boundary value agrees with f on M is that*

$$\lim_{j\to\infty} \int_M f\psi_j \, d\sigma = 0 \tag{21.2}$$

for every sequence $\psi_j \in \mathcal{O}_p^\perp(\partial D)$ such that the restriction of ψ_j to $\partial D \setminus M$ converges to zero in the norm of $\mathcal{L}^q(\partial D \setminus M)$ (that is, $\|\psi_j\|_{\mathcal{L}^q(\partial D\setminus M)} \to 0$ as $j \to \infty$).

Proof. Let $S = \partial D \setminus M$. If f is the boundary value on M of a function $F \in \mathcal{H}^p(D)$, then

$$0 = \int_{\partial D} F\psi_j \, d\sigma = \int_S F\psi_j \, d\sigma + \int_M f\psi_j \, d\sigma, \qquad \text{and}$$

$$\left| \int_S F\psi_j \, d\sigma \right| \le \|F\|_{\mathcal{L}^p(S)} \|\psi_j\|_{\mathcal{L}^q(S)}.$$

Hence (21.2) follows.

Conversely, suppose (21.2) holds. The space $\mathcal{L}^q(\partial D)$ is the direct sum of the subspaces $\mathcal{L}^q(S)$ and $\mathcal{L}^q(M)$, since every function $\psi \in \mathcal{L}^q(\partial D)$ has the form $\psi = \chi_M\psi + \chi_S\psi$ (where χ_m and χ_S are the characteristic functions of the sets M and S). We denote the projection operator from $\mathcal{L}^q(\partial D)$ onto $\mathcal{L}^q(M)$ by π_M, and the projection operator from $\mathcal{L}^q(\partial D)$ onto $\mathcal{L}^q(S)$ by π_S; then $\pi_M + \pi_S = I$ (where I is the identity operator). Suppose f is the boundary value of a function $F \in \mathcal{H}^p(D)$, and T is the linear functional on $\mathcal{L}^q(M)$ induced by f, while T_1 is the linear functional on $\mathcal{L}^q(S)$ induced by the boundary value of F. Then

$$T_1(\pi_S(\psi)) + T(\pi_M(\psi)) = 0 \tag{21.3}$$

for all $\psi \in \mathcal{O}_p^\perp(\partial D)$, since

$$T_1(\pi_S(\psi)) + T(\pi_M(\psi)) = \int_S F\psi \, d\sigma + \int_M f\psi \, d\sigma = \int_{\partial D} F\psi \, d\sigma = 0.$$

If, for the functional T induced by $f \in \mathcal{L}^p(M)$, we have found a functional $T_1 \in [\mathcal{L}^q(S)]'$ for which (21.3) holds for all $\psi \in \mathcal{O}_p^\perp(\partial D)$, then f is the boundary value of a function $F \in \mathcal{H}^p(D)$ (in view of the definition of $\mathcal{O}_p^\perp(\partial D)$). Since M is a uniqueness set for $\mathcal{H}^p(D)$, the extension F will be unique.

Thus, given a functional T induced by $f \in \mathcal{L}^p(M)$ and satisfying (21.2), we must find a functional $T_1 \in [\mathcal{L}^q(S)]'$ for which (21.3) holds. We define T_1 on $\pi_S(\mathcal{O}_p^\perp(\partial D))$ by using (21.3), that is, we set $T_1(\pi_S(\psi)) = -T(\pi_M(\psi))$ for all $\psi \in \mathcal{O}_p^\perp(\partial D)$. This is well defined. Indeed, consider two functions ψ_1 and ψ_2 in $\mathcal{O}_p^\perp(\partial D)$ such that $\pi_S\psi_1 = \pi_S\psi_2$, that is, the function $\psi = \psi_1 - \psi_2$ satisfies the condition $\pi_S\psi = 0$. Choosing the constant sequence $\{\psi\}$, we see that $\|\psi\|_{\mathcal{L}^q(S)} = 0$, so $T(\pi_M(\psi)) = T(\psi) = \int_M f\psi \, d\sigma = 0$ (in view of (21.2)).

This also shows that T_1 is a linear functional on $\pi_S(\mathcal{O}_p^\perp(\partial D))$. We show that T_1 is continuous on $\pi_S(\mathcal{O}_p^\perp)$. It is enough to check continuity at 0. If $\pi_S(\psi_j) \to 0$ in $\mathcal{L}^q(S)$ as $j \to \infty$, then in view of (21.2),

$$T(\pi_M(\psi_j)) = \int_M f\psi_j\, d\sigma \to 0$$

as $j \to \infty$, that is, $T_1(\pi_S(\psi_j)) \to 0$ as $j \to \infty$.

Thus, T_1 is a continuous linear functional on $\pi_S(\mathcal{O}_p^\perp(\partial D))$. By the Hahn-Banach theorem, it extends to all of $\mathcal{L}^q(S)$, and (21.3) holds by construction. □

Theorem 21.2 is one of the theorems in this direction proved by Tarkhanov (see [6, 203]). The simplified proof given here was obtained by Znamenskaya [228].

21.3 Yarmukhamedov's formula

Based on his generalized formula (see Theorem 1.8), Yarmukhamedov [222] constructed the following analogue of Carleman's formula.

Let

$$E_\rho(w) = \sum_{k=0}^{\infty} \frac{w^k}{\Gamma\left(1+\frac{k}{\rho}\right)}$$

be the entire function of Mittag-Leffler. Then E_ρ is real for real w, and $E_\rho(\operatorname{Re} w) \neq 0$, that is, E_ρ satisfies the hypotheses on the entire function K required for the conclusion of Theorem 1.8. Moreover, it is known that E_ρ satisfies, for $\rho > 1/2$, the estimates

$$E_\rho(w) = \begin{cases} \rho e^{w^\rho} + \psi(w), & |\arg w| \leq (\pi/2\rho) + \epsilon, \\ \psi(w), & (\pi/2\rho) + \epsilon \leq |\arg w| \leq \pi, \end{cases}$$

where $\psi(w) = o(1)$ as $|w| \to \infty$, and these estimates may be differentiated (see [46]).

Suppose D is a domain in $\mathbf{C}^n$ bounded by a part S of the boundary surface of a cone with vertex at $z = 0$ of the form

$$\left\{\zeta : \sum_{k=1}^{n-1} |\zeta_k|^2 + (\operatorname{Re}\zeta_n)^2 < \tau^2 \operatorname{Im}\zeta_n^2,\ \operatorname{Im}\zeta_n > 0\right\},$$

where $\tau = \tan(\pi/2\rho) < +\infty$, and by a smooth piece of a surface M lying inside the cone, so that $\partial D = M \cup S$ is a piecewise-smooth surface. If $z \in D$, then

$$\sqrt{\sum_{k=1}^{n-1} |z_k|^2 + (\operatorname{Re} z_n)^2} - \tau \operatorname{Im} z_n = -r < 0.$$

Let $\alpha_0^2 = \sum_{k=1}^{n-1} |z_k|^2 + (\operatorname{Re} z_n)^2$, and $\alpha^2 = s^2 + [\operatorname{Re}(\zeta_n - z_n)]^2 = \sum_{k=1}^{n-1} |\zeta_k - z_k|^2 + [\operatorname{Re}(\zeta_n - z_n)]^2$.

Consider the entire function

$$K(w) = E_\rho \left[\frac{\sigma^{1/\rho}}{r} (\tau w - \alpha_0) \right],$$

where $\sigma > 0$. Then

$$K(i\alpha + \operatorname{Im} \zeta_n) = E_\rho \left[\frac{\sigma^{1/\rho}}{r} (i\alpha\tau + \tau \operatorname{Im} \zeta_n - \alpha_0) \right], \qquad \text{and}$$

$$K(\operatorname{Im} z_n) = E_\rho \left[\frac{\sigma^{1/\rho}}{r} (\tau \operatorname{Im} z_n - \alpha_0) \right] = E_\rho(\sigma^{1/\rho}) \approx \rho e^\sigma$$

as $\sigma \to \infty$.

On the conical part of the boundary,

$$\arg(i\alpha\tau + \tau \operatorname{Im} \zeta_n - \alpha_0) = \arg\left(i\tau\alpha + \sqrt{\sum_{k=1}^{n-1} |\zeta_k|^2 + (\operatorname{Re} \zeta_n)^2} - \alpha_0 \right)$$

$$= \arctan \tau\alpha \bigg/ \sqrt{\sum_{k=1}^{n-1} |\zeta_k|^2 + (\operatorname{Re} \zeta_n)^2} - \alpha_0,$$

so $\pi/2\rho \leq |\arg(i\tau\alpha + \tau \operatorname{Im} \zeta_n - \alpha_0)| \leq \pi$. Therefore $K(i\alpha + \operatorname{Im} \zeta_n) = o(1)$ as $\sigma \to \infty$.

Recall that the kernel in Theorem 1.8 was constructed in the following way: we considered the harmonic function

$$\Phi(\zeta, z) = \frac{\partial^{n-2}}{\partial s^{n-2}} \operatorname{Im} \left[\frac{K(i\alpha + \operatorname{Im} \zeta_n)}{\alpha(i\alpha + \operatorname{Im}(\zeta_n - z_n))} \right] \bigg/ C_n K(\operatorname{Im} z_n),$$

and we proved that $\Phi(\zeta, z) = g(\zeta, z) + h(\zeta, z)$, where $g(\zeta, z)$ is the fundamental solution of Laplace's equation, and $h(\zeta, z)$ is a harmonic function of ζ in $\mathbf{C}^n$ for fixed z. From the form of the function K we obtain that $\Phi(\zeta, z) \to 0$ as $\sigma \to +\infty$ ($\zeta \in S$, and z is a fixed point in D). Moreover, this condition may be differentiated. Denoting

$$\widetilde{U}(\zeta, z) = \sum_{k=1}^{n} (-1)^{k-1} \frac{\partial \Phi}{\partial \zeta_k} (\zeta, z)\, d\bar{\zeta}[k] \wedge d\zeta,$$

we obtain the following assertion.

Theorem 21.3 (Yarmukhamedov). *If $n > 1$, then for $f \in \mathcal{A}(D)$ we have*

$$f(z) = \lim_{\sigma \to +\infty} \int_M f(\zeta)\, \widetilde{U}(\zeta, z), \qquad z \in D. \tag{21.4}$$

Theorem 21.3 was proved in [222].

Formula (21.4) is the first multidimensional Carleman formula constructed for a domain that is not a polycylinder. As a corollary, Yarmukhamedov obtained a two-constant theorem and a Phragmén-Lindelöf principle in several variables.

21.4 Aĭzenberg's formula

By combining the classical Carleman formula (21.1) and the Bochner-Martinelli formula, Aĭzenberg [4] (see also [6]) obtained the following formula.

Let D be a bounded convex domain in $\mathbf{C}^n$ with piecewise-smooth boundary, and fix a point $z \in D$. All complex lines passing through z and not lying in the plane $\{ \zeta : \zeta_1 = z_1 \}$ can be described in the form $\mu = \mu(v, z) = \{ \zeta : \zeta_1 = z_1 + t, \zeta_2 = z_2 + v_2 t, \ldots, \zeta_n = z_n + v_n t, t \in \mathbf{C}^1 \}$. We observed in Lemma 15.8 that if we make the change of variables $\zeta_1 = z_1 + t$, $\zeta_j = z_j + v_j t$, $j \geq 2$, in the kernel $U(\zeta, z)$, we can write it in the form

$$U(\zeta, z) = \frac{dt}{2\pi i t} \wedge \lambda(v)$$

(in Lemma 15.8, we had $\zeta = 1/v_1$, and not $\zeta_1 = z_1 + t$), where

$$\lambda(v) = (-1)^{n-1} \frac{(n-1)!}{(2\pi i)^n} \frac{d\bar{v}[1] \wedge dv[1]}{(1 + |v_2|^2 + \cdots + |v_n|^2)^n}.$$

Integrating $U(\zeta, 0)$ over the sphere $S(0,1)$ in the coordinates t and v, we obtain

$$\begin{aligned} 1 = \int_{S(0,1)} U(\zeta, 0) &= \int_{S(0,1) \setminus \{\zeta : \zeta_1 = 0\}} U(\zeta, 0) \\ &= \frac{1}{2\pi i} \int_{|t|=1} \frac{dt}{t} \int_{\mathbf{C}^{n-1}} \lambda(v) = \int_{\mathbf{C}^{n-1}} \lambda(v), \end{aligned}$$

that is, $\int_{\mathbf{C}^{n-1}} \lambda(v) = 1$.

Suppose $M \subset \partial D$ has positive Lebesgue surface measure. For $z \in D$, we denote by $N(z, M)$ the set of complex lines $\mu(v, z)$ for which the intersection $\mu \cap M$ has positive linear Lebesgue measure. By Fubini's theorem,

$$0 < \int_M U(\zeta, z) = \int_{N(z,M)} \lambda(v) \int_{M \cap \mu(v,z)} \frac{dt}{2\pi i t} \leq \int_{N(z,M)} \lambda(v) \leq 1,$$

that is, $0 < \int_{N(z,M)} \lambda(v) \leq 1$.

The intersection $D \cap \mu(v, z)$ is simply connected for each point $z \in D$ and each line $\mu(v, z)$. We denote by $\varphi_{z,v}(t)$ a holomorphic function in $D \cap \mu(v, z)$ such that $\operatorname{Re} \varphi_{z,v}(t) \to 1$ when t tends to $M \cap \mu(v, z)$, and $\operatorname{Re} \varphi_{z,v}(t) \to 0$ when t tends to $(\partial D \setminus M) \cap \mu(v, z)$ (see Theorem 21.1).

If the linear measure of $M \cap \mu(v, z)$ is positive, then for each $f \in \mathcal{H}^1(D)$ we have by (21.1) that

$$f(z) = \lim_{m \to \infty} \frac{1}{2\pi i} \int_{M \cap \mu(v,z)} \frac{f(z + vt)}{t} e^{m(\varphi_{z,v}(t) - \varphi_{z,v}(0))} \, dt.$$

Integrating both sides of this formula over $N(z, M)$, we obtain

$$f(z) = \frac{1}{\int_{N(z,M)} \lambda(v)} \lim_{m\to\infty} \int_M f(\zeta)\, U(\zeta, z) e^{m(\varphi_{z,v}(t)-\varphi_{z,v}(0))}. \tag{21.5}$$

It remains to show that we can pass to the limit under the integral sign. Indeed

$$\int_{M\cap\mu(v,z)} \frac{f(z+vt)}{t} e^{m(\varphi_{z,v}(t)-\varphi_{z,v}(0))}\, dt$$
$$= 2\pi i f(z) - \int_{(\partial D\setminus M)\cap\mu(v,z)} \frac{f(z+vt)}{t} e^{m(\varphi_{z,v}(t)-\varphi_{z,v}(0))}\, dt,$$

but

$$\left| \int_{(\partial D\setminus M)\cap\mu(v,z)} \frac{f(z+vt)}{t} e^{m(\varphi_{z,v}(t)-\varphi_{z,v}(0))}\, dt \right|$$
$$\le \int_{(\partial D\setminus M)\cap\mu(v,z)} \left| \frac{f(z+vt)}{t} \right| e^{-m \operatorname{Re} \varphi_{z,v}(0)}\, d|t|$$
$$\le C \int_{\partial D\cap\mu(z,v)} |f(z+vt)|\, d|t|,$$

while the function $\int_{\partial D\cap\mu(v,z)} |f(z+vt)|\, d|t|$ is integrable in v since $f \in \mathcal{H}^1(D)$. Thus, we can apply the Lebesgue dominated convergence theorem.

Theorem 21.4 (Aĭzenberg). *If $f \in \mathcal{H}^1(D)$, then (21.5) holds for f.*

22 The Poincaré-Bertrand formula

22.1 The singular Bochner-Martinelli integral depending on a parameter

Suppose D is a bounded domain in $\mathbf{C}^n$, and ∂D is a Lyapunov surface. We will consider functions $f(z, w)$, where $z \in \partial D$ and w is in a compact set K in $\mathbf{C}^m$. Suppose $f \in \mathcal{C}^\alpha(\partial D)$ in z and $f \in \mathcal{C}^\beta(K)$ in w, where $0 < \alpha \le 1$, $0 < \beta \le 1$, and the estimate on f is uniform:

$$|f(z, w) - f(\zeta, w)| \le C|z - \zeta|^\alpha$$

for all $z, \zeta \in \partial D$ and $w \in K$, and

$$|f(z, \zeta) - f(z, w)| \le C|\zeta - w|^\beta$$

for all $z \in \partial D$ and $\zeta, w \in K$.

It follows that $f \in C^\gamma(\partial D \times K)$, where $\gamma = \min(\alpha, \beta)$. Indeed,

$$\begin{aligned}|f(z,w) - f(\zeta,\eta)| &\leq |f(z,w) - f(z,\eta)| + |f(z,\eta) - f(\zeta,\eta)| \\ &\leq C|w-\eta|^\beta + C|z-\zeta|^\alpha \leq C|w-\eta|^\gamma + C|z-\zeta|^\gamma \\ &\leq C_1(\sqrt{|w-\eta|^2 + |z-\zeta|^2}\,)^\gamma\end{aligned}$$

(if $|w-\eta| \leq 1$ and $|z-\zeta| \leq 1$), since on the part of the circle $a^2+b^2=1$ where $a \geq 0$ and $b \geq 0$, the expression $(a^\gamma + b^\gamma)/(a^2+b^2)^{1/2}$ is bounded above and below by positive constants.

Lemma 22.1. *If $f \in C^\alpha(\partial D \times K)$, where $0 < \alpha \leq 1$, then the singular Bochner-Martinelli integral*

$$F(z,w) = \mathrm{P.V.} \int_{\partial D} f(\zeta,w)\,U(\zeta,z), \qquad z \in \partial D, \quad w \in K, \tag{22.1}$$

is a function of class $C^\alpha(\partial D)$ in z and class $C^{\alpha-\epsilon}(K)$ in w for every ϵ such that $0 < \epsilon < \alpha$, and consequently $F \in C^{\alpha-\epsilon}(\partial D \times K)$.

Proof. We have already proved in §2 that F is a function of class $C^\alpha(\partial D)$ in z (see Corollary 2.4). It remains to show that $F \in C^{\alpha-\epsilon}(K)$ in w.

Consider

$$\begin{aligned}&|F(z,w) - F(z,\eta)| \\ &\quad\leq \int_{\partial D \setminus B(z,r)} |f(\zeta,w) - f(\zeta,\eta)|\,|U(\zeta,z)| \\ &\qquad + \int_{B(z,r)\cap\partial D} (|f(\zeta,w) - f(z,w)| + |f(\zeta,\eta) - f(z,\eta)|)\,|U(\zeta,z)| \\ &\qquad + C_1|f(z,w) - f(z,\eta)|.\end{aligned}$$

We have $|\int_{B(z,r)\cap\partial D} U(\zeta,z)| \leq C_1$, since $\mathrm{P.V.}\int_{\partial D} U(\zeta,z) = 1/2$ for $z \in \partial D$ (see Lemma 2.1), and the constant C_1 does not depend on z and r, where $r = |w-\eta|$.

Passing to polar coordinates and using that

$$|U(\zeta,z)|\big|_{\partial D} \leq \frac{(n-1)!}{2\pi^n}\frac{1}{|\zeta-z|^{2n-1}}\,d\sigma(\zeta)$$

$$\left(\text{by Lemma 3.5, } U(\zeta,z)\big|_{\partial D} = \frac{(n-1)!}{2\pi^n}\sum_{k=1}^n \frac{(\bar\zeta_k - \bar z_k)\rho_{\bar k}\,d\sigma}{|\zeta-z|^{2n}},\right.$$

$$\left.\text{and } \left|\sum_{k=1}^n (\bar\zeta_k - \bar z_k)\rho_{\bar k}\right| \leq |\zeta-z|\sqrt{\sum_{k=1}^n |\rho_k|^2} = |\zeta-z|\right),$$

we obtain

$$\int_{B(z,r)\cap\partial D} (|f(\zeta,w)-f(z,w)|+|f(\zeta,\eta)-f(z,\eta)|)\,|U(\zeta,z)|$$
$$\leq C_2\int_{B(z,r)\cap\partial D}|\zeta-z|^{\alpha+1-2n}\,d\sigma(\zeta)\leq C_3r^\alpha=C_3|w-\eta|^\alpha, \qquad \text{and}$$

$$\int_{\partial D\setminus B(z,r)}|f(\zeta,w)-f(\zeta,\eta)|\,|U(\zeta,z)|$$
$$\leq C_4|w-\eta|^\alpha\int_{\partial D\setminus B(z,r)}|\zeta-z|^{1-2n}\,d\sigma(\zeta)$$
$$\leq C_4|w-\eta|^\alpha\int_{B(z,r_0)\cap\partial D\setminus B(z,r)}|\zeta-z|^{1-2n}\,d\sigma(\zeta)+C_5|w-\eta|^\alpha$$
$$\leq C_6|w-\eta|^\alpha\,|\ln|w-\eta||$$

(where r_0 is a sufficiently small fixed number, not depending on z, and in the integral over $B(z,r_0)$, we pass to polar coordinates). Then

$$\int_{S(z,\rho)\cap\partial D}d\omega(\rho,z)\leq C_7\rho^{2n-2},$$

and $d\omega(\rho,z)=d\sigma/d\rho$.

These estimates prove the lemma. In particular, if we consider an integral of the form

$$\Phi(z)=\int_{\partial D_\zeta}f(\zeta,z)\,U(\zeta,z), \qquad z\in\partial D,$$

$f\in\mathcal{C}^\alpha(\partial D\times\partial D)$, then $\Phi\in\mathcal{C}^{\alpha-\epsilon}(\partial D)$ for every ϵ such that $0<\epsilon<\alpha$, since by Lemma 22.1 the function $F(z,w)\in\mathcal{C}^{\alpha-\epsilon}(\partial D\times\partial D)$, and $\Phi(z)=F(z,z)$. □

Lemma 22.1 was proved by Serbin in [178].

22.2 Estimates of some integrals

Let us study the smoothness of the function

$$\Phi(z,w)=\int_{\partial D}\frac{f(\zeta)}{|\zeta-w|^\nu}\,U(\zeta,z), \tag{22.2}$$

where $w,z\in\partial D$, $f\in\mathcal{C}^\alpha(\partial D)$, $0<\alpha\leq 1$, and $0<\nu<2n-1$. First we transform the integral for $\Phi(z,w)$:

$$\Phi(z,w)=I_1+\frac{f(z)}{|z-w|^\nu}\left(I_2+\frac{1}{2}\right),$$

where

$$I_1 = \int_{\partial D} \frac{f(\zeta) - f(z)}{|\zeta - w|^\nu} U(\zeta, z) \qquad \text{and} \qquad I_2 = \int_{\partial D} \frac{|z - w|^\nu - |\zeta - w|^\nu}{|\zeta - w|^\nu} U(\zeta, z).$$

Lemma 22.2. *Let*

$$H(z, w) = \int_{\partial D} \frac{d\sigma(\zeta)}{|\zeta - w|^\nu |\zeta - z|^{2n-1-\alpha}},$$

where $w, z \in \partial D$, $0 < \nu < 2n - 1$, *and* $\alpha > 0$. *Then*

$$\begin{cases} H(z,w) \leq C_1, & \text{if } 0 < \nu < \alpha, \\ H(z,w) \leq C_2 |\ln |z - w|| + C_3, & \text{if } \nu = \alpha, \\ H(z,w) \leq C_4 |z - w|^{\alpha - \nu}, & \text{if } 2n - 1 > \nu > \alpha. \end{cases}$$

Proof. First consider the integral

$$\int_{B(z,r) \cap \partial D} \frac{d\sigma(\zeta)}{|\zeta - w|^\nu |\zeta - z|^{2n-\alpha-1}}$$

where $r = |w - z|/2$. Then $|z - w| \geq |w - \zeta|$ for sufficiently small r, so (passing to polar coordinates in $B(z,r)$) we have

$$\begin{aligned} \int_{\partial D \cap B(z,r)} \frac{d\sigma(\zeta)}{|\zeta - w|^\nu |\zeta - z|^{2n-\alpha-1}} &\leq \frac{1}{|w - z|^\nu} \int_{\partial D \cap B(z,r)} \frac{d\sigma(\zeta)}{|\zeta - z|^{2n-\alpha-1}} \\ &\leq \frac{d_1 r^\alpha}{|w - z|^\nu} = \frac{d_1}{2^\alpha} |z - w|^{\alpha - \nu}. \end{aligned}$$

In the same way,

$$\begin{aligned} \int_{\partial D \cap B(w,r)} \frac{d\sigma(\zeta)}{|\zeta - w|^\nu |\zeta - z|^{2n-\alpha-1}} &\leq \frac{1}{|w - z|^{2n-\alpha-1}} \int_{\partial D \cap B(w,r)} \frac{d\sigma(\zeta)}{|\zeta - w|^\nu} \\ &\leq \frac{d_2 r^{2n-1-\nu}}{|w - z|^{2n-\alpha-1}} \\ &= \frac{d_2}{2^{2n-\nu-1}} |w - z|^{\alpha - \nu}. \end{aligned}$$

On the set $\partial D \setminus (B(z,r) \cup B(w,r))$, the expressions $|\zeta - w|$ and $|\zeta - z|$ are equivalent, since

$$\frac{|\zeta - w|}{|\zeta - z|} \leq \frac{|\zeta - z| + |w - z|}{|\zeta - z|} \leq 1 + \frac{|w - z|}{r} = 3 \qquad (r = |w - z|/2).$$

Therefore

$$\begin{aligned} \int_{\partial D \setminus (B(z,r) \cup B(w,r))} \frac{d\sigma(\zeta)}{|\zeta - w|^\nu |\zeta - z|^{2n-\alpha-1}} &\leq d_3 \int_{\partial D \setminus B(z,r)} \frac{d\sigma(\zeta)}{|\zeta - z|^{2n+\nu-\alpha-1}} \\ &\leq d_3 \int_{B(z,r^0) \cap \partial D \setminus B(z,r)} \frac{d\sigma(\zeta)}{|\zeta - z|^{2n+\nu-\alpha-1}} + d_4. \end{aligned}$$

Passing to polar coordinates in $B(z, r^0)$, we obtain

$$\int_{B(z,r^0)\cap\partial D\setminus B(z,r)} \frac{d\sigma(\zeta)}{|\zeta - z|^{2n+\nu-\alpha-1}} \leq d_5|\ln r| = d_5|\ln|w - z|/2|$$

if $\alpha = \nu$, and

$$\int_{B(z,r^0)\cap\partial D\setminus B(z,r)} \frac{d\sigma(\zeta)}{|\zeta - z|^{2n+\nu-\alpha-1}} \leq d_6 r^{\alpha-\nu} = d_6 2^{\nu-\alpha}|z - w|^{\alpha-\nu}$$

if $\alpha \neq \nu$.

Lemma 22.2 follows. This lemma is well known; it may be found, for example, in [153]. □

Lemma 22.3. *The function $\Phi(z, w)$ defined by (22.2) satisfies the estimate*

$$|\Phi(z, w)| \leq \frac{C}{|z - w|^{\nu+\epsilon}} \quad \textit{for every } \epsilon > 0.$$

Proof. We can estimate the integral I_1 by Lemma 22.2. Indeed,

$$\begin{aligned} |I_1| &\leq \int_{\partial D} \frac{|f(\zeta) - f(z)|}{|\zeta - w|^\nu} |U(\zeta, z)| \\ &\leq C_1 \int_{\partial D} \frac{d\sigma(\zeta)}{|\zeta - w|^\nu |\zeta - z|^{2n-\alpha-1}} \leq C_2|w - z|^{\alpha-\nu-\epsilon}. \end{aligned}$$

We consider the integral I_2. Let m be the least natural number such that $\nu \leq m$. Then

$$\left||z - w|^{\nu/m} - |\zeta - w|^{\nu/m}\right| \leq \left||z - w| - |\zeta - w|\right|^{\nu/m} \leq |\zeta - z|^{\nu/m}.$$

The first inequality follows from the inequality $1 \leq a^\gamma + (1 - a)^\gamma$, if $0 < \gamma \leq 1$ and $0 \leq a \leq 1$, and the second follows because $\left||z - w| - |\zeta - w|\right| \leq |z - \zeta|$. Therefore

$$\left||z - w|^\nu - |\zeta - w|^\nu\right| \leq |z - \zeta|^{\nu/m} \sum_{k=0}^{m-1} |z - w|^{k\nu/m} |\zeta - w|^{(m-1-k)\nu/m}.$$

If $B(w, r)$ is the ball of radius $r = |z - w|$, then

$$\left||z - w|^\nu - |\zeta - w|^\nu\right| \leq m|\zeta - z|^{\nu/m}|z - w|^{(m-1)\nu/m}$$

for $\zeta \in \partial D \cap B(w, r)$, and

$$\left||z - w|^\nu - |\zeta - w|^\nu\right| \leq m|\zeta - z|^{\nu/m}|\zeta - w|^{(m-1)\nu/m}$$

for $\zeta \in \partial D \setminus B(w,r)$. Therefore

$$\begin{aligned}
|I_2| &\le C_3 \int_{\partial D} \frac{\big||z-w|^\nu - |\zeta-w|^\nu\big|\, d\sigma(\zeta)}{|\zeta-w|^\nu |\zeta-z|^{2n-1}} \\
&\le C_4 \int_{\partial D \setminus B(w,r)} \frac{d\sigma(\zeta)}{|\zeta-w|^{\nu/m}|\zeta-z|^{2n-1-(\nu/m)}} \\
&\quad + C_4 \int_{\partial D \cap B(w,r)} \frac{|z-w|^{(m-1)\nu/m}\, d\sigma(\zeta)}{|\zeta-w|^\nu |\zeta-z|^{2n-1-(\nu/m)}} \\
&\le C_5\big|\ln|z-w|\big| + C_6 + \frac{C_7}{|z-w|^\epsilon} \le \frac{C_8}{|z-w|^\epsilon}, \qquad \epsilon > 0,
\end{aligned}$$

by Lemma 22.2. □

Lemma 22.3 was proved by Serbin [178]. It carries over in the obvious way to functions f satisfying the Dini condition.

22.3 Composition of the singular Bochner-Martinelli integral and an integral with a weak singularity

Lemma 22.4. *If $f \in C^\alpha(\partial D \times \partial D)$, then*

$$\int_{\partial D_w} d\sigma(w) \int_{\partial D_\zeta} f(\zeta,w)\, U(\zeta,z) = \int_{\partial D_\zeta} U(\zeta,z) \int_{\partial D_w} f(\zeta,w)\, d\sigma(w)$$

for $z \in \partial D$.

Proof. It is clear that both integrals make sense. Consider

$$\begin{aligned}
I_1(z) &= \int_{\partial D_w} d\sigma(w) \int_{\partial D_\zeta} f(\zeta,w)\, U(\zeta,z) \\
&= \int_{\partial D_w} d\sigma(w) \int_{\partial D_\zeta \setminus B(z,r)} f(\zeta,w)\, U(\zeta,z) \\
&\quad + \int_{\partial D_w} d\sigma(w) \int_{\partial D_\zeta \cap B(z,r)} f(\zeta,w)\, U(\zeta,z), \\
I_2(z) &= \int_{\partial D_\zeta} U(\zeta,z) \int_{\partial D_w} f(\zeta,w)\, d\sigma(w) \\
&= \int_{\partial D_\zeta \setminus B(z,r)} U(\zeta,z) \int_{\partial D_w} f(\zeta,w)\, d\sigma(w) \\
&\quad + \int_{\partial D_\zeta \cap B(z,r)} U(\zeta,z) \int_{\partial D_w} f(\zeta,w)\, d\sigma(w).
\end{aligned}$$

Then

$$
\begin{aligned}
|I_1(z) - I_2(z)| &\leq \left| \int_{\partial D_w} d\sigma(w) \int_{\partial D_\zeta \cap B(z,r)} f(\zeta, w)\, U(\zeta, z) \right| \\
&\quad + \left| \int_{\partial D_\zeta \cap B(z,r)} U(\zeta, z) \int_{\partial D_w} f(\zeta, w)\, d\sigma(w) \right| \\
&\leq \int_{\partial D_w} d\sigma(w) \int_{\partial D_\zeta \cap B(z,r)} |f(\zeta, w) - f(z, w)|\, |U(\zeta, z)| \\
&\quad + \int_{\partial D_\zeta \cap B(z,r)} |U(\zeta, z)| \int_{\partial D_w} |f(\zeta, w) - f(z, w)|\, d\sigma(w) \\
&\quad + 2 \int_{\partial D_w} |f(z, w)|\, d\sigma(w) \left| \int_{\partial D_\zeta \cap B(z,r)} U(\zeta, z) \right| \\
&\leq C_1 \int_{\partial D_\zeta \cap B(z,r)} \frac{d\sigma(\zeta)}{|\zeta - z|^{2n-1-\alpha}} + C_2 \left| \int_{\partial D \cap B(z,r)} U(\zeta, z) \right|.
\end{aligned}
$$

Since the right-hand side tends to zero when $r \to 0$, we have $I_1(z) = I_2(z)$. □

Theorem 22.5 (Serbin). *Suppose $f(z, w) = f_0(z, w)/|z - w|^\nu$, $0 \leq \nu < 2n - 1$, and $f_0 \in \mathcal{C}^\alpha(\partial D \times \partial D)$. Then the following formula holds for interchange of the order of integration:*

$$
\int_{\partial D_w} d\sigma(w) \int_{\partial D_\zeta} f(\zeta, w)\, U(\zeta, z) = \int_{\partial D_\zeta} U(\zeta, z) \int_{\partial D_w} f(\zeta, w)\, d\sigma(w). \tag{22.3}
$$

Proof. First of all, these integrals are well defined. Indeed, $\int_{\partial D_w} f(\zeta, w)\, d\sigma(w)$ is a function that satisfies a Hölder condition with exponent α on ∂D. This is proved the same way as Lemma 22.3 (or Lemma 2.2). Therefore the right-hand integral in (22.3) exists. By Lemma 22.3,

$$
\left| \int_{\partial D_\zeta} f(\zeta, w)\, U(\zeta, z) \right| \leq \frac{C}{|w - z|^{\nu + \epsilon}}, \qquad \epsilon > 0.
$$

If we take ϵ so that $\nu + \epsilon < 2n - 1$, then it is clear that the left-hand integral exists as an improper integral.

We denote

$$
\begin{aligned}
I_1(z) &= \int_{\partial D_w} d\sigma(w) \int_{\partial D_\zeta} f(\zeta, w)\, U(\zeta, z) \\
&= \int_{\partial D_w} d\sigma(w) \int_{\partial D_\zeta \setminus B(w,r)} f(\zeta, w)\, U(\zeta, z) \\
&\quad + \int_{\partial D_w} d\sigma(w) \int_{\partial D_\zeta \cap B(w,r)} f(\zeta, w)\, U(\zeta, z),
\end{aligned}
$$

and

$$\begin{aligned} I_2(z) &= \int_{\partial D_\zeta} U(\zeta,z)\int_{\partial D_w} f(\zeta,w)\,d\sigma(w)\\ &= \int_{\partial D_\zeta} U(\zeta,z)\int_{\partial D_w\setminus B(\zeta,r)} f(\zeta,w)\,d\sigma(w)\\ &\qquad + \int_{\partial D_\zeta} U(\zeta,z)\int_{\partial D_w\cap B(\zeta,r)} f(\zeta,w)\,d\sigma(w). \end{aligned}$$

Now $\partial D_\zeta \times (\partial D_w \setminus B(\zeta,r)) = \partial D_\zeta \times \partial D_w \setminus \{(\zeta,w) : |\zeta - w| \geq r\} = (\partial D_\zeta \setminus B(w,r)) \times \partial D_w$, so by Lemma 22.4 we have

$$\begin{aligned} I_1(z) - I_2(z) &= \int_{\partial D_w} d\sigma(w)\int_{\partial D_\zeta\cap B(w,r)} f(\zeta,w)\,U(\zeta,z)\\ &\qquad - \int_{\partial D_\zeta} U(\zeta,z)\int_{\partial D_w\cap B(\zeta,r)} f(\zeta,w)\,d\sigma(w)\\ &= \int_{\partial D_w} d\sigma(w)\int_{\partial D_\zeta\cap B(w,r)} (f(\zeta,w)-f(z,w))\,U(\zeta,z)\\ &\qquad + \int_{\partial D_w} f(z,w)\,d\sigma(w)\int_{\partial D_\zeta\cap B(w,r)} U(\zeta,z)\\ &\qquad - \int_{\partial D_\zeta} U(\zeta,z)\int_{\partial D_w\cap B(\zeta,r)} (f(\zeta,w)-f(z,w))\,d\sigma(w)\\ &\qquad - \frac{1}{2}\int_{\partial D_w\cap B(\zeta,r)} f(z,w)\,d\sigma(w). \end{aligned}$$

The integral $\int_{\partial D_w\cap B(\zeta,r)} f(z,w)\,d\sigma(w) \to 0$ uniformly in z as $r \to 0$ since $\nu < 2n-1$. By Lemma 22.3,

$$\left|\int_{\partial D_\zeta} f(\zeta,w)\,U(\zeta,z)\right| \leq \frac{C_2}{|z-w|^{\nu+\epsilon}},$$

and $\epsilon > 0$ is such that $\nu + \epsilon < 2n-1$, so

$$\int_{\partial D_w\cap B(\zeta,r)} d\sigma(w)\int_{\partial D_\zeta} f(z,w)\,U(\zeta,z) \to 0$$

when $r \to 0$. The function $\int_{\partial D_w} f(\zeta,w)\,d\sigma(w)$ satisfies a Hölder condition on ∂D with exponent α, and

$$\int_{\partial D_w\cap B(\zeta,r)} f(\zeta,w)\,d\sigma(w) \to 0$$

uniformly in ζ Therefore, given $\epsilon > 0$ we can find $\delta > 0$ such that for $r < \delta$ we have

$$\left|\int_{\partial D_w\cap B(\zeta,r)} (f(\zeta,w)-f(z,w))\,d\sigma(w)\right| \leq \epsilon\, C_3|\zeta - z|^\alpha.$$

Then

$$\left|\int_{\partial D_\zeta} U(\zeta,z)\int_{\partial D_w\cap B(\zeta,r)}(f(\zeta,w)-f(z,w))\,d\sigma(w)\right|$$
$$\le \epsilon C_3\int_{\partial D_\zeta}|U(\zeta,z)|\,|\zeta-z|^\alpha \le \epsilon C_4\int_{\partial D_\zeta}\frac{d\sigma(\zeta)}{|\zeta-z|^{2n-1-\alpha}},$$

and $\left|\int_{\partial D_w} f(z,w)\,d\sigma(w)\right| \le C_5$, while $\int_{\partial D_\zeta\cap B(w,r)} U(\zeta,z)$ converges to zero as $r\to 0$ (for fixed z). We have

$$\left|\int_{\partial D_\zeta\cap B(w,r)}(f(\zeta,w)-f(z,w))\,U(\zeta,z)\right| \le C_6\int_{\partial D_\zeta\cap B(w,r)}\frac{d\sigma(\zeta)}{|\zeta-z|^{2n-1-\alpha}}$$
$$\le C_7 r^\alpha,$$

so
$$\int_{\partial D_w} d\sigma(w)\int_{\partial D_\zeta\cap B(w,r)}(f(\zeta,w)-f(z,w))\,U(\zeta,z)\to 0$$

as $r\to 0$, which concludes the proof of the theorem. □

Lemma 22.4 and Theorem 22.5 were proved in [178]. We remark that a more general theorem on the interchange of a singular integral operator and an integral operator with a weak singularity was proved in [153].

22.4 The Poincaré-Bertrand formula

Theorem 22.6 (Serbin, Song, Zhong, Kytmanov, Prenov, Tarkhanov). *If $f\in\mathcal{C}^\alpha(\partial D\times\partial D)$, then for $z\in\partial D$ we have*

$$\int_{\partial D_w} U(w,z)\int_{\partial D_\zeta} f(\zeta,w)\,U(\zeta,w)$$
$$=\frac{1}{4}f(z,z)+\int_{\partial D_\zeta}\int_{\partial D_w} f(\zeta,w)\,U(w,z)U(\zeta,w). \quad (22.4)$$

Proof. We transform the integral

$$\int_{\partial D_w} U(w,z)\int_{\partial D_\zeta} f(\zeta,w)\,U(\zeta,w)$$
$$=\int_{\partial D_w} U(w,z)\int_{\partial D_\zeta}(f(\zeta,w)-f(w,w))\,U(\zeta,w)$$
$$+\int_{\partial D_w} U(w,z)\int_{\partial D_\zeta}(f(w,w)-f(z,w))\,U(\zeta,w)$$
$$+\int_{\partial D_w} U(w,z)\int_{\partial D_\zeta}(f(z,w)-f(z,z))\,U(\zeta,w)$$
$$+f(z,z)\int_{\partial D_w} U(w,z)\int_{\partial D_\zeta} U(\zeta,w).$$

By Theorem 22.5, we can change the order of integration in the first three terms, while

$$\int_{\partial D_w} U(w,z) \int_{\partial D_\zeta} U(\zeta, w) = \frac{1}{4}.$$

Therefore

$$\int_{\partial D_w} U(w,z) \int_{\partial D_\zeta} f(\zeta,w) U(\zeta,w)$$
$$= \int_{\partial D_\zeta} \int_{\partial D_w} (f(\zeta,w) - f(z,z))\, U(w,z) U(\zeta,w) + \frac{1}{4} f(z,z).$$

When $n = 1$, the term $f(z,z) \int_{\partial D_\zeta} \int_{\partial D_w} U(w,z)U(\zeta,w)$ equals zero, since $\int_{\partial D_w} U(w,z)U(\zeta,w) = 0$ (when $n = 1$, the kernel $U(\zeta,w)$ reduces to the Cauchy kernel $(2\pi i)^{-1}(\zeta - w)^{-1}\, d\zeta$).

It is asserted in [178] that

$$\int_{\partial D_w} U(w,z)U(\zeta,w) = 0, \qquad z, \zeta \in \partial D, \tag{22.5}$$

when $n > 1$ also. We will show that this formula is false already in the case of the ball in $\mathbf{C}^2$. In §5, we computed the Bochner-Martinelli integral of the function

$$f(\zeta,w) = \frac{1}{|w|^2} \frac{1}{|\zeta - w|w|^{-2}|^2} = \frac{1}{|\zeta|^2} \frac{1}{|\zeta|\zeta|^{-2} - w|^2},$$

which differs from the fundamental solution $g(\zeta, z)$ only by a constant factor. The restriction of $U(\zeta,w)$ to ∂D_ζ equals

$$\frac{1}{2\pi^2 |\zeta - w|^4} \sum_{k=1}^{2} (\bar{\zeta}_k - \bar{w}_k) \zeta_k \, d\sigma(\zeta),$$

that is, it consists of derivatives of $f(\zeta,w)$ in ζ. Computing these derivatives and applying the jump formula for them, we obtain

$$\int_{\partial D_w} U(w,z)U(\zeta,w)$$
$$= \frac{1}{2\pi^2} \left[\arctan \frac{|\zeta + z|}{|\zeta - z|} \left(\frac{3(\langle \bar{z}, \zeta\rangle - \langle z, \bar{\zeta}\rangle)^2}{|\zeta - z|^5 |\zeta + z|^5} - \frac{1 + \langle \bar{z}, \zeta \rangle (\langle z, \bar{\zeta}\rangle - \langle \bar{z}, \zeta\rangle)}{|\zeta - z|^3 |\zeta + z|^3} \right) \right.$$
$$+ \frac{(1 - \langle \bar{z}, \zeta\rangle)\langle \bar{z}, \zeta\rangle}{|\zeta + z|^2 |\zeta - z|^4} + \frac{3(1 - \langle \bar{z}, \zeta\rangle)(\langle z, \bar{\zeta}\rangle - \langle \bar{z}, \zeta\rangle)}{|\zeta - z|^4 |\zeta + z|^4}$$
$$\left. + \frac{(1 - \langle \bar{z}, \zeta\rangle)(\langle z, \bar{\zeta}\rangle - \langle \bar{z}, \zeta\rangle)(|\zeta + z|^2 + 2)}{|\zeta - z|^4 |\zeta + z|^2} \right].$$

Thus, formula (22.5) is not true.

That (22.5) is not true can also be shown as follows. In [178] (and also in [177]), an inversion formula for the singular Bochner-Martinelli integral is derived from (22.4) by using (22.5): if $f(z,w) = f(z)$, then

$$\int_{\partial D_w} U(w,z) \int_{\partial D_\zeta} f(\zeta)U(\zeta,w) = \frac{1}{4} f(z). \tag{22.6}$$

But (22.6) also is false, for if we recall the definition of the operator M_σ (which differs from the singular Bochner-Martinelli integral only by the constant 2), formula (22.6) means that $M_\sigma^2 f = f$. But from Corollary 16.10 we then obtain that f extends holomorphically into D. For the ball, this can be seen directly from (5.1) and Corollary 5.3, since

$$M_\sigma P_{s,t} = \frac{(n+s-t-1)}{(n+s+t-1)} P_{s,t},$$

where $P_{s,t}$ is a homogeneous harmonic polynomial of degree s in z and t in $\bar{z}$, and consequently $M_\sigma^2 P_{s,t} = P_{s,t}$ only for holomorphic polynomials $P_{s,t}$ (when $n > 1$).

This error of Serbin was not noticed in a number of works by Chinese mathematicians [34, 140, 148, 229] who also studied the Poincaré-Bertrand formula, and in particular formula (22.6).

Lemma 22.7. *We have*

$$\int_{\partial D_w} U(w,z)U(\zeta,w) = \bar{\partial}_\zeta \int_{\partial D_w} U(w,z) \wedge U_{0,1}(\zeta,w), \qquad z,\zeta \in \partial D, \quad z \neq \zeta$$

(where $U_{0,1}(\zeta,w)$ is the kernel in the Koppelman formula (1.9)).

Proof. When $z \neq \zeta$, we have by definition that

$$\int_{\partial D_w} U(w,z)U(\zeta,w) = \lim_{\epsilon\to 0+} \int_{\partial D \setminus (B(z,\epsilon)\cup B(\zeta,\epsilon))} U(w,z)U(\zeta,w).$$

Consider the integral

$$\int_{\partial D\setminus(B(z,\epsilon)\cup B(\zeta,\epsilon))} U(w,z)U(\zeta,w) = \int_{\partial(D\setminus(B(z,\epsilon)\cup B(\zeta,\epsilon)))} U(w,z)U(\zeta,w)$$
$$+ \int_{D\cap S(z,\epsilon)} U(w,z)U(\zeta,w) + \int_{D\cap S(\zeta,\epsilon)} U(w,z)U(\zeta,w).$$

We have

$$\lim_{\epsilon\to 0+} \int_{D\cap S(z,\epsilon)} U(w,z)U(\zeta,w) = \frac{1}{2} \lim_{\epsilon\to 0+} \int_{S(z,\epsilon)} U(w,z)U(\zeta,w).$$

Therefore, by Lemma 1.13,

$$\lim_{\epsilon\to 0+}\int_{D\cap S(z,\epsilon)} U(w,z)U(\zeta,w)$$
$$= \frac{(n-1)!}{2(2\pi i)^n}\lim_{\epsilon\to 0+}\epsilon^{-2n}\int_{S(z,\epsilon)}\sum_{k=1}^{n}(-1)^{k-1}(\bar w_k-\bar z_k)\,d\bar w[k]\wedge dw\,U(\zeta,w)$$
$$= \frac{1}{2}U(\zeta,z).$$

In the same way,

$$\lim_{\epsilon\to 0+}\int_{D\cap S(\zeta,\epsilon)} U(w,z)U(\zeta,w)$$
$$= \frac{1}{2}\lim_{\epsilon\to 0+}\int_{S(\zeta,\epsilon)} U(w,z)U(\zeta,w)$$
$$= \frac{(n-1)!}{2(2\pi i)^n}\lim_{\epsilon\to 0+}\epsilon^{-2n}\int_{S(\zeta,\epsilon)}\frac{(n-1)!}{(2\pi i)^n}\sum_{k=1}^{n}(-1)^{k-1}\frac{\bar w_k-\bar z_k}{|w-z|^{2n}}\,d\bar w[k]\wedge dw\times$$
$$\times\sum_{j=1}^{n}(-1)^{j-1}(\bar\zeta_j-\bar w_j)\,d\bar\zeta[j]\wedge d\zeta$$
$$= -\frac{(n-1)!}{2n(2\pi i)^n}\sum_{k,j=1}^{n}(-1)^{j-1}\delta_{jk}\frac{\bar\zeta_k-\bar z_k}{|\zeta-z|^{2n}}\,d\bar\zeta[j]\wedge d\zeta = -\frac{1}{2n}U(\zeta,z).$$

By Lemmas 1.12 and 1.15 (and the remark after Lemma 1.15), we have

$$\int_{\partial(D\setminus(B(z,\epsilon)\cup B(\zeta,\epsilon)))} U(w,z)U(\zeta,w)$$
$$= -\int_{D\setminus(B(z,\epsilon)\cup B(\zeta,\epsilon))} U(w,z)\wedge\bar\partial_w U(\zeta,w)$$
$$= \int_{D\setminus(B(z,\epsilon)\cup B(\zeta,\epsilon))} U(w,z)\wedge\bar\partial_\zeta U_{0,1}(\zeta,w)$$
$$= \int_{D\setminus(B(z,\epsilon)\cup B(\zeta,\epsilon))} U(w,z)\wedge\bar\partial_\zeta U_{n,n-2}(w,\zeta)$$
$$\xrightarrow[\epsilon\to 0+]{} \text{P.V.}\int_D U(w,z)\wedge\bar\partial_\zeta U_{n,n-2}(w,\zeta)$$
$$= -\bar\partial_\zeta\int_D U(w,z)\wedge U_{n,n-2}(w,\zeta)-\frac{n-1}{2n}U(\zeta,z)$$
$$= \bar\partial_\zeta\int_D U(w,z)\wedge U_{0,1}(\zeta,w)-\frac{n-1}{2n}U(\zeta,z).$$

Lemma 22.7 follows. It is due to Kytmanov, Tarkhanov, and Prenov [124]. □

We now show that

$$\int_{\partial D_\zeta}\int_{\partial D_w} U(w,z)U(\zeta,w)=0 \tag{22.7}$$

if $z\in\partial D$. By definition and Lemma 22.7,

$$\begin{aligned}\int_{\partial D_\zeta}\int_{\partial D_w} U(w,z)U(\zeta,w) &= \lim_{\epsilon\to 0+}\int_{\partial D_\zeta\setminus B(z,\epsilon)}\int_{\partial D_w} U(w,z)U(\zeta,w)\\ &= \lim_{\epsilon\to 0+}\int_{\partial D_\zeta\setminus B(z,\epsilon)}\overline{\partial}_\zeta\int_{D_w} U(w,z)\wedge U_{0,1}(\zeta,w)\\ &= \lim_{\epsilon\to 0+}\int_{\partial D_\zeta\cap S(z,\epsilon)}\int_{D_w} U(w,z)\wedge U_{0,1}(\zeta,w)\\ &= \lim_{\epsilon\to 0+}\int_{D_w}\int_{\partial D_\zeta\cap S(z,\epsilon)} U(w,z)\wedge U_{0,1}(\zeta,w).\end{aligned}$$

Since the kernels $U_{p,q}$ are invariant with respect to translations and unitary transformations, we may assume that $z=0$, and by making the change of variables $\zeta\to\zeta/\epsilon$ and $w\to w/\epsilon$, we obtain

$$\begin{aligned}&\lim_{\epsilon\to 0+}\int_{D_w}\int_{\partial D_\zeta\cap S(z,\epsilon)} U(w,z)\wedge U_{0,1}(\zeta,w)\\ &=\lim_{\epsilon\to 0+}\int_{(1/\epsilon)D_w}\int_{(1/\epsilon)\partial D_\zeta\cap S(0,1)} U(w,0)\wedge U_{0,1}(\zeta,w)\\ &=\int_{\Pi_w}\int_{T_\zeta\cap S(0,1)} U(w,0)\wedge U_{0,1}(\zeta,w)=\frac{1}{2}\int_{T_\zeta\cap S(0,1)}\int_{\mathbf{C}^n_w} U(w,0)\wedge U_{0,1}(\zeta,w)\end{aligned}$$

(where Π_w is the tangent cone to D_w at the point $z=0$, that is, a half-space, and T_ζ is the tangent plane to ∂D_ζ at the point $z=0$). But the integral

$$\int_{\mathbf{C}^n_w} U(w,0)\wedge U_{0,1}(\zeta,w)$$

consists of terms of the form

$$\int_{\mathbf{C}^n}\frac{\bar{w}_k(\bar{\zeta}_j-\bar{w}_j)-\bar{w}_j(\bar{\zeta}_k-\bar{w}_k)}{|w|^{2n}|\zeta-w|^{2n}}\,d\bar{w}\wedge dw,$$

and such integrals obviously equal zero. Thus formula (22.7) is proved. Other proofs of (22.4) are given in [229] and [200]. □

Corollary 22.8. *If $f\in\mathcal{C}^\alpha(\partial D)$, where $0<\alpha\le 1$, then for $z\in\partial D$ we have*

$$\int_{\partial D_w} U(w,z)\int_{\partial D_\zeta} f(\zeta)\,U(\zeta,w)=\frac{1}{4}f(z)+\int_{\partial D_\zeta} f(\zeta)\,\overline{\partial}_\zeta\int_{D_w} U(w,z)\wedge U_{0,1}(\zeta,w). \tag{22.8}$$

When $n=2$, there is a more revealing formula.

Theorem 22.9 (Vasilevskiĭ, Shapiro). *Let D be a bounded domain in $\mathbf{C}^2$ with smooth boundary ∂D. For $f \in C^\alpha(\partial D)$, where $0 < \alpha \le 1$, define the operator*

$$(M_1 f)(z) = \frac{1}{2\pi^2}\,\mathrm{P.V.}\int_{\partial D} \frac{\bar f(\zeta)}{|\zeta - z|^4}((\bar\zeta_1 - \bar z_1)\,d\zeta_1 + (\bar\zeta_2 - \bar z_2)\,d\zeta_2) \wedge d\bar\zeta, \qquad z \in \partial D.$$

Then $M_\sigma^2 = I + M_1^2$.

Proof. We will give the proof when $D = B(0,1)$ in $\mathbf{C}^2$. First suppose $B(0,1) \subset \mathbf{C}^n$. Consider the integral

$$I(z) = \int_{S(0,1)} f(\zeta)\,dg(\zeta,z) \wedge d\zeta[k,m] \wedge d\bar\zeta, \qquad z \in B(0,1),$$

where $g(\zeta,z)$ is the fundamental solution of Laplace's equation (see §1). Expressing the integrand in terms of the Poisson kernel (as in §5), we obtain

$$I(z) = (-1)^{k+m}\int_{S(0,1)} f(\zeta)P(\zeta,z)\frac{\bar z_m\bar\zeta_k - \bar z_k\bar\zeta_m}{1 - |z|^2}\,d\sigma(\zeta).$$

Let $f = P_{s,t}$ be a homogeneous harmonic polynomial. Since the harmonic extension of $\bar\zeta_k P_{s,t}$ into $B(0,1)$ is given by

$$\bar\zeta_k P_{s,t} + \frac{1 - |\zeta|^2}{n+s+t-1}\frac{\partial P_{s,t}}{\partial \zeta_k},$$

we have

$$I(z) = \frac{(-1)^{k+m}}{n+s+t-1}\left(\bar z_m\frac{\partial P_{s,t}}{\partial z_k} - \bar z_k\frac{\partial P_{s,t}}{\partial z_m}\right). \tag{22.9}$$

The integral $I(z)$ has no jump, so its boundary value coincides with $\mathrm{P.V.}\,I(z)$.

When $n = 2$, we obtain from (22.9) that

$$M_1 P_{s,t} = -2\int_{S(0,1)} \bar P_{s,t}\,dg \wedge d\bar\zeta = \frac{2}{s+t+1}\left(\bar z_2\frac{\partial \bar P_{s,t}}{\partial z_1} - \bar z_1\frac{\partial \bar P_{s,t}}{\partial z_2}\right),$$

so

$$\begin{aligned} M_1^2 P_{s,t} = \frac{-4}{(s+t+1)^2}\Bigg[&\frac{\partial^2 P_{s,t}}{\partial z_2\partial\bar z_2}z_2\bar z_2 + \bar z_1 z_2\frac{\partial^2 P_{s,t}}{\partial\bar z_1\partial z_2} + \frac{\partial^2 P_{s,t}}{\partial z_1\partial\bar z_2}z_1\bar z_2 \\ &+ \frac{\partial^2 P_{s,t}}{\partial z_1\partial\bar z_1}z_1\bar z_1 + \bar z_2\frac{\partial P_{s,t}}{\partial\bar z_2} + \bar z_1\frac{\partial P_{s,t}}{\partial\bar z_1}\Bigg] = -\frac{4t(1+s)}{(s+t+1)^2}P_{s,t}. \end{aligned}$$

Hence, because $M_\sigma^2 P_{s,t} = [(s-t+1)/(s+t+1)]^2 P_{s,t}$, we have $M_\sigma^2 P_{s,t} = P_{s,t} + M_1^2 P_{s,t}$.

Since this equality holds for all polynomials $P_{s,t}$, it holds for arbitrary functions f. □

Theorem 22.9 was given in [207].

23 Problems connected with the possibility of holomorphic extension

23.1 Functions representable by the Cauchy-Fantappiè formula

In §1, we considered the Cauchy-Fantappiè formula (1.6). Let D be a bounded domain in $\mathbf{C}^n$ with smooth boundary ∂D, and let $\eta_k = \eta_k(\zeta, z)$, $\zeta \in \partial D$, $z \in D$, $k = 1, \ldots, n$, be functions, continuously differentiable in ζ, such that $\langle \eta, \zeta - z\rangle = \sum_{k=1}^{n} \eta_k(\zeta_k - z_k) \equiv 1$ for $\zeta \in \partial D$ and $z \in D$. Then

$$f(z) = \frac{(n-1)!}{(2\pi i)^n} \int_{\partial D} f(\zeta)\, \omega'(\eta) \wedge d\zeta, \qquad z \in D, \tag{23.1}$$

where $\omega'(\eta) = \sum_{k=1}^{n} (-1)^{k-1} \eta_k\, d\eta[k]$, and $f \in \mathcal{A}(D)$.

Concerning formula (23.1), we can pose a problem analogous to the problem of §15 for the Bochner-Martinelli integral: namely, if (23.1) holds for $f \in \mathcal{C}(\overline{D})$, will f be holomorphic in D?

If the functions η_k depend holomorphically on z (for example, in the case of the Henkin-Ramirez kernel, or linearly convex domains [10]), then the answer is obvious. If the η_k are arbitrary, then it is easy to give an example where the answer to this question is negative. Let $P(\zeta, z) = (n-1)!\, \pi^{-n}(1 - |z|^2)|\zeta - z|^{-2n}/2$ be the Poisson kernel for the ball, and $D = B(0, 1)$. Then

$$P(\zeta, z) = \frac{1 - \langle \zeta, \bar{z}\rangle}{1 - |z|^2}\, U(\zeta, z)$$

(see §5), and the function $\varphi(\zeta, z) = (1 - \langle \zeta, \bar{z}\rangle)/(1 - |z|^2)$ has the properties that $\varphi(z, z) = 1$ and φ is holomorphic in ζ (for fixed z). As Dautov showed (see Theorem 23.7), the kernel $P(\zeta, z)$ is a Cauchy-Fantappiè kernel. But (23.1) holds with the kernel $P(\zeta, z)$ for all harmonic functions. Therefore our question must be formulated as follows.

Problem 23.1. *For which Cauchy-Fantappiè kernels $\omega'(\eta) \wedge d\zeta$ does the equality (23.1) for a function $f \in \mathcal{C}(\overline{D})$ imply that $f \in \mathcal{A}(D)$?*

One class of Cauchy-Fantappiè kernels can be specified in the following way. The kernel $\omega'(\eta) \wedge d\zeta$ has the form

$$\omega'(\eta) \wedge d\zeta = \sum_{k=1}^{n} \delta_k\, d\bar{\zeta}[k] \wedge d\zeta, \qquad \text{where} \qquad \delta_k = \begin{vmatrix} \eta_1 & \cdots & \eta_n \\ \dfrac{\partial \eta_1}{\partial \bar{\zeta}_1} & \cdots & \dfrac{\partial \eta_n}{\partial \bar{\zeta}_1} \\ \cdots & [k] & \cdots \\ \dfrac{\partial \eta_1}{\partial \bar{\zeta}_n} & \cdots & \dfrac{\partial \eta_n}{\partial \bar{\zeta}_n} \end{vmatrix}.$$

We suppose that η_k can be extended (in ζ) into D so that $\langle \eta, \zeta - z\rangle \equiv 1$ for all $\zeta \neq z$, and that the form $\omega'(\eta) \wedge d\zeta$ has integrable coefficients in $\overline{D}$. Then, as for the Bochner-Martinelli kernel (formula (6.8)), $\overline{\partial}[\omega'(\eta) \wedge d\zeta] = ((2\pi i)^n/(n-1)!)\delta_z\, dv$

in D (in the sense of currents). Therefore

$$\int_{\partial D} f(\zeta)\,\omega'(\eta)\wedge d\zeta - \int_D \overline{\partial} f \wedge \omega'(\eta)\wedge d\zeta = \frac{(2\pi i)^n}{(n-1)!} f(z), \qquad z\in D,$$

if $f \in C^1(\overline{D})$ (see, for example, [18]). (This is the analogue of the Bochner-Martinelli formula for smooth functions.) We can transform the integral over D into an integral over ∂D in the following way: let $\delta_k = (-1)^{k-1} h_k(\zeta)\partial h(\zeta,z)/\partial\zeta_k$. Since $\omega'(\eta)\wedge d\zeta$ is a closed form,

$$\sum_{k=1}^{n} \frac{\partial}{\partial\overline{\zeta}_k}\left(h_k(\zeta)\frac{\partial h(\zeta,z)}{\partial\zeta_k}\right) = 0$$

for $\zeta, z\in D$ and $\zeta\neq z$. Therefore if $f\in C^2(\overline{D})$, we have

$$\begin{aligned} d(f(\zeta)\,\omega'(\eta)\wedge d\zeta) &= d\left(f(\zeta)\sum_{k=1}^{n}(-1)^{k-1} h_k \frac{\partial h}{\partial\zeta_k}\, d\overline{\zeta}[k]\wedge d\zeta\right)\\ &= \sum_{k=1}^{n}\frac{\partial f}{\partial\overline{\zeta}_k} h_k \frac{\partial h}{\partial\zeta_k}\, d\overline{\zeta}\wedge d\zeta, \qquad \text{and} \end{aligned}$$

$$\begin{aligned} &d\left(h\sum_{k=1}^{n}(-1)^{k-1}\frac{\partial f}{\partial\overline{\zeta}_k} h_k\, d\zeta[k]\wedge d\overline{\zeta}\right)\\ &\qquad = \sum_{k=1}^{n}\frac{\partial h}{\partial\zeta_k} h_k \frac{\partial f}{\partial\overline{\zeta}_k}\, d\zeta\wedge d\overline{\zeta} + h\sum_{k=1}^{n}\frac{\partial}{\partial\zeta_k}\left(h_k\frac{\partial f}{\partial\overline{\zeta}_k}\right) d\zeta\wedge d\overline{\zeta}. \end{aligned}$$

Therefore, arguing in the usual way, we obtain the following analogue of Green's formula:

$$\begin{aligned} &\int_{\partial D} f(\zeta)\,\omega'(\eta)\wedge d\zeta - (-1)^n\int_{\partial D} h\sum_{k=1}^{n}(-1)^{k-1}\frac{\partial f}{\partial\overline{\zeta}_k} h_k\, d\zeta[k]\wedge d\overline{\zeta}\\ &\qquad + (-1)^n\int_D h\sum_{k=1}^{n}\frac{\partial}{\partial\zeta_k}\left(h_k\frac{\partial f}{\partial\overline{\zeta}_k}\right) d\zeta\wedge d\overline{\zeta} = \frac{(2\pi i)^n}{(n-1)!} f(z), \qquad z\in D. \end{aligned}$$

(This formula is given in [18].)

If $h_k > 0$, then the operator $\sum_{k=1}^{n}(\partial/\partial\zeta_k)(h_k\partial/\partial\overline{\zeta}_k)$ is elliptic. Solving the Dirichlet problem for this operator for f, we obtain

$$\frac{(2\pi i)^n}{(n-1)!} f(z) = \int_{\partial D} f(\zeta)\,\omega'(\eta)\wedge d\zeta - (-1)^n\int_{\partial D} h(\zeta,z)\sum_{k=1}^{n}(-1)^{k-1} h_k\frac{\partial f}{\partial\overline{\zeta}_k}\, d\zeta[k]\wedge d\overline{\zeta}.$$

The function $h(\zeta,z)$ is, in fact, a fundamental solution for the operator

$$\sum_{k=1}^{n}\frac{\partial}{\partial\overline{\zeta}_k}\left(h_k\frac{\partial}{\partial\zeta_k}\right).$$

If (23.1) holds for f, then by using properties of the fundamental solution h, we obtain that the restriction to ∂D of the form $\sum_{k=1}^{n}(-1)^{k-1}(\partial f/\partial\bar{\zeta}_k)h_k\, d\zeta[k]\wedge d\bar{\zeta}$ equals zero. Multiplying this equality by $\bar{f}$, integrating over ∂D, and applying Stokes's formula, we have

$$\int_D \sum_{k=1}^{n}\left|\frac{\partial f}{\partial\bar{\zeta}_k}\right|^2 h_k\, d\zeta\wedge d\bar{\zeta}=0,$$

that is, $\partial f/\partial\bar{\zeta}_k=0$, and f is holomorphic in D.

Theorem 23.1 (Kytmanov). *Let $\delta_k(\zeta,z)=(-1)^{k-1}h_k(\zeta)\partial h(\zeta,z)/\partial\bar{\zeta}_k$, $k=1,\ldots,n$, for $\zeta\in D$ and $z\in D$, $z\neq\zeta$, where h and $\partial h/\partial\bar{\zeta}_k$ are locally integrable functions on $\overline{D}$, and $h_k\in\mathcal{C}(\overline{D})$, $h_k>0$ in $\overline{D}$. If (23.1) holds for a function $f\in\mathcal{C}^1(\overline{D})$, then f is holomorphic in D.*

The question of completely describing Cauchy-Fantappiè kernels for which this theorem holds remains open.

23.2 Differential criteria for holomorphicity of functions

As before, suppose D is a bounded domain in $\mathbf{C}^n$ with smooth boundary ∂D, and $D=\{\,z:\rho(z)<0\,\}$, where $\rho\in\mathcal{C}^1(\overline{D})$ and $d\rho\neq 0$ on ∂D.

We consider the following problem. Suppose given a vector field $w=w(z)=\sum_{k=1}^{n}w_k(\partial/\partial z_k)$, $w_k\in\mathcal{C}(\partial D)$, such that

$$w(\rho)=\sum_{k=1}^{n}w_k\frac{\partial\rho}{\partial z_k}\neq 0 \text{ on } \partial D, \tag{23.2}$$

that is, for every point $z\in\partial D$ the vector w does not lie in the complex tangent space $T_z^c(\partial D)$.

Problem 23.2. *Suppose $f\in\mathcal{C}^1(\overline{D})$ and f is harmonic in D. If*

$$\bar{w}(f)=\sum_{k=1}^{n}\bar{w}_k\frac{\partial f}{\partial\bar{z}_k}=0 \textit{ on } \partial D, \tag{23.3}$$

will f be holomorphic in D?

This problem was stated in [38, No. 19].

In contrast to the tangential Cauchy-Riemann conditions (6.11) and (6.12), here we require the vanishing of the action of a nontangential vector field $\bar{w}$ on f. Problem 23.2 is an analogue of the oblique derivative problem for real-valued harmonic functions.

If (23.2) does not hold, then it is easy to give an example where (23.3) holds for a function f that is not holomorphic in D. It is enough to consider the ball

$B(0,1)$ in $\mathbf{C}^2$ and the function $f = \bar{z}_1$, with $w = \partial/\partial z_2$. Then $\bar{w}(f) = 0$ on $S(0,1)$ and $w(\rho) = \bar{z}_2$. Therefore inequality (23.2) is violated on the circle $\{ z \in \mathbf{C}^2 : |z| = 1,\ z_2 = 0 \}$.

If $w = \sum_{k=1}^n \bar{\rho}_k(\partial/\partial z_k)$, then Problem 23.2 becomes the homogeneous $\bar{\partial}$-Neumann problem (14.4). The holomorphicity of functions f satisfying (14.4) was proved in §15 (see Theorem 15.1).

We now give a number of equivalent formulations of Problem 23.2. We decompose the field w into a normal and a tangential component:

$$w(z) = \alpha(z) \sum_{k=1}^{n} \bar{\rho}_k \frac{\partial}{\partial z_k} + \tilde{\tau}(z).$$

By hypothesis, $\alpha \neq 0$ on ∂D. The vector field $\tilde{\tau} = \sum_{k=1}^n \tau_k(\partial/\partial z_k) \in T_z^c(\partial D)$. Then condition (23.3) can be rewritten in the form

$$\sum_{k=1}^{n} \frac{\partial f}{\partial \bar{z}_k} \rho_k = \bar{\tilde{\tau}}(f)/\bar{\alpha}. \tag{23.4}$$

As generators of the space $T_z^c(\partial D)$, we may take the vectors

$$\frac{\partial \rho}{\partial z_m} \frac{\partial}{\partial z_k} - \frac{\partial \rho}{\partial z_k} \frac{\partial}{\partial z_m}, \qquad k \neq m, \quad k, m = 1, 2, \ldots, n$$

(see §6). By decomposing $\tilde{\tau}/\alpha$ in terms of these vectors, it is easy to obtain from (23.4) the equality

$$\bar{\partial}_n f = \sum_{k=1}^{n} \frac{\partial f}{\partial \bar{z}_k} \rho_k = \sum_{k>m} \alpha_{m,k}(z) \left[\frac{\partial f}{\partial \bar{z}_k} \bar{\rho}_m - \frac{\partial f}{\partial \bar{z}_m} \bar{\rho}_k \right],$$

where the $\alpha_{k,m}(z)$ are certain continuous functions on ∂D. Multiplying both sides of this equality by $d\sigma$, we obtain

$$\mu_f \big|_{\partial D} = \sum_{k>m} a_{k,m}(z)\, df \wedge d\bar{z}[k,m] \wedge dz \big|_{\partial D} \tag{23.5}$$

(we recall that $\mu_f = \sum_{k=1}^n (-1)^{n+k-1} (\partial f/\partial \bar{z}_k)\, dz[k] \wedge d\bar{z}$ (see §1)).

Problem 23.3. *If $f \in C^1(\overline{D})$ is harmonic in D, and $a_{k,m} \in C(\partial D)$ for $k, m = 1, \ldots, n$, does (23.5) imply that f is holomorphic in D?*

The homogeneous $\bar{\partial}$-Neumann problem is obtained from (23.5) if $a_{k,m} \equiv 0$ for all k and m. When $n = 1$, the right-hand side of (23.5) is always equal to zero.

Suppose $n > 1$. We rewrite (23.5) in integral form by multiplying (23.5) by the fundamental solution $g(\zeta, z)$ of Laplace's equation and integrating over ∂D. We obtain

$$\int_{\partial D} g(\zeta, z)\, u_f(\zeta) = \int_{\partial D} g(\zeta, z) \sum_{k>m} a_{k,m}(\zeta)\, df \wedge d\bar{\zeta}[k,m] \wedge d\zeta, \qquad z \notin \partial D.$$

Using Green's formula (1.3) and Stokes's formula, we get

$$f(z) = \int_{\partial D} f(\zeta)\left[U(\zeta, z) + \sum_{k>m} d(ga_{k,m}) \wedge d\bar{\zeta}[k, m] \wedge d\zeta\right], \qquad z \in D. \tag{23.6}$$

Problem 23.4. *If $f \in \mathcal{C}(\overline{D})$ is harmonic in D, and $a_{k,m} \in \mathcal{C}^1(\partial D)$, does (23.6) imply the holomorphicity of f in D?*

When $a_{k,m} \equiv 0$, we obtain the problem about functions representable by the Bochner-Martinelli integral. If $f \in \mathcal{A}(D)$, then (23.6) is an integral representation of f, since the component $d(ga_{k,m}) \wedge d\bar{\zeta}[k, m] \wedge d\zeta$ is a $\overline{\partial}$-exact form. Thus we again arrive at Problem 23.1.

We can give a solution to Problem 23.3 for some special cases.

When $n = 2$, then (23.5) can be rewritten in the form

$$\mu_f\big|_{\partial D} = a(\zeta)\, df \wedge d\zeta\big|_{\partial D}. \tag{23.7}$$

Theorem 23.2 (Kytmanov). *Suppose $\partial D \in \mathcal{C}^1$ is connected, and $a \equiv$ const. If a harmonic function $f \in \mathcal{C}^1(\overline{D})$ satisfies (23.7), then f is holomorphic in D.*

Proof. We use Corollary 15.7. Suppose $P_{s,t}$ are homogeneous harmonic polynomials of degree s in z and t in $\bar{z}$. By Stokes's formula, we get from (23.7) the chain of equalities

$$\int_{\partial D} f(\zeta) \sum_{k=1}^{2} (-1)^{k-1} \frac{\partial P_{s,t}}{\partial \zeta_k}\, d\bar{\zeta}[k] \wedge d\zeta$$
$$= \int_{\partial D} P_{s,t}\mu_f = \int_{\partial D} aP_{s,t}\, df \wedge d\zeta = -\int_{\partial D} af\, dP_{s,t} \wedge d\zeta.$$

If $t = 0$, that is, $P_{s,t}$ is a holomorphic polynomial, then $dP_{s,t} \wedge d\zeta = 0$, and so

$$\int_{\partial D} f(\star\partial P_{s,t}) = 0.$$

Supposing that $t > 0$, we convert the differential form $dP_{s,t} \wedge d\zeta$ to

$$\frac{\partial \widetilde{P}}{\partial \zeta_1}\, d\bar{\zeta}_2 \wedge d\zeta - \frac{\partial \widetilde{P}}{\partial \zeta_2}\, d\bar{\zeta}_1 \wedge d\zeta.$$

To do this, it suffices to take $\widetilde{P} = -\int (\partial P_{s,t}/\partial \bar{\zeta}_1)\, d\zeta_2$ (here we have in mind formal integration with respect to ζ_2). Then

$$\frac{\partial \widetilde{P}}{\partial \zeta_1} = -\int \frac{\partial^2 P_{s,t}}{\partial \zeta_1 \partial \bar{\zeta}_1}\, d\zeta_2 = \int \frac{\partial^2 P_{s,t}}{\partial \zeta_2 \partial \bar{\zeta}_2}\, d\zeta_2 = \frac{\partial P_{s,t}}{\partial \bar{\zeta}_2}.$$

Therefore

$$\Delta \widetilde{P} = \frac{\partial^2 \widetilde{P}}{\partial \bar{\zeta}_1 \partial \zeta_1} + \frac{\partial^2 \widetilde{P}}{\partial \bar{\zeta}_2 \partial \zeta_2} = \frac{\partial^2 P_{s,t}}{\partial \zeta_1 \partial \bar{\zeta}_2} - \frac{\partial^2 P_{s,t}}{\partial \bar{\zeta}_1 \partial \zeta_2} = 0.$$

The polynomial $\widetilde{P}$ is homogeneous and harmonic, with degree $s+1$ in z and degree $t-1$ in $\bar{z}$. Thus

$$\int_{\partial D} f \sum_{k=1}^{2} (-1)^{k-1} \frac{\partial P_{s,t}}{\partial \zeta_k}\, d\bar{\zeta}[k] \wedge d\zeta = \int_{\partial D} af\, dP_{s,t} \wedge d\zeta$$
$$= \int_{\partial D} af \sum_{k=1}^{2} (-1)^{k-1} \frac{\partial \widetilde{P}}{\partial \zeta_k}\, d\bar{\zeta}[k] \wedge d\zeta.$$

By induction on t, we conclude that $\int_{\partial D} f(\star \partial P_{s,t}) = 0$ for all s and t, and consequently f is holomorphic in D. □

Theorem 23.2 was proved in [111].

Theorem 23.3 (Kytmanov). *Let $D = B(0,1)$ in $\mathbf{C}^n$, let $f \in \mathcal{C}^1(\overline{B})$ be a harmonic function, and suppose $a_{k,m} \in \mathcal{A}(B)$ for $k, m = 1, \ldots, n$. If (23.5) holds for f, then f is holomorphic in B.*

Proof. First consider the case

$$\mu_f \big|_{\partial B} = Q_{l,0}\, df \wedge d\bar{\zeta}[k,m] \wedge d\zeta \big|_{\partial B}, \tag{23.8}$$

where $Q_{l,0}$ is a homogeneous holomorphic polynomial of degree l. We write (23.8) for the ball $B(0,1)$:

$$\sum_{j=1}^{n} \frac{\partial f}{\partial \bar{\zeta}_j} \bar{\zeta}_j = \pm Q_{l,0}(\zeta) \left[\frac{\partial f}{\partial \bar{\zeta}_k} \zeta_m - \frac{\partial f}{\partial \bar{\zeta}_m} \zeta_k \right], \qquad \zeta \in S(0,1). \tag{23.9}$$

We will look for a solution of (23.9) in the form of a series $f(\zeta) = \sum_{s,t} P_{s,t}(\zeta)$ in homogeneous harmonic polynomials. From (23.9) we have

$$\sum_{s,t} t P_{s,t} = \pm Q_{l,0} \sum_{s,t} \left(\zeta_m \frac{\partial P_{s,t}}{\partial \bar{\zeta}_k} - \frac{\partial P_{s,t}}{\partial \bar{\zeta}_m} \zeta_k \right), \qquad \zeta \in S(0,1). \tag{23.10}$$

We now extend the function on the right-hand side of (23.10) into $B(0,1)$ as a harmonic function. To do this, we use Gauss's formula (see, for example, [189, chap. 11]). If $P_{\tilde{k}}$ is an arbitrary homogeneous polynomial of degree $\tilde{k}$, then its harmonic extension $\widetilde{P}_{\tilde{k}}$ from the boundary $S(0,1)$ into the ball $B(0,1)$ is given by the formula $\widetilde{P}_{\tilde{k}} = \sum_{s \geq 0} Z_{\tilde{k}-2s}$, where

$$Z_{\tilde{k}-2s} = \frac{\tilde{k} - 2s + n + 1}{s!\, (\tilde{k} + n - s - 1)!} \sum_{j \geq 0} (-1)^j \frac{(\tilde{k} - j - 2s + n - 2)!}{j!} |z|^{2j} \Delta^{j+s} P_{\tilde{k}}. \tag{23.11}$$

In our case $\tilde{k} = l + s + t$, and

$$P_{\tilde{k}} = Q_{l,0}\left(\zeta_m \frac{\partial P_{s,t}}{\partial \bar{\zeta}_k} - \zeta_k \frac{\partial P_{s,t}}{\partial \bar{\zeta}_m}\right).$$

Therefore we obtain from (23.10) the following system of equations:

$$\begin{aligned} tP_{s,t} &= \sum_{j\geq 0} a_j |\zeta|^{2j} \Delta^j \left[Q_{l,0}\left(\zeta_m \frac{\partial P_{s-l-1,t+1}}{\partial \bar{\zeta}_k} - \zeta_k \frac{\partial P_{s-l-1,t+1}}{\partial \bar{\zeta}_m}\right)\right] \\ &+ \sum_{j\geq 0} b_j |\zeta|^{2j} \Delta^{j+1} \left[Q_{l,0}\left(\zeta_m \frac{\partial P_{s-l,t+2}}{\partial \bar{\zeta}_k} - \zeta_k \frac{\partial P_{s-l,t+2}}{\partial \bar{\zeta}_m}\right)\right] \\ &+ \cdots + \sum_{j\geq 0} c_j |\zeta|^{2j} \Delta^{j+l} \left[Q_{l,0}\left(\zeta_m \frac{\partial P_{s-1,t+l+1}}{\partial \bar{\zeta}_k} - \zeta_k \frac{\partial P_{s-1,t+l+1}}{\partial \bar{\zeta}_m}\right)\right], \end{aligned} \tag{23.12}$$

where a_j, b_j, ..., c_j are certain constants. When $s = 0$, it follows from (23.12) that $tP_{0,t} = 0$ for all t, that is, $P_{0,t} = 0$ for $t \neq 0$. Further, we have $tP_{1,t} = 0$ and so on; generally, $tP_{s,t} = 0$. That is, the series for f contains only holomorphic terms, so f is holomorphic in B.

Now if the $a_{k,m}$ are arbitrary holomorphic functions, then by decomposing them into homogeneous polynomials and applying the preceding argument, we obtain a system of equations of the form (23.11) with a finite number of terms in each equation. □

Theorem 23.3 was proved in [111].

Corollary 23.4. *Let P_k be an arbitrary homogeneous polynomial of degree $k \geq 2$. In order that the restriction of P_k to ∂B extend holomorphically into $B(0,1)$, it is necessary and sufficient that $L_k P_k = 0$ on $S(0,1)$, where*

$$L_k = \sum_{j=1}^{n} \bar{z}_j \frac{\partial}{\partial \bar{z}_j} - \sum_{l\geq 0} \frac{(n+k-2l-3)!\,(2l-1)!!}{(n+k-2)!\,(l+1)!} 2^l \Delta^{l+1}.$$

Proof. Let $P_k = P_{m,t}$, that is,

$$P_k = \sum_{\|\alpha\|=m} \sum_{\|\beta\|=t} a_{\alpha,\beta} z^\alpha \bar{z}^\beta.$$

We extend P_k from ∂B into the ball by Gauss's formula. This extension can be written in the form $\widetilde{P}_k = \sum_{s\geq 0} Z_{k-2s}$, where Z_{k-2s} is defined by (23.11). In order that $\widetilde{P}_k$ be holomorphic, it is necessary and sufficient that

$$\bar{\partial}_n \widetilde{P}_k = \sum_{j=1}^{n} \bar{z}_j \frac{\partial}{\partial \bar{z}_j} \widetilde{P}_k = 0 \text{ on } S(0,1)$$

(see Theorem 15.1). But $\overline{\partial}_n Z_{k-2s} = (t-s)Z_{k-2s}$. Then

$$\overline{\partial}_n \widetilde{P}_k = \sum_{j,s}(t-s)\frac{(-1)^j(k-2s+n-1)(k-j-2s+n-2)!}{s!\,j!\,(k+n-s-1)!}\Delta^{j+s}P_k$$

$$= \sum_{l\geq 0}\Delta^l P_k \sum_{j+s=l}\frac{(-1)^j(k-2s+n-1)(k-l-s+n-2)!\,(t-s)}{s!\,(l-s)!\,(k+n-s-1)!}$$

$$= \sum_{l\geq 0}(-1)^l\Delta^l P_k \sum_{s=0}^{l}\frac{(-1)^s(k-2s+n-1)(k-l-s+n-2)!\,(t-s)}{s!\,(l-s)!\,(k+n-s-1)!}.$$

The proof of Corollary 23.4 will now follow from a lemma. □

Lemma 23.5. *If $k \geq 2l$, then*

$$\sum_{s=0}^{l}\frac{(-1)^s(k-2s+n-1)(k-l-s+n-2)!}{s!\,(l-s)!\,(k+n-s-1)!} = \begin{cases}0, & l>0,\\ 1, & l=0;\end{cases}$$

$$\sum_{s=0}^{l}\frac{s(-1)^s(k-2s+n-1)(k-l-s+n-2)!}{s!\,(l-s)!\,(k+n-s-1)!}$$

$$= \begin{cases}\dfrac{(-1)^l(n+k-2l-1)!\,(2l-2)!}{(n+k-2)!\,l!\,(l-1)!}, & l>0,\\ 0, & l=0.\end{cases}$$

Proof. We compute, for example, the first sum, denoting it by Σ:

$$\Sigma = \frac{(-1)^l(k-2l+n-1)(k-2l+n-2)!}{l!\,(k+n-l-1)!}$$
$$+\frac{(-1)^{l-1}(k-2l+n+1)(k-2l+n-1)!}{(l-1)!\,(k+n-l)!}+\dots$$
$$= \frac{(-1)^{l-1}(k-2l+n-1)!}{(l-1)!\,(k+n-l-1)!}\left(-\frac{1}{l}+\frac{k+n+1-2l}{k+n-l}\right)+\dots$$
$$= \frac{(-1)^{l-1}(k-2l+n-1)!}{l!\,(k+n-l)!}(l-1)(k+n-2l)+\dots$$
$$= \frac{(-1)^{l-1}(k-2l+n)!\,(l-1)}{l!\,(k+n-l)!}+\frac{(-1)^{l-2}(k-2l+n+3)(k-2l+n)!}{(l-2)!\,2!\,(k+n-l+1)!}+\dots$$
$$= \frac{(-1)^{l-2}(k-2l+n+1)!\,(l-2)}{2!\,l(l-2)!\,(k+n-l+1)!}+\dots$$
$$= \frac{(k-l+n-2)!}{(l-1)!\,l(k+n-2)!}+\frac{(k+n-1)(k-l+n-2)!}{l!\,(k+n-1)!}=0.$$

The second sum is computed analogously. □

Corollary 23.4 was given in [111].

Suppose given in the domain D a function $f \in \mathcal{C}^2(\overline{D})$ that satisfies the following elliptic equation:

$$\sum_{j,k=1}^{n} \frac{\partial}{\partial z_j}\left(a_{j,k}(z)\frac{\partial f}{\partial \bar{z}_k}\right) = 0, \tag{23.13}$$

where the matrix $A = \|a_{j,k}\|_{j,k=1}^{n}$ is Hermitian and positive definite on $\overline{D}$, and $a_{j,k} \in \mathcal{C}^1(\overline{D})$, $j,k = 1, \dots, n$.

Theorem 23.6 (Kytmanov). *If the following boundary condition holds for a function $f \in \mathcal{C}^2(\overline{D})$:*

$$\sum_{j,k} a_{j,k}\frac{\partial f}{\partial \bar{z}_k}\rho_j = 0 \textit{ on } \partial D,$$

and if f satisfies (23.13) in D, then f is holomorphic in D.

Proof. Consider the differential form

$$\omega_f = \sum_{j,k} a_{j,k}\frac{\partial f}{\partial \bar{z}_k}(-1)^{j-1}\, dz[j] \wedge d\bar{z}.$$

The restriction of ω_f to ∂D equals zero by the hypothesis of the theorem. By Stokes's formula,

$$\begin{aligned}
0 &= \int_{\partial D} \bar{f}\,\omega_f = \int_D \bar{f}\, d\omega_f + \int_D d\bar{f} \wedge \omega_f \\
&= \int_D \bar{f} \sum_{j,k} \frac{\partial}{\partial z_j}\left(a_{j,k}\frac{\partial f}{\partial \bar{z}_k}\right) dz \wedge d\bar{z} + \int_D \sum_{j,k} a_{j,k}\frac{\partial \bar{f}}{\partial z_j}\frac{\partial f}{\partial \bar{z}_k}\, dz \wedge d\bar{z} \\
&= \int_D \sum_{j,k} a_{j,k}\overline{\frac{\partial f}{\partial \bar{z}_j}}\frac{\partial f}{\partial \bar{z}_k}\, dz \wedge d\bar{z};
\end{aligned}$$

since the Hermitian matrix A is positive definite, it follows from this that

$$\sum_{j,k=1}^{n} a_{j,k}\overline{\frac{\partial f}{\partial \bar{z}_j}}\frac{\partial f}{\partial \bar{z}_k} = 0$$

in D, and hence $\partial f/\partial \bar{z}_j = 0$, $j = 1, \dots, n$, that is, $f \in \mathcal{O}(D)$. □

Theorem 23.6 generalizes Theorem 23.1.

We remark that the vector field $w = \sum_{j,k} \bar{a}_{j,k}\bar{\rho}_j(\partial/\partial z_k)$ does not lie in the complex tangent plane $T_z^c(\partial D)$, since

$$w(\rho) = \sum_{j,k=1}^{n} \bar{a}_{j,k}\bar{\rho}_j\rho_k|\operatorname{grad}\rho|^{-1} > 0 \text{ on } \partial D.$$

Therefore Theorem 23.6 partially solves Problem 23.2, but here the function f is assumed to satisfy the elliptic equation (23.13) instead of being harmonic.

23.3 The generalized $\overline{\partial}$-Neumann problem

Suppose given on the boundary of a bounded domain D a vector field w satisfying condition (23.2) and a function ψ.

Problem 23.5. *Find a harmonic function F in D such that $w(F) = \psi$ on ∂D.*

This problem is analogous to the oblique derivative problem for real-valued harmonic functions. What is required is to determine necessary conditions for solvability of Problem 23.5 and to construct a solution of this problem given by an integral representation, for example, in the ball.

Problem 23.6. *Let L be an elliptic operator of the form (23.13) in a domain D, and suppose given a function ψ on ∂D. Find a solution of the boundary-value problem $\sum_{j,k} a_{j,k}\rho_j(\partial F/\partial \bar{z}_k) = \psi$ on ∂D for a function F satisfying the equation $LF = 0$ in D.*

Problems of this type are not Noetherian or Fredholm because, for example, the set of solutions of the homogeneous problem is infinite-dimensional.

A necessary condition for solvability of Problem 23.6 is that ψ be orthogonal to the holomorphic functions. Indeed, if $f \in \mathcal{O}(\overline{D})$, then

$$\begin{aligned}\int_{\partial D} \psi \bar{f}\, d\sigma &= \int_{\partial D} \sum_{j,k} a_{j,k} \frac{\partial F}{\partial \bar{z}_k} \rho_j \bar{f}\, d\sigma = C \int_{\partial D} \bar{f} \sum_{j,k} a_{j,k} \frac{\partial F}{\partial \bar{z}_k} (-1)^{j-1}\, dz[j] \wedge d\bar{z} \\ &= C \int_{\partial D} \bar{f} d\Big(\sum_{j,k} a_{j,k} \frac{\partial F}{\partial \bar{z}_k} (-1)^{j-1}\, dz[j] \wedge d\bar{z} \Big) = C \int_D \bar{f} LF\, dz \wedge d\bar{z} = 0.\end{aligned}$$

Apparently this condition is also sufficient (for strongly pseudoconvex domains).

In §19, we constructed an integral representation for the solution of the $\overline{\partial}$-Neumann problem in the ball.

Problem 23.7. *Give an integral representation for the solution of Problem (18.4) for forms in the ball.*

We recall that Problem (18.4) consists in finding a form α of type (p,q) in $\overline{D}$ for which $\overline{\partial}^* \overline{\partial} \alpha = 0$ in D and $(\overline{\partial}\alpha)_n = \varphi_n$ on ∂D, where φ is a given form of type $(p, q+1)$ on ∂D.

The Koppelman integral representation lets us find a form α in the ball in terms of its values on ∂D and the values of α and $\overline{\partial}\alpha$ in D. To construct an integral representation for the solution of Problem (18.4), we still need an analogue of the Szegő projection in the ball for forms.

Problem 23.8. *Construct an integral representation for the solution of the $\overline{\partial}$-Neumann problem (14.1) in domains other than the ball.*

23.4 The general form of integral representations in $\mathbf{C}^2$

For completeness, we give a theorem of Dautov from [7, §13].

Theorem 23.7 (Dautov). *Let D be a bounded domain in $\mathbf{C}^2$ with boundary of class $\mathcal{C}^\infty$. Suppose $\mu_z(\zeta)$ is a form of type 3 in ζ which has coefficients of class $\mathcal{C}^\infty(\partial D)$ in ζ and which is a reproducing kernel, that is, for $z \in D$,*

$$f(z) = \int_{\partial D} f(\zeta)\, \mu_z(\zeta)$$

for all $f \in \mathcal{O}(\overline{D})$. In order that each reproducing kernel $\mu_z(\zeta)$ be a Cauchy-Fantappiè kernel, it is necessary and sufficient that D be a domain of holomorphy.

Proof. Suppose D is a domain of holomorphy, and $\mu_z(\zeta)$ is a reproducing kernel for the point $z \in D$. For every point $\zeta \in \partial D$, there exists a neighborhood $B(\zeta, r)$ such that on $\partial D \cap B(\zeta, r)$ all differentials can be expressed in terms of the differentials dz_1, dz_2, $d\bar{z}_1$, or dz_1, dz_2, $d\bar{z}_2$. The sets $B(\zeta, r)$ cover ∂D. We choose a finite subcover and denote its elements by B_1, B_2, $\dots$, B_m. Let $\{\varphi_j\}$ be a partition of unity subordinate to this covering $\{B_j\}$. Then $\mu_z = \mu_1 + \cdots + \mu_m$, where $\mu_j = \varphi_j \mu_z$. In view of the choice of B_j, we have for $\zeta \in B_j \cap \partial D$ that

$$\mu_j = a_j\, d\bar{\zeta}_1 \wedge d\zeta + b_j\, d\bar{\zeta}_2 \wedge d\zeta,$$

where $a_j, b_j \in \mathcal{D}(B_j)$. Therefore this equation holds on all of ∂D, and μ_z has type $(2,1)$ on ∂D.

Consider the form $\alpha_z(\zeta) = \mu_z(\zeta) - U(\zeta, z)$. Since $U(\zeta, z)$ is a reproducing kernel for the point z, we have $\int_{\partial D} f(\zeta)\, \alpha_z(\zeta) = 0$ for all $f \in \mathcal{O}(\overline{D})$.

By Theorem 18.1, we have $\alpha_z\big|_{\partial D} = \overline{\partial}\gamma_z\big|_{\partial D}$, where the form $\gamma_z = \psi_z\, d\zeta$ of type $(2,0)$ has coefficients of class $\mathcal{C}^\infty(\overline{D})$. We set

$$\eta(\zeta, z) = (\eta_1, \eta_2) = \left(\frac{\bar{\zeta}_1 - \bar{z}_1}{|\zeta - z|^4} - (2\pi i)^2 \psi_z (\zeta_2 - z_2), \frac{\bar{\zeta}_2 - \bar{z}_2}{|\zeta - z|^4} + (2\pi i)^2 \psi_z (\zeta_1 - z_1) \right).$$

Then $\langle \eta, \zeta - z \rangle = 1$ and $\eta \in \mathcal{C}^\infty(\partial D)$ in ζ.

Consider the Cauchy-Fantappiè kernel

$$\omega'(\eta) = \sum_{k=1}^{2} (-1)^{k-1} \eta_k\, d\eta[k].$$

Since $\langle \eta, \zeta - z \rangle \equiv 1$, we have

$$\overline{\partial}_\zeta \eta_1 (\zeta_1 - z_1) + \overline{\partial}_\zeta \eta_2 (\zeta_2 - z_2) = 0,$$

so (for $\zeta_1 \neq z_1$)

$$\begin{aligned} \omega'(\eta) \wedge d\zeta &= \eta_1 \overline{\partial}\eta_2 \wedge d\zeta - \eta_2 \overline{\partial}\eta_1 \wedge d\zeta \\ &= \eta_1 \overline{\partial}\eta_2 \wedge d\zeta + \frac{\eta_2}{\zeta_1 - z_1} \overline{\partial}\eta_2 (\zeta_2 - z_2) \wedge d\zeta = \frac{\overline{\partial}\eta_2 \wedge d\zeta}{\zeta_1 - z_1}. \end{aligned}$$

Then

$$\frac{1}{(2\pi i)^2}\,\omega'(\eta)\wedge d\zeta = \frac{1}{(2\pi i)^2}\frac{1}{(\zeta_1 - z_1)}\,\overline{\partial}_\zeta \frac{\bar\zeta_2 - \bar z_2}{|\zeta - z|^2}\wedge d\zeta + \overline{\partial}_\zeta \psi_z(\zeta)\wedge d\zeta$$
$$= U(\zeta, z) + \overline{\partial}\gamma_z = \mu_z(\zeta).$$

By continuity, this equality holds also when $\zeta_1 = z_1$.

We now prove the sufficiency. If η and v are two vector-valued functions such that $\langle \eta, \zeta - z\rangle = 1$ and $\langle v, \zeta - z\rangle = 1$, then the difference of the Cauchy-Fantappiè kernels $\omega'(\eta)\wedge d\zeta - \omega'(v)\wedge d\zeta$ is a $\overline{\partial}$-exact form, since

$$\omega'(\eta)\wedge d\zeta - \omega'(v)\wedge d\zeta = -\overline{\partial}(\eta_1 v_2 - \eta_2 v_1)\wedge d\zeta.$$

Then we obtain from the hypothesis of the theorem that every form $\alpha_z(\zeta)$ of type $(2,1)$ that is orthogonal to holomorphic functions is $\overline{\partial}$-exact.

First we show that ∂D is connected. If ∂D is not connected, we consider a point z^0 lying in a bounded component of the complement $\mathbf{C}^n \setminus \overline{D}$, and let Γ be the boundary of this component. By the Bochner-Martinelli formula,

$$\int_{\partial D} f(\zeta)\, U(\zeta, z^0) = 0$$

for $f \in \mathcal{O}(\overline{D})$, so by the hypothesis of the theorem $U(\zeta, z^0)|_{\partial D} = \overline{\partial}\gamma|_{\partial D}$, where γ is a form of type $(2,0)$. But

$$1 = \int_\Gamma U(\zeta, z^0) = \int_\Gamma \overline{\partial}\gamma = 0.$$

Contradiction.

It remains to prove that D is pseudoconvex. Suppose D is not a domain of holomorphy. Then there exists a point z^0 at which the Levi form is negative definite. Consequently, there is a neighborhood $B(z^0, r)$ of z_0 such that the set $\{\, z \in B(z^0, r) : F(z, z^0) = 0\,\} \setminus \{z^0\} \subset D$, where

$$F(z, z^0) = \sum_{j=1}^{2} \frac{\partial \rho}{\partial z_j}(z^0)(z_j - z_j^0) + \sum_{j,k=1}^{2} \frac{\partial^2 \rho(z^0)}{\partial z_j \partial z_k}(z_j - z_j^0)(z_k - z_k^0)$$

(ρ being a defining function for D).

Suppose $\varphi \in \mathcal{D}(\mathbf{C}^2)$, $\operatorname{supp}\varphi \subset B(z^0, r)$, and $\varphi \equiv 1$ in a neighborhood of z^0. We set $h(z) = \varphi(z)F(z, z^0)^{-1}$ for $z \in \mathbf{C}^2 \setminus D$. Then the form $\overline{\partial}h$ has coefficients of class $\mathcal{C}^\infty(\mathbf{C}^2 \setminus D)$, since $\overline{\partial}h = 0$ outside $B(z^0, r)$, and $\overline{\partial}h = \overline{\partial}\varphi F(z, z^0)^{-1} = 0$ in a neighborhood of z^0. Moreover, if $\alpha = \overline{\partial}h \wedge dz$, then $\int_{\partial D} f\alpha = 0$ for $f \in \mathcal{O}(\overline{D})$. Indeed,

$$\int_{\partial D} f\,\alpha = \int_{\partial D} f\,\overline{\partial}h \wedge dz = \int_{\partial D_{-\epsilon}} f\,\overline{\partial}h \wedge dz = \int_{\partial D_{-\epsilon}} d(fh)\wedge dz = 0,$$

where $D_{-\epsilon} = \{z : \rho(z) < \epsilon\}$ for $\epsilon > 0$. By the hypothesis of the theorem, there is a form $\widetilde{\alpha}$ of type $(2,1)$ with coefficients of class $\mathcal{C}^\infty(\overline{D})$ such that $\overline{\partial}\widetilde{\alpha} = 0$ on $\overline{D}$ and

$$\widetilde{\alpha}\big|_{\partial D} = \alpha\big|_{\partial D}. \tag{23.14}$$

By Koppelman's formula (1.9) applied to the domain $B(0,R) \setminus D$, where $B(0,R)$ is a ball of sufficiently large radius R, we have for $z \in B \setminus \overline{D}$ that

$$\overline{\partial}h \wedge dz = \int_{\partial(B\setminus\overline{D})} \overline{\partial}\eta \wedge d\zeta \wedge U_{2,1}(\zeta,z) - \overline{\partial}\int_{B\setminus\overline{D}} \overline{\partial}\eta \wedge d\zeta \wedge U_{2,0}(\zeta,z).$$

By the same formula applied to D, we have for $z \in B \setminus \overline{D}$ that

$$0 = \int_{\partial D} \widetilde{\alpha} \wedge U_{2,1}(\zeta,z) - \overline{\partial}\int_D \widetilde{\alpha} \wedge U_{2,0}(\zeta,z).$$

Adding these formulas and using (23.14), we obtain

$$\begin{aligned}\overline{\partial}h \wedge dz &= \overline{\partial}\left[-\int_{B\setminus\overline{D}} \overline{\partial}h \wedge d\zeta \wedge U_{2,0}(\zeta,z) - \int_D \widetilde{\alpha} \wedge U_{2,0}(\zeta,z)\right] \\ &= \overline{\partial}h' \wedge dz, \qquad z \in B(0,R) \setminus \overline{D}.\end{aligned}$$

The function h' is continuous in $\mathbf{C}^2$, since the integrand has an integrable singularity. The function $h - h'$ is holomorphic in $B(0,R)\setminus\overline{D}$, since $\overline{\partial}(h-h')\wedge dz = 0$. Therefore $h - h'$ extends holomorphically into D. Then h extends by continuity to z^0. On the other hand, $\lim_{z\to z^0} h(z) = \infty$. □

The necessity in the theorem was proved by Dautov for strongly pseudoconvex domains. But in view of Theorem 18.1, it also holds for every weakly pseudoconvex domain.

When $n > 2$, no such description of reproducing kernels has been obtained, so the following problem is open.

Problem 23.9. *Prove (or disprove) the sufficiency or necessity in Theorem 23.7 when $n > 2$.*

Theorem 18.3 shows that in strongly pseudoconvex domains D, every reproducing kernel $\mu_z(\zeta)$ has the form

$$\mu_z(\zeta) = U(\zeta,z) + \star\partial h_z,$$

where $h_z(\zeta)$ is a harmonic function in D of class $\mathcal{C}^\infty(\overline{D})$.

Chapter 6

Holomorphic Extension of Functions into a Fixed Domain

24 Holomorphic extension of hyperfunctions

24.1 Hyperfunctions as boundary values of harmonic functions

Polking and Wells [160] extended Theorem 6.1 on analytic representation to the class of CR-hyperfunctions. As a corollary, they obtained a Hartogs-Bochner theorem on holomorphic extension of hyperfunctions from the boundary of a domain into the domain. In this section, we prove a generalization of this result in the spirit of Theorems 15.1 and 17.2. All of these results are taken from [126].

First we show that the class of hyperfunctions on the boundary of a domain D can be identified with the class of harmonic functions in D.

Suppose that D is a bounded domain in $\mathbf{R}^m$, where $m > 2$, with connected, real-analytic boundary Γ. We consider the class $H(D)$ of harmonic functions in D with the topology of uniform convergence on compact subsets of D, and the class $H(\mathbf{R}^m \setminus D)$ of harmonic functions in $\mathbf{R}^m \setminus D$ that are equal to zero at infinity, with the inductive limit topology. The Grothendieck duality theorem (see [63], and also [100]) asserts that these spaces are mutually dual, and this duality is given by the nondegenerate pairing

$$(f, g)_H = \int_{\Gamma_\epsilon} \left(f \frac{\partial g}{\partial \nu} - g \frac{\partial f}{\partial \nu} \right) d\sigma, \tag{24.1}$$

where $f \in H(D)$, $g \in H(\mathbf{R}^m \setminus D)$, the domain $D = \{ x : \rho(x) < 0 \}$, the function ρ is a real-valued real-analytic function in $\overline{D}$ with $d\rho \neq 0$ on Γ, the surface $\Gamma_\epsilon = \{ x : \rho(x) = -\epsilon \}$, where $\epsilon > 0$ is small enough that f and g are harmonic in a neighborhood of Γ_ϵ, $\partial/\partial\nu$ is, as usual, the outer normal derivative to Γ_ϵ, and $d\sigma$ is the surface area element of Γ_ϵ.

The same assertion holds for the spaces $H(\overline{D})$ and $H(\mathbf{R}^m \setminus \overline{D})$.

We will view the space $\mathbf{R}^m$ as embedded in $\mathbf{C}^m$ in the following way: $x_j = \operatorname{Re} z_j$, where $x = (x_1, \dots, x_m) \in \mathbf{R}^m$, and $z = (z_1, \dots, z_m) \in \mathbf{C}^m$. Then Γ is a compact set in $\mathbf{C}^m$. We consider the class $B(\Gamma)$ of hyperfunctions with support in Γ. Since Γ is compact, $B(\Gamma)$ can be identified with the space of analytic functionals $A'(\Gamma)$ (see, for example, [78, chap. 9] and [182, chap. 1]). We recall some properties of this space.

Let $A(\Gamma)$ be the class of real-analytic functions on Γ (with respect to the intrinsic structure of the real-analytic manifold Γ). Then each function $f \in A(\Gamma)$ extends holomorphically into some complex neighborhood of Γ. A linear functional T on $A(\Gamma)$ is an analytic functional (that is, $T \in A'(\Gamma)$) if for every complex neighborhood U of Γ there exists a constant $c(U)$ such that

$$|T(f)| \le c(U) \sup_U |f| \tag{24.2}$$

for every entire function f in $\mathbf{C}^m$ (see, for example, [78, chap. 9]). In this definition, we may take holomorphic functions on the closure $\overline{U}$ instead of entire functions (see [78]).

Property (24.2) can be expressed in terms of the function $f \in A(\Gamma)$ and its derivatives on Γ rather than in terms of a holomorphic extension of f into a neighborhood of Γ. Namely, we consider the following expression:

$$\ell(f) = \limsup_{\|\alpha\| \to \infty} \sqrt[\|\alpha\|]{\frac{\sup_\Gamma |D^\alpha f|}{\alpha!}},$$

where $D^\alpha f$ is the derivative of f of order $\alpha = (\alpha_1, \dots, \alpha_m)$, $\|\alpha\| = \alpha_1 + \dots + \alpha_m$, and $\alpha! = \alpha_1! \dots \alpha_m!$. Then $f \in C^\infty(\Gamma)$ is a function in $A(\Gamma)$ if and only if $\ell(f) < +\infty$, and $1/\ell(f)$ is none other than the radius of the polydisc $U(x^0, r)$ with center at x^0 into which f extends holomorphically. The union of all such polydiscs $U(x^0, r)$ for $x^0 \in \Gamma$ gives a complex neighborhood U into which f extends holomorphically.

A sequence of functions f_n converges to f in $A(\Gamma)$ as $n \to \infty$ if f_n converges uniformly with all its derivatives to f on Γ, and all the numbers $\ell(f_n)$ are uniformly bounded (this means exactly that the functions f_n extend holomorphically into the same complex neighborhood U of Γ, and in this neighborhood they converge uniformly to f). Then if $T \in A'(\Gamma)$, we have $T(f_n) \to T(f)$ when $n \to \infty$.

Subsequently we will be interested only in convergence of sequences in $A(\Gamma)$ and weak convergence of functionals in $A'(\Gamma)$.

We may identify the space $A(\Gamma)$ with the space $H(\overline{D})$, or with the space $H(\mathbf{R}^m \setminus D)$, for every function $f \in A(\Gamma)$ can be extended harmonically into D as a function F. This F is a function of class C^∞ up to Γ (see, for example, [29, chap. 1]), and since it is real-analytic on Γ, we have $F \in H(\overline{D})$. The maximum modulus theorem shows that the topologies of $A(\Gamma)$ and $H(\overline{D})$ also coincide. The same argument holds for the spaces $A(\Gamma)$ and $H(\mathbf{R}^m \setminus D)$.

This observation, together with Grothendieck's theorem, shows that the pairing (24.1) gives a functional $T \in A'(\Gamma)$, that is, every function $f \in H(D)$ defines an analytic functional T. We will show that the converse also holds: every analytic functional determines a harmonic function in D.

Suppose $g(x,y)$ is the fundamental solution to Laplace's equation in $\mathbf{R}^m$, that is, $\delta_y = \Delta_x g(x,y)$, where δ_y is the Dirac delta functionat y. We recall that

$$g(x,y) = \frac{c_m}{|x-y|^{m-2}} \qquad \text{if } m > 2.$$

We denote by $V(f)$ the single-layer potential for the function $f \in A(\Gamma)$:

$$V(f)(y) = \int_\Gamma g(x,y) f(x)\, d\sigma(x).$$

We will write $V^+(f)(y)$ if $y \in D$, and $V^-(f)(y)$ if $y \in \mathbf{R}^m \setminus \overline{D}$. We will need the following simple result.

Lemma 24.1. *The function $V^+(f)$ is a function of class $H(\overline{D})$, the function $V^-(f)$ is a function of class $H(\mathbf{R}^m \setminus D)$, and $V^+(f)\big|_\Gamma = V^-(f)\big|_\Gamma \in A(\Gamma)$. Moreover*

$$\ell(V(f)) \le \ell(f). \tag{24.3}$$

Proof. Suppose $x^0 \in \Gamma$, and $U(x^0,r)$ is a polydisc with center x^0 and radius r into which f extends holomorphically. Let $U = \bigcup_{x^0\in\Gamma} U(x^0,r)$. Then we have for f the Cauchy inequalities

$$\left|\frac{D^\alpha f(x^0)}{\alpha!}\right| \le \sup_{x\in U(x^0,r)} |f(x)|/r^{\|\alpha\|} \le \sup_U |f|/r^{\|\alpha\|} = c/r^{\|\alpha\|}, \tag{24.4}$$

so that $\ell(f) \le 1/r$.

The radius r may be taken small enough that there exists a real-analytic diffeomorphism ψ mapping some domain $B \subset \mathbf{R}^{m-1}$ onto $U(x^0,r) \cap \Gamma$. Then

$$\begin{aligned} V(f)(x^0) &= \int_{U(x^0,r)\cap\Gamma} g(x,x^0) f(x)\, d\sigma + \int_{\Gamma\setminus U(x^0,r)} g(x,x^0) f(x)\, d\sigma \\ &= V^1(x^0) + V^2(x^0). \end{aligned}$$

In the first integral, we make the change of variables $x - x^0 = \psi(x')$ and $x^0 = \psi(y')$, where x' and y' are in B. Then by using (24.4), we have

$$\begin{aligned} \left|\frac{D^\alpha V^1(x^0)}{\alpha!}\right| &\le \int_B \left|\frac{D^\alpha f(\psi(y') + \psi(x'))}{\alpha!}\right| |g(\psi(x'))J|\, dx' \\ &\le \frac{c}{r^{\|\alpha\|}} \int_B |g(\psi(x'))J|\, dx', \end{aligned}$$

where J is the Jacobian of the transformation. Thus

$$\left|\frac{D^\alpha V^1}{\alpha!}\right| \le c \cdot \frac{c_1}{r^{\|\alpha\|}}. \tag{24.5}$$

For the integral V^2, we have

$$\left|\frac{D^\alpha V^2(x^0)}{\alpha!}\right| \le \int_{\Gamma\setminus U(x^0,r)} |f(y)| \left|\frac{D^\alpha g(x^0,y)}{\alpha!}\right| d\sigma$$
$$\le c_2 \max_\Gamma |f| \frac{(\|\alpha\|+m-2)!}{\alpha!\,(m-2)!}\frac{1}{r^{\|\alpha\|+m-2}}. \qquad (24.6)$$

It follows from (24.5) and (24.6) that $\ell(V(f)) \le \ell(f)$. □

Suppose $T \in A'(\Gamma)$, and consider the expression $V(T)(y) = T_x(g(x,y))$. Since $g(x,y) \in A(\Gamma)$ for $y \in \Gamma$, the function $V(T)(y)$ is defined outside Γ. Moreover, it is clear that $V(T)$ is a harmonic function outside Γ that equals zero at infinity. We will call it the single-layer potential for the hyperfunction T.

Suppose $\varphi \in A(\Gamma)$, and consider its harmonic extension φ^- into $\mathbf{R}^m \setminus \overline{D}$ that equals zero at infinity. We have already observed that $\varphi^- \in H(\mathbf{R}^m \setminus D)$. Suppose $\epsilon > 0$ is small enough that φ^- and $V^+(T)$ are harmonic in a neighborhood of Γ_ϵ. We take a function $\chi \in C^\infty(\mathbf{R}^m)$ that is equal to one in a neighborhood of $\mathbf{R}^m \setminus D$ and equal to zero in a neighborhood of the set in D where φ^- is not defined. Then $\varphi^-\chi$ belongs to $C^\infty(\mathbf{R}^m)$, and it equals zero at infinity.

Lemma 24.2. *Under the above hypotheses,*

$$T(\varphi) = \int V^+(T)\Delta(\chi\varphi^-)\,dy = \int_{\Gamma_\epsilon}\left(V^+(T)\frac{\partial\varphi^-}{\partial\nu} - \varphi^-\frac{\partial V^+(T)}{\partial\nu}\right) d\sigma. \qquad (24.7)$$

Proof. The support of the function $\Delta(\chi\varphi^-)$ lies in D, so the integral in (24.7) is defined. Consider the integral

$$I = \int g(x,y)\Delta(\chi\varphi^-)(y)\,dy, \qquad dy = dy_1 \wedge \cdots \wedge dy_m.$$

Since $g(x,y)$ is the fundamental solution for the Laplace operator, we have by Stokes's formula that

$$\int g(x,y)\Delta(\chi\varphi^-)(y)\,dy = \langle \Delta_y g(x,y), \chi\varphi^-\rangle = \langle \delta_x, \chi\varphi^-\rangle = \varphi(x)$$

if $x \in \Gamma$ and $y \in \operatorname{supp}\Delta(\chi\varphi^-)$. In an obvious way, the integral I is a holomorphic function in a complex neighborhood U of the compact set Γ. Representing it as the limit of Riemann sums, we see that these sums converge to I uniformly in U, so

$$T_x(I) = \int T_x(g(x,y))\Delta(\chi\varphi^-)\,dy = \int V^+(T)\Delta(\chi\varphi^-)\,dy.$$

If we take χ so that $\chi = 1$ on Γ_ϵ and use the usual Green's formula, we obtain

$$\int V^+(T)\Delta(\chi\varphi^-)\,dy = \int_{\Gamma_\epsilon}\left(V^+(T)\frac{\partial\varphi^-}{\partial\nu} - \varphi^-\frac{\partial V^+(T)}{\partial\nu}\right) d\sigma.$$

□

The same argument gives a representation for $T(\varphi)$ outside $\overline{D}$:

$$T(\varphi) = \int V^-(T)\Delta(\chi\varphi^+)\,dy = \int_{\Gamma_{-\epsilon}} \left(V^-(T)\frac{\partial \varphi^+}{\partial \nu} - \varphi^+ \frac{\partial V^-(T)}{\partial \nu} \right) d\sigma, \tag{24.8}$$

where φ^+ is the harmonic extension of φ into D, which belongs to $H(\overline{D})$, the function $\chi \in C^\infty(\mathbf{R}^m)$ equals 1 in a neighborhood of $\overline{D}$ and equals 0 where φ^+ is not defined, and $\Gamma_{-\epsilon} = \{x : \rho(x) = \epsilon\}$, $\epsilon > 0$.

Now let $P(x, y)$ be the Poisson kernel for D. Then $P(x, y) = \partial G(x, y)/\partial \nu$, where $G(x, y)$ is the Green function for D, $x \in \Gamma$, $y \in D$, and the derivative $\partial/\partial \nu$ is taken in the variable x. Since Γ is real-analytic, $P(x, y)$ is a real-analytic function on Γ for fixed $y \in D$. We define a function $\widetilde{T}$ via $\widetilde{T}(y) = T_x(P(x, y))$, where $T \in A'(\Gamma)$. Then $\widetilde{T} \in H(D)$.

Theorem 24.3 (Kytmanov, Yakimenko). *We have the representation*

$$T(\varphi) = \int \widetilde{T}\Delta(\chi V^-(\varphi))\,dy = \int_{\Gamma_\epsilon} \left(\widetilde{T}\frac{\partial V^-(\varphi)}{\partial \nu} - V^-(\varphi)\frac{\partial \widetilde{T}}{\partial \nu} \right) d\sigma, \tag{24.9}$$

where $\varphi \in A(\Gamma)$, $V^-(\varphi) \in H(\mathbf{R}^m \setminus D)$, and Γ_ϵ and the function χ are the same as in Lemma 24.2.

Proof. It suffices to show that

$$\begin{aligned} \varphi(x) &= \int P(x, y)\Delta(\chi V^-(\varphi))(y)\,dy \\ &= \int_{\Gamma_\epsilon} \left(P(x, y)\frac{\partial V^-(\varphi)}{\partial \nu}(y) - V^-(\varphi)\frac{\partial P}{\partial \nu_y}(x, y) \right) d\sigma(y), \qquad x \in \Gamma. \end{aligned}$$

As remarked in [199, Proposition 1.6], the kernel $P(x, y)$ has finite order of growth near Γ, that is,

$$|P(x, y)| \le C/\operatorname{dist}^k(y, \Gamma)$$

for some positive constants C and k when y is sufficiently close to Γ and x is fixed. Then, using the equality

$$\int_{\Gamma_\epsilon} \left(P(x, y)\frac{\partial V^+(\varphi)}{\partial \nu} - \frac{\partial P(x, y)}{\partial \nu_y} V^+(\varphi) \right) d\sigma(y) = 0,$$

the properties of harmonic functions of finite order of growth (see Chap. 3), and the jump theorem for the normal derivative of the single-layer potential of φ, we have that

$$\int P(x, y)\Delta(\chi V^-(\varphi))\,dy = \lim_{\epsilon \to 0+} \int_{\Gamma_\epsilon} P(x, y)\varphi^+(y)\,d\sigma(y).$$

It remains to use Proposition 1.6 from [199] to obtain the required equality. □

Formula (24.9) shows that the boundary value of an arbitrary harmonic function $f \in H(D)$ can be suitably defined not by using (24.1), but in the following way: we regard the boundary value $bf \in A'(\Gamma)$ of $f \in H(D)$ as the expression

$$bf(\varphi) = \int f\Delta(\chi V^-(\varphi))\,dy = \int_{\Gamma_\epsilon} \left(f\frac{\partial V^-(\varphi)}{\partial \nu} - V^-(\varphi)\frac{\partial f}{\partial \nu} \right) d\sigma, \tag{24.10}$$

where $\varphi \in A(\Gamma)$, and χ and Γ_ϵ are the same as in (24.9).

Formula (24.10) is equivalent to (24.1), since the operator V is one-to-one and continuous in $A(\Gamma)$, as follows from Lemma 24.1 and the jump formula for the normal derivative of the single-layer potential.

Thus we can identify the space $A'(\Gamma)$ with $H(D)$ via formulas (24.9) and (24.10) and Grothendieck's theorem. In the case when D is a half-space in $\mathbf{R}^m$, this assertion was proved by Hörmander [78, chap. 9].

If T is a distribution on Γ, then it follows from (24.9) and (24.10) in the usual way that

$$T(\varphi) = \lim_{\epsilon \to 0+} \int_{\Gamma_\epsilon} \widetilde{T}(y)\varphi^+(y)\,d\sigma. \tag{24.11}$$

Indeed, by Green's formula

$$T(\varphi) = \int_{\Gamma_\epsilon} \left[\widetilde{T}\left(\frac{\partial V^-(\varphi)}{\partial \nu} - \frac{\partial V^+(\varphi)}{\partial \nu} \right) - \frac{\partial \widetilde{T}}{\partial \nu}\left(V^-(\varphi) - V^+(\varphi)\right) \right] d\sigma.$$

From the properties of harmonic functions of finite order of growth (see §11), we have

$$\int_{\Gamma_\epsilon} \frac{\partial \widetilde{T}}{\partial \nu}\left(V^-(\varphi) - V^+(\varphi)\right) \to 0 \qquad \text{when } \epsilon \to 0^+, \quad \text{and}$$

$$\lim_{\epsilon \to 0+} \int_{\Gamma_\epsilon} \widetilde{T}\left(\frac{\partial V^-(\varphi)}{\partial \nu} - \frac{\partial V^+(\varphi)}{\partial \nu} \right) d\sigma = \lim_{\epsilon \to 0+} \int_{\Gamma_\epsilon} \widetilde{T}\varphi^+\,d\sigma.$$

This proves (24.11).

Subsequently we shall need a formula for the jump of the normal derivative of the single-layer potential of a hyperfunction.

Theorem 24.4 (Kytmanov, Yakimenko). *If $T \in A'(\Gamma)$, then*

$$T = \frac{\partial V^-(T)}{\partial \nu} - \frac{\partial V^+(T)}{\partial \nu} \qquad \text{on } \Gamma.$$

Here the normal derivative of a harmonic function $f \in H(D)$ is defined as follows: first compute the derivatives $\partial f/\partial x_j$, then take their boundary values $b(\partial f/\partial x_j)$ by (24.10); then

$$\left.\frac{\partial f}{\partial \nu}\right|_\Gamma = \sum_{j=1}^m \nu_j b\frac{\partial f}{\partial x_j}, \qquad \nu = (\nu_1, \dots, \nu_m).$$

The normal derivative of $f \in H(\mathbf{R}^m \setminus \overline{D})$ is defined analogously.

Proof. First we show that

$$\frac{\partial f}{\partial \nu}(\varphi) = bf\left(\frac{\partial \varphi^+}{\partial \nu}\right), \tag{24.12}$$

if $f \in H(D)$, where φ^+ is the harmonic extension of $\varphi \in A(\Gamma)$ into D. By (24.10),

$$bf\left(\frac{\partial \varphi^+}{\partial \nu}\right) = \int f\Delta\left(\chi V^-\left(\frac{\partial \varphi^+}{\partial \nu}\right)\right) dy, \qquad \text{while}$$

$$V^-\left(\frac{\partial \varphi^+}{\partial \nu}\right)(y) = \int_\Gamma g(x,y)\frac{\partial \varphi^+}{\partial \nu}(x)\, d\sigma(x), \qquad y \in \mathbf{R}^m \setminus \overline{D}.$$

By Green's formula,

$$\int_\Gamma g(x,y)\frac{\partial \varphi^+}{\partial \nu}(x)\, d\sigma = \int_\Gamma \varphi(x)\frac{\partial g}{\partial \nu_x}(x,y)\, d\sigma.$$

On the other hand,

$$\begin{aligned}\frac{\partial f}{\partial \nu}(\varphi) &= \sum_{j=1}^m b\frac{\partial f}{\partial x_j}(\nu_j\varphi) = \sum_{j=1}^m \int \frac{\partial f}{\partial y_j}\Delta\left(\chi V^-(\nu_j\varphi)\right)(y)\, dy \\ &= -\sum_{j=1}^m \int f\Delta\frac{\partial}{\partial y_j}\left(\chi V^-(\nu_j\varphi)\right) dy = -\sum_{j=1}^m \int f\Delta\left[\chi\frac{\partial}{\partial y_j}V^-(\nu_j\varphi)\right] dy.\end{aligned}$$

The integral

$$\int f\Delta\left[\frac{\partial \chi}{\partial y_j}V^-(\nu_j\varphi)\right] dy = 0$$

by Green's formula, while

$$\begin{aligned}\sum_{j=1}^m \frac{\partial}{\partial y_j}V^-(\nu_j\varphi) &= \sum_{j=1}^m \frac{\partial}{\partial y_j}\int_\Gamma g(x,y)\nu_j(x)\varphi(x)\, d\sigma \\ &= -\sum_{j=1}^m \int_\Gamma \frac{\partial g}{\partial x_j}\nu_j\varphi\, d\sigma = -\int_\Gamma \varphi(x)\frac{\partial g(x,y)}{\partial \nu_x}\, d\sigma.\end{aligned}$$

Formula (24.12) is proved. Moreover, by (24.10) we have

$$(bV^+(T))(\varphi) = \int V^+(T)\Delta(\chi V^-(\varphi))\, dy.$$

Comparing this expression with (24.7), we obtain that $[bV^+(T)](\varphi) = T(V^-(\varphi))$. In the same way, using (24.8), we find $[bV^-(T)](\varphi) = T(V^+(\varphi))$. Since $V^+(\varphi) = V^-(\varphi)$ on Γ, we have $bV^+(T) = bV^-(T)$. By (24.12) and (24.10), we have

$$\frac{\partial V^+}{\partial \nu}(\varphi) = bV^+\left(\frac{\partial \varphi^+}{\partial \nu}\right)$$
$$= \int_{\Gamma_\epsilon}\left[V^+(T)\frac{\partial}{\partial \nu}V^-\left(\frac{\partial \varphi^+}{\partial \nu}\right) - \frac{\partial V^+(T)}{\partial \nu}V^-\left(\frac{\partial \varphi^+}{\partial \nu}\right)\right] d\sigma.$$

We have already seen that

$$V^-\left(\frac{\partial \varphi^+}{\partial \nu}\right) = \int_\Gamma \varphi\frac{\partial g}{\partial \nu}\,d\sigma = S^-(\varphi),$$

where S is the double-layer potential of $\varphi \in A(\Gamma)$.

Lemma 24.1 shows that $S^-(\varphi) \in H(\mathbf{R}^m \setminus D)$, while $S^+(\varphi) \in H(\overline{D})$. Thus

$$\frac{\partial V^+(T)}{\partial \nu}(\varphi) = \int_{\Gamma_\epsilon}\left[V^+(T)\frac{\partial S^-(\varphi)}{\partial \nu} - \frac{\partial V^+(T)}{\partial \nu}S^-(\varphi)\right] d\sigma,$$

so by (24.7) we have

$$\frac{\partial V^+(T)}{\partial \nu}(\varphi) = T(S^-(\varphi)).$$

In the same way,

$$\frac{\partial V^-(T)}{\partial \nu}(\varphi) = T(S^+(\varphi)).$$

Using the jump formula for the double-layer potential, we have the required result

$$\frac{\partial V^-(T)}{\partial \nu}(\varphi) - \frac{\partial V^+(T)}{\partial \nu}(\varphi) = T(S^+(\varphi) - S^-(\varphi)) = T(\varphi).$$

□

Similarly, we can prove a jump theorem for the double-layer potential of hyperfunctions.

Theorem 24.4 shows that the operator V is one-to-one in $A'(\Gamma)$.

24.2 Holomorphic extension of hyperfunctions into a domain

Now suppose that the domain D with connected real-analytic boundary Γ lies in $\mathbf{C}^n$, where $n > 1$. If D has the form $\{\, z : \rho(z) < 0 \,\}$, where ρ is a real-valued function of class $A(\overline{D})$, and $d\rho \neq 0$ on $\Gamma = \partial D$, then by Lemma 3.5 we have

$$U(\zeta, z)\big|_\Gamma = M(\zeta, z)\,d\sigma(\zeta), \quad \text{where} \quad M(\zeta, z) = \frac{(n-1)!}{2\pi^n}\sum_{k=1}^n \frac{\bar{\zeta}_k - \bar{z}_k}{|\zeta - z|^{2n}}\rho_{\bar{k}}(\zeta).$$

Since ρ_k and $\rho_{\bar{k}}$ are real-analytic functions, we can define the Bochner-Martinelli transform of an analytic functional T via

$$M(T)(z) = T_\zeta(M(\zeta, z)).$$

The function $M(T)$ is harmonic outside Γ and equals zero at infinity. We are interested in the following question: if $M^+(T) = \widetilde{T}$ (or, what is the same thing, $bM^+(T) = T$), will $\widetilde{T}$ be holomorphic in D? (In other words, will T be a CR-hyperfunction on Γ? See the theorem of Polking and Wells in [160].) This problem is a generalization of the problem from §14 about functions represented by the Bochner-Martinelli integral.

We recall that the operators $\overline{\partial}_n$, $\overline{\partial}_{-n}$, ∂_n, ∂_{-n} have the following form (see §16):

$$\overline{\partial}_n \varphi = \frac{1}{2}\left(\frac{\partial \varphi^+}{\partial \nu} + i\frac{\partial \varphi^+}{\partial \tau}\right),$$

where $\varphi \in A(\Gamma)$, ν is the outer unit normal to Γ, as before, and $\tau = i\nu$;

$$\partial_n \varphi = \frac{1}{2}\left(\frac{\partial \varphi^+}{\partial \nu} - i\frac{\partial \varphi^+}{\partial \tau}\right), \quad \overline{\partial}_{-n} \varphi = -\frac{1}{2}\left(\frac{\partial \varphi^-}{\partial \nu} + i\frac{\partial \varphi^-}{\partial \tau}\right),$$

and finally

$$\partial_{-n} \varphi = -\frac{1}{2}\left(\frac{\partial \varphi^-}{\partial \nu} - i\frac{\partial \varphi^-}{\partial \tau}\right),$$

where φ^+ is the harmonic extension of φ into D, and φ^- is the harmonic extension of φ into $\mathbf{C}^n \setminus \overline{D}$ that equals zero at infinity.

Also, just as we defined the operator $\partial/\partial\nu$ on functionals $T \in A'(\Gamma)$, we may define the operators $\overline{\partial}_n$, ∂_n, $\overline{\partial}_{-n}$, and ∂_{-n} on T. For example,

$$\overline{\partial}_n \varphi = \sum_{k=1}^n \frac{\partial \varphi^+}{\partial \bar{z}_k}\rho_k, \qquad \text{so} \qquad (\overline{\partial}_n T)(\varphi) = \sum_{k=1}^n \left(b\frac{\partial \widetilde{T}}{\partial \bar{z}_k}\right)(\rho_k \varphi).$$

Since

$$U(\zeta, z) = \sum_{k=1}^n (-1)^{k-1} \frac{\partial g(\zeta, z)}{\partial \zeta_k}\, d\bar{\zeta}[k] \wedge d\zeta,$$

we have $M(\zeta, z) = \partial_n g(\zeta, z)$ if $z \notin \overline{D}$, and $M(\zeta, z) = -\partial_{-n} g(\zeta, z)$ if $z \in D$.

We will show that a jump formula holds for the Bochner-Martinelli transform of a functional $T \in A'(\Gamma)$.

Theorem 24.5 (Kytmanov, Yakimenko). *If $T \in A'(\Gamma)$, then*

$$bM^+(T) - bM^-(T) = T \qquad \text{on } \Gamma.$$

Proof. First of all we note the analogues of formula (24.12), that is,

$$\begin{aligned}(\overline{\partial}_n T)(\varphi) &= T(\partial_n \varphi^+),\\ (\partial_n T)(\varphi) &= T(\overline{\partial}_n \varphi^+).\end{aligned} \tag{24.13}$$

Their proofs are the same as for (24.12), except that we need to use the complex analogue of Green's formula (for $z \notin \overline{D}$)

$$\int_\Gamma \partial_n g(\zeta, z)\varphi(\zeta)\, d\sigma = \int_\Gamma g(\zeta, z)\overline{\partial}_n \varphi^+(\zeta)\, d\sigma$$

and the conjugate formula

$$\int_\Gamma \overline{\partial}_n g(\zeta, z)\varphi(\zeta)\, d\sigma = \int_\Gamma g(\zeta, z)\partial_n \varphi^+(\zeta)\, d\sigma$$

(see Corollary 11.5).

Moreover, the complex analogue of Theorem 24.4 holds, that is,

$$\begin{aligned}T &= -\overline{\partial}_{-n} V(T) - \overline{\partial}_n V(T),\\ T &= -\partial_{-n} V(T) - \partial_n V(T).\end{aligned} \tag{24.14}$$

Then, using (24.7), (24.10), and (24.13), we have

$$\begin{aligned} bM^+(T)(\varphi) &= \int M^+(T)\Delta(\chi V^-(\varphi))\, dv(z)\\ &= -\sum_{k=1}^n \int \frac{\partial}{\partial z_k} T_\zeta\left(\rho_{\bar{k}} g(\zeta, z)\right)\Delta\left(\chi V^-(\varphi)\right)\, dv(z)\\ &= \sum_{k=1}^n \int V^+(\rho_{\bar{k}} T)\Delta\left(\chi \frac{\partial}{\partial \bar{z}_k} V^-(\varphi)\right)\, dv(z) = \sum_{k=1}^n \rho_{\bar{k}} T\left(\frac{\partial}{\partial z_k} V^-(\varphi)\right)\\ &= -T(\partial_{-n} V(\varphi)).\end{aligned}$$

In the same way, $bM^-(T)(\varphi) = T(\partial_n V(\varphi))$, and it remains to use formulas (24.14). □

In the course of the proof, we have showed that

$$\begin{aligned} bM^+(T)(\varphi) &= -T(\partial_{-n} V(\varphi)),\\ bM^-(T)(\varphi) &= T(\partial_n V(\varphi)).\end{aligned} \tag{24.15}$$

Theorem 24.6 (Kytmanov, Yakimenko). *In order that $bM^+(T) = T$ on Γ, it is necessary and sufficient that $\overline{\partial}_n T = 0$ on Γ.*

This is an analogue of Theorem 14.1.

Proof. From (24.13)–(24.15), we have

$$[bM^+(T)](\varphi) - T(\varphi) = T(-\partial_{-n}V(\varphi) - \varphi) = T(\partial_n V(\varphi)) = \overline{\partial}_n T(V(\varphi)) = 0$$

for all $\varphi \in A(\Gamma)$, but since V is a one-to-one operator, $\overline{\partial}_n T = 0$. The converse is obvious. □

Lemma 24.7. *If the domain D is such that the iterates $M^k\varphi$ converge in the metric of $C^\infty(\partial D)$ for every $\varphi \in C^\infty(\partial D)$, then the $M^k\psi$ converge in $A(\Gamma)$ to some function ψ_0 of class $\mathcal{O}(\overline{D})$ for every $\psi \in A(\Gamma)$.*

Proof. By Lemma 24.1, we have $M\psi \in H(\overline{D})$, and moreover $\ell(M(\psi)) \leq \ell(\psi)$, so we can apply the operator M again to consider $M^2\psi$, and so on. Lemma 24.1 shows that

$$\ell(M^k(\psi)) \leq \ell(\psi). \tag{24.16}$$

From (24.16), the hypothesis of the lemma, and the definition of convergence in $A(\Gamma)$, we find that $\psi_0 \in \mathcal{O}(\overline{D})$, and $M^k\psi \to \psi_0$ in $A(\Gamma)$ as $k \to \infty$. □

Theorem 24.8 (Kytmanov, Yakimenko). *Suppose D satisfies the hypotheses of Theorem 17.1, and $T \in A'(\Gamma)$. Then the sequence $(bM^+)^kT$ converges weakly to some CR-hyperfunction S as $k \to \infty$.*

Proof. For simplicity, we denote bM^+ by M. Basically we need to repeat the proof of Theorem 17.1, using Theorems 24.5 and 24.6 and Lemma 24.7. We briefly sketch the proof.

Consider the bilinear form

$$B(T, \varphi) = V^{-1}[T(\bar{\varphi})] = T[V^{-1}(\bar{\varphi})],$$

where $T \in A'(\Gamma)$ and $\varphi \in A(\Gamma)$. It is easy to show by using (24.14) and (24.15) that the operators M and V are self-adjoint with respect to the form $B(\, , \,)$. Then

$$M^k(T(\bar{\varphi})) = B(VM^kT, \varphi) = B(T, M^kV\varphi) = T[V^{-1}(M^k(V(\bar{\varphi})))].$$

Now $M^kV\bar{\varphi} \to h$ in $A(\Gamma)$ by Lemma 24.7, so a finite limit $\lim_{k\to\infty} M^k(T(\bar{\varphi}))$ exists.

By applying a corollary of the Banach-Steinhaus theorem (see [44, Corollary 7.1.4]), we deduce that this limit defines an analytic functional S. It remains to show that S is a CR-hyperfunction. By the theorem of Polking and Wells [160], we need to verify that $S(\overline{\partial_\tau\varphi}) = 0$ for all functions $\partial_\tau\varphi$ of the following form:

$$\partial_\tau\varphi \, d\sigma = \partial\varphi \wedge d\zeta[k, s] \wedge d\bar{\zeta}\big|_\Gamma. \tag{24.17}$$

Lemma 24.7 shows that $H(\overline{D})$ decomposes into the direct sum of subspaces $\mathcal{O}(\overline{D})$ and $Y(D)$, orthogonal in the sense of the bilinear form B. If $\varphi \in H(\overline{D})$, then $V[\partial_\tau\varphi] \in Y$ (a consequence of (24.14) and (24.15)), so

$$S(\overline{\partial_\tau\varphi}) = \lim_{k\to\infty} M^k[T(\overline{\partial_\tau\varphi})] = \lim_{k\to\infty} B(VM^kT, \partial_\tau\varphi) = \lim_{k\to\infty} B(T, M^kV\partial_\tau\varphi).$$

Then $M^k[V(\partial_\tau\varphi)] \in Y(D)$, since $V(\partial_\tau\varphi) \in Y(D)$. It follows from Lemma 24.7 and the hypothesis of the theorem that $M^k[V(\partial_\tau\varphi)] \to 0$ in $A(\Gamma)$ as $k \to \infty$, that is, $S(\overline{\partial_\tau\varphi}) = 0$. □

Theorem 24.8 was given in [126] without the additional hypothesis of convergence of the iterates of the operator M in $W_2^s(D)$.

Theorem 24.9 (Kytmanov, Yakimenko). *Suppose D is a pseudoconvex domain, and $bM^+(T) = T$ on Γ. Then T is a CR-hyperfunction, that is, the harmonic extension $\widetilde{T}$ of the functional T is holomorphic.*

Proof. We again apply the theorem of Polking and Wells [160]. We need to show that $T(\overline{\partial_\tau\varphi}) = 0$, where $\partial_\tau\varphi$ is a function of the form (24.17). These functions are orthogonal to holomorphic functions for integration over Γ. Therefore we can apply Theorem 18.3 to them, so that $\overline{\partial_\tau\varphi} = \overline{\overline{\partial}_n \overline{F}} = \partial_n F$, where F is a function that is harmonic in D. (We have already observed that Theorem 18.3 is valid not only for strongly pseudoconvex domains, but also for pseudoconvex domains with real-analytic boundaries.) Moreover, F will belong to $H(\overline{D})$ since the Neumann operator preserves real analyticity. Thus we obtain from Theorems 24.5 and 24.6 and formula (24.13) that $T(\overline{\partial_\tau\varphi}) = T(\partial_n F) = 0$, that is, T is a CR-hyperfunction on Γ. □

Corollary 24.10. *Under the hypotheses of Theorem 24.9, if $bM^-(T) = 0$ on Γ, then T is a CR-hyperfunction on Γ.*

Corollary 24.11. *Under the hypotheses of Theorem 24.9, if $\overline{\partial}_n T = 0$, then T is a CR-hyperfunction on Γ.*

Apparently Theorem 24.9, like Theorem 17.2, holds for all domains.

25 Conditions for holomorphic extension of functions into a fixed domain

25.1 Holomorphic extension using the Bochner-Martinelli integral

We consider the problem of describing those functions on a hypersurface that can be extended holomorphically into a fixed domain. We do not assume that this domain is the envelope of holomorphy of the hypersurface, as was the case in §8. We remark that this problem cannot be solved by requiring the vanishing of a certain family of continuous linear functionals. A survey of previously known results may be found in the book [6, §27]. Our method consists in using the Bochner-Martinelli integral. Here we give results from [8, 9].

Let Ω be a domain in $\mathbf{C}^n$, where $n > 1$, and suppose the hypersurface Γ has the form $\Gamma = \{z \in \Omega : \rho(z) = 0\}$, where $\rho \in \mathcal{C}^k(\Omega)$, $k \geq 1$, ρ is real-valued, and $d\rho \neq 0$ on Γ. Thus, Γ is a smooth (class $\mathcal{C}^k$), orientable, relatively closed hypersurface in Ω. We consider the open sets $\Omega^+ = \{ z \in \Omega : \rho(z) > 0 \}$ and $\Omega^- = \{ z \in \Omega : \rho(z) < 0 \}$. The orientation of Γ is taken to be compatible with Ω^+. For $\epsilon > 0$, we denote $\Omega_\epsilon^+ = \{ z \in \Omega : \rho(z) > \epsilon \}$, $\Omega_\epsilon^- = \{ z \in \Omega : \rho(z) < -\epsilon \}$, and $\Gamma_{\pm\epsilon} = \{ z \in \Omega : \rho(z) = \pm\epsilon \}$. For every relatively compact domain $G \subset \Omega$, there is a positive ϵ_0 such that for all ϵ in the range $0 < \epsilon \leq \epsilon_0$, the surfaces $\Gamma_{\pm\epsilon} \cap G$ are also smooth. Let $\nu(z)$ be the unit normal to Γ at the point $z \in \Gamma$, directed to the side of increasing ρ. Then for every compact set $K \subset \Gamma$, there is a positive ϵ_0 such that for all ϵ in the range $0 < \epsilon \leq \epsilon_0$, the point $z + \epsilon\nu(z) \in \Omega^+$ and the point $z - \epsilon\nu(z) \in \Omega^-$, where $z \in K$.

We recall that the classes $\mathcal{G}(\Omega^\pm)$ of harmonic functions in $\Omega^\pm$ having finite order of growth at Γ are defined as follows (see §11): a function f is in $\mathcal{G}(\Omega^+)$ if for every ball $B(z^0, r)$ with center at $z^0 \in \Gamma$ and radius r, there are positive constants c and m for which

$$|f(z)| \leq c\rho^{-m}(z), \qquad z \in \Omega^+ \cap B(z^0, r). \tag{25.1}$$

If the constant m in (25.1) does not depend on the point z^0, then we will call f a function of growth order (near Γ) not greater than m; we denote the class of such functions by $\mathcal{G}_m(\Omega^+)$.

We define the classes $\mathcal{G}(\Omega^-)$ and $\mathcal{G}_m(\Omega^-)$ of harmonic functions analogously, except that ρ must be replaced by $-\rho$ in (25.1). It is clear that these classes do not depend on the choice of the defining function ρ. If $k \geq 2$ (where k is the order of smoothness of Γ), then (locally) we may take for ρ the distance function $d(z, \Gamma) = \inf_{w \in \Gamma} |z - w|$. Therefore, we may replace $\rho(z)$ by $d(z, \Gamma)$ in (25.1).

As we showed in §11, if $\Gamma \in C^\infty$, then a function $f \in \mathcal{G}(\Omega^+)$ determines a distribution (denoted f_0) on Γ by the formula

$$\langle f_0, \varphi \rangle = \lim_{\epsilon \to 0+} \int_\Gamma f(z + \epsilon\nu(z))\varphi(z)\, d\sigma(z), \qquad \varphi \in \mathcal{D}(\Gamma).$$

If $f \in \mathcal{G}_m(\Omega^+)$, then f determines a distribution f_0 on Γ with order of singularity no greater than $[m] + 1$. This is easy to show by using the arguments from Theorem 11.1. Consequently, f will determine a distribution f_0 even on hypersurfaces of finite smoothness $C^{[m]+1}$.

Subsequently we shall need Corollary 11.5 for hypersurfaces with finite order of smoothness.

Theorem 25.1 (Kytmanov). *Suppose* $\Gamma \in C^\infty$, $f^- \in \mathcal{G}(\Omega^-)$, *and* $f^+ \in \mathcal{G}(\Omega^+)$. *If* $f_0^+ = f_0^-$ *on* $\Gamma \cap B(z^0, r)$, *and* $(\overline{\partial}_n f^+)_0 = (\overline{\partial}_n f^-)_0$ *on* $\Gamma \cap B(z^0, r)$, *for some point* $z^0 \in \Gamma$, *then there exists a harmonic function* F *in* $\Omega^+ \cup \Omega^- \cup B(z^0, r)$ *such that* $F = f^+$ *in* Ω^+ *and* $F = f^-$ *in* Ω^-. *If* $\Gamma \in C^{[m]+2}$, *then the theorem holds for functions* $f^\pm \in \mathcal{G}_m(\Omega^\pm)$. *If the harmonic functions* $f^\pm$ *extend continuously to* $\Omega^+ \cup (\Gamma \cap B(z^0, r))$, *then we may take the hypersurface* Γ *to be class* C^1.

Proof. The proof is an exact repetition of the proof of Corollary 11.5. □

As shown in §6, a necessary condition for the holomorphic extension of a function f given on Γ is the Cauchy-Riemann condition, that is, f must be a CR-function. We are interested in the question of when a CR-function f does extend holomorphically into the fixed domain Ω^+. We will consider a holomorphic (or harmonic) extension from Γ in the weak sense, that is, there must exist a function $F \in \mathcal{G}(\Omega^+)$ such that $F_0 = f$ on Γ and F is holomorphic in Ω^+. If $\Gamma \in \mathcal{C}^{[m]+1}$, then we will suppose that the extension F belongs to the class $\mathcal{G}_m(\Omega^+)$.

Corollary 25.2. *Suppose $\Gamma \in \mathcal{C}^\infty$, and f is a CR-function (or CR-distribution) on Γ. If there exists a harmonic extension $F \in \mathcal{G}(\Omega^+)$ of f (in the sense that $F_0 = f$ on Γ) such that $(\bar{\partial}_n F)_0 = 0$ on Γ, then this extension F is a holomorphic extension of f into Ω^+. Moreover*

(a) if $\Gamma \in \mathcal{C}^k$, $f \in \mathcal{L}^p_{\text{loc}}(\Gamma)$, $p \geq 1$, $k \geq 3 + [(2n-1)/p]$, then the extension F satisfies the condition that for each point $z^0 \in \Gamma$ there is a ball $B(z^0, r) \subset \Omega$ for which

$$\lim_{\epsilon \to 0^+} \int_{\Gamma \cap B(z^0,r)} |F(z + \epsilon\nu(z)) - f(z)|^p \, d\sigma = 0, \tag{25.2}$$

and $F(z + \epsilon\nu(z)) \to f(z)$ at the Lebesgue points of f when $\epsilon \to 0^+$;

(b) if $\Gamma \in \mathcal{C}^k$, $k \geq 1$, $f \in \mathcal{C}^r(\Gamma)$, $0 \leq r \leq k$, then $F \in \mathcal{C}^r(\Gamma \cup \Omega^+)$, and $F = f$ on Γ.

Proof. First suppose that Ω is a ball in $\mathbf{C}^n$. If f is a CR-distribution on Γ, then by Theorem 6.1 on analytic representation, $f = h_0^+ - h_0^-$ on Γ, where the $h^\pm$ are holomorphic in $\Omega^\pm$, and $h^\pm \in \mathcal{G}(\Omega^\pm)$. If F is a harmonic extension of f into Ω^+, $F \in \mathcal{G}(\Omega^+)$, and $(\bar{\partial}_n F)_0 = 0$ on Γ, then we have $h_0^- = (h^+ - F)_0$ on Γ, and $(\bar{\partial}_n h^-)_0 = 0 = [\bar{\partial}_n(h^+ - F)]_0$ on Γ. By Theorem 25.1, there is a harmonic function h in Ω that agrees with h^- in Ω^- and with $h^+ - F$ in Ω^+. Since h^- is holomorphic in Ω^-, the function h is holomorphic in Ω, that is, F is holomorphic in Ω^+.

Now suppose $f \in \mathcal{L}^p_{\text{loc}}$, where $p \geq 1$. By Theorem 6.1, the local behavior of $h^\pm$ near Γ is determined by the Bochner-Martinelli integral of f. It is easy to show that the Bochner-Martinelli integral of f has order of growth near Γ at most $(2n-1)/p$. Thus, for Theorem 25.1 to be applicable to the function $f \in \mathcal{L}^p$, we need to require that the smoothness of Γ be greater than $3 + [(2n-1)/p]$, and the order of growth of F also must not exceed $(2n-1)/p$. It follows from Theorem 3.4 on the jump of the Bochner-Martinelli integral for integrable functions that

$$\lim_{\epsilon \to 0^+} \int_{\Gamma \cap B(z^0,r)} |h^+(z + \epsilon\nu(z)) - h^-(z - \epsilon\nu(z)) - f(z)|^p \, d\sigma = 0,$$

and since h^- and $h^+ - F$ extend holomorphically into Ω, we obtain (25.2).

If $\Gamma \in \mathcal{C}^k$ and $f \in \mathcal{C}^r(\Gamma)$, where $k \geq 1$ and $0 \leq r \leq k$, then we can also apply Theorem 25.1. It follows from Theorem 6.1 that in this case, $h^\pm \in \mathcal{C}^{r-0}(\Gamma \cup \Omega^\pm)$.

But since h^- extends holomorphically into Ω, we have $h^+ \in \mathcal{C}^r(\Gamma \cup \Omega^+)$ (see Corollary 4.8).

Now if Ω is an arbitrary domain in $\mathbf{C}^n$, we apply the preceding argument to some ball $B(z^0, r) \subset \Omega$ to obtain that F extends holomorphically into this ball. Consequently, F will be holomorphic in Ω^+. □

Remark. If the distribution f has order of singularity on Γ at most k, then the Bochner-Martinelli integral of f has order of growth near Γ at most $2n+k-1$, so the surface Γ in Corollary 25.2 may be taken in class $\mathcal{C}^{2n+k+1}$, and the harmonic extension F in class $\mathcal{G}_{2n+k-1}(\Omega^+)$.

Example. If f is not a CR-function on Γ, then Corollary 25.2 is false in general. Indeed, consider the hypersurface $\Gamma = \{z : \operatorname{Re} z_1 = 0\}$. Then $\overline{\partial}_n f = \partial f/\partial \bar{z}_1$. Therefore every harmonic function f in $\mathbf{C}^n$ that is independent of z_1 (for example, $f = \operatorname{Re} z_2$) satisfies the condition $\overline{\partial}_n f = 0$ on Γ, but in general such a function does not extend holomorphically into any domain abutting Γ.

Nonetheless, if Γ is the boundary of a bounded domain $D \subset \Omega$, then it follows from the conditions $f \in \mathcal{G}(D)$ and $\overline{\partial}_n f = 0$ on Γ that f extends holomorphically into D; see Chapter 4.

Corollary 25.2 is an assertion that is hard to verify. Presently we shall give more constructive results. For the moment we prove an auxiliary assertion.

Let D be a bounded domain in $\mathbf{C}^n$ with boundary $\Gamma = \partial D$ of class $\mathcal{C}^m$. We recall that the Bochner-Martinelli transform is defined for $f \in \mathcal{D}'(\Gamma)$ via (see §17)

$$(Mf)(z) = -\frac{(n-2)!}{2\pi^n}\langle f_\zeta, (\partial_n)_\zeta |\zeta - z|^{2-2n}\rangle, \qquad z \notin \Gamma.$$

Lemma 25.3. *If $\Gamma \in \mathcal{C}^m$, where $m \geq 2$, and f is a distribution on Γ with order of singularity at most $m-2$, then*

$$\lim_{\epsilon \to 0+} \int_\Gamma \left[(\overline{\partial}_n M^+ f)(z + \epsilon\nu(z)) - (\overline{\partial}_n M^- f)(z - \epsilon\nu(z))\right] \varphi \, d\sigma = 0$$

for all $\varphi \in \mathcal{C}^m(\mathbf{C}^n)$. If $\Gamma \in \mathcal{C}^2$, $f \in \mathcal{L}^p(\Gamma)$, $p \geq 1$, then

$$\lim_{\epsilon \to 0+} \int_\Gamma \left|(\overline{\partial}_n M^+ f)(z + \epsilon\nu(z)) - (\overline{\partial}_n M^- f)(z - \epsilon\nu(z))\right|^p d\sigma = 0,$$

and at the Lebesgue points of f we have

$$\lim_{\epsilon \to 0+} \left[(\overline{\partial}_n M^+ f)(z + \epsilon\nu(z)) - (\overline{\partial}_n M^- f)(z - \epsilon\nu(z))\right] = 0.$$

Proof. The last part of the lemma follows from the remark after Theorem 4.10. Suppose $f \in \mathcal{L}^p(\Gamma)$, where $p \geq 1$. In order to prove convergence in the $\mathcal{L}^p$ norm, we need to apply the arguments of Theorems 4.10 and 3.4 to obtain what is required.

If $\Gamma \in \mathcal{C}^m$, where $m \geq 2$, and $f \in \mathcal{C}^s(\Gamma)$, where $0 \leq s \leq m-2$, then by using smoothness properties of derivatives of the Bochner-Martinelli integral (see §4), we obtain that the difference $(\overline{\partial}_n M^+ f)(z^+) - (\overline{\partial}_n M^- f)(z^-)$ tends to zero in the metric of $\mathcal{C}^s(\Gamma)$ when $\epsilon \to 0^+$, where $z^+ = z + \epsilon\nu(z)$ and $z^- = z - \epsilon\nu(z)$. Then

$$\begin{aligned}\int_\Gamma \left[\overline{\partial}_n M^+ f(z^+) - \overline{\partial}_n M^- f(z^-)\right] \varphi(z)\, d\sigma(z) \\ = c\int_\Gamma \langle f_\zeta, (\overline{\partial}_n)_z(\partial_n)_\zeta |\zeta - z^+|^{2-2n} - (\overline{\partial}_n)_z(\partial_n)_\zeta|\zeta - z^-|^{2-2n}\rangle \varphi(z)\, d\sigma \\ = c\left\langle f_\zeta, (\partial_n)_\zeta \int_\Gamma \left[(\overline{\partial}_n)_z|\zeta - z^+|^{2-2n} - (\overline{\partial}_n)_z|\zeta - z^-|^{2-2n}\right] \varphi(z)\, d\sigma\right\rangle.\end{aligned}$$

Replacing $|\zeta - z^+|$ by $|\zeta^+ - z|$ in the last expression (their difference is $o(|\zeta - z|)$ as $\zeta \to z$), we obtain that $c(\overline{\partial}_n)_z|\zeta^\pm - z|^{2-2n}\, d\sigma(z) = \overline{U(z,\zeta^\pm)}\big|_\Gamma$. Consequently, $\partial_n \overline{M}^+ \varphi(\zeta^+) - \partial_n \overline{M}^- \varphi(\zeta^-) \to 0$ in the metric of $\mathcal{C}^{m-2}(\Gamma)$ when $\epsilon \to 0^+$. □

We approximate the domain Ω from inside by a sequence of bounded domains, that is, $\Omega = \cup_{s\geq 1}\Omega_s$, $\Omega_s \Subset \Omega_{s+1}$. Suppose $\varphi_s \in \mathcal{D}(\Omega)$ and $\varphi_s \equiv 1$ on Ω_s.

Theorem 25.4 (Aĭzenberg, Kytmanov). *Suppose $\Gamma \in \mathcal{C}^\infty$ and f is a CR-distribution on Γ. Then for f to extend holomorphically into Ω^+ (as a function $F \in \mathcal{G}(\Omega^+)$), it is necessary and sufficient that the Bochner-Martinelli transform $M^-(\varphi_s f)$ extend to a harmonic function in Ω_s for every s.*

If $\Gamma \in \mathcal{C}^{k+2n+3}$, where $k \geq 0$, then the theorem holds for CR-distributions with singularity k. If $\Gamma \in \mathcal{C}^k$, and the CR-function $f \in \mathcal{L}^p_{\text{loc}}(\Gamma)$, $p \geq 1$, $k \geq 3 + [(2n-1)/p]$, then the holomorphic extension F will satisfy condition (25.2). Finally, if $\Gamma \in \mathcal{C}^2$ and the CR-function $f \in \mathcal{C}(\Gamma)$, then the holomorphic extension F will be continuous up to Γ.

Proof. Necessity. Suppose f extends holomorphically into Ω^+ as a function $F \in \mathcal{G}(\Omega^+)$. Consider the Bochner-Martinelli transform $M(\varphi_s f)$. Lemma 25.3 implies $\left[\overline{\partial}_n M^+(\varphi_s f)\right]_0 = \left[\overline{\partial}_n M^-(\varphi_s f)\right]_0$ on $\Gamma \cap \Omega_s$, while $[M^+(\varphi_s f)]_0 - [M^-(\varphi_s f)]_0 = f$ on $\Gamma \cap \Omega_s$ (see Theorem 3.6). Then $[M^-(\varphi_s f)]_0 = [M^+(\varphi_s f) - F]_0$ on $\Gamma \cap \Omega_s$, and $\left[\overline{\partial}_n M^-(\varphi_s f)\right]_0 = \left[\overline{\partial}_n M^+(\varphi_s f) - \overline{\partial}_n F\right]_0$ on $\Gamma \cap \Omega_s$. By Theorem 25.1, the function $M^-(\varphi_s f)$ extends harmonically into Ω_s. The necessity of the other parts of Theorem 25.4 also follows from Theorem 25.1 and Lemma 25.3.

Sufficiency. Suppose $M^-(\varphi_s f)$ extends to a harmonic function h_s in Ω_s. Then by Lemma 25.3, we have $[\overline{\partial}_n M^-(\varphi_s f)]_0 = [\overline{\partial}_n M^+(\varphi_s f)]_0$ on $\Gamma \cap \Omega_s$. Therefore $(\overline{\partial}_n h_s)_0 - [\overline{\partial}_n M^+(\varphi_s f)]_0 = 0$ on $\Gamma \cap \Omega_s$, while $[h_s - M^+(\varphi_s f)]_0 = f$ on $\Gamma \cap \Omega_s$. We then obtain from Corollary 25.2 that $h_s - M^+(\varphi_s f)$ is holomorphic in Ω_s^+. This extension (by the uniqueness theorem) also defines a holomorphic extension of f into Ω^+. □

If the distribution f defines a continuous linear functional on the functions of class $\mathcal{C}^\infty(\Gamma)$ that are bounded (together with all derivatives) (that is, $f \in \mathcal{O}_0'(\Gamma)$), then we have the following.

Corollary 25.5. *Suppose Ω and Γ satisfy the hypotheses of Theorem 25.4, and f is a CR-function of class $\mathcal{O}_0'(\Gamma)$ (in particular, $f \in \mathcal{L}^p(\Gamma)$, $p \geq 1$). In order that f extend holomorphically into Ω^+, it is necessary and sufficient that the Bochner-Martinelli transform $M^- f$ extend to a harmonic function in Ω^+.*

Theorem 25.4 and Corollary 25.5 say that a CR-function f extends holomorphically into that domain Ω^+ to which the Bochner-Martinelli integral $M^- f$ extends harmonically.

Remark. For the proof of the necessity in Theorem 25.4, it is not necessary to require that f be a CR-function, but the sufficiency is false without this requirement. Consider the hyperplane $\Gamma = \{ z : \operatorname{Re} z_1 = 0 \}$ in the ball $B(0,1)$ and the function $f = \operatorname{Re} z_2$. Then f does not extend holomorphically into either B^+ or B^-. The Bochner-Martinelli integral Mf does satisfy the condition $\overline{\partial}_n M^+ f = \overline{\partial}_n M^- f$ on Γ (by Lemma 25.3), and $M^+ f - M^- f = f$ on Γ. Besides, $\overline{\partial}_n f = 0$ on Γ, so (by Theorem 25.1) the function $M^- f + f$ extends harmonically into the ball $B(0,1)$, that is, $M^- f$ also extends harmonically into $B(0,1)$.

Now suppose $\Omega = B(0,1)$ and the domain Ω^- contains 0. We will be interested in the question of when the Bochner-Martinelli integral $M^- f$ extends harmonically into Ω.

Consider in Ω a set of homogeneous harmonic polynomials that forms a complete orthonormal system of functions in $\mathcal{L}^2(\partial\Omega)$ with respect to Lebesgue measure $d\sigma$ (see §5). We denote these polynomials by $P_{k,s}$, where k is the degree of homogeneity, $k = 0, 1, 2, \dots$, and s indexes the polynomials of degree k contained in the basis, $s = 1, 2, \dots, \sigma(k) = (2n - 2k - 2)(k + 2n - 3)!/k!\,(n-2)!$ (see, for example, [189, chap. 10]). It is clear that $\sigma(k)$ is a polynomial (in k) of degree $2n - 2$ with leading coefficient $-2/(n-2)!$.

Theorem 25.6 (Kytmanov). *If $\{P_{k,s}\}$ is a complete orthonormal system of homogeneous harmonic polynomials in $\mathcal{L}^2(\partial\Omega)$, then the Bochner-Martinelli kernel has the following expansion:*

$$U(\zeta, z) = -\sum_{k,s} \frac{P_{k,s}(z)}{n+k-1} \left[\star\partial \frac{\overline{P_{k,s}(\zeta)}}{|\zeta|^{2n+2k-2}} \right], \tag{25.3}$$

where the series in (25.3) converges uniformly on compact sets in the domain $\{ (z,\zeta) \in \mathbf{C}^{2n} : |\zeta| > |z| \}$.

Proof. We denote the restriction of $P_{k,s}$ to $\partial\Omega$ by $Y_{k,s}$. Then $\{Y_{k,s}\}$ is a basis in $\mathcal{L}^2(\partial\Omega)$ consisting of spherical harmonics. Let $\zeta \in \partial\Omega$, $z \in \Omega$, and $|\zeta - z|^{2-2n} = \sum_{k,s} c_{k,s} \overline{Y}_{k,s}(\zeta)$ (here z is fixed), where

$$c_{k,s} = \int_{\partial\Omega} |\zeta - z|^{2-2n} Y_{k,s}(\zeta)\, d\sigma(\zeta).$$

If we express $|\zeta - z|^{2-2n}$ in terms of the Poisson kernel for the ball

$$P(\zeta, z) = \frac{(n-1)!}{2\pi^n} \frac{1 - |z|^2}{|\zeta - z|^{2n}},$$

we obtain

$$c_{k,s} = \frac{2\pi^n}{(n-1)!} \int_{\partial\Omega} P(\zeta, z) \frac{1 - \langle \zeta, \bar{z} \rangle - \langle \bar{\zeta}, z \rangle + |z|^2}{1 - |z|^2} Y_{k,s}\, d\sigma.$$

It is easy to verify that the functions

$$\zeta_j P_{k,s}(\zeta) - \frac{1}{n+k-1} \frac{\partial P_{k,s}}{\partial \bar{\zeta}_j}(|\zeta|^2 - 1) \qquad \text{and}$$
$$\bar{\zeta}_j P_{k,s}(\zeta) - \frac{1}{n+k-1} \frac{\partial P_{k,s}}{\partial \zeta_j}(|\zeta|^2 - 1)$$

are harmonic extensions into Ω of the functions $\zeta_j Y_{k,s}$ and $\bar{\zeta}_j Y_{k,s}$ given on $\partial\Omega$. Therefore

$$\begin{aligned} c_{k,s} &= \frac{2\pi^n}{(n-1)!} \frac{1}{1 - |z|^2} \Bigg[(1 + |z|^2) P_{k,s}(z) - 2|z|^2 P_{k,s}(z) \\ &\qquad + \frac{|z|^2 - 1}{n+k-1} \sum_{j=1}^{n} \left(\bar{z}_j \frac{\partial P_{k,s}}{\partial \bar{z}_j} + z_j \frac{\partial P_{k,s}}{\partial z_j} \right) \Bigg] \\ &= \frac{2\pi^n}{(n-2)!\,(n+k-1)} P_{k,s}(z). \end{aligned}$$

Thus

$$|\zeta - z|^{2-2n} = \frac{2\pi^n}{(n-2)!} \sum_{k,s} \frac{1}{n+k-1} P_{k,s}(z) \overline{Y}_{k,s}(\zeta). \tag{25.4}$$

The series (25.4) converges in ζ in the sense of $\mathcal{L}^2(\partial\Omega)$, and in z uniformly on compact subsets of Ω. The harmonic extension in ζ of (25.4) into Ω reduces to the equality

$$\left(|\zeta| \left| \frac{\zeta}{|\zeta|^2} - z \right| \right)^{2-2n} = \frac{2\pi^n}{(n-2)!} \sum_{k,s} \frac{1}{n+k-1} P_{k,s}(z) \overline{P_{k,s}(\zeta)}. \tag{25.5}$$

Applying the Kelvin transform in ζ to both sides of (25.5), we obtain

$$|\zeta - z|^{2-2n} = \frac{2\pi^n}{(n-2)!} \sum_{k,s} \frac{P_{k,s}(z)}{n+k-1} \frac{\overline{P_{k,s}(\zeta)}}{|\zeta|^{2n+2k-2}},$$

where the series converges uniformly on compact subsets of the domain $\{ (z,\zeta) \in \mathbf{C}^{2n} : |\zeta| > |z| \}$. Hence, since

$$U(\zeta,z) = -\frac{(n-2)!}{2\pi^n} \star \partial_\zeta |\zeta - z|^{2-2n},$$

equation (25.3) follows.

Consider a function $f \in \mathcal{L}^1(\Gamma)$ (or $f \in \mathcal{O}_0'(\Gamma)$). It follows from (25.3) that

$$\begin{aligned}(M^- f)(z) &= \int_\Gamma f(\zeta)\, U(\zeta,z) \\ &= -\sum_{k,s} \frac{P_{k,s}(z)}{n+k-1} \int_\Gamma f(\zeta) \left[\star\partial \frac{\overline{P_{k,s}(\zeta)}}{|\zeta|^{2n+2k-2}} \right]. \qquad (25.6)\end{aligned}$$

Thus, the coefficients of the expansion of $M^- f$ into a series in terms of the system $\{P_{k,s}\}$ have the form

$$a_{k,s} = -\frac{1}{n+k-1} \int_\Gamma f(\zeta) \left[\star\partial \frac{\overline{P_{k,s}(\zeta)}}{|\zeta|^{2n+2k-2}} \right], \qquad (25.7)$$

$$M^- f = \sum_{k,s} a_{k,s} P_{k,s}(z), \qquad (25.8)$$

where the series in (25.8) converges uniformly in some neighborhood of the origin. The orientation of Γ is compatible with Ω^+; if instead it is compatible with Ω^-, then the minus sign in (25.7) disappears. □

We now discuss the question of convergence of the series (25.8) in the ball Ω. This series converges in $\mathcal{L}^2(\partial\Omega)$ if and only if

$$\sum_{k,s} |a_{k,s}|^2 < \infty,$$

in which case the series will converge uniformly on compact subsets of Ω. Therefore, if $M^- f$ extends harmonically from Ω^- into Ω, then the series

$$\sum_{k,s} |a_{k,s}|^2 r^{2k}$$

converges for all $r < 1$. Conversely, if this convergence holds, then $M^- f$ extends harmonically from Ω^- into Ω. We obtain the following result.

Theorem 25.7 (Aĭzenberg, Kytmanov). *Suppose Γ is such that $\Omega \setminus \Gamma$ consists of two connected components, and the CR-function $f \in \mathcal{L}^1(\Gamma)$ (or $f \in \mathcal{O}_0'(\Gamma)$). A necessary and sufficient condition for f to extend holomorphically into Ω^+ is that*

$$\limsup_{k\to\infty} \max_s \sqrt[k]{|a_{k,s}|} \le 1, \qquad (25.9)$$

where the $a_{k,s}$ *are given by (25.7). The smoothness of* Γ *is the same as in Corollary 25.5. If it is known that* $\Gamma \in \mathcal{C}^2$ *and* $f \in \mathcal{C}(\Gamma)$, *then the extension lies in* $\mathcal{C}(\Omega^+ \cup \Gamma)$.

If Γ is the boundary of a bounded domain in Ω, then the requirement that f be a CR-distribution on Γ is unnecessary.

Corollary 25.8. *Suppose* $\Gamma \in \mathcal{C}^\infty$ *is the boundary of a bounded domain in* Ω *(that is,* $\Gamma = \partial\Omega^-$*),* $\Omega^- = \Omega \setminus \overline{\Omega^+}$, *and* $0 \in \Omega^-$. *If* $f \in \mathcal{D}'(\Gamma)$ *satisfies (25.9), then* f *extends holomorphically into* Ω^+. *If* $f \in \mathcal{C}(\Gamma)$ *and* $\Gamma \in \mathcal{C}^2$, *then the extension lies in* $\mathcal{C}(\overline{\Omega^+})$.

Proof. If (25.9) holds for f, then the series (25.8) for $M^- f$ converges in Ω. On the other hand, $M^- f$ will be defined in $\mathbf{C}^n \setminus \overline{\Omega^+}$, and $(M^- f)(z) \to 0$ when $|z| \to \infty$. Hence $M^- f$ extends to a harmonic function in $\mathbf{C}^n$, so $M^- f \equiv 0$. Now we get the result by using Theorems 15.4 and 17.2. □

Example. Suppose Ω is the whole space $\mathbf{C}^n$, and Γ is a hypersurface of class $\mathcal{C}^2$ that divides $\mathbf{C}^n$ into two domains. Suppose that the points $\pm b$ lie in different domains. We denote by $P^{\pm}_{k,s}$ the spherical harmonics of Theorem 25.6 for the unit spheres with centers at the points $\pm b$. Consider $f \in \mathcal{C}(\Gamma) \cap \mathcal{L}^1(\Gamma)$, and set

$$a^{\pm}_{k,s} = \int_\Gamma f(\zeta) \left[\star \partial \frac{\overline{P^{\pm}_{k,s}(\zeta)}}{|\zeta \pm b|^{2n+2k-2}} \right].$$

From Theorem 25.7, we obtain the following necessary and sufficient condition for the CR-function f to extend holomorphically to an entire function:

$$\lim_{k\to\infty} \max_s \sqrt[k]{|a^{+}_{k,s}|} = \lim_{k\to\infty} \max_s \sqrt[k]{|a^{-}_{k,s}|} = 0.$$

Example. Suppose Γ is a convex surface, $\Gamma \cap \partial\Omega$ is a smooth surface, $0 \in \Gamma$, the planes $T_c = \{ z : \operatorname{Re} z_1 = c \}$ do not intersect Γ for $c < 0$, $T_c \cap \Omega \subset \Omega^-$, and the plane T_0 is tangent to Γ at 0. Consider the differential form

$$U_1(\zeta, z) = \frac{(n-2)!}{(2\pi i)^n} \frac{1}{\zeta_1 - z_1} \sum_{k=2}^{n} \frac{(-1)^k (\bar\zeta_k - \bar z_k)\, d\bar\zeta[k] \wedge d\zeta}{|\zeta - z|^{2n-2}}.$$

Then $\bar\partial_\zeta U_1(\zeta, z) = U(\zeta, z)$ if $\zeta_1 \neq z_1$. If $f \in \mathcal{L}^1(\Gamma)$ is a CR-function on Γ, then

$$(M^- f)(z) = \int_\Gamma f(\zeta)\, \bar\partial_\zeta U_1 = \int_{\Gamma\cap\partial\Omega} f(\zeta)\, U_1(\zeta, z)$$

if $\operatorname{Re} z_1 < 0$. The integral $\int_{\Gamma\cap\partial\Omega} f(\zeta)\, U_1(\zeta, z)$ realizes a real-analytic extension of the integral $M^- f$ to a neighborhood of 0. This extension is also a harmonic extension. That is, the CR-function f on Γ extends holomorphically into a one-sided neighborhood of 0 in Ω^+. We can prove in the same way, for example, a well-known result on local holomorphic extension of CR-functions from a hypersurface Γ with nondegenerate Levi form (see, for example, [75]).

25.2 Holomorphic extension using Cauchy-Fantappiè integrals

We consider a ball $B \subset \overline{B} \subset \Omega^-$ and a real-valued function $\chi \in \mathcal{C}^\infty(\Omega)$ such that $\chi < 0$ in B, $\chi > 0$ outside some neighborhood of $\overline{B}$, and $\Omega_\epsilon = \{ z \in \Omega : \chi(z) < \epsilon \} \Subset \Omega$ for $\epsilon > 0$. We choose an increasing sequence $\epsilon(s) \to \infty$ so that the $\Omega_{\epsilon(s)} = \Omega_s$ are domains (if necessary, we take in place of $\Omega_{\epsilon(s)}$ a connected component), and a sequence of functions $\varphi_s \in \mathcal{D}(\Omega)$, where $\varphi_s = 1$ in a neighborhood of the closure $\overline{\Omega_s}$. Furthermore, we assume that there exist vector-valued functions $P(\zeta, z) = (P_1(\zeta, z), \dots, P_n(\zeta, z))$ with the following properties:

(a) $P(\zeta, z)$ is defined on the set $W = \{ (\zeta, z) \in \mathbf{C}^{2n} : \zeta \in \Omega \setminus \overline{B},\ z \in \Omega(\zeta) \}$, where $\Omega(\zeta) = \{ z \in \Omega : \chi(\zeta) < \chi(z) \}$;

(b) $P \in \mathcal{C}^\infty(W)$;

(c) $P(\zeta, z)$ is a real-analytic function in $z \in \Omega(\zeta)$ for fixed $\zeta \in \Omega \setminus B$;

(d) $\sum_{j=1}^n P_j(\zeta, z)(\zeta_j - z_j) \neq 0$ in W.

Here we have required that χ and P be class $\mathcal{C}^\infty$ because we want to consider CR-distributions f acting on χ and P. It is understood that if we consider distributions with a finite order of singularity, or functions of class $\mathcal{L}^p$, $\mathcal{C}^\lambda$, etc., then the smoothness of χ and P can be correspondingly reduced.

We consider the Cauchy-Fantappiè differential form

$$\omega(\zeta, z, P) = \frac{(n-1)!}{(2\pi i)^n} \frac{\sum_{k=1}^n (-1)^{k-1} P_k \, dP[k] \wedge d\zeta}{\left[\sum_{j=1}^n P_j(\zeta, z)(\zeta_j - z_j)\right]^n},$$

where $dP[k] = dP_1 \wedge \cdots \wedge dP_{k-1} \wedge dP_{k+1} \wedge \cdots \wedge dP_n$. Here P is viewed under differentiation as a differential in ζ with z fixed. The form $\omega(\zeta, z, P)$ is a $\overline{\partial}$-closed differential form in ζ of type $(n, n-1)$ on W.

When f is a CR-distribution on Γ, we set

$$\Phi_s(z) = \left\langle (\varphi_s f)_\zeta, \frac{\omega(\zeta, z, P)}{d\sigma(\zeta)} \right\rangle, \tag{25.10}$$

where $d\sigma(\zeta)$ is the surface area element of Γ. Since $\overline{B} \subset \Omega^-$, the function $\Phi_s(z)$ is real-analytic in B.

Lemma 25.9. *If $\Gamma \in \mathcal{C}^\infty$ and f is a CR-distribution on Γ, then a necessary and sufficient condition for f to extend holomorphically into Ω^+ (as a function $F \in \mathcal{G}(\Omega^+)$) is that $\Phi_s(z)$ extend from B into Ω_s for each s as a real-analytic function. If $\Gamma \in \mathcal{C}^{k+2n+\epsilon}$, where $k \geq 0$ and $\epsilon > 0$, then the lemma holds for CR-distributions with order of singularity k. If $\Gamma \in \mathcal{C}^k$ and the CR-function $f \in \mathcal{L}^p_{\mathrm{loc}}(\Gamma)$, where $p \geq 1$ and $k \geq 3 + [(2n-1)/3]$, then the holomorphic extension F will satisfy the condition that for each point $z^0 \in \Gamma$ there is a ball $B(z^0, r) \subset \Omega$ for which*

$$\lim_{\epsilon \to 0+} \int_{\Gamma \cap B(z^0, r)} |F(z + \epsilon\nu(z)) - f(z)|^p \, d\sigma = 0.$$

Moreover, $F(z+\epsilon\nu(z)) \to f(z)$ as $\epsilon \to 0^+$ at the Lebesgue points of f. Finally, if $\Gamma \in \mathcal{C}^2$ and the CR-function $f \in \mathcal{C}(\Gamma)$, then the holomorphic extension F will be continuous up to Γ.

Proof. Consider the difference

$$U(\zeta, z) - \omega(\zeta, z, P). \tag{25.11}$$

Then the form (25.11) is $\overline{\partial}_\zeta \mu \wedge d\zeta$ (see [7, Lemma 1.3]), where

$$\mu = \frac{1}{(2\pi i)^n} \sum_{j=0}^{n-2} \det(\bar\zeta - \bar z, P, \underbrace{\overline{\partial}_\zeta(\bar\zeta - \bar z), \dots, \overline{\partial}_\zeta(\bar\zeta - \bar z)}_{j}, \underbrace{\overline{\partial}_\zeta P, \dots, \overline{\partial}_\zeta P}_{n-j-2}).$$

Here the determinant has as columns the vector-valued functions $\bar\zeta - \bar z$, P, $\overline{\partial}_\zeta(\bar\zeta - \bar z)$, and $\overline{\partial}_\zeta P$, and it is computed by the usual rule for computing determinants, taking into account the properties of the exterior product. The form μ is defined in W.

Now suppose $\tilde\varphi_s \in \mathcal{D}(\Omega)$ equals 1 on $\overline{\Omega_s}$, and $\operatorname{supp}\varphi_s \subset \{\, z \in \Omega : \varphi_s = 1 \,\}$. Then

$$\begin{aligned}\langle(\varphi_s f)_\zeta, U(\zeta,z) - \omega(\zeta,z,P)\rangle &= \langle \varphi_s f, \overline{\partial}_\zeta \mu \wedge d\zeta\rangle \\ &= \langle \varphi_s f, \overline{\partial}_\zeta(\tilde\varphi_s \mu \wedge d\zeta)\rangle + \langle \varphi_s f, \overline{\partial}_\zeta[(1-\tilde\varphi_s)\mu \wedge d\zeta]\rangle \\ &= \langle f, \overline{\partial}_\zeta(\tilde\varphi_s \mu \wedge d\zeta)\rangle + \langle \varphi_s f, \overline{\partial}_\zeta[(1-\tilde\varphi_s)\mu \wedge d\zeta]\rangle \\ &= \langle \varphi_s f, \overline{\partial}_\zeta[(1-\tilde\varphi_s)\mu \wedge d\zeta]\rangle.\end{aligned}$$

In view of the properties of P, the last function is real-analytic in Ω_s. It remains to use Theorem 25.4 and the fact that the real-analytic extension is a harmonic extension. □

Remark. If $f \in \mathcal{O}'_0(\Gamma)$, then the holomorphic extendibility of f from Γ into Ω^+ is equivalent to the real-analytic extendibility from B into Ω of the function

$$\Phi(z) = \left\langle f_\zeta, \frac{\omega(\zeta, z, P)}{d\sigma(\zeta)} \right\rangle. \tag{25.12}$$

Example. For the vector-valued function

$$P(\zeta, z) = (|\zeta_1 - z_1|^{k_1}(\bar\zeta_1 - \bar z_1), \dots, |\zeta_n - z_n|^{k_n}(\bar\zeta_n - \bar z_n)),$$

we have

$$\sum_{j=1}^{n} P_j(\zeta, z)(\zeta_j - z_j) = \sum_{j=1}^{n} |\zeta_j - z_j|^{k_j+2} \neq 0$$

for $\zeta \neq z$, so the function P satisfies all the above requirements for every domain Ω.

Now we observe that in the case of a domain of holomorphy Ω, the vector-valued function $P(\zeta, z)$ may be taken to be holomorphic in z. By Theorem 2.6.11 in [79], the function χ my be taken to be strictly plurisubharmonic in Ω. Using the Henkin-Ramirez construction, we can construct a vector-valued function $P(\zeta, z)$ satisfying all the necessary conditions that is in addition holomorphic in z in $\Omega(\zeta)$ for fixed ζ (see also the construction in [74, Theorem 3.10]).

Theorem 25.10 (Aĭzenberg, Kytmanov). *If Ω is a domain of holomorphy, and Γ and the CR-distribution f satisfy the hypotheses of Lemma 25.9, then f extends holomorphically into Ω^+ if and only if the functions $\Phi_s(z)$ given in (25.10) extend holomorphically from B into Ω_s for every s.*

Corollary 25.11. *If Ω is a domain of holomorphy, and f is a CR-distribution in $\mathcal{O}_0'(\Gamma)$, then a necessary and sufficient condition for f to extend holomorphically into Ω^+ is that the function $\Phi(z)$ defined by (25.12) extend holomorphically into Ω.*

Example. Suppose Ω is a convex domain. Then the exhaustion function $\chi \in \mathcal{C}^\infty(\Omega)$ may be chosen to be convex. In this case, we may take as P the vector-valued function $\operatorname{grad}\chi = (\partial\chi/\partial\zeta_1, \dots, \partial\chi/\partial\zeta_n)$. By Sard's theorem, $\operatorname{grad}\chi \neq 0$ on $\partial\Omega_\epsilon$ for almost all $\epsilon > 0$. We will assume that $\operatorname{grad}\chi \neq 0$ on Γ. Suppose $0 \in B \Subset \Omega^-$. Then $|(\operatorname{grad}\chi, \bar\zeta)| > |(\operatorname{grad}\chi, \bar z)|$ for z sufficiently close to zero, where $(a, b) = a_1\bar b_1 + \cdots + a_n\bar b_n$. Therefore

$$\begin{aligned}\frac{1}{(\operatorname{grad}\chi(\zeta), \bar\zeta - \bar z)^n} &= \sum_{k=0}^{\infty} \frac{(k+n-1)!}{k!\,(n-1)!}\frac{(\operatorname{grad}\chi, \bar z)^k}{(\operatorname{grad}\chi, \bar\zeta)^{n+k}} \\ &= \sum_{\|\alpha\|\geq 0} \frac{(\|\alpha\|+n-1)!}{\alpha!\,(n-1)!}\frac{z^\alpha(\operatorname{grad}\chi)^\alpha}{(\operatorname{grad}\chi, \bar\zeta)^{n+\|\alpha\|}},\end{aligned}$$

where $\alpha = (\alpha_1, \dots, \alpha_n)$, $\|\alpha\| = \alpha_1 + \cdots + \alpha_n$, $\alpha! = \alpha_1! \dots \alpha_n!$, $z^\alpha = z_1^{\alpha_1} \dots z_n^{\alpha_n}$. Now from (25.10) it follows that

$$\Phi_s(z) = \sum_{\|\alpha\|\geq 0} c_\alpha(s) z^\alpha, \tag{25.13}$$

where

$$c_\alpha(s) = \frac{(\|\alpha\|+n-1)!}{\alpha!\,(n-1)!}\left\langle (\varphi_s f)_\zeta, \left(\frac{\operatorname{grad}\chi}{(\operatorname{grad}\chi, \bar\zeta)}\right)^\alpha \frac{\omega(\zeta, \operatorname{grad}\chi)}{d\sigma(\zeta)}\right\rangle.$$

By Theorem 25.10, a necessary and sufficient condition for f to extend holomorphically into Ω^+ is that the sum of the series (25.13) extend holomorphically into Ω_s for every s. If Ω is a bounded, convex, n-circular domain, then we can give simple conditions for this holomorphic extension. Let $\psi(z) = \psi(|z|) = \psi(|z_1|, \dots, |z_n|)$ be the Minkowski functional of the domain Ω. Then $\Omega = \{z : \psi(z) < 1\}$. We set $d_\alpha(\Omega) = \max_{\overline{\Omega}} |z|^\alpha$. Then $d_\alpha(\Omega_{(r)}) = r^{\|\alpha\|} d_\alpha(\Omega)$, where

$\Omega_{(r)} = r\Omega = \{z : \psi(z) < r\}$. Now from Theorem 25.10, Corollary 25.11, and the condition for convergence of power series in a bounded, complete, n-circular domain (see, for example, [209, §14]), we obtain the following (where we set $\Omega_s = \Omega_{(1-1/s)}$).

Theorem 25.12 (Aĭzenberg, Kytmanov). *The following conditions are necessary and sufficient for a distribution f to extend holomorphically into Ω^+, where Ω is a bounded, convex, n-circular domain.*

1. *f is a CR-distribution on Γ;*

2. $\limsup_{\|\alpha\|\to\infty} \sqrt[\|\alpha\|]{|c_\alpha(s) d_\alpha(\Omega)|} \le (1 - 1/s)^{-1}$ *for all s.*

If $f \in \mathcal{L}^1(\Gamma)$, then the second condition may be replaced by

3. $\limsup_{\|\alpha\|\to\infty} \sqrt[\|\alpha\|]{|c_\alpha| d_\alpha(\Omega)} \le 1$, *where*

$$c_\alpha = \frac{(\|\alpha\| + n - 1)!}{\alpha!} \int_\Gamma f(\zeta) \left(\frac{\operatorname{grad} \psi(\zeta)}{(\operatorname{grad} \psi, \bar{\zeta})} \right)^\alpha \omega(\zeta, \operatorname{grad} \chi).$$

We further consider the special case of this theorem when $\Omega = B$ is the unit ball with center at the origin. Then it is convenient to consider $\psi^2(z) = |z|^2$, $\operatorname{grad} \psi^2 = \bar{z}$, and

$$d_\alpha(B) = \frac{\prod_{j=1}^n \alpha_j^{\alpha_j/2}}{\|\alpha\|^{\|\alpha\|/2}} = \sqrt{\frac{\alpha^\alpha}{\|\alpha\|^{\|\alpha\|}}}.$$

Corollary 25.13. *If $\Omega = B$, then necessary and sufficient conditions for the existence of a holomorphic extension of $f \in \mathcal{L}^1(\Gamma)$ into Ω^+ are*

1. *f is a CR-function on Γ;*

2. $$\limsup_{\|\alpha\|\to\infty} \sqrt{\|\alpha\|} \sqrt[\|\alpha\|]{\frac{|c_\alpha|}{\sqrt{\alpha^\alpha}}} \le 1, \tag{25.14}$$

   where $c_\alpha = \int_\Gamma f(\zeta)(\bar{\zeta}/|\zeta|^2)^\alpha \omega(\zeta, \bar{\zeta})$.

It is interesting to compare Corollary 25.13 with Theorem 25.7. There we considered a complete orthonormal system of polynomials $\{P_{k,s}\}$, and instead of (25.14) we had the condition (25.9)

$$\limsup_{k\to\infty} \max_s \sqrt[k]{|a_{k,s}|} \le 1.$$

The holomorphic monomials z^α and z^β are orthogonal in $\mathcal{L}^2(\partial B)$ if $\alpha \neq \beta$. The norm of z^α in $\mathcal{L}^2(\partial B)$ equals (compare with Proposition 1.4.9 from [175])

$$\|z^\alpha\| = \sqrt{\frac{\alpha!\,(n-1)!\,s_n}{(\|\alpha\|+n-1)!}}, \tag{25.15}$$

where s_n is the volume of the sphere ∂B. Furthermore, if as $P_{k,s}$ we consider the monomial $z^\alpha/\|z^\alpha\|$, where $\|\alpha\| = k$, then

$$\begin{aligned}
\star\partial \frac{\overline{P}_{k,s}}{|z|^{2n+2k-2}} &= \frac{2^{1-n}i^n}{\|z^\alpha\|}\sum_{j=1}^{n}(-1)^{j-1}\frac{\partial}{\partial z_j}\frac{\overline{P}_{k,s}}{|z|^{2n+2k-2}}\,d\bar z[j]\wedge dz \\
&= -\frac{2^{1-n}i^n}{\|z^\alpha\|}(n+k-1)\bar z^\alpha \sum_{j=1}^{n}(-1)^{j-1}\frac{\bar z_j\,d\bar z[j]\wedge dz}{|z|^{2n+2k}} \\
&= \frac{2(-1)^{n-1}(n+k-1)\pi^n}{(n-1)!\,\|z^\alpha\|}\left(\frac{\bar z}{|z|^2}\right)^\alpha \omega(z,\bar z).
\end{aligned}$$

Now it is clear that the quantities entering in (25.14) are equivalent to those figuring in (25.9), but only in the case that we are considering not all harmonic polynomials $P_{k,s}$, but just the holomorphic ones. We recall that in (25.9), there appear the coefficients of the series expansion of the harmonic function M^-f in terms of the system $\{P_{k,s}\}$ in a neighborhood of zero, and (25.9) is equivalent to the convergence of this series in the whole ball $\Omega = B$. At the same time, (25.14) guarantees convergence in the ball of a subseries of the given series, its so-called "holomorphic" part. The reason that the existence of a harmonic extension of the Bochner-Martinelli integral into B is equivalent to the existence of an extension of its holomorphic part is made clear by the following argument. We introduce the form

$$\begin{aligned}
U_k(\zeta,z) = \frac{(-1)^k(n-1)!}{(2\pi i)^n}\Bigg(&\sum_{j=1}^{k-1}(-1)^j\frac{\bar\zeta_j-\bar z_j}{|\zeta-z|^{2n}}\,d\bar\zeta[j,k] \\
&+\sum_{j=k+1}^{n}(-1)^{j-1}\frac{\bar\zeta_j-\bar z_j}{|\zeta-z|^{2n}}\,d\bar\zeta[k,j]\Bigg)\wedge d\zeta.
\end{aligned}$$

It is easy to check that

$$\frac{\partial}{\partial \bar z_k}U(\zeta,z) = \overline{\partial}_\zeta U_k(\zeta,z), \qquad k = 1,\dots,n.$$

Next we compute the derivatives of the Bochner-Martinelli transform:

$$\begin{aligned}
\frac{\partial}{\partial \bar z_k}M(\varphi_s f) &= \langle(\varphi_s f)_\zeta, \overline{\partial}_\zeta U_k(\zeta,z)\rangle \\
&= \langle f, \overline{\partial}_\zeta(\tilde\varphi_s U_k(\zeta,z))\rangle + \langle \varphi_s f, \overline{\partial}_\zeta((1-\tilde\varphi_s)U_k(\zeta,z))\rangle.
\end{aligned}$$

In the last expression, the first term equals zero, since f is a CR-distribution, and the second term is harmonic in Ω_s. Thus, all the functions $\partial M(\varphi_s f)/\partial \bar{z}_k$ are harmonic not only in $\Omega_s \setminus \Gamma$, but even in all of Ω_s. The operator $D_a M(\varphi_s f)$ has the same property, where

$$D_a = \sum_{k=1}^{n} (\bar{z}_k - \bar{a}_k) \frac{\partial}{\partial \bar{z}_k}.$$

This consideration lets us constructively select the holomorphic and the nonholomorphic part in the Bochner-Martinelli transform when Ω is a complete, bounded, n-circular domain. It is easy to see that in such domains, there is a complete orthonormal system (in $\mathcal{L}^2(\partial\Omega)$) of homogeneous harmonic polynomials $\{P^*_{k,l}\}$ generalizing the system in the case of the ball. The basis $\{P^*_{k,l}\}$ may be chosen so that each polynomial $P^*_{k,l}$ is homogeneous of degree p in z and degree q in $\bar{z}$, where $p + q = k$ (compare §5 in [175]). The function $M^-(\varphi_s f)$ can be expanded in some neighborhood of the origin in a series

$$M^-(\varphi_s f) = \sum_{k,l} a_{k,l} P^*_{k,l} = \sum_{k=0}^{\infty} \sum_{l=0}^{\delta(k)} a_{k,l} P^*_{k,l}.$$

Moreover,

$$D_0 M^-(\varphi_s f) = \sum_{k=0}^{\infty} \sum_{\substack{p+q=k \\ q>0}} a_{k,l}\, q(P^*_{k,l}) P^*_{k,l}, \tag{25.16}$$

where $q(P)$ denotes the degree of the polynomial P in $\bar{z}$.

The sum of the series (25.16) is harmonic in Ω_s, so this series converges in Ω_s (in this case Ω_s can be taken to be a homothety of Ω), but then the series

$$\sum_{k=0}^{\infty} \sum_{\substack{l=0 \\ q(P^*_{k,l})>0}}^{\delta(k)} a_{k,l} = P^*_{k,l} \tag{25.17}$$

also converges in Ω_s. Its sum is the nonholomorphic part in the Bochner-Martinelli transform, and the sum of the series

$$\sum_{k=0}^{\infty} \sum_{\substack{l=0 \\ q(P^*_{k,l})=0}}^{\delta(k)} a_{k,l} = P^*_{k,l} \tag{25.18}$$

is the holomorphic part. If $f \in \mathcal{L}^1(\Gamma)$, then we obtain in place of the Bochner-Martinelli transform the Bochner-Martinelli integral, which can be expanded in a neighborhood of the origin in a series in terms of $\{P^*_{k,l}\}$. This series divides into two parts: the sum of the series (25.17) is the nonholomorphic part of M^-f, which extends harmonically into the whole domain Ω, and the sum of the series (25.18) is the holomorphic part of M^-f, which in general does not extend into Ω. We have obtained the following.

Proposition 25.14. *A harmonic extension of M^-f from a neighborhood of zero into the whole of a complete, bounded, n-circular domain exists if and only if there exists a harmonic (and consequently holomorphic) extension of its holomorphic part, given by the series (25.18).*

In the same way we can obtain a new proof of Corollary 25.13 from the proposition and Theorem 25.7.

We now show that Corollary 25.13 is a special case of a considerably more general situation. Let Ω be a complete, bounded, n-circular domain, and let $|\partial\Omega|$ be the image of $\partial\Omega$ under the mapping $z \mapsto (|z_1|, \dots, |z_n|)$. Given a finite measure λ on $|\partial\Omega|$, a necessary and sufficient condition for the existence of a Szegő integral representation

$$f(z) = \frac{1}{(2\pi i)^n} \int_{|\partial\Omega|} d\lambda \int_{\Delta_{|\zeta|}} f(\zeta) h(z_1\bar{\zeta}_1, \dots, z_n\bar{\zeta}_n) \frac{d\zeta}{\zeta}, \tag{25.19}$$

where

$$h(z\bar{\zeta}) = \sum_{\|\alpha\| \geq 0} a_\alpha z^\alpha \bar{\zeta}^\alpha, \qquad a_\alpha = \left[\int_{|\partial\Omega|} |\zeta^{2\alpha}| \, d\lambda \right]^{-1},$$

$$\Delta_{|\zeta|} = \{ z : |z_j| = |\zeta_j|, j = 1, \dots, n \}, \qquad \frac{d\zeta}{\zeta} = \frac{d\zeta_1}{\zeta_1} \wedge \dots \wedge \frac{d\zeta_n}{\zeta_n},$$

is that the measure λ be massive on $|S(\Omega)|$, where $S(\Omega)$ is the Shilov boundary of Ω (see [10, §11]), that is, for each set $E \subset |\partial\Omega|$ of zero λ measure we have the inclusion $\overline{|\partial\Omega| \setminus E} \supset |S(\Omega)|$. In a number of cases, the Szegő kernel has been computed in closed form (see [10, §11]), but for our purposes this is not necessary. We rewrite (25.19) in the form

$$f(z) = \frac{1}{(2\pi i)^n} \int_{\partial\Omega} f(\zeta) h(z\bar{\zeta}) \, d\lambda \, \frac{d\zeta}{\zeta}$$

and suppose that $d\lambda$ is some differential form $(2\pi i)^n b(|\zeta|) \in \mathcal{C}^2$, that is, we will assume that the Szegő integral representation has the form

$$f(z) = \int_{\partial\Omega} f(\zeta) h(z\bar{\zeta}) \, b(|\zeta|) \wedge \frac{d\zeta}{\zeta} \tag{25.20}$$

for strictly logarithmically convex (in other words, strongly pseudoconvex) n-circular domains Ω with boundary $\partial\Omega \in \mathcal{C}^2$. In such domains, $S(\Omega) = \partial\Omega$, and the requirement of massiveness means that the support of the form b is all of $|\partial\Omega|$. Now if we consider a homothety $\Omega_{(r)}$, then we obtain from (25.20) the Szegő integral representation

$$f(z) = \int_{\partial\Omega_{(r)}} f(\zeta) h\left(z\frac{\bar{\zeta}}{r^2}\right) b\left(\frac{|\zeta|}{r}\right) \wedge \frac{d\zeta}{\zeta}. \tag{25.21}$$

Further, as in the proof of Lemma 25.9, we consider the difference

$$U(\zeta, z) - h\left(z\frac{\bar{\zeta}}{r^2}\right) b\left(\frac{|\zeta|}{r}\right) \wedge \frac{d\zeta}{\zeta}, \tag{25.22}$$

where $r = \psi(|\zeta|) = \psi(|\zeta_1|, \ldots, |\zeta_n|)$ is the Minkowski functional of the domain Ω. The form (25.22) is orthogonal to holomorphic functions for integration over $\partial\Omega_{(r)}$, and by a theorem of Dautov (see [10, §25], and also §18 of this book), it is representable on $\partial\Omega_{(r)}$ in the form $\bar{\partial}_\zeta \mu$, where μ has an explicit integral representation and consequently depends real-analytically on the parameter z. Now repeating the arguments from the proof of Lemma 25.9 and Theorem 25.12, we obtain the following result (we give it in the classical style in the spirit of Corollary 25.13, but it extends naturally to distributions).

Theorem 25.15 (Aĭzenberg, Kytmanov). *If Ω is a strongly pseudoconvex n-circular domain, and $b(|\zeta|) \in \mathcal{C}^2$, $|\zeta| \in |\partial\Omega|$, is a form of degree $(n-1)$ with support $|\partial\Omega|$, then necessary and sufficient conditions for the existence of a holomorphic extension of $f \in \mathcal{L}^1(\Gamma)$ into Ω^+ are*

1. *the function f is a CR-function on Γ;*
2. *we have the inequality*

$$\limsup_{\|\alpha\|\to\infty} \sqrt[\|\alpha\|]{\frac{|c_\alpha|}{d_\alpha(\Omega)}} \le 1, \tag{25.23}$$

where

$$c_\alpha = \int_\Gamma f(\zeta) \left(\frac{\bar{\zeta}}{\psi^2(|\zeta|)}\right)^\alpha b\left(\frac{|\zeta|}{\psi(|\zeta|)}\right) \wedge \frac{d\zeta}{\zeta}. \tag{25.24}$$

Proof. The proof was already indicated before the statement of the theorem. As in the proof of Theorem 25.12, we obtain the condition

$$\limsup_{\|\alpha\|\to\infty} \sqrt[\|\alpha\|]{|\tilde{c}_\alpha| d_\alpha(\Omega)} \le 1 \tag{25.25}$$

for extendibility, where

$$\tilde{c}_\alpha = \frac{\displaystyle\int_\Gamma f(\zeta) \left(\frac{\bar{\zeta}}{\psi^2(|\zeta|)}\right)^\alpha b\left(\frac{|\zeta|}{\psi(|\zeta|)}\right) \wedge \frac{d\zeta}{\zeta}}{\displaystyle\int_{\partial\Omega} |\zeta^{2\alpha}|\, b(|\zeta|) \wedge \frac{d\zeta}{\zeta}}. \tag{25.26}$$

It remains to note that under our hypotheses, we have the equality

$$\lim_{\|\alpha\|\to\infty} \sqrt[\|\alpha\|]{\frac{d_\alpha^2(\Omega)}{\ell_\alpha}} = 1$$

(where ℓ_α is the denominator of the right-hand side of (25.26)), so (25.25) is equivalent to (25.23). □

We remark that if $\Omega = B$ is the unit ball, then

$$b(|\zeta|) = \sum_{k=1}^{n} (-1)^{k-1} |z_k|^2 \, d|z|^2[k],$$

where $d|z|^2[k] = d|z_1|^2 \wedge \cdots \wedge d|z_{k-1}|^2 \wedge d|z_{k+1}|^2 \wedge \cdots \wedge d|z_n|^2$, and taking into account that $\psi(|z|) = |z|$, we obtain that

$$b\left(\frac{|\zeta|}{\psi(|\zeta|)}\right) \wedge \frac{d\zeta}{\zeta} = \omega(\zeta, \bar{\zeta}),$$

that is, Corollary 25.13 is a special case of Theorem 25.15 (compare c_α in (25.14) and (25.24)).

In Theorem 25.13, we did not need to know an explicit form for the Szegő kernel of Ω and the measure $b(|\zeta|) \wedge d\zeta/\zeta$, for the requirement of strong pseudoconvexity let us use the result of Dautov to establish that the difference of the differential forms (25.14) is real-analytic in the parameter z. We accomplished the same end in the proof of Lemma 25.19 without the requirement of strong pseudoconvexity, but by using that the difference (25.11) is the difference of two Cauchy-Fantappiè forms. Therefore the analogue of Theorem 25.15 holds for complete n-circular domains Ω if we know for the domains $\Omega_{(r)}$ a Cauchy-Fantappiè integral representation with a kernel that is holomorphic in z and real-analytic in ζ.

In conclusion, we remark that Theorems 25.12 and 25.13 appear to be the simplest and most natural conditions for the existence of a holomorphic extension in the multidimensional case.

26 The Cauchy problem for holomorphic functions of a Lebesgue class in a domain

26.1 Statement of the problem

We continue the study of conditions for the existence of a holomorphic extension of a function into a given domain, but from a rather different direction. Let S be a set of positive $(2n-1)$-dimensional Lebesgue measure on the boundary of a Lyapunov domain $D \Subset \mathbf{C}^n$.

Problem 1. Given a function $f_0 \in \mathcal{L}^q(S)$, where $1 \leq q \leq \infty$, find a function f, holomorphic in the domain D, whose nontangential boundary value on S agrees almost everywhere with f_0.

From the jump theorem for the Bochner-Martinelli integral with density in some particular class, it follows that the behavior of the extension f near S is completely determined by the smoothness of f_0. In particular, if f_0 is continuous inside S (that is, in $\overset{\circ}{S}$), then f is continuous on $D \cup \overset{\circ}{S}$, etc. The behavior of the extension f near the remaining part of the boundary of D is determined by the class of functions in which a solution of Problem 1 is sought. This problem was investigated in the previous section in the class of functions without any restrictions on the behavior near $\partial D \setminus S$. Here we study Problem 1 in the space of holomorphic functions of class $\mathcal{L}^2(D)$.

26.2 Some additional information on the Bochner-Martinelli integral

When $1 \leq q \leq \infty$, we denote by $h^q(D)$ the Hardy class of harmonic functions in the domain D (see [75]), that is, the harmonic functions $h(z)$ for which

$$\limsup_{\epsilon \to 0+} \int_{\partial D} |h(z - \epsilon\nu(z))|^q \, d\sigma < \infty.$$

Theorem 26.1 (Shlapunov, Tarkhanov). *If D is a bounded domain in $\mathbf{C}^n$ with smooth boundary, and $f \in \mathcal{L}^q(\partial D)$, where $1 < q < \infty$, then $M^+ f \in h^q(D)$, and $M^- f \in h^q(\mathbf{C}^n \setminus \overline{D})$.*

Proof. We will need a result that is actually a reformulation of Theorem 2.6 (Privalov's lemma for the Bochner-Martinelli integral). For a point $z^0 \in \partial D$, we denote by $V(z^0)$ a cone with vertex at z^0, with axis along the normal to ∂D at z^0, and with angle α between the axis and the generator less than $\pi/2$ (α does not depend on z^0).

Lemma 26.2. *If $f \in \mathcal{L}^1(\partial D)$, and z^0 is a Lebesgue point of f, then*

$$M^{\pm} f(z) = \pm\frac{1}{2} f(z^0) + \int_{\partial D \setminus B(z^0, |z - z^0|)} f(\zeta)\, U(\zeta, z^0) + r(z^0, z), \qquad z \notin \partial D, \tag{26.1}$$

where $r(z^0, z)$ tends uniformly to zero if $z \to z^0$ and $z \in V(z^0)$.

Proof. Represent $\frac{1}{2} f(z^0) = \mathrm{P.V.} \int_{\partial D} f(z^0)\, U(\zeta, z^0)$, and apply Privalov's lemma (Theorem 2.6). □

It follows from Lemma 26.2 that at all Lebesgue points $z^0 \in \partial D$ of f at which the singular integral $\mathrm{P.V.} \int_{\partial D} f(\zeta)\, U(\zeta, z^0)$ exists, that is, almost everywhere on ∂D (see [193, §4]), the functions $M^{\pm} f(z)$ have nontangential boundary values $M_b^{\pm} f(z^0)$ on ∂D. Moreover

$$M_b^{\pm} f(z^0) = \pm\frac{1}{2} f(z^0) + \text{P.V.} \int_{\partial D} f(\zeta)\, U(\zeta, z^0). \tag{26.2}$$

Since $f \in \mathcal{L}^q(\partial D)$, where $1 < q < \infty$, and singular integral operators are bounded in these spaces (see, for example, [193, §4]), we may conclude from (26.2) that $M_b^{\pm} f \in \mathcal{L}^q(\partial D)$. Moreover, we have the following generalization of Theorem 3.4.

Lemma 26.3. *Under the above hypotheses on f,*

$$\lim_{\epsilon\to 0+} \int_{\partial D} |M^{\pm} f(z \mp \epsilon\nu(z)) - M_b^{\pm} f(z)|^q \, d\sigma = 0. \tag{26.3}$$

Proof. Indeed, for $z \in \partial D$ and $\epsilon > 0$, we obtain from (26.1) and (26.2) that

$$\begin{aligned} & M^+ f(z - \epsilon\nu(z)) - M_b^+ f(z) \\ &= \int_{\partial D \setminus B(z,\epsilon)} f(\zeta)\, U(\zeta, z) - \text{P.V.} \int_{\partial D} f(\zeta)\, U(\zeta, z) + r(z, z - \epsilon\nu(z)). \end{aligned}$$

It is known from the theory of singular integral operators (see [193, §4]) that

$$\lim_{\epsilon\to 0+} \int_{\partial D} \left| \int_{\partial D \setminus B(z,\epsilon)} f(\zeta)\, U(\zeta, z) - \text{P.V.} \int_{\partial D} f(\zeta)\, U(\zeta, z) \right|^q d\sigma(z) = 0. \tag{26.4}$$

On the other hand, since $r(z, z - \epsilon\nu(z))$ tends uniformly to zero as $\epsilon \to 0^+$, we have further that

$$\lim_{\epsilon\to 0+} \int_{\partial D} |r(z, z - \epsilon\nu(z))|^q \, d\sigma = 0. \tag{26.5}$$

To deduce (26.3) from (26.4) and (26.5), it suffices to use the triangle inequality for the norm in $\mathcal{L}^q(\partial D)$. □

Theorem 26.1 follows immediately from (26.3). □

Theorem 26.1 was given in [187].

Corollary 26.4. *If D is a bounded domain in $\mathbf{C}^n$ with class C^2 boundary, and $f \in \mathcal{L}^q(\partial D)$, where $1 < q < \infty$, then $Mf \in \mathcal{L}^q(D)$.*

Proof. For $\epsilon > 0$, let $D_\epsilon = \{ z \in D : \operatorname{dist}(z, \partial D) > \epsilon \}$, and choose ϵ so that for each point in the shell $D \setminus D_\epsilon$ there is just one closest point of ∂D (concerning the possibility of such a choice for domains with boundary of class C^2 see [61, 213]). Then $D \setminus D_\epsilon = \{ \zeta = z - \delta\nu(z) : z \in \partial D,\ 0 < \delta \le \epsilon \}$, and the volume form in the shell $D \setminus D_\epsilon$ is $dv = \varphi\, d\delta\, d\sigma$. Therefore

$$\int_D |Mf(z)|^q \, dv \le \int_{D_\epsilon} |Mf(z)|^q \, dv + \int_0^\epsilon \int_{\partial D} |Mf(z - \epsilon\nu(z))|^q \, d\sigma d\delta < \infty$$

in view of Theorem 26.1. Consequently, $Mf \in \mathcal{L}^q(D)$. □

This corollary is contained in [187].

26.3 Weak boundary values of holomorphic functions of class $\mathcal{L}^q(D)$

We have already seen that holomorphic (or more generally, harmonic) functions f of finite order of growth near the boundary ∂D have weak boundary values f_0 on ∂D (see §11).

A further elementary, but useful, observation is that f can be represented in terms of its weak boundary value f_0 by the Bochner-Martinelli formula

$$(\chi_D f)(z) = \langle (f_0)_\zeta, U(\zeta, z)/d\sigma(\zeta)\rangle, \qquad z \notin \partial D,$$

where χ_D is the characteristic function of D. It follows at once that the condition of finite-order growth near ∂D is necessary for the existence of weak boundary values. Indeed, f_0 necessarily has a finite order of singularity on ∂D, so to obtain the necessity we may use the structure of the kernel $U(\zeta, z)$.

Furthermore, it follows from Theorem 26.1 that a necessary and sufficient condition for f_0 to belong to $\mathcal{L}^q(\partial D)$ is that f belong to the Hardy class $\mathcal{H}^q(D)$ (also see Theorem 1.5).

Now what can be said about the boundary values on ∂D of holomorphic functions in class $\mathcal{L}^q(D)$, that is, $f \in \mathcal{O}(D) \cap \mathcal{L}^q(D)$, when $1 \le q \le \infty$? It is known that when $q < \infty$, this class is larger than $\mathcal{H}^q(D)$. Moreover, it is not even clear a priori if functions $f \in \mathcal{O}(D) \cap \mathcal{L}^q(D)$ must have finite order growth at ∂D, that is, if they must have weak boundary values on ∂D.

Thus, we fix $f \in \mathcal{O}(D) \cap \mathcal{L}^q(D)$. Not worrying for the moment about whether it is well defined, we associate to f a distribution $f_0 \in \mathcal{D}'(\partial D)$ in the following way: given $g \in C^1(\partial D)$, we choose any differential form β of bidegree $(n, n-1)$ of class $C^1(\overline{D})$ whose restriction to ∂D has the form $\beta|_{\partial D} = g\, d\sigma$, and we set

$$\langle f_0, g\rangle = \int_D f\, \overline{\partial}\beta. \tag{26.6}$$

Lemma 26.5. *Formula (26.6) is well defined, that is, it does not depend on the choice of the $(n, n-1)$-form β of class $C^1(\overline{D})$ for which $\beta/d\sigma = g$.*

Proof. It is enough to see that if the restriction of β to ∂D equals zero, then $\int_D f\,\overline{\partial}\beta = 0$. First of all, it is easy to show by using Koppelman's formula for exterior differential forms (Theorem 1.11) that if the (p, q)-form β of class $C^1(\overline{D})$ has zero tangential part on ∂D, then there is a (p, q)-form $\tilde{\beta}$ of the same class on $\overline{D}$ that equals zero on ∂D coefficientwise, and such that $\overline{\partial}\tilde{\beta} = \overline{\partial}\beta$. Consequently, we may assume without loss of generality that β vanishes coefficientwise on ∂D. We then use Bochner's lemma asserting that for any $\epsilon > 0$ there exists a function $\varphi_\epsilon \in \mathcal{D}(\mathbf{C}^n)$, $0 \le \varphi_\epsilon \le 1$, with support in an ϵ-neighborhood of ∂D, equal to 1 in some small neighborhood of ∂D, for which $|D^\alpha \varphi_\epsilon| \le c_\alpha \epsilon^{-\|\alpha\|}$ everywhere in $\mathbf{C}^n$, where the constant c_α does not depend on ϵ (see, for example, [78, chap. 1]). We have

$$\int_D f\,\overline{\partial}\beta = \int_D f\,\overline{\partial}((1-\varphi_\epsilon)\beta) + \int_D f\,\overline{\partial}(\varphi_\epsilon\beta). \tag{26.7}$$

Since the form $(1-\varphi_\epsilon)\beta$ has compact support in D, the first term on the right-hand side of (26.7) vanishes by Stokes's formula. Concerning the second term, denoting the set $\{ z \in D : \operatorname{dist}(z, \partial D) > \epsilon \}$ by D_ϵ, we can write

$$\int_D f\,\overline{\partial}(\varphi_\epsilon \beta) = \int_{D\setminus D_\epsilon} f(\overline{\partial}\varphi_\epsilon \wedge \beta) + \int_{D\setminus D_\epsilon} f(\varphi_\epsilon\,\overline{\partial}\beta). \tag{26.8}$$

To estimate the first term on the right-hand side of (26.8), we use Hölder's inequality. For this purpose, let us agree that the modulus of the differential form β (at the point z) means the square root of the sum of the squares of the moduli of its coefficients (at the point z).

Let q' be the number such that $(1/q) + (1/q') = 1$. Taking into account the estimate for the derivatives of φ_ϵ, we obtain for some constant $C_1 > 0$ not depending on ϵ that

$$\begin{aligned} \left| \int_{D\setminus D_\epsilon} f(\overline{\partial}\varphi_\epsilon \wedge \beta) \right| &\leq \|f\|_{\mathcal{L}^q(D\setminus D_\epsilon)} \left\| \frac{\overline{\partial}\varphi_\epsilon \wedge \beta}{dv} \right\|_{\mathcal{L}^{q'}(D\setminus D_\epsilon)} \\ &\leq C_1 \|f\|_{\mathcal{L}^q(D\setminus D_\epsilon)} \left[\frac{1}{\epsilon} \|\, |\beta|\, \|_{\mathcal{L}^{q'}(D\setminus D_\epsilon)} \right]. \end{aligned} \tag{26.9}$$

Since $|\beta| \in C^1(\overline{D})$, and $|\beta| = 0$ on ∂D, it is easy to see by using a localization procedure and the Newton-Leibniz formula that there is a constant $C_2 > 0$ such that for all sufficiently small $\delta > 0$, we have the estimate

$$\|\, |\beta|\, \|_{\mathcal{L}^{q'}(\partial D_\delta)} \leq C_2 \|\, |\beta|\, \|_{\mathcal{W}^1_{q'}(D\setminus D_\delta)}. \tag{26.10}$$

Here $\mathcal{W}^1_{q'}(D \setminus D_\delta)$ is the Sobolev space of functions $f \in \mathcal{L}^{q'}(D \setminus D_\delta)$ with first derivatives of the same class, provided with the usual norm. We now take $\epsilon > 0$ sufficiently small and integrate inequality (26.10) in δ from 0 to ϵ. Using Fubini's theorem, just as in the proof of Corollary 26.4, we arrive at the inequality

$$\|\, |\beta|\, \|_{\mathcal{L}^{q'}(D\setminus D_\epsilon)} \leq C_2 \epsilon \,\|\, |\beta|\, \|_{\mathcal{W}^1_{q'}(D\setminus D_\epsilon)}.$$

Substituting this estimate into (26.9), we obtain

$$\left| \int_{D\setminus D_\epsilon} f\,(\overline{\partial}\varphi_\varsigma \wedge \beta) \right| \leq C_1 C_2 \|f\|_{\mathcal{L}^q(D\setminus D_\epsilon)} \|\, |\beta|\, \|_{\mathcal{W}^1_{q'}(D\setminus D_\epsilon)}.$$

The analogous estimate for the second term on the right-hand side of (26.8) is evident. Thus, there is a constant $C > 0$ such that for all sufficiently small $\epsilon > 0$, we have

$$\left| \int_D f\,\overline{\partial}\beta \right| \leq C \|f\|_{\mathcal{L}^q(D\setminus D_\epsilon)} \|\, |\beta|\, \|_{\mathcal{W}^1_{q'}(D\setminus D_\epsilon)}. \tag{26.11}$$

From the property of absolute continuity of the Lebesgue integral with respect to the integration set, it follows that whatever q is in the range $1 \le q \le \infty$, the expression on the right-hand side of (26.11) tends to zero as $\epsilon \to 0^+$. Therefore $\int_D f\,\overline{\partial}\beta = 0$, as required. □

Now we can state the main result of this subsection. If $s \in \mathbf{Z}_+$ and $1 \le q \le \infty$, we denote by $\mathcal{W}_q^s(D)$ the space of all functions f in D whose derivatives through order s are qth power Lebesgue summable, that is,

$$\sum_{\|\alpha\| \le s} \int_D |D^\alpha f(z)|^q \, dv < \infty.$$

We also need the Sobolev space $\mathcal{W}_q^s(D)$ for nonintegral indices. Let $s = [s]+h$, where $[s] \ge 0$ is the integral part of s, and $0 < h < 1$. A function f belongs to $\mathcal{W}_q^s(D)$ if f and all its derivatives through order $[s]$ satisfy a Hölder condition of order h in D in the metric of $\mathcal{L}^q$, that is,

$$\sum_{\|\alpha\| \le s} \int_D \int_D \frac{|D^\alpha f(z) - D^\alpha f(w)|^q}{|z-w|^{2n+hq}} \, dv_z \, dv_w < \infty.$$

It is known that when $1 < q < \infty$, the space $\mathcal{W}_q^s(D)$ is reflexive, which makes it possible to extend the scale even to nonintegral negative indices s. Namely, for nonintegral s we set by definition $\mathcal{W}_q^s = (\mathcal{W}_{q'}^{-s})'$, where the space of functionals on the right-hand side is given the strong topology (see, for example, [106]).

Theorem 26.6 (Kudryavtsev, Nikolskiĭ [106]). *If $\partial D \in \mathcal{C}^{s+1}$, where $s \in \mathbf{Z}_+$ and $s - 1/q > 0$, then each function $f \in \mathcal{W}_q^s(D)$ has a trace on ∂D that lies in the space $\mathcal{W}_q^{s-1/q}(\partial D)$, and the corresponding mapping is surjective, continuous, and has a continuous inverse branch.*

Theorem 26.7 (Shlapunov, Tarkhanov). *Formula (26.6) defines a continuous linear functional on the space $\mathcal{C}^1(\partial D)$. Moreover, $f_0 \in \mathcal{W}_q^{-1/q}(\partial D)$.*

Proof. It is clear that only the second part requires proof. Using Hölder's inequality, we obtain from (26.6) the estimate

$$|\langle f_0, g\rangle| \le C \|f\|_{\mathcal{L}^q(D)} \| \, |\beta| \, \|_{\mathcal{W}_{q'}^1(D)}. \tag{26.12}$$

According to Lemma 26.5, the expression on the left-hand side does not depend on the choice of the $(n, n-1)$-form β of class $\mathcal{C}^1(\overline{D})$ for which $\beta/d\sigma = g$. In particular, we can make this choice in correspondence with the trace Theorem 26.6 so that

$$\| \, |\beta| \, \|_{\mathcal{W}_{q'}^1(D)} \le C_1 \|g\|_{\mathcal{W}_{q'}^{1-1/q'}}(\partial D),$$

where the constant $C_1 > 0$ does not depend on g. Substituting this estimate into (26.12) we conclude that

$$|\langle f_0, g\rangle| \leq C_f \|g\|_{\mathcal{W}^{1/q'}_{q'}(\partial D)} \tag{26.13}$$

for all $g \in \mathcal{C}^1(\partial D)$. Consequently, f_0 extends to be a continuous linear functional on the space $\mathcal{W}^{1/q'}_{q'}(\partial D)$, that is $f_0 \in \mathcal{W}^{-1/q}_q(\partial D)$. □

Corollary 26.8. *Suppose $D \Subset \mathbf{C}^n$ is a Lyapunov domain, and $f \in \mathcal{O}(D) \cap \mathcal{L}^q(D)$, where $1 < q < \infty$. Then f has a weak boundary value $f_0 \in \mathcal{W}^{-1/q}_q(\partial D)$, which is defined by (26.6).*

Theorem 26.6 and Corollary 26.8 are contained in [106].

26.4 Doubly orthogonal bases in spaces of harmonic functions

Suppose Ω is a bounded domain in $\mathbf{C}^n$, and B is some relatively compact subdomain of Ω. We denote by $\mathcal{G}_2(B)$ the (obviously closed) subspace of $\mathcal{L}^2(B)$ consisting of harmonic functions in B. We will construct a special basis $\{b_j\}$ in the space $\mathcal{G}_2(B)$. What we want precisely is for the system $\{b_j\}$ to be an orthonormal basis in $\mathcal{G}_2(\Omega)$ and an orthogonal basis in $\mathcal{G}_2(B)$. Such bases are called doubly orthogonal.

In the following theorem, we state a condition on B under which a basis with these properties exists, and we show how to find it.

Theorem 26.9 (Shlapunov, Tarkhanov). *If the boundary of B is regular in the sense of potential theory, and the complement of B has no compact connected component in Ω, then there exists an orthonormal basis $\{b_j\}$ in the space $\mathcal{G}_2(\Omega)$ whose restriction to B is an orthogonal basis in $\mathcal{G}_2(B)$.*

Proof. We use a result of Krasichkov [105]. Suppose $\mathcal{L}^2$ is a Hilbert space with scalar product $(\,,\,)$, $\mathcal{G}_2$ is a subspace of $\mathcal{L}^2$, P is the orthogonal projection onto $\mathcal{G}_2$, and $M : \mathcal{L}^2 \to \mathcal{L}^2$ is a self-adjoint linear operator. We want to find a system $\{b_j\}$ such that

1. $\{b_j\}$ is dense in $\mathcal{G}_2$,
2. $\{b_j\}$ is orthonormal in $\mathcal{L}^2$,
3. $(Mb_j, b_k) = 0$ if $k \neq j$.

It is proved in [105] that if the restriction of the operator PM to $\mathcal{G}_2$ is one-to-one and compact, then such a system $\{b_j\}$ exists, is unique, and coincides with a system of eigenvectors of the operator PM, that is, $PMb_j = \lambda_j b_j$.

We use this result and consider $\mathcal{G}_2(\Omega)$ as a subspace of the Hilbert space $\mathcal{L}^2(\Omega)$. As M we take the operator of multiplication by the characteristic function of the set B, that is, $Mf = \chi_B f$. It is clear that M is a self-adjoint linear operator.

Lemma 26.10. *The restriction of M to $\mathcal{G}_2(\Omega)$ is compact.*

Proof. We need to show that M takes any bounded set to a precompact set.

Suppose $K \subset \mathcal{G}_2(\Omega)$ is a bounded set, that is, there is a constant $C > 0$ such that $\|f\|_{\mathcal{L}^2(\Omega)} \le C$ for all f in K. The image $M(K)$ of K under the mapping M is precompact if we can extract from each sequence $\{F_j\} \subset M(K)$ a subsequence $\{F_{j_k}\}$ that converges in $\mathcal{L}^2(\Omega)$. If $\{F_j\} \subset M(K)$, then $F_j = \chi_B f_j$, where $\{f_j\} \subset K$. The sequence $\{f_j\}$ is bounded in $\mathcal{D}'(\Omega)$, the space of distributions in Ω. Since $\mathcal{D}'$ is a Montel space, we can extract from $\{f_j\}$ a subsequence $\{f_{j_k}\}$ that converges in $\mathcal{D}'(\Omega)$ to $f \in \mathcal{D}'(\Omega)$. Now we use the Stieltjes-Vitali theorem, in view of which f is harmonic in Ω, and $\{f_{j_k}\}$ converges to f in the topology of the space $C^\infty(\Omega)$. We set $F = \chi_B f$, and $F_{j_k} = \chi_B f_{j_k}$. Then $F \in \mathcal{L}^2(\Omega)$, and $\{f_{j_k}\}$ converges to F in $\mathcal{L}^2(\Omega)$. □

We denote by P the operator of orthogonal projection onto the subspace $\mathcal{G}_2(\Omega)$ of $\mathcal{L}^2(\Omega)$.

Lemma 26.11. *The restriction of the operator PM to $\mathcal{G}_2(\Omega)$ is compact and one-to-one.*

Proof. We remark that P is a continuous linear operator, so the restriction of PM to $\mathcal{G}_2(\Omega)$ is compact, being the composition of the compact operator M and the bounded operator P.

We now show that the restriction of PM to $\mathcal{G}_2(\Omega)$ is one-to-one. Suppose $f \in \mathcal{G}_2(\Omega)$, and $PMf = 0$. Then Mf lies in the orthogonal complement of $\mathcal{G}_2(\Omega)$, that is, $(Mf, f) = 0$, or in other words $\int_B |f|^2\, dv = 0$. Consequently $f = 0$ almost everywhere in B. Since f is a harmonic function, $f \equiv 0$ in Ω. □

Now we finish the proof of Theorem 26.9. In view of the theorem of Krasichkov [105], there exists a system $\{b_j\} \subset \mathcal{L}^2(\Omega)$ such that $\{b_j\}$ is dense in $\mathcal{G}_2(\Omega)$, orthonormal in $\mathcal{L}^2(\Omega)$, and orthogonal in $\mathcal{G}_2(B)$, with $PMb_j = \lambda_j b_j$. In particular, $\{b_j\}$ is an orthonormal basis in $\mathcal{G}_2(\Omega)$. Furthermore, it follows from results of [22, 92] that if the boundary of B is regular and the complement of B has no compact connected component in Ω, then $\mathcal{G}_2(\Omega)$ is dense in $\mathcal{G}_2(B)$. This means that $\{b_j\}$ is an orthogonal basis in $\mathcal{G}_2(B)$. □

Theorem 26.9 was given in [106].

26.5 Criteria for solvability of Problem 1

Suppose D is a bounded domain in $\mathbf{C}^n$, where $n > 1$, and S is a domain in the boundary of D. Concerning the smoothness of S, we suppose that it is a piece of a Lyapunov surface. As before, we denote the space of holomorphic functions in D by $\mathcal{O}(D)$. We consider solvability of the following variant of Problem 1.

Problem 1'. Under what conditions on the function $f_0 \in \mathcal{L}^2(S)$ is there a function $f \in \mathcal{O}(D) \cap \mathcal{L}^2(D)$ such that $f|_S = f_0$?

To formulate a solvability condition for this problem, we turn to the construction of §25. Let $\Omega \Subset \mathbf{C}^n$ be a domain larger than D, where $\Omega \setminus S$ consists of two connected components: $\Omega^+ = D$ and $\Omega^- = \Omega \setminus \overline{D}$. Now we choose a domain $B \subset \Omega^-$ with piecewise-smooth boundary such that the complement of B has no compact connected component in Ω. Then in view of Theorem 26.9, there exists an orthonormal basis $\{b_j\}$ in the space $\mathcal{G}_2(\Omega)$ whose restriction to B is an orthogonal basis in $\mathcal{G}_2(B)$. Furthermore, we can consider f_0 as an element of the space $\mathcal{L}^2(\partial D)$, extending it by zero to $\partial D \setminus S$. We will denote this extension also by f_0. Let Mf_0 be the Bochner-Martinelli integral of f_0. The restriction of Mf_0 to B lies in the space $\mathcal{G}_2(B)$. We denote by c_j the Fourier coefficients of the function Mf_0 with respect to the orthogonal system $\{b_j\}$: namely,

$$c_j = \int_B Mf_0(z)\overline{b_j(z)}\, dv \Big/ \int_B |b_j(z)|^2\, dv. \tag{26.14}$$

Theorem 26.12 (Shlapunov, Tarkhanov). *If $S \in C^{n+2}$, then necessary and sufficient conditions for solvability of Problem 1′ are*

1. $\sum_{j=1}^{\infty} |c_j|^2 < \infty$,
2. *f_0 is a CR-function on S.*

Necessity. Suppose that there is a function $f \in \mathcal{O}(D) \cap \mathcal{L}^2(D)$ whose weak limit on S agrees with f_0. Consider the following function:

$$F(z) = \begin{cases} Mf_0(z), & \text{if } z \in \Omega^-, \\ Mf_0(z) - f(z), & \text{if } z \in \Omega^+ = D. \end{cases}$$

We will show that $F \in \mathcal{L}^2(\Omega)$. Indeed, by our hypotheses on S we may assume that D is contained in some Lyapunov domain $\widetilde{D}$ having S as a part of its boundary. Consider the function $\tilde{f}_0 \in \mathcal{L}^2(\partial\widetilde{D})$ that agrees with f_0 on S and equals 0 on $\partial\widetilde{D}\setminus S$. Then $Mf_0 = \int_{\partial\widetilde{D}} \tilde{f}_0(\zeta)\, U(\zeta, z)$, so $Mf_0 \in \mathcal{L}^2(\Omega^+)$ by Corollary 26.4. Arguing analogously for Ω^-, we obtain that $Mf_0 \in \mathcal{L}^2(\Omega^-)$. Consequently, $F \in \mathcal{L}^2(\Omega)$.

We now verify that F is harmonic in Ω, that is, $F \in \mathcal{G}_2(\Omega)$. For this purpose we denote the restriction of F to $\Omega^\pm$ by $F^\pm$. Then $F^\pm$ is harmonic in $\Omega^\pm$. It follows from Hölder's inequality and the structure of the Bochner-Martinelli kernel that $F^\pm(z)$ has finite growth order k near S. Consequently $F^\pm$ has weak boundary values on S (see §11). It follows from Lemma 26.3 that when $\epsilon \to 0^+$, the function $M^\pm f_0(\zeta \mp \epsilon\nu(\zeta))$ converges to the function $\mp\frac{1}{2}f_0(\zeta) + \mathrm{P.V.}\, Mf_0(\zeta)$ in the norm of the space $\mathcal{L}^2(S)$. On the other hand, f has a weak boundary value f_0 on S. Therefore the boundary values of the $F^\pm$ on S agree and equal $-\frac{1}{2}f_0(\zeta) + \mathrm{P.V.}\, Mf_0(\zeta)$.

Furthermore, $\overline{\partial}_n f = 0$ since $f \in \mathcal{O}(D)$. On the other hand, by Lemma 25.3,

$$\lim_{\epsilon \to 0+} \int_S \left(\overline{\partial}_n Mf_0(z + \epsilon\nu(z)) - \overline{\partial}_n Mf_0(z - \epsilon\nu(z))\right) \varphi\, d\sigma = 0 \tag{26.15}$$

for all $\varphi \in \mathcal{D}(\Omega)$. This means that the weak boundary values of the complex normal derivatives agree on S.

Now we can use Theorem 25.1 to conclude that the $F^{\pm}$ will be harmonic in Ω after being suitably defined on S. Thus $F \in \mathcal{G}_2(\Omega)$ is a harmonic extension of Mf_0 from B into Ω. Therefore the expansion $F(z) = \sum_{j=1}^{\infty} c_j b_j(z)$ still converges in Ω in the norm of $\mathcal{L}^2(\Omega)$. From Bessel's inequality $\sum_{j=1}^{\infty} |c_j|^2 \leq \|F\|_{\mathcal{L}^2(\Omega)}$ it follows that the series $\sum_{j=1}^{\infty} |c_j|^2$ converges. We have proved that condition (1) of the theorem holds.

For the proof of condition (2), we fix a differential form γ of bidegree $(n, n-2)$ of class $\mathcal{D}(\mathbf{C}^n)$ such that $\operatorname{supp}\gamma \cap \partial D \subset S$. Then we obtain from Corollary 26.8 and definition (26.6) that

$$\int_S f_0\, \overline{\partial}\gamma = \int_D f\, \overline{\partial}(\overline{\partial}\gamma) = 0,$$

that is, $\overline{\partial}_\tau f_0 = 0$. □

Sufficiency. Suppose conditions (1) and (2) of the theorem hold for a function $f_0 \in \mathcal{L}^2(S)$. By condition (1) and the Riesz-Fischer theorem, there is a function $F \in \mathcal{G}_2(\Omega)$ such that $F(z) = \sum_{j=1}^{\infty} c_j b_j(z)$ in Ω. In particular, $F(z) = Mf_0(z)$ in B, that is, $F(z)$ is a harmonic extension of Mf_0 from Ω^- into Ω. Consider the function $f(z) = Mf_0(z) - F(z)$ for $z \in D$. It is harmonic in D with growth order k near S, where $k > (2n-1)/2$. This means that $f(z)$ has a weak boundary value on S. It follows from (26.3) that the weak boundary value of M^+f_0 on S equals $\frac{1}{2}f_0 + \mathrm{P.V.}\, Mf_0$. On the other hand, the weak boundary value on S of the function $F(z)$, $z \in \Omega^-$, equals the weak boundary value on S of the function Mf_0 in Ω^-, that is, $-\frac{1}{2}f_0 + \mathrm{P.V.}\, Mf_0$. Consequently, the weak boundary value on S of f equals f_0. Furthermore, since F agrees with Mf_0 in Ω^-, it follows from (26.15) that the weak boundary value on S of the complex normal derivative $\overline{\partial}_n f$ equals zero. Using Corollary 25.2, we find that a function $f(z)$ with such properties is holomorphic in D. Consequently, f is the desired solution of Problem 1′. □

Theorem 26.12 was given in [106]. It is a generalization of Theorem 25.7 for the given class of functions. In Theorem 25.7, the domain Ω is a ball, and the domain B is actually a ball of smaller radius with center at zero, while the system $\{b_j\}$ consists of the homogeneous harmonic polynomials $\{P_{k,s}\}$.

Bibliography

[1] M. L. Agranovskiĭ and R. E. Val′skiĭ, Maximality of invariant algebras of functions, *Sibirsk. Mat. Zh.* **12** (1971), no. 1, 3–12; English translation in *Siberian Math. J.* **12** (1971).

[2] M. L. Agranovskiĭ and A. M. Semenov, Boundary analogues of Hartogs's theorem, *Sibirsk. Mat. Zh.* **32** (1991), no. 1, 168–170; English translation in *Siberian Math. J.* **32** (1991), no. 1, 137–139.

[3] R. A. Aĭrapetyan and G. M. Henkin, Integral representations of differential forms on Cauchy-Riemann manifolds, and the theory of CR functions, *Uspekhi Mat. Nauk* **39** (1984), no. 3, 39–106; English translation in *Russian Math. Surveys* **39** (1984), no. 3, 41–118.

[4] L. A. Aĭzenberg, A multidimensional analogue of Carleman's formula, *Dokl. Akad. Nauk SSSR* **277** (1984), no. 6, 1289–1291; English translation in *Soviet Math. Dokl.* **30** (1984), no. 1, 241–243.

[5] L. A. Aĭzenberg, Multidimensional residues and their applications, in *Several Complex Variables II*, Encyclopaedia of Mathematical Sciences, vol. 8, Springer-Verlag, 1994, pp. 24–39. (translated from the Russian)

[6] L. A. Aĭzenberg, *Carleman's formulas in complex analysis*, Kluwer, Dordrecht, Boston, London, 1993. (translated from the Russian)

[7] L. A. Aĭzenberg and Sh. A. Dautov, *Differential forms orthogonal to holomorphic functions or forms, and their properties*, American Mathematical Society, Providence, RI, 1983. (translated from the Russian)

[8] L. A. Aĭzenberg and A. M. Kytmanov, On the possibility of holomorphic extension into a domain of functions defined on a connected piece of its boundary, *Mat. Sb.* **182** (1991), no. 4, 490–507; English translation in *Math. USSR Sb.* **72** (1992), no. 2, 467–483.

[9] L. A. Aĭzenberg and A. M. Kytmanov, On the possibility of holomorphic extension into a domain of functions defined on a connected piece of its

boundary, II, *Ross. Akad. Nauk Matem. Sbornik* **184** (1993), no. 1, 3–14; English translation in *Russian Akad. Sci. Sb. Math.* **78** (1994), no. 1, 1–10.

[10] L. A. Aĭzenberg and A. P. Yuzhakov, *Integral representations and residues in multidimensional complex analysis*, American Mathematical Society, Providence, RI, 1983. (translated from the Russian)

[11] N. I. Akhiezer and I. M. Glazman, *Theory of linear operators in Hilbert space*, Pitman, Boston, 1981. (translated from the Russian)

[12] W. Alt, Singuläre Integrale mit gemischten Homogenitäten auf Mannigfaltigkeiten und Anwendungen in der Funktionentheorie, *Math. Z.* **137** (1974), no. 3, 227–256.

[13] A. Andreotti and C. D. Hill, E. E. Levi convexity and Hans Lewy problem, I. Reduction to vanishing theorems, *Ann. Scuola Norm. Sup. Pisa* **26** (1972), no. 2, 325–363.

[14] A. Andreotti and F. Norguet, Problème de Levi pour les classes de cohomologie, *C. R. Acad. Sci. Paris* **258** (1964), 778–781.

[15] Piotr Antosik, Jan Mikusiński, and Roman Sikorski, *Theory of distributions: the sequential approach*, Elsevier, Amsterdam, 1973.

[16] A. M. Aronov, Functions that can be represented by a Bochner-Martinelli integral, *Properties of holomorphic functions of several complex variables*, Inst. Fiz. Sibirsk. Otdel. Akad. Nauk SSSR, Krasnoyarsk, 1973, 35–39. (Russian)

[17] A. M. Aronov, On the boundary values of derivatives of an integral of Bochner-Martinelli type, *Holomorphic functions of several complex variables*, Inst. Fiz. Sibirsk. Otdel. Akad. Nauk SSSR, Krasnoyarsk, 1976, pp. 20–28. (Russian)

[18] A. M. Aronov, The boundary behavior of an integral of Cauchy-Fantappiè type, *Combinatorial and asymptotic analysis, No. 2*, Krasnoyarsk State Univ., Krasnoyarsk, 1977, 114–117. (Russian)

[19] A. M. Aronov and Sh. A. Dautov, On a result of B. Weinstock, *Properties of holomorphic functions of several complex variables*, Inst. Fiz. Sibirsk. Otdel. Akad. Nauk SSSR, Krasnoyarsk, 1973, 193–195. (Russian)

[20] A. M. Aronov and A. M. Kytmanov, The holomorphy of functions that are representable by the Martinelli-Bochner integral, *Funktsional. Anal. i Prilozhen.* **9** (1975), no. 3, 83–84; English translation in *Functional Anal. Appl.* **9** (1975).

[21] A. M. Aronov and A. M. Kytmanov, Criterion for the existence of a holomorphic continuation of a smooth function given on the boundary of a domain in $\mathbf{C}^n$, *Holomorphic functions of several complex variables*, Inst. Fiz. Sibirsk. Otdel. Akad. Nauk SSSR, Krasnoyarsk, 1976, pp. 13–19. (Russian)

[22] T. Bagby, Approximation in the meanby solutions of elliptic equations, *Trans. Amer. Math. Soc.* **281** (1984), no. 2, 761–784.

[23] V. A. Baĭkov, On a boundary theorem for holomorphic functions of two complex variables, *Uspekhi Mat. Nauk* **24** (1969), no. 5, 221–222. (Russian)

[24] M. S. Baouendi and L. P. Rothschild, Extension of holomorphic functions in generic wedges and their wave front sets, *Comm. Partial Differential Equations* **13** (1988), no. 11, 1441–1466.

[25] D. Barrett and J. E. Fornæss, Uniform approximation of holomorphic functions on bounded Hartogs domains in $\mathbf{C}^2$, *Math. Zeit.* **191** (1986), 61–72.

[26] H. P. Boas, A geometric characterization of the ball and the Bochner-Martinelli kernel, *Math. Ann.* **248** (1980), no. 3, 275–278.

[27] H. P. Boas and Mei-Chi Shaw, Sobolev estimates for the Lewy operator on weakly pseudoconvex boundaries, *Math. Ann.* **274** (1986), no. 2, 221–231.

[28] S. Bochner, Analytic and meromorphic continuation by means of Green's formula, *Ann. of Math.* **44** (1943), 652–673.

[29] M. Brelot, *Éléments de la théorie classique du potentiel*, fourth edition, Paris, 1969.

[30] Hans Bremermann, *Distributions, complex variables, and Fourier transforms*, Addison-Wesley, Reading, MA, 1965.

[31] J. Bros and D. Iagolnitzer, Causality and local analyticity: mathematical study, *Ann. Inst. H. Poincaré Sect. A* **18** (1973), no. 2, 147–184.

[32] Yu. A. Brychkov and A. P. Prudnikov, *Integral transforms of generalized functions*, Nauka, Moscow, 1977; English translation, Gordon and Breach, New York, 1989.

[33] Jianxin Chen, Cauchy type integral for differential forms of type $(p, n-1)$, *J. Xiamen Univ. Nat. Sci.* **22** (1983), no. 4, 398–405. (Chinese with English summary)

[34] Jianxin Chen, Singular integral equations for exterior differential forms, *J. Math. Wuhan Univ.* **5** (1985), 357–370. (Chinese with English summary)

[35] Shu Jin Chen, The Sokhotskiĭ-Plemelj formula for a Cauchy-Fantappiè type integral, *J. Xiamen Univ. Natur. Sci.* **23** (1984), no. 3, 267–273. (Chinese with English summary)

[36] E. M. Chirka, Analytic representation of CR functions, *Mat. Sb.* **98** (1975), no. 4, 591–623; English translation in *Math. USSR Sb.* **27** (1975).

[37] E. M. Chirka, Currents and some of their applications, in the Russian edition of *Holomorphic chains and their boundaries* by Reese Harvey, Mir, Moscow, 1979, 122–158. (Russian)

[38] E. M. Chirka, ed., Some unsolved problems of multidimensional complex analysis, Inst. Fiz. Sibirsk. Otdel. Akad. Nauk SSSR, Krasnoyarsk, 1987, 38 pp. (Russian)

[39] E. M. Chirka and E. L. Stout, Removable singularities in the boundary, *Astérisque* (1994), to appear.

[40] Sh. A. Dautov, Forms orthogonal to holomorphic functions with respect to integration over the boundary of strictly pseudoconvex domains, *Dokl. Akad. Nauk SSSR* **203** (1972), 16–18; English translation in *Soviet Math. Dokl.* **13** (1972).

[41] Sh. A. Dautov and A. M. Kytmanov, The boundary values of an integral of Martinelli-Bochner type, *Properties of holomorphic functions of several complex variables*, Inst. Fiz. Sibirsk. Otdel. Akad. Nauk SSSR, Krasnoyarsk, 1973, 49–54. (Russian)

[42] J.-P. Demailly and C. Laurent-Thiébaut, Formules intégrales pour les formes différentielles de type (p, q) dans les variétés de Stein, *Ann. Sci. École Norm. Sup.* **20** (1987), no. 4, 579–598.

[43] P. Dolbeault, Théorème de Plemelj en plusieurs variables, *Riv. Mat. Univ. Parma* (4) **10** (1984), 47–54.

[44] R. E. Edwards, *Functional analysis: theory and applications*, Dover, New York, 1994.

[45] Yu. V. Egorov, M. A. Shubin, and R. V. Gamkrelidze, editors, *Partial differential equations I. Foundations of the classical theory*, Encyclopaedia of Mathematical Sciences 30, Berlin, Springer-Verlag, 1992. (translated from the Russian)

[46] M. A. Evgrafov, *Asymptotic estimates and entire functions*, Gordon and Breach, New York, 1961. (translated from the Russian)

[47] G. Fichera, Caratterizzazione della traccia, sulla frontiera di un campo, di una funzione analitica di più variabili complesse, *Atti. Accad. Naz. Lincei. Rend. Cl. Sci. Fis. Mat. Nat.* **22** (1957), 706–715.

[48] G. Fichera, Unification of global and local existence theorems for holomorphic functions of several complex variables, *Atti Accad. Naz. Lincei Mem. Cl. Sci. Fis. Mat. Natur. VIII Ser. Sez. I* **18** (1986), 61–83.

[49] B. Fisher, Neutrices and the product of distributions, *Studia Math.* **57** (1976), no. 3, 263–274.

[50] G. B. Folland and J. J. Kohn, *The Neumann problem for the Cauchy-Riemann complex*, Princeton Univ. Press, Princeton, NJ, 1972.

[51] B. A. Fuks, *Introduction to the theory of analytic functions of several complex variables*, second printing, Amer. Math. Soc., Providence, RI, 1965. (translated from the Russian)

[52] B. A. Fuks, *Special chapters in the theory of analytic functions of several complex variables*, Amer. Math. Soc., Providence, RI, 1965. (translated from the Russian)

[53] A. Gaziev, Limits of the Martinelli-Bochner integral, *Izv. Vyssh. Uchebn. Zaved. Matematika* **1978**, no. 9, 25–30. (Russian)

[54] A. Gaziev, Behavior of a Martinelli-Bochner type integral near the surface of integration, *Questions of mathematical analysis and their applications*, Samarkand State Univ., Samarkand, 1981, pp. 111–119. (Russian)

[55] A. Gaziev, Necessary and sufficient conditions for continuity of the Martinelli-Bochner integral, *Izv. Vyssh. Uchebn. Zaved. Mat.* (1983), no. 9, 13–17; English translation in *Sov. Math.* **27** (1983), no. 9.

[56] A. Gaziev, Some properties of a Martinelli-Bochner integral with continuous density, *Izv. Akad. Nauk UzSSR Ser. Fiz.-Mat. Nauk* **1985**, no. 1, 16–22. (Russian)

[57] J. Globevnik and E. L. Stout, Boundary Morera theorems for holomorphic functions of several complex variables, *Duke Math. J.* **64** (1991), no. 3, 571–615.

[58] V. D. Golovin, *Homology of analytic sheaves and duality theorems*, Consultants Bureau, New York, 1989. (translated from the Russian)

[59] Sheng Gong, A remark of integral of Cauchy type in several complex variables, *Proceedings of the 1980 Beijing Symposium on Differential Geometry and Differential Equations*, vol. 3, Gordon & Breach, New York, 1982, pp. 1183–1189.

[60] Sheng Gong (=Kung) and Ji-Huai Shi, Singular integrals in several complex variables, *Several complex variables: Proceedings of the 1981 Hangzhou conference*, Birkhäuser, Boston, 1984, pp. 181–185.

[61] N. Yu. Gorenskiĭ, Some applications of the differential properties of the distance function to open sets in $\mathbf{C}^n$, *Properties of holomorphic functions of several complex variables*, Inst. Fiz. Sibirsk. Otdel. Nauk SSSR, Krasnoyarsk, 1973, 203–208. (Russian)

[62] Phillip Griffiths and Joseph Harris, *Principles of algebraic geometry*, Wiley, New York, 1978.

[63] A. Grothendieck, Sur les espaces de solutions d'une classe générale d'équations aux dérivées partielles, *J. Analyse Math.* **2** (1953), 243–280.

[64] N. M. Günther, *Potential theory and its applications to basic problems of mathematical physics*, English translation of the Russian translation of the French, Ungar, New York, 1967.

[65] Reese Harvey, Holomorphic chains and their boundaries, *Proc. Symp. Pure Math.* **XXX, Part 1** (1977), 309–382.

[66] F. R. Harvey and H. B. Lawson, On boundaries of complex analytic varieties. I, *Ann. of Math.* **102** (1975), no. 2, 223–290.

[67] F. R. Harvey and J. Polking, Removable singularities of solutions of linear partial differential equations, *Acta Math.* **125** (1970), no. 1–2, 39–56.

[68] F. R. Harvey and J. C. Polking, The $\overline{\partial}$-Neumann kernel in the ball in $\mathbf{C}^n$, *Proc. Symp. Pure Math.* **41**, Amer. Math. Soc., Providence, RI, 1984, 117–136.

[69] F. R. Harvey and J. C. Polking, The $\overline{\partial}$-Neumann solution to the inhomogeneous Cauchy-Riemann equation in the ball in $\mathbf{C}^n$, *Trans. Amer. Math. Soc.* **281** (1984), no. 2, 587–613.

[70] T. E. Hatziafratis, An explicit Koppelman type integral formula on analytic varieties, *Michigan Math. J.* **33** (1986), no. 3, 335–341.

[71] W. K. Hayman and P. B. Kennedy, *Subharmonic functions*, vol. 1, Academic Press, London, 1976.

[72] G. M. Henkin, Analytic representation for CR-functions on submanifolds of codimension 2 in $\mathbf{C}^n$, *Lecture Notes in Math.* **798** (1980), 169–191.

[73] G. M. Henkin, The Hartogs-Bochner effect on CR manifolds, *Dokl. Akad. Nauk SSSR* **274** (1984), no. 3, 553–558; English translation in *Soviet Math. Dokl.* **29** (1984), no. 1, 78–82.

[74] G. M. Henkin, The method of integral representations in complex analysis, *Several Complex Variables I*, Encyclopaedia of Mathematical Sciences, Vol. 7, Springer-Verlag, 1990, pp. 19–116. (translated from the Russian)

[75] G. M. Henkin and E. M. Chirka, Boundary properties of holomorphic functions of several complex variables, *Current problems in mathematics*, vol. 4, Akad. Nauk SSSR Vsesoyuz. Inst. Nauchn. i Tekhn. Informatsii, Moscow, 1975, 12–142; English translation in *J. Soviet Math.* **5** (1976), no. 5.

[76] G. M. Henkin and J. Leiterer, Global integral formulas for solving the $\overline{\partial}$-equation on Stein manifolds, *Ann. Polon. Math.* **39** (1981), 93–116.

[77] G. M. Henkin and J. Leiterer, *Theory of functions on complex manifolds*, Birkhäuser, Basel, Boston, Stuttgart, 1984.

[78] L. Hörmander, *The analysis of linear partial differential operators I*, Springer-Verlag, Berlin, Heidelberg, New York, Tokyo, 1983.

[79] L. Hörmander, *An introduction to complex analysis in several variables*, third edition, North-Holland, Amsterdam, New York, 1989.

[80] M. Itano, On the multiplicative products of distributions, *J. Sci. Hiroshima Univ.*, Ser. A-I Math. **29** (1965), no. 1, 51–74.

[81] V. K. Ivanov, Multiplication of distributions and regularization of divergent integrals, *Iav. Vyssh. Uchebn. Zaved. Matematika* **1971**, no. 3, 41–49. (Russian)

[82] V. K. Ivanov, Hyperdistributions and multiplication of Schwartz distributions, *Dokl. Akad. Nauk SSSR* **204** (1972), 1045–1048; English translation in *Soviet Math. Dokl.* **13** (1972).

[83] V. K. Ivanov, The operation of multiplication for distributions of slow growth, *Izv. Vyssh. Uchebn. Zaved. Matematika* **1972**, no. 3, 10–19. (Russian)

[84] V. K. Ivanov, The algebra of a class of generalized functions, *Dokl. Akad. Nauk SSSR* **237** (1977), no. 4, 779–781. (Russian)

[85] V. K. Ivanov, An associative algebra of elementary generalized functions, *Sibirsk. Mat. Zh.* **20** (1979), 731–740; English translation in *Siberian Math. J.* **20** (1979).

[86] V. K. Ivanov, An algebra of elementary generalized functions, *Dokl. Akad. Nauk SSSR* **246** (1979), no. 4, 805–808. (Russian)

[87] V. K. Ivanov, Asymptotic approximations to the product of generalized functions, *Izv. Vyssh. Uchebn. Zaved. Matematika* **1981**, no. 1, 19–26. (Russian)

[88] V. K. Ivanov, Multiplication of homogeneous generalized functions of several variables, *Dokl. Akad. Nauk SSSR* **257** (1981), no. 1, 29–33. (Russian)

[89] B. Jöricke, Removable singularities of CR functions, *Dokl. Akad. Nauk SSSR* **296** (1987), no. 5, 1038–1041; English translation in *Soviet Math. Dokl.* **36** (1988), no. 2, 340–342.

[90] B. Jöricke, Removable singularities of CR-functions, *Ark. Mat.* **26** (1988), no. 11, 117–143.

[91] V. A. Kakichev, Character of continuity of the boundary values of the Martinelli-Bochner integral, *Oblast. Ped. Inst. Uchen. Zap.* **96** (1960), 145–160. (Russian)

[92] M. V. Keldysh and M. A. Lavrent′ev, Sur les suites convergentes de polynomes harmoniques, *Trudy Mat. Ins. Grus. Filial Akad. Nauk SSSR* **1** (1937), 165–184.

[93] K. Keller, Irregular operations in quantum field theory. I. Multiplication of distributions, *Rep. Math. Phys.* **14** (1978), no. 3, 285–309.

[94] K. Keller, Analytic regularizations, finite part prescriptions and products of distributions, *Math. Ann.* **236** (1978), no. 1, 49–84.

[95] N. Kerzman, Singular integrals in complex analysis, *Proc. Symp. Pure Math.* **XXXV, Part 2**, Amer. Math. Soc., Providence, RI, 1979, 3–41.

[96] N. Kerzman and E. M. Stein, The Szegő kernel in terms of Cauchy-Fantappiè kernels, *Duke Math. J.* **45** (1978), no. 2, 197–224.

[97] Kh. Ya. Khristov and B. P. Damyanov, Asymptotic functions: a new class of generalized functions. I. General statement and definitions, *Bulg. Fiz. Zh.* **5** (1978), no. 6, 543–556. (Russian)

[98] K. Kimura, Kernels for the $\overline{\partial}$-Neumann problem on the unit ball in $\mathbf{C}^2$, *Comm. Partial Differential Equations* **12** (1987), no. 9, 967–1028.

[99] J. J. Kohn, Subellipticity of the $\overline{\partial}$-Neumann problem on pseudoconvex domains: sufficient conditions, *Acta Math.* **142** (1979), no. 1–2, 79–122.

[100] H. Komatsu, Microlocal analysis in Gevrey classes and in complex domains, *Lecture Notes in Math.* **1495** (1991), 161–236.

[101] A. V. Kopaev, On a generalization of the Martinelli-Bochner integral formula, *Mathematical analysis and the theory of functions*, Moskov. Oblast. Ped. Inst., Moscow, 1980, 42–46. (Russian)

[102] W. Koppelman, The Cauchy integral for differential forms, *Bull. Amer. Math. Soc.* **73** (1967), no. 4, 554–556.

[103] A. P. Kopylov, Stability of classes of multidimensional holomorphic mappings. II. Stability of classes of holomorphic mappings, *Sibirsk. Mat. Zh.* **23** (1982), no. 4, 65–89; English translation in *Siberian Math. J.* **23** (1984), no. 4, 500–519.

[104] S. G. Krantz, *Function theory of several complex variables*, second edition, Wadsworth & Brooks/Cole, Pacific Grove, CA, 1992.

[105] I. F. Krasichkov, Systems of functions with the property of double orthogonality, *Mat. Zametki* **4** (1968), no. 5, 551–556. (Russian)

[106] L. D. Kudryavtsev and S. M. Nikol′skiĭ, Spaces of differentiable functions of several variables and embeddings, in *Analysis III: spaces of differentiable functions*, Encyclopaedia of Mathematical Sciences, vol. 26, Springer-Verlag, 1991. (translated from the Russian)

[107] A. M. Kytmanov, A criterion for the holomorphy of an integral of Martinelli-Bochner type, *Combinatorial and asymptotic analysis*, Krasnoyarsk State Univ., Krasnoyarsk, 1975, 169–177. (Russian)

[108] A. M. Kytmanov, A certain characteristic property of $\overline{\partial}$-closed exterior differential forms, *Uspekhi Mat. Nauk* **31** (1976), no. 2, 217–218. (Russian)

[109] A. M. Kytmanov, On an integral characteristic property of $\overline{\partial}$-closed complex differential forms, *Sibirsk. Mat. Zh.* **19** (1978), no. 4, 788–792; English translation in *Siberian Math. J.* **19** (1978).

[110] A. M. Kytmanov, The representation and product of distributions of several variables by means of harmonic functions, *Izv. Vyssh. Uchebn. Zaved. Matematika* **1978**, no. 1, 36–42; English translation in *Soviet Math.* **22** (1978).

[111] A. M. Kytmanov, Some differential criteria for holomorphy of functions in $\mathbf{C}^n$, *Some problems of multidimensional complex analysis*, Akad. Nauk SSSR Sibirsk. Otdel., Inst. Fiz., Krasnoyarsk, 1980, 51–64. (Russian)

[112] A. M. Kytmanov, A class of multidimensional distributions, *Izv. Vyssh. Uchebn. Zaved. Matematika* **1980**, no. 10, 23–28; English translation in *Soviet Math.* **24** (1980).

[113] A. M. Kytmanov, Multiplication of multidimensional distributions, *Generalized functions and their applications in mathematical physics*, Akad. Nauk SSSR, Vychisl. Tsentr, Moscow, 1981, pp. 316–322. (Russian)

[114] A. M. Kytmanov, Exact calculation of an integral of Martinelli-Bochner type in a ball in $\mathbf{C}^n$, *Uspekhi Mat. Nauk* **36** (1981), no. 3, 217–218; English translation in *Russian Math. Surveys* **36** (1981).

[115] A. M. Kytmanov, Calculation of a Martinelli-Bochner type integral in a ball and some of its applications, *Izv. Vyssh. Uchebn. Zaved. Mat.* **1983**, no. 3, 59–66; English translation in *Soviet Math.* **27** (1983).

[116] A. M. Kytmanov, On the removal of singularities of CR functions, *Uspekhi Mat. Nauk* **42** (1987), no. 6, 197–198; English translation in *Russian Math. Surveys* **42** (1987), no. 6, 239–240.

[117] A. M. Kytmanov, On the $\bar{\partial}$-Neumann problem for smooth functions and distributions, *Mat. Sb.* **181** (1990), no. 5, 656–668; English translation in *Math. USSR Sb.* **70** (1991), no. 1, 79–92.

[118] A. M. Kytmanov, On the removal of singularities of integrable CR functions, *Mat. Sb.* **136** (1988), no. 2, 178–186; English translation in *Math. USSR Sb.* **64** (1989), no. 1, 177–185.

[119] A. M. Kytmanov, Holomorphic extension of integrable CR functions from part of the boundary of a domain, *Mat. Zametki* **48** (1990), no. 2, 64–71; English translation in *Math. Notes* **48** (1990), no. 2, 761–765.

[120] A. M. Kytmanov, Generalized Fourier transform of distributions of slow growth, *Sibirsk. Mat. Zh.* **31** (1990), no. 2, 94–103; English translation in *Siberian Math. J.* **31** (1990), no. 2, 264–272.

[121] A. M. Kytmanov, Holomorphic extension of CR functions with singularities on a hypersurface, *Izv. Akad. Nauk SSSR, Ser. Mat.* **54** (1990), no. 6, 1320–1330; English translation in *Math. USSR Izv.* **37** (1991), no. 3, 681–691.

[122] A. M. Kytmanov, The analogue of Privalov lemma for the Bochner-Martinelli integral, *Work Collect. Equations of Mathematical Physics and Function Theory*, Krasnoyarsk, Krasnoyarsk State Univ. (1991), pp. 81–85. (Russian)

[123] A. M. Kytmanov and L. A. Aĭzenberg, The holomorphy of continuous functions that are representable by the Martinelli-Bochner integral, *Izv. Akad. Nauk Armyan. SSR Ser. Mat.* **13** (1978), no. 2, 158–169. (Russian)

[124] A. M. Kytmanov, B. B. Prenov, and N. N. Tarkhanov, The Poincaré-Bertrand formula for the Bochner-Martinelli integral, *Izv. Vyssh. Uchebn. Zaved. Matematika* **1992**, no. 11, 29–34; English translation in *Soviet Math.* **36** (1992), no. 11.

[125] A. M. Kytmanov and C. Rea, Elimination of $\mathcal{L}^1$ singularities on Hölder peak sets for CR functions, *Ann. Scuola Norm. Sup. Pisa* (1994), to appear.

[126] A. M. Kytmanov and M. Sh. Yakimenko, On holomorphic extension of hyperfunctions, *Sib. Mat. Zh.* **34** (1993), no. 6; English translation in *Siberian Math. J.* **34** (1993), no. 6, 1101-1109.

[127] N. S. Landkof, *Foundations of modern potential theory*, Nauka, Moscow, 1966; English translation, Springer-Verlag, 1972.

[128] C. Laurent-Thiébaut, Formules intégrales et théorèmes du type "Bochner" sur une variété de Stein, *C. R. Acad. Sci. Paris Sér. I Math.* **295** (1982), no. 12, 661–664.

[129] C. Laurent-Thiébaut, Théorème de Bochner sur une variété de Stein, *Lecture Notes in Math.* **1094** (1984), 151–161.

[130] C. Laurent-Thiébaut, Formules intégrales de Koppelman sur une variété de Stein, *Proc. Amer. Math. Soc.* **90** (1984), no. 2, 221–229.

[131] C. Laurent-Thiébaut, Transformation de Bochner-Martinelli dans une variété de Stein, *Lecture Notes in Math.* **1295** (1987), 96–131.

[132] C. Laurent-Thiébaut, Extension de formes différentielles CR, *C. R. Acad. Sci. Paris Sér. I Math.* **306** (1988), no. 13, 539–542.

[133] C. Laurent-Thiébaut, Sur l'extension des fonctions CR dans une variété de Stein, *Ann. Mat. Pura Appl.* **150** (1988), 141–151.

[134] Hsuan-Pei Lee and John Wermer, Orthogonal measures for subsets of the boundary of the ball in $\mathbf{C}^2$, *Recent developments in several complex variables*, Princeton Univ. Press, Princeton, NJ, 1981, 277–289.

[135] J. Leray, Fonction de variables complexes: sa représentation comme somme de puissances négatives de fonctions linéaires, *Atti Accad. Naz. Lincei. Rend. Cl. Sci. Fis. Mat. Natur.* **20** (1956), no. 5, 589–590.

[136] J. Leray, Le calcul différentiel et intégral sur une variété analytique complexe (Problème de Cauchy III), *Bull. Soc. Math. France* **87** (1959), 81–180.

[137] Bang He Li, Non-standard analysis and multiplication of distributions, *Sci. Sinica* **21** (1978), no. 5, 561–585.

[138] Liang Yu Lin, Integral of Cauchy type in $\mathbf{C}^n$ space, *J. Math. Res. Expo.* **4** (1984), no. 4, 105–106.

[139] Liang Yu Lin, Limit values of Cauchy type integral in several complex variables at angular points, *J. Xiamen Univ. Nat. Sci.* **25** (1986), no. 2, 128–134. (Chinese with English summary)

[140] Liang Yu Lin, The transformation formulas for angular point of Cauchy type integral of several complex variables, *J. Xiamen Univ. Nat. Sci.* **27** (1988), no. 3, 267–273. (Chinese with English summary)

[141] C. H. Look and T. D. Zhong, An extension of Privalov theorem, *Acta Math. Sinica* **7** (1957), no. 1, 144–165.

[142] G. Lupacciolu, Some global results on extensions of CR-objects in complex manifolds, *Trans. Amer. Math. Soc.* **321** (1990), no. 2, 761–774.

[143] G. Lupacciolu, A theorem on holomorphic extension of CR-functions, *Pacific J. Math.* **124** (1986), no. 1, 177–191.

[144] G. Lupacciolu, Holomorphic continuation in several complex variables, *Pacific J. Math.* **128** (1987), no. 1, 117–126.

[145] G. Lupacciolu, Valeurs au bord de fonctions holomorphes dans des domaines non bornés de $\mathbf{C}^n$, *C. R. Acad. Sci. Paris Sér. I Math.* **304** (1987), no. 3, 67–69.

[146] G. Lupacciolu and E. L. Stout, Removable singularities for $\overline{\partial}_b$, *Several complex variables: proceedings of the Mittag-Leffler Institute, 1987–1988*, Princeton Univ. Press, Princeton, NJ, 1993, 507–518.

[147] G. Lupacciolu and G. Tomassini, An extension theorem for CR-functions, *Ann. Mat. Pura Appl.* **137** (1984), 257–263. (Italian; English summary)

[148] Daowei Ma, Two problems on singular integrals in $\mathbf{C}^n$, *J. Wuhan Univ. Nat. Sci.* (1983), no. 3, 7–22. (Chinese with English summary)

[149] E. Martinelli, Alcuni teoremi integrali per le funzioni analitiche di piú variabili complesse, *Mem. R. Accad. Ital.* **9** (1938), 269–283.

[150] E. Martinelli, Sopra una dimonstrazione de R. Fueter per un theorema di Hartogs, *Comment. Math. Helv.* **15** (1943), 340–349.

[151] E. Martinelli, Sopra una formula di Andreotti-Norguet, *Boll. Un. Mat. Ital.* **11** (1975), no. 3, 455–457.

[152] E. Martinelli, Onalche riflessione sulla rappresentazione integrale di massima dimensione per le funzioni di più variabili complesse, *Atti Accad. Naz. Lincei. Rend. Cl. Sci. Fis. Mat. Natur.* **76** (1984), no. 4, 235–242.

[153] S. G. Mikhlin, *Multidimensional singular integrals and integral equations*, Pergamon, Oxford, New York, 1965. (translated from the Russian)

[154] S. G. Mikhlin, *Linear partial differential equations*, Vyssh. Shkola, Moscow, 1977. (Russian)

[155] E. S. Mkrtchyan, A logarithmic residue formula in $\mathbf{C}^n$ and its application to the study of univalent mappings, *Dokl. Akad. Nauk Armyan. SSR* **68** (1979), no. 1, 14–16. (Russian)

[156] N. I. Muskhelishvili, *Singular integral equations*, Nauka, Moscow, 1968; English translation, Dover, New York, 1992.

[157] A. Nagel and W. Rudin, Moebius-invariant function spaces on balls and spheres, *Duke Math. J.* **43** (1976), 841–865.

[158] M. Naser Shafii, Generalizations of Severi's theorem, *Mat. Zametki* **26** (1979), no. 5, 687–690; English translation in *Math. Notes* **26** (1979).

[159] M. Passare, Residues, currents, and their relation to ideals of holomorphic functions, *Math. Scand.* **62** (1988), no. 1, 75–152.

[160] J. C. Polking and R. O. Wells, Jr., Boundary values of Dolbeault cohomology classes and a generalized Bochner-Hartogs theorem, *Abh. Math. Semin. Univ. Hamburg* **47** (1978), 3–24.

[161] B. B. Prenov and N. N. Tarkhanov, A remark on the jump of the Martinelli-Bochner integral for domains with piecewise smooth boundary, *Sibirsk. Mat. Zh.* **30** (1989), no. 1, 199–201; English translation in *Siberian Math. J.* **30** (1989), 153–155.

[162] B. B. Prenov and N. N. Tarkhanov, On the singular Bochner-Martinelli integral, *Sib. Mat. Zh.* **33** (1992), no. 2, 202–205. (Russian)

[163] I. I. Privalov, *Randeigenschaften analytischer Funktionen*, second edition, Deutscher Verlag, Berlin, 1956.

[164] A. P. Prudnikov, Yu. A. Brychkov, and O. I. Marichev, *Integrals and series: Elementary functions*, Nauka, Moscow, 1981; English translation, Gordon & Breach, 1986.

[165] A. P. Prudnikov, Yu. A. Brychkov, and O. I. Marichev, *Integrals and series: Special functions*, Nauka, Moscow, 1983; English translation, Gordon & Breach, 1986.

[166] R. M. Range, The $\overline{\partial}$-Neumann operator on the unit ball in $\mathbf{C}^n$, *Math. Ann.* **266** (1984), no. 4, 449–456.

[167] Ya. A. Roĭtberg, On boundary values of generalized solutions of elliptic equations, *Mat. Sb.* **86** (1971), no. 2, 248–267; English translation in *Math. USSR Sb.* **15** (1971), no. 2, 241–260.

[168] A. V. Romanov, Spectral analysis of the Martinelli-Bochner operator for the ball in $\mathbf{C}^n$ and its applications, *Funkt. Anal. i Prilozhen.* **12** (1978), no. 3, 86–87; English translation in *Functional Anal. Appl.* **12** (1978), 232–234.

[169] A. V. Romanov, Convergence of iterates of the Bochner-Martinelli operator and the Cauchy-Riemann equation, *Dokl. Akad. Nauk SSSR* **242** (1978), no. 4, 780–783; English translation in *Soviet Math. Dokl.* **19** (1978), no. 5, 1211–1215.

[170] G. Roos, L'integrale de Cauchy dans $\mathbf{C}^n$, *Lecture Notes in Math.* **409** (1974), 176–195.

[171] G. Roos, Cocycles de noyaux de Martinelli, *Lecture Notes in Math.* **670** (1978), 365–369.

[172] G. Roos, Fonctiones de plusieurs variables complexes et formulas de representation integrale, *Lecture Notes in Math.* **1188** (1986), 45–182.

[173] J.-P. Rosay, Équation de Lewy—résolubilité globale de l'équation $\bar{\partial}_b u = f$ sur la frontière de domaines faiblement pseudoconvexes de $\mathbf{C}^2$ (ou $\mathbf{C}^n$), *Duke Math. J.* **49** (1982), no. 1, 121–128.

[174] J.-P. Rosay and E. L. Stout, Radó's theorem for CR-functions, *Proc. Amer. Math. Soc.* **106** (1989), no. 4, 1017–1026.

[175] Walter Rudin, *Function theory in the unit ball of* $\mathbf{C}^n$, Springer-Verlag, New York, Heidelberg, Berlin, 1980.

[176] L. Schwartz, Sur l'impossibilité de la multiplication des distributions, *C. R. Acad. Sci. Paris* **239** (1954), 847–848.

[177] A. I. Serbin, On functions representable by the Martinelli-Bochner integral, *Dokl. Akad. Nauk SSSR* **196** (1971), no. 6, 1276–1279; English translation in *Soviet Math. Dokl.* **12** (1971), no. 1, 357–361.

[178] A. I. Serbin, Permutation of the order of integration in an iterated integral with the Martinelli-Bochner kernel, *Izv. Vyssh. Uchebn. Zaved. Matematika* **1973**, no. 12, 64–72. (Russian)

[179] A. I. Serbin, Conditions for the representation of functions by the Martinelli-Bochner integral, *Izv. Vyssh. Uchebn. Zaved. Matematika* **1975**, no. 6, 136–142. (Russian)

[180] B. V. Shabat, *Introduction to complex analysis: Part II. Functions of several variables*, Amer. Math. Soc., Providence, 1992. (translated from the Russian)

[181] B. A. Shaimkulov, On an analogue of the F. and M. Riesz theorem, *Inst. Fiz. Sibirsk. Otdel. Akad. Nauk SSSR*, Krasnoyarsk, 1988, 17 pp. (Russian)

[182] P. Schapira, *Théorie des hyperfonctions* (Lecture Notes in Math., vol. 126), Springer-Verlag, 1970.

[183] Mei-Chi Shaw, L^2 estimates and existence theorems for the tangential Cauchy-Riemann complex, *Invent. Math.* **82** (1985), 133–150.

[184] G. E. Shilov, *Mathematical analysis. Second special course*, Nauka, Moscow, 1965; English translation, Gordon and Breach, New York, 1968.

[185] Yu. M. Shirokov, Algebra of one-dimensional generalized functions, *Teoret. Mat. Fiz.* **39** (1979), no. 3, 291–301. (Russian)

[186] Yu. M. Shirokov, Algebra of three-dimensional generalized functions, *Teoret. Mat. Fiz.* **40** (1979), no. 3, 348–354. (Russian)

[187] A. A. Shlapunov and N. N. Tarkhanov, On the Cauchy problem for holomorphic functions of Lebesgue class $\mathcal{L}^2$ in a domain, *Sib. Mat. Zh.* **33** (1992), no. 5, 186–195. (Russian)

[188] Z. Słodkowski, Analytic set-valued functions and spectra, *Math. Ann.* **225** (1981), 363–386.

[189] S. L. Sobolev, *Introduction to the theory of cubature formulas*, Nauka, Moscow, 1974; English translation *Cubature formulas and modern analysis: an introduction*, Gordon and Breach, Philadelphia, 1992.

[190] F. Sommen, Martinelli-Bochner type formulae in complex Clifford analysis, *Z. Anal. Anwendungen* **6** (1987), no. 1, 75–82.

[191] G. Sorani, Integral representations of holomorphic functions, *Amer. J. Math.* **88** (1966), no. 4, 737–746.

[192] L. N. Sretenskiĭ, *Theory of the Newtonian potential*, Gostekhizdat, Moscow, 1946. (Russian)

[193] E. M. Stein, *Singular integrals and differentiability properties of functions*, Princeton Univ. Press, Princeton, NJ, 1970.

[194] E. M. Stein, *Boundary behavior of holomorphic functions of several complex variables*, Princeton Univ. Press, Princeton, NJ, 1972.

[195] E. M. Stein and G. Weiss, *Introduction to Fourier analysis on Euclidean spaces*, second printing, with corrections, Princeton Univ. Press, Princeton, NJ, 1975.

[196] E. L. Stout, The boundary values of holomorphic functions of several complex variables, *Duke Math. J.* **44** (1977), no. 1, 105–108.

[197] E. L. Stout, Analytic continuation and boundary continuity of functions of several complex variables, *Proc. Edinburg Royal Soc.* **89A** (1981), 63–74.

[198] E. L. Stout, Removable singularities for the boundary values of holomorphic functions, *Several complex variables: proceedings of the Mittag-Leffler Institute, 1987–1988*, Princeton Univ. Press, Princeton, NJ, 1993, 600–629.

[199] E. J. Straube, Harmonic and analytic functions admitting a distribution boundary value, *Ann. Scuola Norm. Sup. Pisa Cl. Sci.* **11** (1984), no. 4, 559–591.

[200] Ji Guang Sun, Singular integral equations on a closed smooth manifold, *Acta Math. Sinica* **22** (1979), 675–692. (Chinese; English summary)

[201] N. N. Tarkhanov, On integral representation of solutions of systems of linear partial differential equations of first order and some of its applications, *Some questions of multidimensional complex analysis*, Inst. Fiz. Sibirsk. Otdel. Akad. Nauk SSS, Krasnoyarsk, 1980, pp. 147–160. (Russian)

[202] N. N. Tarkhanov, Theorems on the jump and exterior product of currents from $\mathcal{D}'_q(M)$ by means of their harmonic representations, *Sibirsk. Mat. Zh.* **24** (1983), no. 2, 203–204. (Russian)

[203] N. N. Tarkhanov, *The parametrix method in the theory of differential complexes*, Nauka, Novosibirsk, 1990. (Russian)

[204] H. G. Tillmann, Darstellung der Schwartzschen Distributionen durch analytische Funktionen, *Math. Z.* **77** (1961), 106–124.

[205] N. L. Vasilevskiĭ and M. V. Shapiro, An analogue of monogeneity in the sense of Moisil-Teodorescu and some applications in the theory of boundary value problems, Reports of the extended sessions of a seminar of the I. N. Vekua Institute of Applied Mathematics, Vol. I, no. 1, pp. 63–66, Tblis. Gos. Univ., Tblisi, 1985. (Russian)

[206] N. L. Vasilevskiĭ and M. V. Shapiro, Some questions of hypercomplex analysis, *Complex analysis and applications*, conference proceedings (Varna 1987), Publ. House Bulg. Acad. Sci, Sofia, 1989, pp. 523–531.

[207] N. L. Vasilevskiĭ and M. V. Shapiro, Integrals with Martinelli-Bochner kernel and quaternionic theory of functions, Abstracts of School-seminar "Complex analysis and mathematical physics," Krasnoyarsk, 1987, p. 18. (Russian)

[208] V. S. Vinogradov, Analogue of a Cauchy type integral for analytic functions of several complex variables, *Dokl. Akad. Nauk SSSR* **178** (1968), no. 2, 282–285; English translation in *Soviet Math. Dokl.* **9** (1968), no. 1, 73–76.

[209] V. S. Vladimirov, *Methods of the theory of functions of many complex variables*, MIT Press, Cambridge, MA, 1966. (translated from the Russian)

[210] V. S. Vladimirov, Problems of linear conjugacy of holomorphic functions of several complex variables, *Izv. Akad. Nauk SSSR Ser. Mat.* **29** (1965), 807–835; English translation in *Amer. Math. Soc. Transl.* (2) **71** (1968), 203–232.

[211] V. S. Vladimirov, *Equations of mathematical physics*, Mir, Moscow, 1984. (revised English translation of the 1981 Russian edition)

[212] V. S. Vladimirov, *Generalized functions in mathematical physics*, Nauka, Moscow, 1976; English translation, Mir, Moscow, 1979.

[213] E. A. Volkov, Boundaries of subdomains, Hölder weight classes and the solution of the Poisson equation in these classes, *Trudy Mat. Inst. Steklov* **117** (1972), 75–99; English translation in *Proc. Steklov Inst. Math.* **117** (1972).

[214] B. M. Weinstock, Continuous boundary values of analytic functions of several complex variables, *Proc. Amer. Math. Soc.* **21** (1969), no. 2, 463–466.

[215] B. M. Weinstock, An approximation theorem for $\overline{\partial}$-closed forms of type $(n, n-1)$, *Proc. Amer. Math. Soc.* **26** (1970), no. 4, 625–628.

[216] B. M. Weinstock, Uniform approximation by solutions of elliptic equations, *Proc. Amer. Math. Soc.* **41** (1973), no. 2, 513–517.

[217] R. O. Wells, *Differential analysis on complex manifolds*, Springer-Verlag, 1980.

[218] Dejun Wu and Zhongluan Zhong, On the integral representation on the bounded convex domain and the integral representation of Bochner-Martinelli, *J. Xiamen Univ. Nat. Sci.* **22** (1983), no. 4, 405–415. (Chinese with English summary)

[219] Zong Yuan Yao, An integral representation on bounded domains in $\mathbf{C}^n$, *J. Xiamen Univ. Nat. Sci.* **25** (1986), no. 3, 260–269. (Chinese with English summary)

[220] Zong Yuan Yao, A Bochner-Martinelli integral formula of type I in $\mathbf{C}^n$, *J. Xiamen Univ. Nat. Sci.* **26** (1987), no. 4, 390–398. (Chinese with English summary)

[221] Sh. Yarmukhamedov, Generalization of the Martinelli-Bochner integral representation, *Mat. Zametki* **15** (1974), no. 5, 739–747; English translation in *Math. Notes* **15** (1974).

[222] Sh. Yarmukhamedov, The Martinelli-Bochner integral formula and the Phragmén-Lindelöf principle, *Dokl. Akad. Nauk SSSR* **243** (1978), no. 6, 1414–1417; English translation in *Soviet Math. Dokl.* **19** (1978), 1592–1595.

[223] A. P. Yuzhakov and A. V. Kuprikov, The logarithmic residue in $\mathbf{C}^n$, *Properties of holomorphic functions of several complex variables*, Inst. Fiz. Sibirsk. Otdel. Akad. Nauk SSSR, Krasnoyarsk, 1973, 181–191. (Russian)

[224] P. Zappa, Remarks on the Bochner-Martinelli kernels, *Atti Accad. Naz. Lincei Rend. Cl. Sci. Fis. Mat. Natur.* **67** (1979), no. 1–2, 21–26. (Italian)

[225] A. I. Zaslavskiĭ, Distributions, closed differential forms and Fourier transforms, Ural State Univ., Sverdlovsk, 1983, 8 pp. (Russian)

[226] A. I. Zaslavskiĭ, Multipliable distributions, *Izv. Vyssh. Uchebn. Zaved. Matematika* **1983**, no. 5, 26–34. (Russian)

[227] I. V. Zhuravlev, Existence of solutions to a multidimensional analogue of the Beltrami equation, *Sibirsk. Mat. Zh.* **30** (1989), no. 1, 103–113; English translation in *Siberian Math. J.* **30** (1989), 79–87.

[228] L. N. Znamenskaya, Criterion for holomorphic extension of functions of class $\mathcal{L}^p$ given on a part of the Shilov boundary of a circular strictly starlike domain, *Sibirsk. Mat. Zh.* **31** (1990), no. 5, 175–177; English translation in *Siberian Math. J.* **31** (1990), no. 5, 848–850.

[229] Tongde Zhong, Transformation formulae of multiple singular integrals with Bochner-Martinelli kernel, *Acta Math. Sinica* **23** (1980), no. 4, 554–565. (Chinese with English summary)

[230] Tongde Zhong, Some applications of Bochner-Martinelli integral representation, *J. Xiamen Univ. Nat. Sci.* **22** (1983), no. 2, 127–132. (Chinese with English summary)

[231] Tongde Zhong, Singular integrals and singular integral equations on the smooth boundary of an unbounded domain in the space $\mathbf{C}^n$, *J. Xiamen Univ. Nat. Sci.* **23** (1984), no. 1, 1–13. (Chinese with English summary)

[232] Tongde Zhong, Some applications of Bochner-Martinelli integral representation, *Several complex variables: Proceedings of the 1981 Hangzhou conference*, Birkhäuser, Boston, 1984, pp. 217–225.

[233] Tongde Zhong, The Andreotti-Norguet formula on Stein manifolds and its generalizations, *J. Xiamen Univ. Nat. Sci.* **26** (1987), no. 6, 641–644. (Chinese with English summary)

[234] Tongde Zhong, Singular integrals and integral representations in several complex variables, *Several complex variables in China*, Contemporary Mathematics Vol. 142, American Mathematical Society, 1993, pp. 151–173.

Index

♯ defined, 65
⋆ defined, 155
□ defined, 156

$\mathcal{A}(U)$ defined, 1
$\mathcal{A}_2^s$ defined, 169
a priori estimates, 159
Abel mean, 138
absolute continuity, 266
Agranovskiĭ, M. L.
  and Semenov, 168
  and Val′skiĭ, 165
Alexander, H., 88
Alexander-Pontryagin duality, 90
algebra
  generated by finite products, 116, 119, 124, 135
  of Vladimirov, 116
  written as direct sum, 137
algebraic
  geometry, X
  hypersurface, 91
Almansi, E., 50, 52
analytic
  disc, 103
  functional, 122–124, 126, 234–235, 243
    Bochner-Martinelli transform of, 241
    multiplication for, 124
  hypersurface, 7
  real, *see* real-analytic
  representation, 55–70
    and local extension, 70
    and multiplication of distributions, 116
    application of, 61, 85, 89, 246
    problem of, 61–64
    theorem on, 64–69
  subset, 74
  wave front set, 138
Andreotti, A., 9, 64
angle, solid, 13, 197
antiholomorphicity, 127, 183
applications, 189–219
  of analytic representation of CR-functions, 85, 89, 246
  of analytic representation of distributions, 61
  of Bochner-Martinelli in the ball, 51–52
  of Carleman's formula, 201
  of Cauchy-Fantappiè, 7
  of Hartogs-Bochner theorem, 73
  of Lupacciolu's theorem, 90
  of residue formula, 199
  of Roos-Yuzhakov formula, 192
  of theorem on functions represented by Bochner-Martinelli, 165
approximation
  asymptotic, 277
  by bounded domains, 248
  by harmonic functions, 166
  by holomorphic functions, X, 69, 85, 86
  by polynomials, 70, 98
  by pseudoconvex domains, 85, 86, 88–90
  by smooth domains, 191
  by smooth functions, 4
  in the mean, 273
  of harmonic functions, 166
  to the identity, 115, 126
  Weinstock's theorem, 287
arc, 61

Aronov, A. M., 39, 160
and Dautov, 71, 73
and Kytmanov, 161
associated
function, 135
system, 135
associative, 115, 277
asymptotic
approximation, 277
decomposition, 116
estimates, 274
expansion, 136
functions, 278
average of Cauchy kernel, 164
axioms for multiplication of distributions, 115
Aĭzenberg, L. A., X, 201
analogue of Riesz theorem, 165, 177
and Dautov, 30
and Kytmanov
holomorphicity theorem, 162
jump theorem, 40
theorem on holomorphic extension of distributions, 248, 251–252, 255–256
theorem on holomorphic extension of integrable functions, 260
and Poisson summation formula, 199–200
Carleman type formula, 205–206

$B(z, \epsilon)$ defined, 2
B_s defined, 170
ball
$\bar{\partial}$-Neumann problem in, 183–188
Bochner-Martinelli integral in, 44–54, 249–252
characterization of, 53–54
counterexample in, 23–24, 162, 215, 220, 222, 249
holomorphic extension in, 256–258
holomorphicity of functions in, 225–227
Hörmander, 21
integral representation in, 183–188, 229
iterates of Bochner-Martinelli integral in, 172
Banach-Steinhaus theorem, 243
Bang He Li, 116
Baouendi, M. S., 70
Barrett, D., example of, 172
basis, *see* orthonormal basis
doubly orthogonal, 267–268
Baĭkov, V. A., 71
behavior, boundary
of Bochner-Martinelli integral, IX, 5, 13–21, 69
of derivatives, 33–44
of Green function, 5
of holomorphic extension, 75, 83, 176, 262
Bergman projection, 48
Bessel
function, 148
inequality, 270
bidegree, 32, 58, 59, 264, 270
bidimension, 59
bidisc, 63, 88
biholomorphic mapping, 8, 91, 163, 190, 195–197
bilinear form, 170–171, 174–175, 180, 243
Boas, H. P., 53, 178
Bochner transform, 145
Bochner, S., IX, 4, 5, 71
lemma of, 264
theorem of, *see* Hartogs-Bochner theorem
Bochner-Martinelli field, 67, 68
Bochner-Martinelli formula
analogue of, 221
applications of, IX
for differential forms, 9
for holomorphic functions, 4–6
for smooth functions, 4
for Stein manifolds, 8–9
for $\mathcal{W}_2^s$, 169
generalization of, 7–13, 191
in terms of currents, 60–61
proof of, 4–6, 45
Bochner-Martinelli integral, IX, 1–54
and holomorphic extension, 72

and the theorem of Hartogs, 70
boundary behavior, IX, 13–21, 69
computed in ball, 49–51
derivatives of, 33–44
functions represented by, X, 159–168, 172, 224, 241
holomorphic extension via, 244–252
in ball, 44–54, 249–252
inversion formula, 216
iterates of, 168–176
jump of, IX, 16–17, 21–33, 36–44, 81, 262
of polyharmonic function, 50
of polynomials, 48, 49
properties of, 5, 262–263
singular, X, 13–17, 20, 47, 53, 192–195, 206–219

Bochner-Martinelli kernel, IX, 2–5
$\overline{\partial}$-problem for, 77–80
adjoint of, 53
current defined by, 60
distributions acting on, 27
in terms of potentials, 8
invariance of, 18, 193, 218
restriction to boundary, 14, 27, 40, 45, 194, 215
series expansion of, 249

Bochner-Martinelli operator, 159
and characterization of ball, 53–54
bounded in C^α, 16
eigenvalues of, 47, 172
iterates of, 48
on spherical harmonics, 46–48
positivity of, 171
self-adjointness of, 47, 53, 171, 175, 243
spectrum of, 44–49, 172

Bochner-Martinelli transform, 247
jump of, 241
of analytic functional, 241

Bochner-Martinelli-Koppelman integral, 10, 67

Borel measure, 102, 138, 176

boundary
behavior, *see* behavior, boundary
class C^k, 1
integration over whole, 5
of ball, 2
of complex manifold, characterized, 73–74
orientation of, 2
piecewise-smooth, defined, 2
real-analytic, 181, 233, 240, 244
singularities on, 195–200
smooth, 1
values
nontangential, 262, 263
normal, 101
of harmonic functions, 107–108, 233–240
of polyharmonic functions, 128–129
weak, 264–267, 269, 270

boundary-value problem, 229

branch, inverse, 266

Bremermann, H., 116

Bros, J., 138

Bros-Iagolnitzer transform, 138, 142

bundle
cotangent, 8
tangent, 8

C^k defined, 1

C^r defined, 1

canonical solution, 180, 187

Carleman's formula, 200–206

Cartan's Theorem B, 8

Cauchy
formula, 5, 146, 201
inequality, 235
integral, IX, 5, 61, 192
kernel, IX, 2, 215
average of, 164
operator, 44
principal value, 13, 192
problem for holomorphic functions, 261–270
theorem, 64
type integral, IX, 16, 19, 23, 44, 47, 61

Cauchy-Bochner transform, 145

Cauchy-Fantappiè
formula, 6–7, 9, 13, 21

functions represented by, 220–222
integral, IX, 33
holomorphic extension via, 253–261
kernel, 220, 230, 231, 253
singular integral, 21
Cauchy-Green formula, 4, 7, 64
Cauchy-Riemann equation
tangential, 62–63, 74, 85, 89, 93, 158, 177–179, 182, 222, 246
weak, 62
chain, X, 28, 57, 74
characteristic function, 60, 97, 144, 191, 201, 202, 264, 267
characterization
of $\bar\partial$-closed forms, 165–168
of ball, 53–54
of boundaries of complex manifolds, 73–74
of harmonic functions of finite order of growth, 107–108
Chinese mathematicians, 216
Chirka, E. M., 17, 28, 64, 69, 71
and Stout, XI, 91
chord, 53–54
circle, 207, 223
circular
cone, 17, 21, 94
domain, 255, 256, 258–261
Clifford analysis, 9
closed current, 56
cohomology, 64, 67, 71, 85, 86, 89, 90
commutative, 115
compact
cohomology, 71, 86
curve, 74
operator, 267, 268
singularity, IX, 71, 75
complete
circular domain, 256, 258, 259, 261
orthonormal system, 249, 256, 258
completeness, weak, 106
complex
curve, 73
dimension, 73, 74
Laplacian, 157
line, 164, 168, 205
manifold, *see* manifold, complex
normal derivative, 270
plane, two-dimensional, 97
projective space, 90
structure, 58
tangent space, 63, 70, 73, 222
complexity, maximal, 73, 74
composition of singular integrals, 211–214
computation
of Bochner-Martinelli integral in ball, 49–51
of Fourier transform, 139–141, 146–154
concave, 75, 93
cone
Cauchy-Bochner transform for, 145
circular, 17, 21, 94
dual, 146
future light, 148
tangent, 13, 194, 195, 197, 198, 218
conical, 194, 204
connected component, 88
continuity
absolute, 266
modulus of, 20, 37, 193
continuous functions, 21–24
represented by Bochner-Martinelli, 162–163
contour, 199, 200
contraction, convolution, 65, 68
convex
domain, 205, 255, 256
holomorphically, 83–85, 87, 92
linearly, 220
logarithmically, 259
polynomially, 76
rationally, 103
surface, 252
convolution, 97, 126, 144
contraction, 65, 68
of distributions, 65, 115

counterexample
- in the ball, 23–24, 98–99, 162, 215, 220, 222, 249
- in the bidisc, 88–89
- Itano's, 118
- to Lupacciolu's theorem, 84–85
- to preservation of angle, 197
- to theorem on iterates, 172

Cousin problem, first, 64
covector, 9
CR-distribution
- approximated by holomorphic functions, 69
- approximated by polynomials, 70
- defined, 63
- holomorphic extension of, 87, 248, 251, 255, 256
- removable singularities of, 102
- uniqueness theorem for, 174–177

CR-form, 70
CR-function, 55–103
- analytic representation of, 55–70
- bounded, 91–94
- defined, 61
- holomorphic extension of, 70–91, 244–261
- integrable, 94–96
- removable singularities of, 97, 91–97, 103
- Riemann's theorem for, 97–103

CR-hyperfunction, *see* hyperfunction
CR-manifold, 70, 75, 103
critical points, 76
currents, 28, 55–60
- $\bar\partial$-problem for, 67
- closed, 56
- convolution contraction of, 65
- definition of, 55
- exact, 56
- examples of, 56–58
- Grothendieck's lemma for, 67
- of measure type, 56, 57
- singularity of, 55–56, 58
- support of, 56

curve
- compact, 74
- complex, 73
- smooth, 73, 164

cycle, 7, 194
cylinder, 164

$\bar\partial$-exact, 64, 67, 85, 94, 224, 231
∂-exact, 175
$\bar\partial_n$ defined, 39
$\bar\partial$-Neumann operator, kernel of, 184–188
$\bar\partial$-Neumann problem, 155–188
- for distributions, 158, 181–182
- for forms, 158, 182–183
- for smooth functions, 179–181
- generalized, 229
- homogeneous, 159–160, 223
- in the ball, 183–188
- integral representation for, 183–188
- series solution, 180–181
- solvability, 177–183
- statement of, 155–160
- uniqueness for, 174–177

$\bar\partial$-problem for Bochner-Martinelli kernel, 77–80
$\bar\partial^*$ defined, 156
∂^*-exact, 179, 183
$\bar\partial_\tau$ defined, 177
Δ defined, 2
$\mathcal{D}^p$ defined, 55
$\mathcal{D}'_p$ defined, 55
$\mathcal{D}^*$ defined, 117
dv defined, 1
$d\bar\zeta[k]$ defined, 2
Damyanov, B. P., 116, 278
Dautov, Sh. A., 178, 220, 230, 232, 260, 261
- and Aronov, 71, 73
- and Aĭzenberg, 30
- and Kytmanov, 21

decomposition
- asymptotic, 116
- Dolbeault, 59
- Hefer, 78
- of d, 59
- spectral, 44, 48, 172

decreasing, rapidly, 137
defining function, 1, 5, 40, 45, 99
definite

negative, 231
positive, 228
degree zero, 194
Δ defined, 2
delta function, 60, 115, 122, 135, 189, 235
Fourier transform of, 139
density
of fractions, 160, 166
of harmonic functions, 113
of harmonic polynomials, 45
of holomorphic functions, 69, 92, 172
of orthogonal system, 267, 268
of smooth functions, 172
derivative
boundary behavior of, 33–44
complex normal, 270
formulas for finding, 33–36
jump of, 36–44
oblique, 222, 229
of Bochner-Martinelli integral, 33–44
of Bochner-Martinelli transform, 257
of double-layer potential, 44
of finite order, 138
of fundamental solution, IX, 31, 121
of Green function, 5
of harmonic function, X, 238
of single-layer potential, IX, 5, 8, 34–35, 237, 238
determinant, 128, 254
De Rham currents, 55
diffeomorphism, 235
difference quotient, 160
differential condition, 71
differential criteria for holomorphicity, 222–228
differential equation
arbitrary linear, 75
Euler, 52
partial, 115
differential form
Bochner-Martinelli formula for, 9
characterization of $\overline{\partial}$-closed, 165–168
current associated to, 58
$\overline{\partial}$-Neumann problem for, 182–183
double, 10
exterior, 2, 9, 10, 142
jump of, 30–33
restriction to boundary, 27, 111
spaces of, 55
differentiation, exterior, 56
dilation, 53
Dini condition, 20, 192, 193, 195–198, 211
Dirac delta function, *see* delta function
direct
product, 120
sum, 58, 137, 175, 202, 243
of currents, 59
direction cosine, 27
Dirichlet problem, 201, 221
disc, 44
analytic, 103
unit, 64
discrete
set, 189
spectrum, 49
distance function, 98, 102, 106, 245
distribution
$\overline{\partial}$-Neumann problem for, 158, 181–182
acting on Bochner-Martinelli kernel, 27
analytic representation of, 61
coefficients, 56
convolution of, 65, 115
CR, *see* CR-distribution
current identified as, 56
Fourier transform of, 149–154
given on a hypersurface, 105–154
harmonic representation of, 105–115
homogeneous, 120–121
jump of, 27–30
multipliable, 126–135
multiplication of, *see* multiplication of distributions
of slow growth, 137, 141–142
open question for, 172

solution of $\overline{\partial}$, 86
with point singular support, 116
Dolbeault
decomposition, 59
theorem, 67
domain
circular, 255, 256, 258–261
convex, 255, 256
double differential forms, 10
double-layer
operator, 48, 49
potential, 44, 114, 240
jump for, 49, 240
jump of normal derivative, 44
of hyperfunctions, 240
singular, 49
summand of Bochner-Martinelli kernel, IX, 5, 8
doubly orthogonal bases, 267–268
dual
cone, 146
formal, 156
space, 144, 233
duality
Alexander-Pontryagin, 90
Grothendieck, 233
Dynin, 9

$\mathcal{E}$ defined, 27
E_f defined, 189
$\mathcal{E}^p$ defined, 55
$\mathcal{E}'$ defined, 56
$\mathcal{E}'$ defined, 27
E_ρ defined, 203
eigenfunction, 49
eigenvalue
of Bochner-Martinelli operator, 47, 172
of layer potentials, 49
of Levi form, 75, 93
eigenvector, 267
ellipsoid, 194
elliptic
complex, 21
equation, 228
operator, 108, 221, 229
embedding theorem, 175, 181
entire function, 7, 199, 203, 204, 252
linear forms on, 122
Mittag-Leffler, 203
envelope
$\widehat{K}$, 76
of holomorphy, XI, 72, 75, 76, 84, 87, 90, 92, 201, 244
rational, 84
error of Serbin, *see* Serbin
essential support, 58
estimates, a priori, 159
Euclidean
metric, 155
volume, 63
Euler
constant, 150
equation, 52
even-dimensional, 73
exact
$\overline{\partial}$, 64, 67, 85, 94, 224, 231
∂, 175
∂^*, 179, 183
current, 56
form, 72
exhaustion function, 255
extension
harmonic, 44, 46, 49, 52, 110–115, 133, 169, 170, 219
holomorphic, 75–91, 201–203, 220–270
of hyperfunctions, 233–244
via Bochner-Martinelli, 244–252
via Cauchy-Fantappiè, 253–261
impossibility of, 70
local, 70, 110
one-dimensional, 163–165, 168
real-analytic, 252, 254
theorem of Hartogs-Bochner, 70–74
theorem of Weinstock, 72–73
uniqueness of, 81
exterior
differential form, IX, 2, 55, 56, 264
differentiation, 56, 58
product, 58, 254

$F^{\pm}$ defined, 13
FBI transform, 138, 142
fiber-preserving map, 8

Fichera, G., 71
first Cousin problem, 64
Fisher, B., 116
Folland, G. B., 161
form, bilinear, *see* bilinear form
Fornæss, J. E., example of, 172
Fourier coefficients, 269
Fourier transform, 115, 137
  examples, 139–141
  generalized, 137–154
  inversion formula, 142–144
  nonlinear, 138
  of characteristic function, 144–149
  of delta function, 139
  of distribution, 149–154
  properties of, 139
Fredholm problem, 229
Fubini's theorem, 64, 76, 77, 96, 103, 190, 205, 265
function
  associated, 135
  characteristic, 60, 97, 144, 191, 201, 202, 264, 267
  class $\mathcal{L}^p$, 24–26
  continuous, *see* continuous function
  CR, *see* CR-function
  defining, 1, 5, 40, 45, 99
  delta, 60, 115, 122, 135, 189, 235
  distance, 98, 102, 106, 245
  harmonic, *see* harmonic function
  holomorphic, *see* holomorphic function
  of slow growth, 137–141
  polyharmonic, *see* polyharmonic function
  rapidly decreasing, 137
  represented by Cauchy-Fantappiè formula, 220–222
  smooth, *see* smooth function
  subharmonic, 101, 102
functional
  analytic, *see* analytic functional
  Minkowski, 255, 260
fundamental solution, 2, 7, 33, 99, 166, 204, 215, 219, 223, 235, 236
  derivatives of, IX, 31, 121
  for elliptic operator, 221, 222
  of $\overline{\partial}$, 60–61
future light cone, 148

$g(\zeta, z)$ defined, 2
$\mathcal{G}_2^s$ defined, 168
γ_k defined, 27
$\mathcal{G}$ defined, 105
$\mathcal{G}^*$ defined, 116
Gauss
  formula, 132, 153, 225, 226
  representation, 132
Gaziev, A., 20, 24
geometry, algebraic, X
global representation, 63
Goluzin formula, 200–201
grad ρ defined, 27
gradient, 127
graph, 74, 107
Green
  formula
    analogue of, 221
    applied, 110, 123, 153, 174, 224, 236, 238, 239
    classical, 100
    complex form, 1–4, 110, 242
  function, 6, 237
    boundary behavior, 5
    classical, 99
    properties of, 99–100
Griffiths, P., 164
Grothendieck
  duality, 233, 234, 238
  lemma, 67, 68
    for currents, 67
growth
  finite order, 105, 107–108, 128–129
  order, 245
  slow, 137–142

$\mathcal{H}$ defined, 111
$\mathcal{H}^p$ defined, 5
$\mathcal{H}_s$ defined, 144
$\mathcal{H}^*$ defined, 116
$\mathcal{H}_0$ defined, 114
Hadamard finite part, 133, 135, 149
Hahn-Banach theorem, 203
Hardy

class, 262, 264
space, 5–6, 176
harmonic
continuation, 110–115
extension, 44, 49, 52
of polynomial, 46, 49
forms, 157
function
boundary values of, 107–108, 233–240
of finite order of growth, 105, 107–108
spaces of, 267–268
with harmonic product, 127
majorant, 101, 102
representation, 105–118
Harris, J., 164
Hartogs theorem on removable singularities, IX, 71, 78, 83, 90, 97, 162
Hartogs-Bochner theorem, 70–74, 99, 175, 183
application of, 73
counterexample to, 72
for hyperfunctions, 233
for Stein manifold, 71
Harvey, F. R.
and Lawson, X, 17, 23, 71
theorem, 66, 73–74
and Polking, 75
Hausdorff measure, 57, 74, 76, 98
Hefer
decomposition, 78
theorem, 78
Henkin, G. M., 70, 75, 93, 178
and Leiterer, 8, 9
proof of Cauchy-Fantappiè formula, 7
Henkin-Ramirez kernel, 21, 220, 255
Hermitian
form, 193
matrix, 223
Hilbert space, 144, 267
Hill, C. D., 64
Hodge
scalar product, 156, 179
star operator, 155–157
Hölder
condition
and removable singularities, 97
for data in analytic representation, 65
for derivatives of Green function, 5
for integral, 212, 213
Plemelj formula for functions satisfying, 13–17
density, 21
inequality, 265, 266, 269
holomorphic
extension, 75–91, 220–270
from part of the boundary, 201–203
of hyperfunctions, 233–244
one-dimensional, 163–165, 168
via Bochner-Martinelli, 244–252
via Cauchy-Fantappiè, 253–261
function
Cauchy problem for, 261–270
represented by Bochner-Martinelli, 161–168
mapping, 192, 193, 195, 198
holomorphically convex, 83–85, 87, 92
holomorphicity, differential criteria for, 222–228
homogeneous
$\overline{\partial}$-Neumann problem, 159–160
coordinates, 91
distribution, 120–121, 129–135
function, 120, 121, 135, 146
homothety, 258, 260
homotopy, 7, 66
Hörmander, L., 71, 86, 238
balls, 21
multiplication of distributions, 115, 135
theorem on harmonic representation, 122, 123
hyperdistribution, 117, 124, 137
equal to a distribution, 118
equal to a function, 118
space of, 117
support of, 118
hyperfunction

as boundary value of harmonic function, 233–240
double-layer potential of, 240
holomorphic extension of, 233–244
multiplication of, 122–124
single-layer potential of, 236, 238
hyperplane at infinity, 90
hypersurface, XI
analytic, 7

Iagolnitzer, D., 138, 142
ideal, 137
identity, 48
improper integral, X, 13, 32, 131, 192, 198, 212
inductive limit topology, 233
inequality
Bessel, 270
Cauchy, 235
Hölder, 265, 266, 269
Jensen, 25, 101
Schwarz, 37
triangle, 263
integral
equations, 51–52
estimated, 208–211
representation, 6–13
in $\mathbf{C}^2$, 230–232
in the ball, 183–188, 229
with weak singularity, 211–214
integration by parts, 109
inverse branch, 266
inversion formula
for Bochner-Martinelli integral, 216
for Fourier transform, 142–144
for generalized Fourier transform, 142–144
irreducible, 74
isomorphism, 124, 137, 168, 171
Itano, M., 116, 118
iterates of Bochner-Martinelli integral, 48, 168–176
Ivanov, V. K., 116, 124, 135, 137
theorem of, 136, 137

Jacobian, 96, 190, 193, 235
Jensen's inequality, 25, 101
Jöricke, B., 97
jump
across a hypersurface, 88–90
for Bochner-Martinelli transform of a functional, 241
for derivatives, 36–44
for double-layer potential, 49
for hyperfunctions, 238, 240
of Bochner-Martinelli integral, IX, 16–17, 21–33, 81
of differential form, 30–33
of holomorphic functions, 55–70
of normal derivative, 40–44
problem, *see* analytic representation
problem for distributions, 105–115

$\widehat{K}$ defined, 75
$\widetilde{K}$ defined, 84
Kakichev, V. A., 17
Keldysh-Lavrent′ev theorem, 160
Keller, K., 115
Kelvin transform, 51, 129, 250
kernel
Bochner-Martinelli, *see* Bochner-Martinelli kernel
Cauchy, 2, 215
average of, 164
Henkin-Ramirez, 21, 220, 255
Koppelman, 10, 30, 31
of $\overline{\partial}$-Neumann operator, 184–188
Poisson, *see* Poisson kernel
reproducing, 230, 232
Szegő, 185
Khenkin, *see* Henkin, G. M.
Khristov, Kh. Ya., 116
Kohn, J. J., 159, 161, 178
Koppelman's integral representation, IX, 9–13, 216, 229, 232, 264
and characterization of $\overline{\partial}$-closed forms, 166, 168
deduced from $\overline{\partial}$-homotopy formula, 67
for elliptic complexes, 21
kernel of, 30, 31
Koppelman, W., 4
Krasichkov, I. F., 267, 268
Krasnoyarsk, X

Kronecker symbol, 12, 43
Krylov formula, 200–201
Kudryavtsev, L. D., 266
Kytmanov, A. M., XI, 110, 160, 161, 165, 172, 217
Kytmanov, A. M., theorems of
  analogue of Privalov's theorem, 17
  analytic representation, 89, 90, 237, 238, 241
  Bochner-Martinelli integral in the ball, 49
  classification of multipliable distributions, 133, 134
  $\bar{\partial}$-Neumann problem, 179, 181, 182, 185
  Fourier transform, 138, 143, 145, 154
  functions represented by the Bochner-Martinelli integral, 161 162, 166, 167, 176, 242, 244
  functions represented by the Cauchy-Fantappiè formula, 222
  harmonic extension, 245
  harmonic representation, 111
  holomorphic extension, 248, 251, 255, 256, 260
  holomorphic extension of CR-functions, 87, 88
  holomorphicity of harmonic functions, 224, 225, 228
  iterates of the Bochner-Martinelli integral, 174, 243
  jump of Bochner-Martinelli integral, 21, 24
  jump of normal derivative, 40
  multiplication of distributions, 117, 118, 120
  Poincaré-Bertrand formula, 214
  removable singularities, 91, 93, 96, 98, 99, 102, 103
  residues, 197
  series expansion of the Bochner-Martinelli kernel, 249

Laplace
  equation, IX, 2, 7, 33, 99, 121, 122, 166, 204, 219, 223, 235
  operator, 2, 107, 129, 131, 236
    invariance of, 107
  transform, 138
Laplacian, complex, 157
Laurent-Thiébaut, C., 83
Lavrent′ev, *see* Keldysh-Lavrent′ev
Lawson, H. B., *see* Harvey and Lawson
Lebesgue
  class, 261
  dominated convergence, 26, 32, 92, 101, 126, 133, 136, 206
  integral, 266
  measure, linear, 200, 205
  measure, surface, 202, 205
  point, 17–21, 23, 39, 44, 76, 246, 247, 254, 262
  summable, 266
Lee, H.-P., 103
Leibniz's rule, 120, 265
Leiterer, J., 8, 9
Leray, J., IX, 6
level
  set, 63, 77, 94
  surface, 28, 63
Levi form
  eigenvalue of, 75, 93
  negative definite, 231
  nondegenerate, 90, 96, 103, 252
  restriction on, 70
Levi-Civita, theorem of, 73
Lewy, H., example of, 177
Li, Bang He, 116
light cone, 148
limit
  inductive, 233
  nontangential, 20, 65, 71, 76, 87, 89
  of Riemann sums, 123, 160, 236
  weak, 103, 158
line, complex, 164, 168, 205
Liouville's theorem, 71, 124
localization, 122, 265
logarithmically convex, 259
Lu Qi-Keng, 17
Lupacciolu, G., 75, 76, 84, 88

theorem of, 75–77, 84, 91
proof, 81–84
Lyapunov
domain, 261, 267, 269
surface, 5, 41, 71, 96, 99, 201, 206, 268
theorem, 44

$[M]$ defined, 57
$\mathfrak{M}$ defined, 129
$\mathfrak{M}_\delta$ defined, 133
Mf defined, 46
$\mathfrak{M}_h$ defined, 133
M_σ defined, 47
μ_f defined, 2
majorant, harmonic, 101, 102
manifold
complex, 65, 71
characterization of boundary, 73–74
CR, 70, 75, 103
even-dimensional, 73
1-concave, 93
one-dimensional, 74
real-analytic, 234
Stein, *see* Stein manifold
mapping
biholomorphic, *see* biholomorphic mapping
holomorphic, *see* holomorphic mapping
nondegenerate, 197
Martinelli, E., IX, 4, 5, 8, 71, 78
massiveness, 259
mathematical physics, 138
maximal
chord, 53, 54
complexity, 73, 74
distance, 54
function, 100, 101
length, 53
maximum principle, 83, 103, 234
mean
Abel, 138
approximation in, 273
mean-value
property, 132
theorem, 3, 10, 60
measure
of slow growth, 138
probability, 101
Mikusiński, J., 115
multiplication in the sense of, 124–126
minimality, 103
Minkowski functional, 255, 260
minor, 18
mistake, *see* error
Mittag-Leffler's function, 203
modulus
maximum, 234
of continuity, 20, 37, 193
moment condition, 73, 74
monomial, 45, 130, 188, 257
Montel space, 268
μ_f defined, 2
multipliable distributions, 126–135
multiplication
in the sense of Mikusiński, 124–126
of analytic functionals, 124
of distributions, 115–137
axioms for, 115
different approaches to, 115–116
Hörmander, 115
Mikusiński, 115
properties of, 118–122
via analytic representation, 116
via harmonic representations, 116–118
of hyperfunctions, 122–124
multiplicity
infinite, 47
of a zero, 189, 190, 199
multiplier, 115, 118

Naser Shafii, 72
negative definite, 231
Neumann problem, 157
Newton-Leibniz formula, 265
Nikolskiĭ, S. M., 266
Noetherian problem, 229
nondegenerate
Levi form, 90, 96, 103, 252
mapping, 197
pairing, 233

nonholomorphic part, 258, 259
nonintegral indices, 266
nonlinear Fourier transform, 138
nonnegative operator, 171
nontangential
  boundary value, 262, 263
  limit, *see* limit, nontangential
  path, 20, 200, 202
  vector field, 222
nontrivial
  forms, 74
  intersection, 78
  solution, 52
Norguet, F., 9
norm
  in a fiber, 9
  in C^α, 16
  in $\mathcal{G}_2^s$, 171
  in $\mathcal{H}_s$, 144
  in $\mathcal{L}^2$, 257, 269, 270
  in $\mathcal{L}^p$, 72, 247
  in $\mathcal{L}^q$, 202, 263
  in $\mathcal{W}_2^s$, 144
  of $\overline{\partial}$-Neumann operator, 184
normal
  boundary values, 101
  field, 8

$\mathcal{O}(U)$ defined, 1
$\Omega^\pm$ defined, 61
$\mathcal{O}_p^\perp$ defined, 201
oblique derivative problem, 222, 229
odd
  degree, 131
  dimension, 134
  dimensional manifold, X, 73
  function, 114, 122, 123
  power, 133
Oka-Weil theorem, 85
$\Omega^\pm$ defined, 61
ω' defined, 6
one-dimensional holomorphic extension
    property, 163–165, 168
open problems, 220–232
operator measure, 172
ordinary differential equation, 52
orientation
  of $\mathbf{C}^n$, 1
  of boundary, 2, 14, 57
  of surface, 61, 68, 105, 245, 251
orthogonal transformation, 139, 140
orthogonality condition, 72, 158, 163
orthonormal basis
  in $\mathcal{G}_2$, 267–269
  in $\mathcal{L}^2$, 46, 162, 184
Osgood, W. F., IX

$\mathcal{P}_{s,t}$ defined, 45
$\mathcal{P}^*$ defined, 135
pairing, nondegenerate, 233
parallelogram, 200
partial sum, 113, 199
partition of unity, 230
peak
  function, 85, 102
  point, 84
  set, 99, 102
Phragmén-Lindelöf principle, 204
physics, 138
Plancherel's theorem, 142
Plemelj, *see* Sokhotskiĭ-Plemelj
plurisubharmonic function, 85, 255
Poincaré-Bertrand formula, 48, 206–219
Poisson
  equation, 107
  formula, 6, 45, 48
  integral, 52, 100, 176
    connected to
      Bochner-Martinelli, 6
  kernel, 6, 114, 137, 219
    and Green function, 99, 237
    for ball, 44, 49, 149, 185, 220, 250
    for half-plane, 146
    for half-space, 19, 23, 121, 122, 124, 135, 140
  summation formula, 199–200
Polking, J.
  and Harvey, 75
  and Wells, 233, 241, 243, 244
polycylinder, 204
polydisc, 67, 68, 234, 235
polyharmonic function, 129, 141
  Bochner-Martinelli integral of, 50
  boundary values of, 128–129
polyhedron, 2

polynomially convex, 76
polytope, 200
positive definite, 228
precompact set, 268
Prenov, B. B.
  example of, 197
  lemma of, 217
  theorem of, 192, 195, 214
prime on summation sign, 10, 55
primitive, 107
principal
  part, 116
  value, 13, 21, 142, 192
Pringsheim, convergence in the sense of, 200
Privalov
  lemma, 262
  theorem, 17–20
probability measure, 101
product of distributions, *see* multiplication of distributions
projective space, 90, 91
pseudoconvex domain
  strongly, *see* strongly pseudoconvex domain
  weakly, *see* weakly pseudoconvex domain
pure monomial, 188

Qi-Keng Lu, 17
quaternionic analysis, 9
quotient group, 67, 85

Radó's theorem, 96, 97
Ramirez, E., 21, 220, 255
rapidly decreasing function, 137
rational
  convexity, 103
  envelope, 84
  number, 47
ray, 194
real-analytic
  boundary, 181, 233, 240, 244
  dependence on parameter, 260, 261
  diffeomorphism, 235
  extension, 252, 254
  function, 234, 241, 253
    Poisson summation formula for, 199
  manifold, 234
reflexive, 266
regularization, 85, 86, 132
removable singularities, 75, 97–103
  of CR-functions, 91–97
reproducing kernel, 230, 232
residue
  logarithmic, 189–200
  theorem, 140
restriction
  of Bochner-Martinelli to boundary, 14, 27, 40, 45, 194, 215
  of forms to boundary, 14, 27, 30, 33, 62, 63, 111, 165, 167, 180, 222
  operator, 168
Riemann
  removable singularities theorem, 92, 97–103
  sum, 123, 160, 236
Riesz, theorem of brothers, 165, 177
Riesz-Fischer theorem, 270
rigidity, X
Roĭtberg, Ya. A., 108
Romanov, A. V., 47, 48, 169, 171
Roos, G., 191
Roos-Yuzhakov formula, 196, 198, 199
Rosay, J.-P., 96
rotation, 53, 54, 198
Rothschild, L. P., 70

$S(z,\epsilon)$ defined, 2
$\mathcal{S}$ defined, 137
$\sum'$ defined, 10
$\mathcal{S}'$ defined, 137
saltus, *see* jump
Sard's theorem, 76, 164, 190, 255
scalar product
  for forms, 155
  Hodge, 156
  in G^s, 170
  in Hilbert space, 267
  in L^2, 44, 168
  in $\mathbf{R}^n$, 105
  in W^s, 168

schlicht, 75, 76, 84, 87, 91
Schwartz, L., 115
Schwarz inequality, 37
self-adjointness
 and doubly orthogonal bases, 267
 of the Bochner-Martinelli operator, 47, 53, 171, 175, 243
 of the single-layer operator, 174
Semenov, A. M., 168
seminorm, 137
Serbin, A. I.
 error of, 162, 215–216
 lemma of, 208, 211
 theorem of, 212, 214
Shafii, Naser, 72
Shaimkulov, B. A., 165
Shapiro, M. V., 219
Shaw, Mei-Chi, 178
shell, 263
Shilov boundary, 259
Shirokov, Yu. M., 115
Shlapunov, A. A., 262, 266, 267, 269
$\sum'$ defined, 10
single-layer
 operator, 48–49, 170, 174
 potential, IX, 36
  continuity of, 111
  derivative of, 34–35
  of distribution, 119
  of hyperfunction, 236, 238
  of real-analytic function, 235
  smoothness of, 119
  summand of Bochner-Martinelli kernel, 5, 8
singular
 chain, 57
 integral, 53
  equation, 52
  interchange of, 214
  operator, 20, 263
singularity
 compact, IX, 71, 75
 finite order, 28, 29, 56, 112, 253
 integrable, 133, 153, 232
 integral with weak, 211–214
 of current, 55–56, 58
 on boundary, 195–200
 removable, 75, 91–97
Słodkowski, Z., 97
slow growth, 137–142
 measure of, 138
 space of functions, 144
Smirnov's theorem, analogue of, 101
smooth functions represented by Bochner-Martinelli, 161–162
Sobolev
 embedding theorem, 175, 181
 space, XI, 108, 168, 265, 266
Sokhotskiĭ-Plemelj formula, 13–17, 20, 21, 24, 61, 173
 applied, 47, 52
solid angle, 13, 197
solution, canonical, 180, 187
spanning, X, 74
spectral decomposition, 44, 48, 172
spectrum of Bochner-Martinelli operator, 44–49, 172
Spencer, D. C., 159
spherical harmonics, 45–46, 162, 177, 184, 249, 252
star operator, 155–157
Stein manifold, 87
 Andreotti-Norguet formula for, 9
 Bochner-Martinelli formula for, 8–9
 Hartogs-Bochner theorem for, 71
 Lupacciolu's theorem for, 83, 86
 Plemelj formula for, 21
Stieltjes-Vitali theorem, 268
Stout, E. L., 84, 91, 165
 and Alexander, 88
 and Chirka, XI, 91
 and Lupacciolu, 91
 and Rosay, 96
Straube, E. J., XI, 108, 172
strictly pseudoconvex domain, *see* strongly pseudoconvex domain
strong operator topology, 48, 169, 172, 174
strong topology, 266
strongly pseudoconvex domain, 21, 84, 88, 92, 97, 159, 178–183, 229,

232, 244, 259–261
approximating by, 85, 89
subalgebra, 137
subharmonic function, 101, 102
submanifold
current associated to, 57–60
with boundary, 74
sum of squares, 193
supersingular support, 118
support
essential, 58
of current, 56
of hyperdistribution, 118
singular, 116, 118
supersingular, 118
surface
convex, 252
hyper, *see* hypersurface
level, 28, 63
Lyapunov, 5, 41, 71, 96, 99, 201, 206, 268
measure, 27, 28, 202, 205, 249, 261
orientation of, 61, 68, 105, 245, 251
piecewise-smooth, 195, 203
surface area element, 5, 23, 34, 57, 76, 108, 111, 233, 253
of plane, 38, 42
of sphere, 24, 42, 130
surjective mapping, 266
system, associated, 135
Szegő
integral representation, 259
kernel, 185, 259, 261
projection, 180, 181, 229

$\tau(z)$ defined, 13
T defined (single-layer operator), 48
T_v defined, 111
T^c_ζ defined, 63
tangent
bundle, 8
cone, 13, 194, 195, 197, 198, 218
plane, 21, 25, 95, 218, 252
complex, 228
space, complex, 63, 70, 73, 222
vector, complex, 74
tangential
Cauchy-Riemann equation, 62–63, 74, 85, 89, 93, 158, 177–179, 182, 222
derivative, IX, 5, 8
part of form, 30, 62, 157, 167, 179, 182, 183, 264
vector field, 28, 29, 37, 112, 223
Tarkhanov, N. N., X, 12, 114, 168, 202, 203
and Prenov, 192, 195, 214, 217
and Shlapunov, 262, 266, 267, 269
Taylor
expansion, 197, 198
polynomial, 136, 149, 153
termwise integration, 9, 199
Tillmann, H. G., 116
Tongde Zhong, 17, 214
topology
strong, 266
strong operator, *see* strong operator topology
weak, *see* weak topology
totally real, 70
trace theorem, 168–169, 266
transform
Bochner-Martinelli, *see* Bochner-Martinelli transform
Cauchy-Bochner, 145
FBI, 138, 142
Fourier, *see* Fourier transform
Kelvin, 51, 129, 250
Laplace, 138
translation, 17, 18, 21, 41, 107, 163, 218
transversality, 94, 95
Trépreau, J.-M., 70
triangle inequality, 263
two-constant theorem, 204
two-dimensional complex planes, 97

$U(\zeta, z)$ defined, 2
$U_{p,q}(\zeta, z)$ defined, 10
unbounded
component, 163, 164
domain, 7, 8, 71, 72, 91
unique solution, 178, 267
uniqueness
for $\overline{\partial}$-Neumann problem, 174–177
of extension, 81, 82

set, 200–202
theorem, 72, 82, 88, 99, 162, 182, 243
unitary
matrix, 18
transformation, 17, 18, 21, 41, 43, 107, 193, 218
upper half-plane, 146, 199

Val'skiĭ, R. E., 165
Vasilevskiĭ, N. L., 219
vertex, 17, 21, 94, 146, 203, 262
Vinogradov, V. S., 71
Vitali theorem, 268
Vladimirov, V. S.
algebra of, 116
theorem of, 144–149
volume
element, 189
form, 1, 59, 156, 263
of sphere, 257
potential, 158

$\mathcal{W}_2^s$ defined, 144, 168
W defined (double-layer operator), 49
W_σ defined, 49
$\mathcal{W}_q^s$ defined, 266
wave front set, 70, 115, 135, 138
weak
boundary values, 264–267, 269, 270
Cauchy-Riemann equations, 62
completeness, 106
convergence, 55, 71, 181, 234, 243
limit, 103, 158
sense, 246
singularity, 211–214
topology, 55, 69, 70, 87, 115, 116, 129
weakly pseudoconvex domain, 178, 181, 232
wedge, 70
Weinstock, B. M., 71, 73, 166, 272
theorem of, 72–73
Wells, R. O., Jr., 233, 241, 243, 244
Wermer, J., 103
Weyl's lemma, 110

Yakimenko, M. Sh., 237, 238, 241–244
Yarmukhamedov, Sh., 7, 33
construction, 7
formula, 203–204
representation, 20
theorem, 8
Yuzhakov, A. P., 191
formula, 196, 198, 199
theorem, 191

Zaslavskiĭ, A. I., 124, 127, 138
zero
at infinity, 124, 169–171, 174, 233, 236, 241
distribution, 116, 117, 120, 142
finite number of, 195
Hausdorff measure, 74, 98
isolated, 197, 199
jump, 37, 39, 40
measure, 259
multiplicity of, 189, 190, 199
normal part, 158, 179, 183
set, 91, 92, 96, 99, 189, 191
simple, 190, 193, 195, 199
tangential part, 30, 62, 167, 179, 183, 264
Zhong Tongde, 17, 214
Znamenskaya, L. N., 203

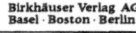